CIVIL ENGINEERING MATERIALS

CIVIL ENGINEERING MATERIALS

Shan Somayaji
California Polytechnic State University, San Luis Obispo

PRENTICE HALL, Upper Saddle River, New Jersey 07458

Library of Congress Cataloging-in-Publication Data

Somayaji, Shan.
Civil engineering materials / Shan Somayaji.—2nd ed.
p. cm.
Includes bibliographical references and index.
ISBN 0-13-083906-X
1. Materials. I. Title.

TA403 .S598 2001
624.1'8—dc21

00-051640

Vice president and editorial director, ECS: **Marcia J. Horton**
Acquisitions editor: **Laura Curless**
Supervisor: **Dolores Mars**
Marketing manager: **Holly Stark**
Production editor: **Chanda Wakefield**
Executive managing editor: **Vince O'Brien**
Managing Editor: **David A. George**
Art director: **Jayne Conte**
Cover design: **Bruce Kenselaar**
Art editor: **Adam Velthaus**
Manufacturing manager: **Trudy Pisciotti**
Manufacturing buyer: **Dawn Murrin**
Vice president of production and manufacturing, ESM: **David W. Riccardi**

Upper Saddle River, New Jersey 07458

Printed in the United States of America

10 9 8 7

ISBN 0-13-083906-X

Prentice-Hall International (UK) Limited, London
Prentice-Hall of Australia Pty. Limited, Sydney
Prentice-Hall Canada Inc., Toronto
Prentice-Hall Hispanoamericana, S.A., Mexico
Prentice-Hall of India Private Limited, New Delhi
Prentice-Hall of Japan, Inc., Tokyo
Pearson Education Asia Pte. Ltd., Singapore
Editora Prentice-Hall do Brasil, Ltda., Rio de Janeiro

Contents

PREFACE TO THE SECOND EDITION xiii

PREFACE TO THE FIRST EDITION xv

1 INTRODUCTION 1

1.1 Materials and Methods—A Historical Perspective 1

1.2 Materials and Types 5

1.3 Civil Engineering Materials 14

1.4 Properties of Engineering Materials 15
1.4.1 Forces, Loads, and Stresses, 17
1.4.2 Strain, 20
1.4.3 Stiffness, 22
1.4.4 Ductile and Brittle Materials, 26

1.5 Selection of Materials 29

1.6 Standards 31

Problems 34

2 AGGREGATES 35

2.1 Rocks 35
2.1.1 Some Common Rocks and Rock Minerals, 36

2.2 Types of Aggregates 40
2.2.1 Based on Source or Method of Manufacture, 42
2.2.2 Based on Size, 44
2.2.3 Based on Density, 44

2.3 Properties of Aggregates 47
2.3.1 Specific Gravity and Moisture Content, 48
2.3.2 Bulk Density and Voids, 53
2.3.3 Modulus of Elasticity and Strength, 55
2.3.4 Gradation and Fineness Modulus, 56
2.3.5 Other Properties, 65

2.4 Examples 68

2.5 Testing 70

Problems 77

3 CONCRETE AND OTHER CEMENTITIOUS MATERIALS 79

3.1 Types of Cement 79
3.1.1 Hydraulic Cement, 80
3.1.2 Nonhydraulic Cement, 82

3.2 Various Cementitious Materials 84

3.3 Uses of Concrete 85

3.4 Portland Cement 87
3.4.1 Manufacture, 88
3.4.2 Cement Chemistry, 90
3.4.3 Types of Portland Cement, 91
3.4.4 Fineness, 93
3.4.5 Strength of Cement, 94
3.4.6 Consistency of Cement, 95

3.5 Hydration 95
3.5.1 Setting, 95
3.5.2 Hardening, 96

3.6 Properties of Concrete 98
3.6.1 Properties of Fresh Concrete, 100
3.6.2 Factors Affecting Consistency and Workability, 105
3.6.3 Segregation and Bleeding, 106

3.7 Mixing, Placing, and Curing 107
3.7.1 Pumping and Placing, 112
3.7.2 Concrete Tools, 114
3.7.3 Finishing and Types of Finishes, 116
3.7.4 Curing, 117

3.8 Properties of Hardened Concrete 123
3.8.1 Compressive Strength, 124
3.8.2 Tensile Strength, 135
3.8.3 Flexural Strength, 136
3.8.4 Stress-strain Diagram and Modulus of Elasticity, 138
3.8.5 Examples, 142
3.8.6 Shrinkage, 144
3.8.7 Creep, 153
3.8.8 Carbonation, 154

3.9 Durability 155
3.9.1 Alkali-aggregate Reaction, 156
3.9.2 Sulfate Attack, 156
3.9.3 Freeze-thaw Cycle, 157
3.9.4 Corrosion, 158

3.10 Mix Proportioning and Design 160
3.10.1 Mix Design, 161
3.10.2 Mix Design Procedure, 163
3.10.3 Examples of Mix Design, 172

3.11 Admixtures 175
3.11.1 Chemical Admixtures, 175
3.11.2 Mineral Admixtures or Pozzolans, 181

3.12 Types of Concrete 187
3.12.1 Reinforced Concrete, 187
3.12.2 Prestressed and Precast Concrete, 192
3.12.3 Fiber-reinforced Concrete, 193
3.12.4 Lightweight Concrete, 197
3.12.5 High-strength and High-performance Concrete, 198
3.12.6 Examples, 201

3.13 Other Cementitious Materials 202
3.13.1 Plaster and Stucco, 202
3.13.2 Mortar, 204
3.13.3 Grout, 205
3.13.4 Shotcrete, 206
3.13.5 Soil Cement, 207
3.13.6 Pervious Concrete and Cement-bonded Particleboard, 207

3.14 Testing 208

Problems 220

4 MASONRY ***223***

4.1 Masonry Units 227
4.1.1 Clay Bricks and Structural Clay Tiles, 228
4.1.2 Manufacture of Bricks, 230

4.1.3 *Grades and Types, 232*
4.1.4 *Sizes of Bricks, 234*
4.1.5 *Properties of Bricks, 235*
4.1.6 *Concrete Masonry Units, 239*
4.1.7 *Types and Grades of Concrete Masonry Units, 241*
4.1.8 *Properties of Concrete Masonry Units, 244*

4.2 Mortar, Grout, and Plaster 245
4.2.1 *Lime and Lime Mortar, 246*
4.2.2 *Mortar for Unit Masonry, 249*
4.2.3 *Mixing and Properties of Mortar, 251*
4.2.4 *Grout and Its Uses, 256*
4.2.5 *Plaster, 257*

4.3 Masonry Construction 258
4.3.1 *Types of Bond, 263*
4.3.2 *Types of Joints, 265*
4.3.3 *Control Joints and Expansion Joints, 266*

4.4 Properties of Masonry 271
4.4.1 *Efflorescence, 276*

4.5 Reinforced Masonry 279

4.6 Mix Proportioning and Examples 282

4.7 Testing 284

Problems 291

5 WOOD AND WOOD PRODUCTS 292

5.1 Structure of Wood 292
5.1.1 *Chemical Composition of Wood, 293*

5.2 Types of Wood 295

5.3 Physical Properties of Wood 297
5.3.1 *Moisture Content, 298*
5.3.2 *Density and Specific Gravity, 299*
5.3.3 *Examples, 302*
5.3.4 *Defects, 302*

5.4 Shrinkage and Seasoning 304
5.4.1 *Shrinkage, 305*
5.4.2 *Seasoning, 307*

5.5 Treatment and Durability 308
5.5.1 *Decay and Destruction, 309*

5.6 Lumber Sizes 311

5.7 Use Classification and Lumber Grading 313
5.7.1 Dimension, Decking, and Timbers, 316

5.8 Moisture Content Classification and Grade Stamp 318
5.8.1 Moisture Content Classification, 318
5.8.2 Grading, 318

5.9 Mechanical Properties and Allowable Values 321
5.9.1 Effect of Slope of Grain, 327
5.9.2 Strength Ratio and In-grade Testing, 328
5.9.3 Examples, 329

5.10 Wood Products 331
5.10.1 Glulam, 333
5.10.2 Plywood, 335
5.10.3 Other Panel Products, 339
5.10.4 Manufactured Components, 342

5.11 Creep 343

5.12 Wood Construction 344

5.13 Testing 348

Problems 353

6 BITUMINOUS MATERIALS AND MIXTURES 354

6.1 Tars and Pitches 355

6.2 Asphalts 356

6.3 Petroleum Asphalts 357
6.3.1 Asphalt Cement, 358
6.3.2 Cutback Asphalts, 359
6.3.3 Emulsified and Blown Asphalt, 360

6.4 Properties of Asphalt 361
6.4.1 Consistency, 362
6.4.2 Specific Gravity, 364
6.4.3 Durability, 365
6.4.4 Rate of Curing, 365
6.4.5 Resistance to Action of Water, 366
6.4.6 Ductility and Adhesion, 366
6.4.7 Temperature Susceptibility, 367
6.4.8 Hardening and Aging, 369

6.5 Asphalt Grades 370
6.5.1 Viscosity and Penetration Grading, 370
6.5.2 Performance-based Grading, 372
6.5.3 Cutback Asphalt Grades, 374

6.6 Asphalt Concrete 374
6.6.1 Aggregates, 375
6.6.2 Types of Asphalt Concrete, 378

6.7 Asphalt Pavement 380
6.7.1 Elements of Flexible Pavement, 381
6.7.2 Stabilization, 383

6.8 Spray Applications 384

6.9 Testing 386

Problems 392

7 IRON AND STEEL **393**

7.1 Iron 394
7.1.1 Cast Iron and Wrought Iron, 397

7.2 Steel 399
7.2.1 Steel Products, 400

7.3 Properties of Steel 402

7.4 Structural Steel 406
7.4.1 Structural Grades, 408

7.5 Reinforcing Steel 415
7.5.1 Grades and Types, 417
7.5.2 Handling, 421

7.6 Welded Wire Fabric 422
7.6.1 Grades, 422
7.6.2 Sizes, 423
7.6.3 Uses, 425

7.7 Epoxy-coated Reinforcing Steel 426

7.8 Testing 426

Problems 427

8 PLASTICS AND SOILS **428**

8.1 Plastics 428
8.1.1 Types of Plastics, 429
8.1.2 Properties of Plastics, 432
8.1.3 General Properties, 435

8.2 Modified Plastics 436

8.3 Uses of Plastics 438

8.4 Soils 439
8.4.1 Types and Properties of Rocks, 441
8.4.2 Types of Soil, 441
8.4.3 Soil Classification Systems, 445

8.5 Strength Property of Soils 449

8.6 Constituent Properties of Soils 449
8.6.1 Particle Size, 450
8.6.2 Unit Weight, 452
8.6.3 Moisture Content, 453
8.6.4 Degree of Saturation, 453
8.6.5 Void Ratio, 454
8.6.6 Porosity, 455
8.6.7 Particle Shape, 455
8.6.8 Specific Gravity, 456
8.6.9 Uses of Soil, 457

REFERENCES AND FURTHER READING ***458***

INDEX ***464***

Preface to the Second Edition

In revising, updating, and expanding several chapters and sections from the first edition of this text, the primary goal has remained the same—to assist a civil engineering or construction engineering student in understanding how construction materials differ in their physical, mechanical and chemical properties. The three key features of the text—properties, testing and standards, and application of materials—have been kept intact. The chemical nature of a material, a topic that was intentionally ignored in the first edition, has appeared—although in a minor role—in the chapters on cement, concrete, masonry, and wood. As in the first edition, the thrust of the text is to submit information that is essential for material selection and elementary design—the wedge between a course in solid mechanics and the subject of structural design.

Several chapters have been reshaped and rewritten to make them more readable and comprehensive; these include Chapter 2 ("Aggregates"), Chapter 3 ("Concrete and Other Cementitious Materials"), Chapter 6 ("Bituminous Materials and Mixtures"), and Chapter 8 (renamed "Plastics and Soils" in this edition). A number of new tables and figures are expected to increase familiarity. A brief introduction to several important new engineering materials, from semiconductors to thermosetting plastics, is hoped to help a civil engineering or construction engineering student identify and understand the basic nature of the new products that are constantly being introduced into the market.

A detailed discussion of the types and properties of rocks has been added to Chapter 2, to help in appreciating the importance of these natural materials (which make up for a major portion of mortar, portland cement concrete, and asphalt concrete) in the production of a constructed facility. The section on the effects of variables on the compressive strength of portland cement concrete has been expanded to cover the effects of several aspects of concrete manufacture—ingredients, mix proportions, environment, and construction practice—on the strength properties. Several new sections (such as "Acceptance Criteria," "Durability," and "Corrosion") are expected to benefit the student in the study of construction procedures and material evaluation. It is hoped that some of the rewritten sections such as "Mortar (Sections 3.13 and 4.2)," "Structure of Wood (Section 5.1)," "Physical Properties of Wood (Section 5.3)," and "Reinforcing Steel (Sections 7.5 and 7.6)," will have additional appeal to the reader. A detailed presentation of the performance-grading system of asphalt binders will help the student recognize the new trends in asphalt technology and pavement construction. A separate chapter on soils (along with plastics), in its revised and expanded form, is expected to suggest the importance of soil both as a construction and foundation material. The additional examples and end-of-the-chapter problems are the added features of the new edition.

San Luis Obispo *Shan Somayaji*

Preface to the First Edition

This text is intended for an introductory course in civil engineering materials, usually taken by students of engineering, construction, architecture, and technology in the junior year. It is assumed that the student has completed a basic course on solid mechanics or strength of materials.

A course on the properties of engineering materials generally forms a link between the study of solid mechanics and courses on structural design concepts and methods. The ideas of equilibrium and deformation from the course on solid mechanics are needed to describe material behavior and its properties, which form the foundation of structural design principles. Design is a process, iterative in nature, of designing a product or a structure that is yet to be built and constructed. The construction of a structure requires a thorough understanding of fabrication, application, field testing, and properties of materials. Thus, a thorough knowledge of properties and performance of materials and construction practices are required to develop, design, and build a safe, economical, and durable structure.

An introductory course on materials of construction or civil engineering is commonly set on the following aspects:

- Physical, mechanical, and other important properties of materials
- Fabrication or method of manufacture
- Durability and long-term performance
- Specifications and standards
- Application or methods of use

- Laboratory testing procedures
- Material testing procedures history

This text is designed to embrace all these aspects, although the extent of coverage of each may vary from one material to another. Very little, if any, importance is given to chemical properties or characteristics of microstructure of materials, as civil and construction engineering professionals are, in general, not heavily interested in them. Essentially, the focus is on explaining properties and behavior that emphasize concepts from solid mechanics and are required in the understanding of design principles.

Common materials of construction that are dealt with in the text are portland cement concrete, masonry, wood and wood products, bituminous materials, iron, steel, plastics, and soils. Chapter 3 deals with concrete and other cementitious materials (mortar, grout, and so on). A brief introduction to various types of concrete (such as fiber-reinforced, reinforced, and precast concrete) is also given in this chapter. Various types of masonry units and masonry are discussed in Chapter 4. Methods of construction, properties, and specifications are also described. Chapter 5 describes properties of wood, wood products such as glulam and plywood, and lumber, grading of lumber and wood products, and wood construction. Bituminous materials such as tars, pitches and asphalts, grades of asphalt, and asphalt pavements are described in Chapter 6. Iron, reinforcing steel, and structural steel are dealt with in Chapter 7. Properties and uses of plastics and soils are covered in Chapter 8.

Many of these materials are made as composites, by combining various other materials, and a study of the primary material requires an understanding of the basic or raw materials. Thus, aggregates (which are required in the construction of portland cement concrete, mortar, grout, masonry, and asphalt concrete) are described in Chapter 2, before introducing the latter. A brief review of the properties of soil as a construction material is given in Chapter 8.

As this is an introductory text on materials, the author felt that the presentation should be as simple as possible and devoid of research ambiguities or summaries. Thus, the writing style came from a need to generalize material behavior as much as possible, so that the student understands the "general characteristics" in terms of properties and uses. Of course, properties of civil engineering materials depend on a number of variables originating from several sources, such as raw materials, method of manufacture, environment, workmanship, method of testing, and age. Due to this, one could question the validity of generalization—and the outcome. But all can agree that an introductory course on materials should be able to present a general picture of overall material behavior, and it is this understanding that prompted the style of presentation.

Information on properties, specifications, and uses of a few of the materials is repeated at several places within the text. The purpose of this design is two-fold:

First is the pedagogical notion that students understand better through repetition; second is the fact that there exist many variations in subject matter as well as teaching style in various schools, which demands that each chapter or section be complete with all relevant information without the need to go back and forth within the text.

To illustrate the second point, details about reinforcing steel can be found in Chapter 3 ("Concrete and other Cementitious Materials") and in Chapter 7 ("Iron and Steel"). When the concepts of reinforced concrete are introduced, the material is described in the text as a composite material made from concrete and reinforcing steel; thus, data on the properties and grades of reinforcing steel are required, which are given in Chapter 3. These details are repeated in Chapter 7, wherein physical properties, mechanical properties, types, and grades of reinforcing and structural steel are further described. The author did not find it proper to require the student to turn back to Chapter 3 in order to understand the details about steel.

Laboratory testing procedures for selected tests are described at the end of most chapters, instead of as an adjunct to the text. This is done to drive home a point that laboratory testing is an essential part of the study of materials of civil engineering or construction. All the test procedures are based on relevant ASTM specifications, but are presented in a manner that makes them more useful and convenient in a classroom atmosphere.

Although a great deal of care is taken to see that the details are current, noncontroversial, and free of error, it is natural to expect errors and omissions due to the vastness in scope and breadth of this material. The author would be very grateful if these are brought to his notice at the following address:

Civil and Environmental Engineering Dept.
California Polytechnic State University
San Luis Obispo
CA 93407

The author also wishes to acknowledge the valuable contributions of Wei-Chou V. Ping of Florida A&M University/Florida State University and Mumtaz A. Usmen of Wayne State University, whose insightful comments in reviewing this manuscript are greatly appreciated.

Shan Somayaji

1

Introduction

Civil engineering consists of the design, construction, maintainance, inspection, and management of characteristically diverse public works projects, from railroads to high-rise buildings to sewage treatment centers. Their construction may be under or above ground, offshore or inland, over mile-deep valleys or flat terrains, and upon rocky mountains or clayey soils. The thought that all these creative efforts are made possible through the marvelous innovative spirits of civil engineers is in itself comforting and appealing, as well as challenging for prospective civil and construction engineers. The gigantic achievements of the past stand as a flashing beacon to promote the potential of civil engineering.

Although the profession of civil engineering per se is of fairly recent origin—the American Society of Civil Engineers (ASCE), the oldest national engineering society in the United States, was founded in 1852—the work of civil engineering is as old as humankind. The most ambitious and historically significant projects throughout the history of civilization were accomplished to satisfy the fundamental human needs for transportation, water, shelter, and disaster control. Nonetheless, the systematic approach to planning for the community's future, by training young minds professionally in all facets of civil engineering, is quite new.

1.1 MATERIALS AND METHODS—A HISTORICAL PERSPECTIVE

At the core of civil engineering rests the investigation of materials and methods that can satisfy the needs of the community. For example, shelter is provided for through housing; dwellings are built in accordance with a *method* that is appropriate for the *material* selected, the method of construction changing with the material.

Remnants of the methods and materials of civil engineering can be found in plenty among the records of ancient civilizations. In addition to fighting wars and conquering other kingdoms, rulers all over the world were involved in constructing facilities and building programs that catered to the public spirit. The first Babylonian dynasty of King Hammurabi (c. 1800 B.C.) initiated sweeping reforms and construction programs that were documented in contemporary manuscripts. King Sennacherib of Assyria (c. 700 B.C.), who was called a great engineer-king, built a dam across the river Tebitu, and from the reservoir thus created constructed many canals. The canal walls were built from cubes of stone and the canal floor had a layer of concrete or mortar under a top course of stone to prevent leakage. Nearly all Mesopotamian cities around that time were paved with slabs of stone and brick. The first emperor of the Chin dynasty in China, between 259 and 210 B.C., started the building of the Great Wall for protection from the Huns. The great Roman emperor Constantine I, after his conversion to Christianity, built the city of Constantinople and dedicated it as his capital (A.D. 330).

The handling of materials in construction necessitated proper tools. Stones had to be cut to proper size and shape before they could be used to build masonry. Trees had to be felled, cut, and shaped before they could be used in construction. Soil and mineral deposits had to be dug up prior to making bricks and cement. Metals provided the base material from which tools could be fashioned, making it possible to advance public work facilities with relative ease and safety.

The use of metals came in before 4000 B.C. with the advent of native copper—which was naturally occurring—for ornaments and utensils. During this period, it was learned that heat softens metals, promoting the technique of forming. The period 3000–1000 B.C. is known as the Bronze Age, due to the increased use of bronze, the first manmade alloy, formed by mixing molten copper and tin. Around the same time, elaborate techniques were developed for forming gold and silver jewelry. Following the Bronze Age came the Iron Age. Iron ore—mostly iron oxide—was heated in a charcoal hearth. The carbon in the charcoal reacted with the iron oxide, releasing carbon dioxide and producing a spongy mass of iron. Pure iron has a high melting temperature of around 1600°C. Though facilities for melting at this high temperature were not yet available, iron containing large amounts of carbon could be melted and cast into shapes, marking the origin of *cast iron*.

History shows us that ancient engineers were innovative and efficient in terms of the materials they chose and the methods of construction they employed. Sumerians, for example, around 3000 B.C., built houses with mud bricks joined by locally available bitumen. They are supposed to have constructed 18-ft (5.4-m) thick walls along a circumference of about 6 miles (9.6 km) to provide refuge for people and cattle. Mesopotamians built mud-brick huts without windows to keep out the sizzling heat of the summer sun. People in south India and Sri Lanka had houses made of wooden frames and removable reed mat for walls. Such houses, which are cheap and practical, are still being built.

Historical sites and ruins show us the skills of ancient builders. During the period of the Eastern Chou Dynasty in China, 770–250 B.C., a number of cities were built, usually rectangular or square in a north-south axis, surrounded by double walls and a moat. At Harappa and Mohenjo Daro in Pakistan, along the rich alluvial banks of the Indus river, are the remains of two large and expertly constructed cities dating back to 3000–1500 B.C. The cities were planned around a central citadel and constructed of good quality burned bricks. Elaborate municipal drainage and complex irrigation systems were without parallel during ancient times. Remains of domestic utensils and jewelry made from carved ivory, silver, copper, bronze, and earthenware indicate a well-developed technology of metals.

The Assyrians of Mesopotamia around 1100–750 B.C. knew how to construct buildings that could not be destroyed by fire. The building walls were made of stone, so that the fire burned off only the roof. One of the most important technological discoveries of ancient engineers—which brought revolutionary changes to twentieth-century civil engineering construction—was the introduction of hydraulic cement by the Romans, around 145 B.C. They discovered that the local sandy volcanic ash, when added to lime mortar, made a material that became as strong as rock when dried. They called this mortar *pulvis puteolanus*, and used it for gigantic endeavors like the Colosseum, and to build aqueducts, bridges, and roads, some of which are still in use. Their standard roads were 15 ft (4.5 m) wide and had a 4-ft (1.2-m) deep foundation formed of layers of stone, rubble, and concrete, and were topped with a surface of concrete, stone, and powdered gravel. Underground sewage facilities were installed because their cities were located in valleys between sharp hills. The walls of their buildings had thin facings of brick, stone, or marble. They built apartment houses of five or more stories, and provided public latrines that were flushed with water delivered from baths and industrial establishments. The Romans are also credited with building semicircular arch bridges using stones.

In Patiliputra, India, the houses around 300 B.C. were constructed of wood, and to protect them from fire an elaborate system of fire protection was enforced. Records show the construction of suspension bridges held by iron chains. Ingots of steel made in India were taken to Damascus, where they were converted into sword blades.

The historical records available all over the globe show us that the basic materials used in construction were either derived from the earth or made from plants. Every continent of the world possesses three basic types of surface characteristics: hard crystalline rocks, such as granite; mountainous belts of folded sedimentary rocks; and plainland basins filled with sediments. This means that the core of civil engineering construction or material technology is the same in every part of the world, though different materials have been used in different places, depending on the local availability and need.

In Mesopotamia, for example, the most abundant natural resource of the land was mud. Hence the city walls were made of clayey mud. Molding the clay into bricks made it possible to build straight walls without visible weak spots. The brick mold is believed to have been a Sumerian invention—around 3000 B.C.—and the use of molds made it possible to manufacture bricks that were flat on all six sides. The bricks were dried from a few days up to 5 years depending on the strength required. But independent of the extent of the drying period, the bricks softened and crumbled when they became wet. This led to the discovery of a new kind of brick—burned brick. The chemical changes in the clay during burning resulted in strong and durable bricks. But these bricks were costly due to the scarcity of fuel, and thus were employed only for the outside of important buildings.

Egyptian temple buildings had stone-paved floors supporting colossal (hollow) stone columns, holding up the loads from massive stone lintels. Stones were used exclusively in temple buildings and for tombs, for the Egyptians considered a tomb a house of eternity and a temple a house of a million years. In contrast, the houses for people—mortals—were built of mud and wood, and thus were not durable. The vast cedar forests of Lebanon supplied timber to Egypt and Mesopotamia, which had little good building timber of their own. Assyrians made use of swamp reeds as structural material for house construction. A bundle of reeds tied together served as a pillar to hold up a house of light construction. In building construction, the post-and-beam framing technique using timber owes its development to the Greeks. Into the Mediterranean basin they brought a tradition of using wood, featuring a sloping and pitched roof.

In the Babylon of King Nabopolassar, around 600 B.C., double city walls were built, the space between outer and inner filled with rubble, generally up to ground level. The Ishtar Gate of Babylon, built during the rule of King Nebuchadrezzar, around 580 B.C., was finished with enameled bricks, of blue for the towers, and green and pink for the connecting walls. The Babylonian roads were paved with massive stone blocks set in asphalt. Geologists have identified a 7.5-mile (12-km) stretch of road paved with slabs of sandstone and limestone, about 43 miles southwest of Cairo, Egypt, that may have been the world's first paved road, built roughly 4600 years ago, and used to transport heavy stones for the building of the pyramids.

This brief historical perspective on materials and methods of civil engineering shows that construction materials were, for the most part, of native origin and satisfied environmental compatibility as well as financial constraints. This statement applies to most (but not all) of the basic materials used in today's civil engineering facilities. Advances in engineering techniques, resource constraints, and cost-cutting measures are responsible for the introduction of a significant number of new materials into today's construction market. Although it is beyond the scope of this basic textbook on the properties and use of materials in civil engineering or construction, the aspect of material selection, is central in importance. The choice of a construction material should be made only after a detailed review of its long-term performance, its potential to effect the durability of other materials in the structure, and its environmental compatibility. For example, asbestos may appeal as a good construction material due to one or more favorable properties—fire resistance, in this case—as well as for financial reasons, but its long-term potential to cause environmental hazards and human discomfort must outweigh the immediate utility and financial dividends.

1.2 MATERIALS AND TYPES

An introduction to various types of engineering materials is given in this section. A *material* is defined as a substance or thing from which something else can be made. Cloth, cement, sugar, brick, aluminum, soil, and water are all examples of materials. In engineering, materials are employed to design and build structures, elements, or products. Buildings are made of concrete, tennis rackets are molded out of reinforced plastics, boats are carved out of wood, and roads are made from asphalt concrete. The subject of *materials science* examines the whys and hows of materials, making it possible to advance the development of new materials. The term *materials engineering* refers to the understanding and review of properties and uses of materials commonly used in engineering.

Materials can be divided into several categories; some of the common groups of materials are introduced in the following.

Amorphous Materials. Materials in which the atoms are arranged almost randomly, or those that do not have crystalline structure, are called amorphous materials. Generally these materials are strong but brittle. *Examples:* Soot (or, impure carbon) and glass. A crystalline material can be converted into an amorphous material by quenching—heating the material to its melting temperature followed by rapid cooling so that the material has no time to return to its crystalline arrangement. For example, sugar candy is produced from crystalline sugar. Similarly, an amorphous material, like glass, can be converted into a crystalline material, like glass ceramic, by a process called seeding.

Brittle Materials. Brittleness denotes relatively little or no elongation or increase in length at fracture. A material that exhibits brittleness is called a brittle material. *Examples:* Cast iron, concrete, and glass.

Building Materials. Materials that are used in the building industry, such as cement, steel, brick, plastics, wood, glass, ceramics, and concrete, are called building materials.

Cementitious Materials. Materials in which the principal binder is portland cement or another type of hydraulic cement are called cementitious materials. Concrete, mortar, grout, and roller-compacted concrete, which are obtained by combining cement, aggregates, and water, are the most common cementitious materials. The products of the reaction between cement and water form compounds that bind the aggregate particles together, so that the resulting material can be considered homogeneous. The aggregates are of two types, fine and coarse; and both contain particles of various sizes, from large to small. All cementitious materials are porous, the porosity depending upon many factors, such as the amount and type of cement, and the amount of water.

Ceramic Materials. The word "ceramic" comes from the Greek, meaning "burned earth" When something is burned, it combines with oxygen in the air; ceramic materials are nonmetallic materials often based on clay (silicate mineral).

They are usually crystalline and brittle, do not conduct heat or electricity very well, and can withstand high temperatures. When loaded, they remain mostly elastic and exhibit practically no plastic flow. Many ceramic materials are used for insulation—thermal (firebricks), building (fiberglass), and electrical. *Examples:* glass, cement, china, stone, and brick.

Clay Brickwork. Brick is a burned clay masonry unit, generally rectangular and solid. The term "brickwork" refers to masonry built with bricks and mortar, primarily as vertical members subjected to compressive and bending forces. The coefficient of thermal expansion of brickwork is approximately $5–7 \times 10^{-6}$ per °C, which is about half that of concrete and twice that of limestone. The expansion of clay brick from moisture is about one-fifth that of concrete. Even with their low coefficients of expansion, materials such as clay brickwork may present problems if used in large unbroken lengths, like in factory walls with ends restrained as from transverse walls or adjacent buildings. They should be provided with suitable expansion joints.

Composite Materials. When two or more separate materials are combined in macroscopic structural units, the resulting elements are called composite materials. Many materials that have two or more constituents, such as metallic alloys and polymer blends, are not composites, as the structural unit is formed at the microscopic rather than macroscopic level. Examples of composite materials are straw brick, paper, reinforced concrete, wood, and polymers reinforced with graphite (carbon) or glass fibers. It is also possible to form a composite by joining together two or more members. A steel plate glued onto a wood beam makes a wood-steel composite. The most common composite material in building construction is plywood, which is made from gluing together three or more layers of ply or veneer. But the term composite is more often understood to mean a material made from combining a fiber, a matrix, and some fillers. Wood is a natural composite.

Construction Materials. A construction material is any material used in the construction industry. *Examples:* Concrete, cement, soil, stones, aggregates, plastics, and asphalt.

Crystalline Materials. Materials in which the atoms are arranged in a discernible repeated pattern in three dimensions are called crystalline materials. The repeated unit is called the unit cell. *Examples:* talc, ceramics, quartz, and glass ceramics. Two materials can have the same atomic structure but different crystal structure, and vice versa. *Examples:* glass (noncrystalline) and quartz, and diamond and graphite (identical atomic structure); silicon and diamond (identical crystal structure). When a crystalline material is heated, the atoms receive the energy and begin to vibrate. With increase in temperature, the movement (or vibration) becomes very energetic and the material melts. In the liquid form, the atoms move very freely and are arranged quite randomly, and the material is said to be amorphous.

Ductile Materials. Ductility is the property that makes it possible for a material to be drawn out or stretched to a considerable extent, from a significant sustained load, before rupture. It is usually measured as the percentage of elongation (increase in length), or as the percentage of the reduction in the cross-sectional area, when the material is subjected to tension. *Examples:* mild steel, aluminum, and wood.

Elastic Materials. Elasticity is the ability of a material to deform under a load, without a permanent set or deformation upon the release of the load. Springs, rubber bands, and cricket balls behave elastically. Elasticity can also be defined as that property of a material by virtue of which deformations from a load or stress disappear after the removal of the load. Some materials (gases, for example) possess elasticity of volume only (that is, the volume is the only characteristic that remains unchanged), but solids such as metals may possess elasticity of form and shape as well. As an example, the top of a metal desk will not deform in shape, form or volume from the stresses caused by a stack of books piled on the desk. A perfectly elastic material should recover completely its original shape and dimensions when loads are removed. None of the materials known today remain perfectly elastic throughout the range of stress leading up to failure, but all exhibit elastic properties up to some stress level. Metals such as steel remain elastic over very high stress levels, whereas some materials such as polymers and concrete can be considered elastic only at low stress levels. An elastic material behaves inelastically when the stresses exceed the elastic limit, beyond which changes in the volume, shape, and form are permanent. In addition to the characteristics of the material, the elasticity depends also on the type of load: tensile, compressive, or shear. Temperature is also a factor that determines the range of elasticity. The high elasticity of a material, such as asphalt, at ordinary temperatures may diminish to a low value at elevated temperatures. In materials such as concrete and wood, a rapid loading rate generally increases elasticity whereas a slow rate of loading lessens it. A long-term load may produce a permanent set whereas a short-term load may not.

Elastomeric Materials. A polymer having elastic properties is called an elastomer, which means it can be stretched by large amounts, and that it will return to its original shape when the load is removed. Rubber (latex) and polybutadiene (also called synthetic rubber) are examples of elastomers.

Electronic Materials. An atom has a nucleus consisting of two types of particles: neutrons, which have no electrical charge (are neutral), and protons, which have a positive electrical charge. The nucleus is surrounded by electrons, which have a negative electrical charge. The number of protons in an atom is equal to the number of electrons, and the atom is said to be electrically neutral. Like the planets that go around the sun, the electrons go around the nucleus in fixed orbits, called electron shells. The size of a shell and the number of electrons in it become larger as the shell moves farther away from the nucleus. The maximum number of electrons permitted in each shell is $2n^2$, where n represents the shell number. Of the two lightest elements, hydrogen and helium, the former has one electron, and the latter, two, in their single outer shell. A carbon atom has six electrons, two in the first and four in the second. The second shell contains up to a maximum of 8 electrons, and the third up to a maximum of 18. Atoms that have $2n^2$ electrons in each shell are stable (or less reactive). When this is not possible, the atom becomes stabilized or unreactive if it contains 8 electrons in the outer shell. On the other hand, if the outer shell has only seven electrons, as in fluorine, iodine, chlorine, and bromine, the atom is reactive. Hydrogen is very reactive as it has only one electron in the first shell. Unstable atoms most effectively capture electrons from any atoms nearby to complete the shell. Elements such as oxygen, with two electrons missing, are not as reactive as flourine, with one missing electron in the outer shell.

Metals are elements that have a tendency to release their electrons. When the electrons from the outer shell of a pure metal atom are removed, leaving a unit with a net positive charge, the atom is called an ion, or cation because it is positively charged. If, instead, electrons are added, it is called an anion, for it is negatively charged. An electron current consists of charged particles—electrons—for the most part moving through a conductor. Electrons in some elements such as copper and aluminum are free to move and jump from one atom to another, and these materials are called conductors. Others, such as plastics and wood, which do not contain as many moving electrons, are insulators. When a material is neither completely a conductor nor an insulator it is called a semiconductor.

Insulating Materials. Materials that are provided for sound, thermal (heat), or electrical insulation are called insulating materials. The spread of molecular movement through bodies that are in direct contact constitutes the flow of heat, and the rate of such movement is described by the thermal conductivity of the material. Thermal conductivity is the ability of a material to conduct heat, and is defined as the ratio of the flux of heat to the temperature gradient. It is measured in joules per second per square meter of area of a body when the temperature difference is 1°C per meter thickness of the body (or in Btu per hour per square foot when the temperature difference is 1°F per foot thickness). The conductivity of air is lower than that of water; and the conductivity of a material increases sharply with rise in moisture. For thermal insulating materials the conductivity is very low, and for metals it is very high. Higher density materials normally have higher conductivity, but there is no direct relationship between the density and the conductivity. Concrete, which has higher density than wool, has a higher conductivity value; but steel, which is nearly three times denser than aluminum, has lower conductivity. If air in a porous material is replaced by water, the conductivity increases. Impurity atoms present in metals decrease their thermal as well as electrical conduction. Thus, when we want a good conductor, we select a pure material with large crystals. Ends of electrical wires in electronic equipment are often gold-plated, for gold does not react with the oxygen in the air or another surface. Heating a metal lowers its conductivity, whereas cooling improves it.

Thermal diffusivity represents the rate at which the temperature changes within a mass. Diffusivity is directly related to thermal conductivity, and is inversely proportional to specific heat, which is the heat capacity, and to the density of the material. Specific heat is the amount of heat energy necessary to cause a unit increase in temperature of a unit mass of the material. The more heat a material absorbs, the higher is the specific heat. Of all materials, water has the highest specific heat, and the specific heat of air is about 30 percent of that of water. Thermal resistance of a material is inversely proportional (is reciprocal) to thermal conductivity. The thermal insulation value of a material is obtained by calculating the U-value of that material, which is a measure of the thermal remittance through the material or through a member element made from it. It is expressed as the rate of transmission of energy—through a unit area of an element—per degree difference in temperature between the internal and external surfaces. The location of the material and the penetration of moisture will also influence the U-value. A thermal insulation material is a material with very low thermal conductivity, or a low U-value.

The thermal conductivity of metals is a hundred to a thousand times that of stone or brick. Rankings of some construction materials in terms of their thermal conductivity, in decreasing order, are as follows:

Metals: Silver (high), copper, aluminum, steel, lead (low)

Stones: Granite, limestone, sandstone

Others: Bricks, concrete, asphalt

Common materials of high thermal resistivity (low thermal conductivity) are ranked in decreasing order as follows:

Air, asbestos, cork, blanket, wool, foam (polystyrene), wood

When the energy from a sound wave hits a wall, it is partly absorbed and partly reflected. The sound absorption coefficient is a measure of the proportion of the sound energy striking a surface that is absorbed by that surface. It measures the ability of a surface to reduce the reflection of incident sound. Porous or irregular surfaces have high sound absorption, and smooth surfaces have very low absorption. The materials used in construction for reducing the transmission of noise from one room or space to another through walls, ceilings, and floors are called sound insulating materials. The sound transmission loss is the difference between the incident and the transmitted sound energy (that radiated into the next room). The loss, which is measured in decibels (dB), depends on the unit mass of the partition wall per unit area. Resistance of a material to airborne sound transmission increases with increasing density or mass of the material, but the effectiveness of the material for thermal insulation is reduced. Increase in the frequency of sound waves also increases the transmission resistance.

An unwanted sound from a vibrating body is generally termed noise, and this definition is subjective. The sources of noise are many: traffic, industry, aircraft, loudspeakers, machinery, and so on. Acoustics is the science of sound. The sound originating in a room will be reflected, absorbed, or transmitted through the walls, and it is desirable to limit both the amount reflected and the amount transmitted. The sound absorption potential of the wall affects the reduction of sound within a room. Increases in the porosity and roughness of the surface increase the amount of sound absorbed, decreasing the amount reflected. Thus, using porous units with a rough surface or providing surface undulations for walls can decrease the reflected noise within a room. Hard smooth plaster or painting may produce an elevated level of reflected sound. Changing the acoustical properties of walls, ceiling, floor, and furnishings can effectively modify the sound levels within a room.

Sound transmission through a material can be accomplished through two mechanisms: forced vibration and porosity. Sound from impact is transmitted through forced vibrations, which can be decreased through cushioning. Airborne sound passes through a wall by both forced vibration and porosity mechanisms. The sound absorption coefficient generally indicates the amount of sound not reflected (the part that is absorbed and transmitted) from a wall. The airborne sound emitted by a source can be decreased at the source by enclosing it within a sound insulating material, such as a heavy box with absorptive linings on the inside. Placing a screen between the source and the listener, and having the screen close to the source, can reduce the direct noise. Using absorbent materials can decrease the reverberant (re-reflected) noises. The sound transmission coefficient expresses the proportion of sound energy transmitted. The sound transmission loss (also called sound reduction index) gives the sound-reduction effect of an element. Porous materials are best absorbents at higher frequencies of sound—frequencies in excess of 1000 Hz. A singer's low note has a frequency of about 100 Hz, and the top note in a flute around 4000 Hz, where 1 Hz is equal to one wave per second. Membrane or panel materials such as plywood and felt are best absorbers at low frequencies. A combination of resonant and porous absorbers, such as a perforated panel, is best in medium frequencies. When a room needs to be treated for sound, the most likely surface to receive the treatment—the most critical element—is the ceiling, for it causes multiple reflections.

Magnetic Materials. When a material has a net magnetic moment or spin it is called a magnetic material. In certain atoms, the magnitude and direction of the magnetic moments associated with various electrons may be such that they all balance out to give a zero net magnetic moment. In certain other atoms, however, there are more electrons with spins in one direction than in the other, and the balancing may not be complete. The atom as a whole may therefore exhibit a net magnetic field. When such a material is placed in an external magnetic field, each atomic dipole experiences a torque, the effect of which is to align it in the direction of the field. The bulk of the matter, therefore, acquires a net magnetic moment in this direction, forming a magnetic material. Magnetic materials are classified into three types: diamagnetic, paramagnetic, and ferromagnetic materials. Diamagnetic materials are those repelled by an external magnetic field. *Examples:* bismuth, copper, mercury, and sodium. Paramagnetic materials are those whose atoms have a net intrinsic permanent magnetic moment because of the spin and orbital motion of their electrons. *Examples:* titanium, aluminum, magnesium, tungsten, and platinum. Ferromagnetic materials are derived from iron. The atoms of ferromagnetic materials also have a net intrinsic magnetic dipole moment, primarily due to the spin of the electrons. In such materials, the interaction between the neighboring atomic magnetic dipoles is described by quantum mechanics and is very strong. *Examples:* iron, cobalt, and nickel.

Manufacturing Materials. These are materials used in machinery or in manufacturing industries—industries that make products. *Examples:* metals, plastics, ceramics, and rubber.

Masonry Materials. A mason is one who builds with bricks, stones, and blocks. Masonry is the part of a building or structure that is made from combining the masonry units: stone, block or brick, and mortar. Egyptians built their pyramids (called *mastabas*) first using mud brick masonry and later (around 2500 B.C.) with stone masonry using gypsum mortar. Romans employed a type of masonry construction for walls in which the space between two parallel layers of burned brick was filled with concrete. Mortar from bitumen was used to bond the bricks in some early masonry construction. Masonry was also used for building columns and towers, such as the Tower of Pisa, and arches, such as the 83-ft span semicircular arch in the Basilica of Constantine (A.D. 313). Masonry walls are erected today using the same two types of materials: masonry units and mortar. The common masonry units are clay bricks and concrete blocks, although stones, mud bricks, and fly ash bricks can also be used. Masonry units can be solid (such as burned clay bricks) or hollow (such as hollow concrete blocks). The hollow spaces, called cells, in hollow-block masonry can be kept hollow or filled with grout. When a wall is built using a single layer of brick it is called a single-wythe construction, and the thickness of the wall is equal to the width of the brick. When a wall has two layers of brick separated by a certain distance it is called double-wythe construction, and the thickness of the wall is equal to the thickness of the two layers plus the space between the two, which is filled with grout. When the wall has reinforcing bars, running both horizontally and vertically, located in the cells or in the space between the two layers, the construction is called reinforced masonry. The arrangement used in laying masonry units is called the bond. The mortar serves as the binder required for the masonry units to be attached to each other, without which the masonry wall is nothing more than a stack of bricks—like a stack of books or cards. Mortar is made by combining portland cement, lime, sand, and water.

Metallic Materials (Metals). Generally classified as ferrous and nonferrous, metals are used in construction and manufacturing. Ferrous metals are principally iron-carbon alloys containing small amounts of sulfur, phosphorous, silicon, and manganese. Nickel, chromium, molybdenum, and vanadium are some of the metallic alloys of iron. Ferrous metals come in three forms: steel, cast iron, and wrought iron. The principal nonferrous metals—which are more than 70 in number—are copper, nickel, zinc, aluminum, and magnesium. More than 90 percent of the world consumption of metals is in the form of steels and cast iron.

Polymeric Materials. The word polymer is from the Greek: *poly* means many and *meros* means parts, and polymer means a material with many parts. A polymer is defined as any material that belongs to a group of carbon-containing (organic) materials, natural or manmade, that have macromolecular structure: very large molecules made up of repeating molecular units. Monomers are the small molecules from which polymer molecules are made. A polymer can be represented as a large molecule containing hundreds or thousands of atoms formed by combining one, two, or (occasionally) more types of monomers into a chain structure. *Examples:* timber, rubber, polyethylene, polyester, polyvinyl chloride, wool, cotton, silk, starch, and cellulose. Some polymers (such as polyethylene, polypropylene, nylon, and polycarbonate) are crystalline or semicrystalline, and others (such as acrylic and polystyrene) are amorphous. Most polymers are plastic, which means that they can be easily bent into different shapes. The first synthetic polymer produced in the United States was Bakelite, commonly used to make electrical plugs and switches.

Plastic Materials (Plastics). Plastics are organic-based materials derived primarily from the petrochemical industry, which are capable of being formed into any shape. They are also defined as synthetic organic materials that can be molded under heat and pressure into shapes that will be retained after the removal of heat and pressure. Plastics are often referred to as resins, and consist of few basic polymeric materials often mixed with dyes, fillers, additives, or reinforcement such as glass fibers. Divided into two types, thermoplastic and thermosetting, all plastics are combustible and their mechanical and physical properties vary. They have relatively large coefficients of thermal expansion (around three to ten times that of steel) and low moduli of elasticity. Ultraviolet light can break their large molecules into smaller ones, resulting in discoloration, loss of gloss, and loss of ductility. Mechanical stress may also cause molecular fracture. Plastics are used as floor coverings (vinyl tiles and carpets), surface finishes (polishes and sealers), adhesives (formaldehyde), wallpaper, pipes, window frames, shutters, claddings, and so on. Polyester, glass-reinforced polyester (GRP), nylon, phenol-formaldehyde, polycarbonate, polystyrene, and polyvinyl chloride (PVC) are some of the common plastics. The basic raw materials for making plastics, supplied by only a few companies around the world, are acquired by industrial organizations to formulate final products with specific characteristics needed for use. It is quite common to use several types of polymers in combination to obtain optimum results.

Raw Materials. Natural products or materials that are transformed through manufacturing processes are called raw materials. *Examples:* Coal, petroleum, iron ore, and limestone.

Repair Materials. These are materials used to repair a deteriorating structure of concrete, masonry, or steel. They may include several classes of materials such as fillers (materials used as the base for the sealant in full-movement joints), sealants (to seal the joints), waterproofing compounds, and materials for general repair work. *Examples:* rubber (filler), cork (filler), mastics such as asphalt (sealant) and hot-applied rubber-bitumen compound (sealant), polyurethane (sealant and repair mortar), cement mortar, and concrete.

Semiconductors. Conduction of electricity in materials is caused by the motion of electrons. Electrons can be visualized as particles that carry charge, and are commonly called charge carriers. The conduction of electricity improves with an increase in the number of carriers. The movement of electrons in a material is controlled by the effective speed of movement, called electron mobility. In general, the conduction of electricity depends on the number of electrons that are available, the speed of movement, and the method (or ease) of travel. In every material that conducts electricity, there is a basic speed at which electrons can travel. In silicon, this speed is more than 2000 mph. Electrical conductivity is the measure of a material's ability to conduct electricity. At one end, some materials can insulate from a very high voltage gradient (they let no electrons go through); at the other, superconductors allow electrons to keep on traveling forever, which means that a current set to move in a closed loop continues to flow indefinitely. Many metals, chemical compounds, and alloys exhibit superconductivity at temperatures very near absolute zero.

In metals, the electrons are naturally free and move about, with just as many going in any one direction as in another, resulting in no electron flow. To produce a net flow, or current, the speed of movement should be increased so that the energy of some electrons is increased. Applying a voltage through a metal wire lifts the electrons to higher levels, and the electrons accept this energy by increasing their speed and producing a current. Metals are electrical conductors, for their electrons can take on the energies offered by even a small voltage source. For both insulators and semiconductors, the energy levels required for the electrons to move at a higher speed do not exist at normal voltage gradients. As the electrons cannot increase their velocity, no current is produced, making these materials poor conductors.

Though, metals can take on the energies offered by even a small voltage source and set up a net flow of electrons, so that an electrical current flows, in semiconductors there are no such available energies, and therefore something else is needed to lift the electrons up the large gap between the energy bands. The energy from light can be used to make an electron jump over this gap, and this event causes the material to have a metallic luster and look. Thus, a silicon chip in a calculator is encapsulated, for exposure to light will turn the semiconductors into good conductors, causing shorts. However, light does not have enough energy to lift the electrons over the energy gap in an insulator. Thus an insulator, which is colorless and transparent, remains an insulator even when exposed to light. But we can make any material conduct electricity by exposing it to x-rays or by heating it to a high enough temperature. Semiconductors, which have a conductivity intermediate between that of metals and insulators, are used in electronic components, solar cells, transistors, radar detectors, and so on. *Examples:* silicon, germanium, and gallium arsenide.

Thermoplastic Materials. Materials that turn plastic (soft) when subjected to heat are thermoplastic. Petroleum pitch, which is the black viscous liquid that remains after gasoline has been removed from crude oil, is a thermoplastic material. It is a hard and brittle solid at room temperature, but becomes a viscous liquid at higher temperatures. The term thermoplastic material, however, is more commonly applied to the kind of polymer that becomes soft at higher temperatures so that it can be shaped. *Examples:* Polyethylene, polyvinyl chloride, polypropylene, and nylon.

Thermoset Materials. Synthetic polymeric materials that become more rigid when heated are called thermosets. *Examples:* phenol-formaldehyde, urea-formaldehyde, epoxy, and unsaturated polyester.

Waste Materials. The waste matter produced by a manufacturing facility is called a waste material. *Examples:* fly ash (from coal-burning power plants), carbon dioxide (from heating limestone), and sludge (from sewage treatment facilities).

1.3 CIVIL ENGINEERING MATERIALS

The basic materials used in civil engineering applications or in construction projects are:

- Wood
- Cement and concrete
- Bitumens and bituminous materials
- Structural clay and concrete units
- Reinforcing and structural steels

These are sometimes called *structural materials*. Added to these are plastics, soils, and aluminum. All these materials are employed in a variety of civil engineering structures such as dams, bridges, roads, foundations, liquid-retaining structures, waterfront construction, buildings, and retaining walls. The basic materials most common to highway construction are soils, aggregates, bituminous binders, lime, and cement.

Wood is derived from trees, and can be put to use directly, as pieces of lumber cut from a log, or as a raw material in the manufacture of various wood products or manufactured components. Plywood, glue-laminated timber, and oriented strandboard are some of the wood products most commonly found in the construction of buildings and bridges.

Concrete is one of the most common construction materials, in which portland cement is the essential ingredient. Portland cement (and other types of hydraulic cement) is also a key ingredient in the manufacture of many other cementitious products, such as masonry blocks, soil-cement bricks, and plaster. In combination with other materials, such as reinforcing bars, polypropylene fibers, and high-strength strands or wires, different types of concrete are produced, such as reinforced, fiber, and prestressed concrete.

Bitumen, which comes in a variety of forms, is mixed with other raw materials for the construction of pavements, roof shingles, waterproofing compounds, and many other materials. Structural clay and concrete masonry units, commonly called bricks and blocks, are the principal elements in the construction of masonry walls. Structural steel, which is fabricated in many forms and shapes, is employed in the construction of railroad ties, high-rise buildings, roof trusses, and many more structural elements.

These basic materials or products are selected for their properties, performance, availability, aesthetics, and cost. Knowledge of all these aspects is essential in selecting a suitable material for a particular situation.

In addition to the materials mentioned above, there are a significant number of secondary construction materials common to engineering projects. Sealants, adhesives, floor and wall coverings, fasteners, and doors and windows fall into this category. Most of these, also called *nonstructural materials,* are chosen based on quality guidelines and aesthetic considerations.

1.4 PROPERTIES OF ENGINEERING MATERIALS

Materials for engineering applications are selected so as to perform satisfactorily during service. The material for a highway bridge should possess adequate strength, rough surface, and sufficient rigidity. A water-retaining structure would be built with a material that is impermeable, crack-free, strong, and does not react with water. A road surface needs such materials that show little movement under the impact of loads, are water-resistant, and are easy to repair.

Performance requirements, or property specifications, are not the same for all structures or structural materials. What is expected of a material used for the construction of a liquid-retaining structure is not the same as that chosen for a pavement. To evaluate the performance characteristics of engineering materials, and to assist an engineer in the selection of the most appropriate and economical material for a particular application, one needs to study the properties of the materials of construction. In general, the common properties of engineering materials are grouped under three major headings:

- Physical properties
- Mechanical properties
- Chemical properties

Physical properties are those derived from the properties of matter or attributed to the physical structure. They include density, porosity, void content, moisture content, specific gravity, permeability, and structure (micro or macro). In addition, properties such as texture, color, and shape fall under this classification. Physical properties are helpful in evaluating a material in terms of the appearance, weight, permeability, and water retention of a structure.

By knowing the specific gravity, the density (or mass per unit volume) of a material can be established. Perhaps the earliest use of this term was by Archimedes (287–212 B.C.), who discovered that a comparison of the specific gravity of the material used in the king's crown with that of a block of pure gold is enough to establish the purity of the gold in the crown. The knowledge of permeability of materials lets us compare them in terms of their effectiveness as moisture barriers. An understanding of porosity and moisture content of construction materials is essential in assessing the performance of structures during service.

Mechanical properties measure the resistance of a material to applied loads or forces. Some reflect the strength of the material, whereas others measure the deformation capacity or stiffness. Strength is a measure of the maximum load per unit area, and can be in relation to tension, compression, shear, flexure, torsion, or impact. If we compare the physical strength of one individual with the emotional strength of another, we know that the two "strengths" are not the same and that the comparison is inappropriate. The same reasoning can be applied when describing the strength of a material; it is important to specify the type of strength.

It should be noted that a material might be subjected to a combination of loads that would result in more than one type of stress, or pressure, acting on it. For example, a floor joist may be subjected to loads that cause a bending moment and a shear force as well as torsion at a certain section. In this type of situation, measurement of the inherent strength—relative to the forces acting on it—becomes analytically and experimentally complex; in many situations, it cannot be done. As in the maintenance of good health in the human body, for which both emotional and physical strengths are critical, so in the performance of a construction material, which requires adequate strengths of all types.

The deformation capacity of a material (its stiffness) is assessed through its elastic modulus, defined in subsection 1.4.2. A knowledge of the strength (or various types of strength) and the deformation characteristics of materials is absolutely essential in the selection process. A high-strength material may not necessarily possess adequate deformation capacity or stiffness, and vice versa.

In addition to strength and deformation capacities, the mechanical properties include other measurements such as brittleness, plasticity, and ductility each of which reflects the deformation behavior of the material.

Chemical properties are those pertaining to the composition and potential reaction of a material. The compounds of composition, such as oxides and carbonates, describe the chemical nature of the material, and the way it would behave in a certain environment. For example, by reviewing the proportions of the principal compounds in various cements, we will be able to choose the right type of cement for a particular application. Knowledge of the chemical composition of clays is indispensable in evaluating the characteristics expected in burned bricks. Chemical properties such as acidity, alkalinity, and resistance to corrosion of materials are especially noteworthy.

In addition to the physical, chemical, and mechanical properties, the thermal, electrical, magnetic, acoustical, and optical properties of materials are also of relevance in civil engineering. For example, the coefficient of thermal expansion of concrete, which is a thermal property, is fundamental in assessing the expansion potential of concrete slabs. Thermal properties, customarily, represent the behavior of a material under heat and temperature. Acoustical properties such as sound transmission and sound reflection are critical in choosing materials that should offer sound resistance and function as sound barriers. Optical properties such as color, light transmission, and light reflection are considered in determining the energy consumption capacity of a material. Measures of other properties, such as electrical conductivity and magnetic permeability, are needed in materials used in electrical works.

In civil engineering construction, though some materials are selected primarily for their physical properties or characteristics, most are chosen because of their mechanical properties and durability. For example, lightweight aggregates, such as pumice and shale, are selected for the manufacture of lightweight concrete floors due primarily to their low density. In areas of high seismic activity, structural steel is preferred for the columns and beams of high-rise buildings over reinforced concrete, for its high tensile strength and ductility.

Thus a proper understanding of the environment and the constraints within which a particular project is to be developed is crucial in the material selection process. The goal of engineering design should be to select the most appropriate material for a particular job. A general knowledge of all relevant properties of the various materials that are available, and an appreciation of their performance characteristics, are fundamental in achieving this goal. A brief description of general strength and deformation properties and the definitions of some common terms in solid mechanics are presented in the following sections.

1.4.1 Forces, Loads, and Stresses

When a body is pulled or pushed, it is said to be acted upon by a *force*. If a chair is pushed downwards by sitting on it, it does not move, for the solid floor pushes upward against it with an equal force. If the chair is pulled horizontally with sufficient force, however, it moves. Thus, when a body that is free to move is pushed or pulled the force—which is external—causes the movement. And we can conclude that force can cause motion. This leads us to Newton's first law, which states that every body continues in its state of rest or of uniform motion in a straight line unless it is acted on by a force. Newton's third law states that if one body exerts a force on a second, the second body always exerts upon the first a force that is equal to the first force, and that acts in the opposite direction.

When a solid body is subjected to external forces, called *loads,* the body is deformed and internal forces are produced (Fig. 1.1). The internal forces or internal stresses that act between consecutive particles are said to be proportional to the external loads. This "balancing act" between the internal and external forces is stated to maintain the equilibrium condition in the body.

To illustrate this theory, let us take a rubber band. When the two ends of the rubber band are pulled apart, it stretches, which is the deformation or the change in length from the original length—extension in this case. The change in length gets larger with the increase in pull, which is the applied or external force. But each time we increase the force or the pull, the rubber band elongates a little bit more and then stays at that length. As long as the force is held constant at the two ends, the extension remains unchanged at that moment in time. This balancing act between the pull and the extension illustrates the equilibrium condition.

Finally, when the force applied at the ends exceeds a certain limit, the rubber band snaps, which illustrates the limitation of the equilibrium theory. However, if we

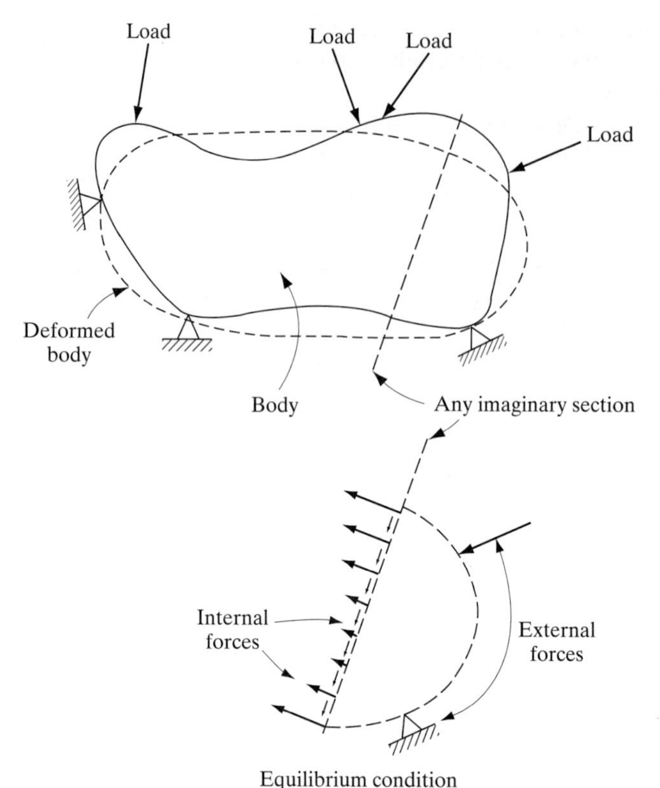

Figure 1.1 Solid body subjected to external forces.

can now rejoin the broken ends of the band with glue or cement, the "new" rubber band can once again be used for the same experiment. The test can be repeated, ending in the fracture of the band a second time, perhaps at a different level of pull. Just before the failure, the external pull on the rubber band is at its maximum ("at failure"), and the extension is the largest.

Therefore it can be stated that the internal forces, or stresses, acting on any imaginary section through a body of a material—rubber band in this example—must be in equilibrium with the external forces acting on either side of this section. This theory provides us with a procedure to calculate the stresses and deformations in a material subjected to external forces. However, it should be noted that this principle of equilibrium applies to a body at rest. When a body is in motion, additional forces from the acceleration must be considered, as explained in Newton's second law of motion.

In the example of the rubber band, what we are able to observe or measure in the material is its extension when subjected to the pull, and not the stress. Materials undergo extension or contraction depending upon the type of load and the direction of measurement. This change in length immediately following the application of a unidirectional force is called *deformation* or *linear deformation*. The deformation per unit length is called *strain*.

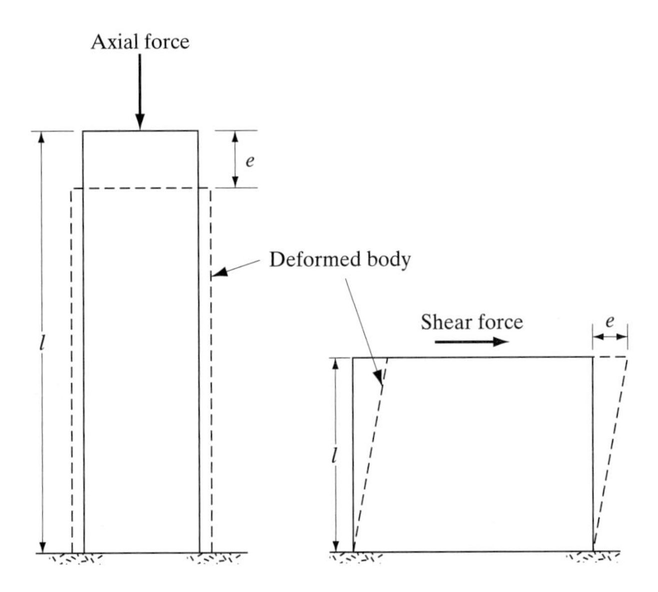

Figure 1.2 Deformation.

Strain is therefore a ratio between the change in length and the length over which the change is measured, called *gage length*. It is a dimensionless quantity, sometimes expressed as inches per inch length, or m per m (Fig. 1.2).

$$\text{Strain } e' = \frac{e}{l}$$

where e is the change in length over a length l.

When a 2-in. long rubber band stretches to a length of 3 in., the deformation is equal to 1 in., the gage length is equal to 2 in., and the average strain (over the gage length) is equal to (1 in./2 in.), or 0.5 in. per in. This means that a 4-in. long rubber band has to be stretched to a length of 6 in. to have the same strain as the stretched 2-in. long rubber band. In other words, the 4-in. long rubber band with 2-in. extension has the same strain as the 2-in. long rubber band with 1-in. extension. It can thus be seen that the concept of strain is helpful in comparing the effects of a certain load or force on bodies of different shapes and sizes.

The term *stress* describes the measure of a force acting on a unit area of an imaginary section through a body. As indicated earlier, there are as many types of stresses as there are forces. A force acting along the axis of a member causes an *axial stress* or *direct stress* (Fig. 1.3a). When an axial force is in tension, the resulting stress is a *tensile stress,* and when it is in compression, the stress is a *compressive stress.*

Shearing stress is produced by forces that tend to slide one particle upon another, and it acts along or parallel to the cross-sectional plane (as shown on the square element in Fig. 1.3b). A tensile or compressive stress acts perpendicular to a plane and is thus normal to the shear stress.

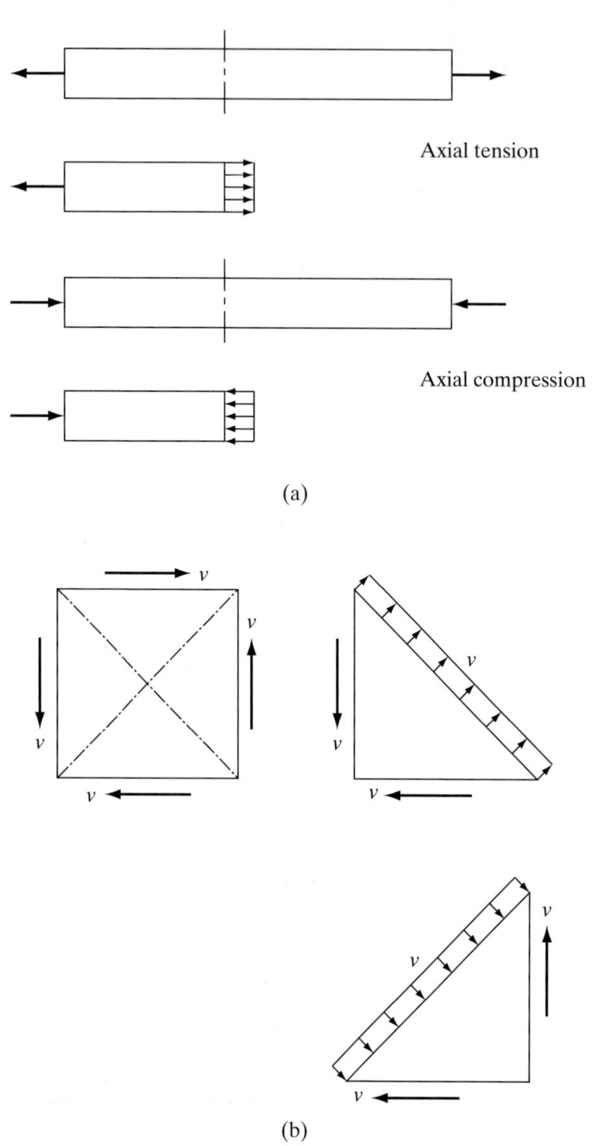

Figure 1.3 (a) Axial and (b) shear stresses.

A *bending, normal,* or *flexural stress* is produced by external forces that create a bending moment. Due to the action of the moment one side of a cross-section has tensile stresses (the bottom of the beam in Fig. 1.4), and the other side has compressive stresses (the top of the beam in Fig. 1.4).

1.4.2 Strain

It was pointed out that the change in linear dimension is called deformation. But this term—deformation—is also used to indicate the change in form of a body, and

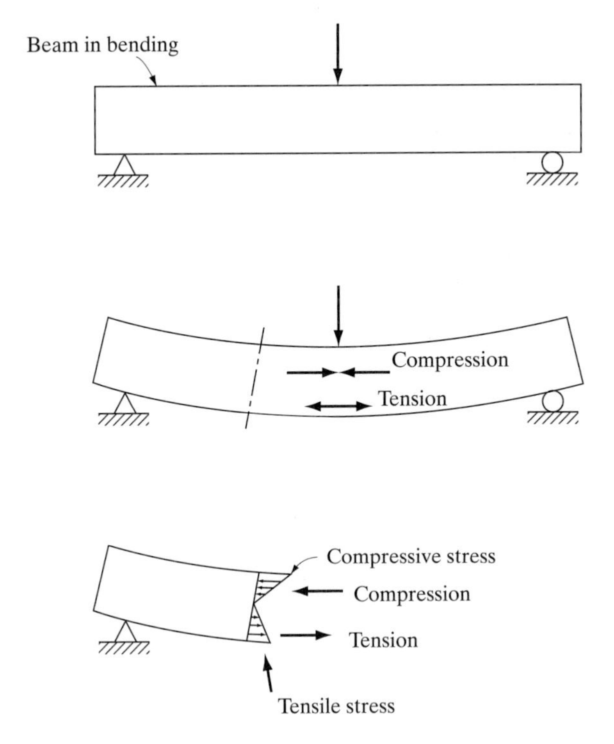

Figure 1.4 Bending stresses.

can be the net result from more than a single cause. Thermal changes, moisture loss, and applied loads all cause deformation in the body. When explaining the effects of direct stresses, the term deformation is taken to mean the change in linear dimension; while discussing the effects of shear forces, *shear deformation* is the change in length measured parallel to the direction of the shear forces. The *shear strain* is computed as the shear deformation per unit length measured perpendicular to the direction of the shear forces (Fig. 1.2). Therefore, shear strain is the change in the angle between the two adjacent sides of the element, and is expressed in radians.

When a body is subjected to an axial stress (tensile or compressive), deformation takes place not only in the axial direction but also in other directions perpendicular to the axis. The ratio of lateral strain (normal to the axial direction) to axial strain (when loaded along the axis) is called *Poisson's ratio,* named after the French scientist who defined it in 1811. When a body is subjected to an axial tensile force along its length, it stretches over its length and contracts along the sides; similarly, when it is compressed along its length (as in a pillow, sponge, or eraser), it stretches out along the sides.

The Poisson's ratio of most materials of construction ranges between 0.15 to 0.40. For glass, its value is 0.24, and for granite the value is 0.25. The Poisson's ratio for concrete is between 0.1 and 0.18, depending on the proportions between ingredients. Increase in cement content increases the Poisson's ratio. Cement mortar has a Poisson's ratio of about 0.16. The Poisson's ratios of common materials are listed in the following table.

Material	Poisson's ratio
Aluminum	0.25
Brass	0.32–0.35
Cast iron	0.23–0.27
Concrete	0.1–0.18
Copper	0.31–0.34
Glass	0.24
Lead	0.43
Steel	0.27–0.30
Stone	0.20–0.34
Wrought iron	0.27–0.29
Metals in plastic range	0.50

The change in the volume of a material is called the *volumetric deformation*. The change in volume per unit volume, or the *volumetric strain,* is calculated as the ratio between the change in volume and the initial volume. In an element of rectangular cross-section (cubic element) subjected to an axial force, the volumetric strain can be computed using the axial strain and the Poisson's ratio:

$$\text{Change in volume} = l \times b \times d \times (1 - 2p_r)e'$$

where l, b, and d are the dimensions of the member, p_r is the Poisson's ratio, and e' is the axial strain. Since e' is a small quantity, the terms involving $(e')^2$ and $(e')^3$ are omitted in the derivation.

Using the longitudinal or axial strain and the Poisson's ratio, the volumetric strain or volumetric deformation per unit deformation can be calculated using the following equation.

$$\text{Volumetric strain} = \frac{lbd(1 - 2p_r)e'}{lbd}$$

$$= (1 - 2p_r)e'$$

1.4.3 Stiffness

Stiffness is a relative measure of the deformability of a material under load. A material that develops a high level of strain at a given stress is less stiff than a material showing less strain under the same stress. In other words, the greater the stress needed to produce a known strain, the stiffer the material. A rubber band requires only a slight load to stretch it by an inch, whereas a paper clip is nearly impossible to elongate, even by a fraction of an inch. Steel is stiffer than cast iron, which is stiffer than concrete.

The stiffness of a material is measured in terms of its modulus of elasticity. *Elasticity* is that property of a material that enables it to change its length, volume, or form in direct response to an applied force, and to recover its original size or form when the load is completely removed. Some materials, such as steel and rubber, can endure large amounts of strain or deformation and still return to their original form when the loads are taken off. Concrete (in compression) is an example of a material that will recover its original size and shape only when the applied stresses are low; when the stress exceeds a certain limit and then is brought back to zero, the shape and size are permanently altered.

The *elastic limit* is the maximum stress below which a material will fully recover its original form upon the removal of applied forces. It is also defined as the greatest stress that can be applied without causing a permanent deformation. In some materials, such as concrete, brick, and stone, the elastic limit is low. For most metals the elastic limit is high. To establish the value of the elastic limit experimentally, a load should be applied and released a number of times, the magnitude of the load at each application being somewhat larger than at the preceding one. This experimental procedure is tedious and is not generally carried out. Instead, elastic limit is approximated as the proportional limit or yield point of the material by plotting its stress-strain diagram.

The *proportional limit* is the maximum stress below which the ratio between the stress and the strain is constant. It is ascertained by plotting the stress-strain relationship of a material subjected to uniaxial tension or compression. Most materials exhibit the same elastic properties in both tension and compression.

Figure 1.5 shows typical stress-strain diagrams for two materials, labeled type A and type B. The ordinate of the graph represents the stress and the abscissa the strain. The basic procedure followed in drawing such a plot is to collect a series of extension or strain readings against the data on load or stress. If loads and extensions are recorded, they can be converted to stresses and strains by knowing the element's cross-sectional area and the gage length. At a given stress, the strain in material A is smaller than that in material B. Thus, per the definition of stiffness given above, it can be concluded that material A is stiffer than material B.

In both diagrams we note that the stress is linearly proportional to the strain, up to a certain stress. The stress at which the plot deviates from the straight line is the proportional limit. When the element is loaded up to a stress that is less than the proportional limit, and then unloaded, the unloading plot follows exactly the same straight line as the loading plot. When the load is taken off completely, the element regains its initial size and shape, and the deformation is zero. Because of this characteristic, the proportional limit is generally used to mark the elastic limit of a material.

A short distance above the proportional limit in material A, the curve stays horizontal over a brief portion of the strain, indicating that the material is stretching without increase in load. The corresponding stress is called the *yield point.* A common meaning of the word "yield" is to concede under some pressure but not to surrender totally (e.g., a yielding mattress). In a material, yielding indicates, in a similar fashion, that the material is flexing a little but not giving up completely. The yield point is the first unit stress at which deformation continues without increase in load. For most materials, the yield point is higher than the elastic limit or the proportional limit. In steel, this difference is about 5 to 15 percent of the yield point. But the exact

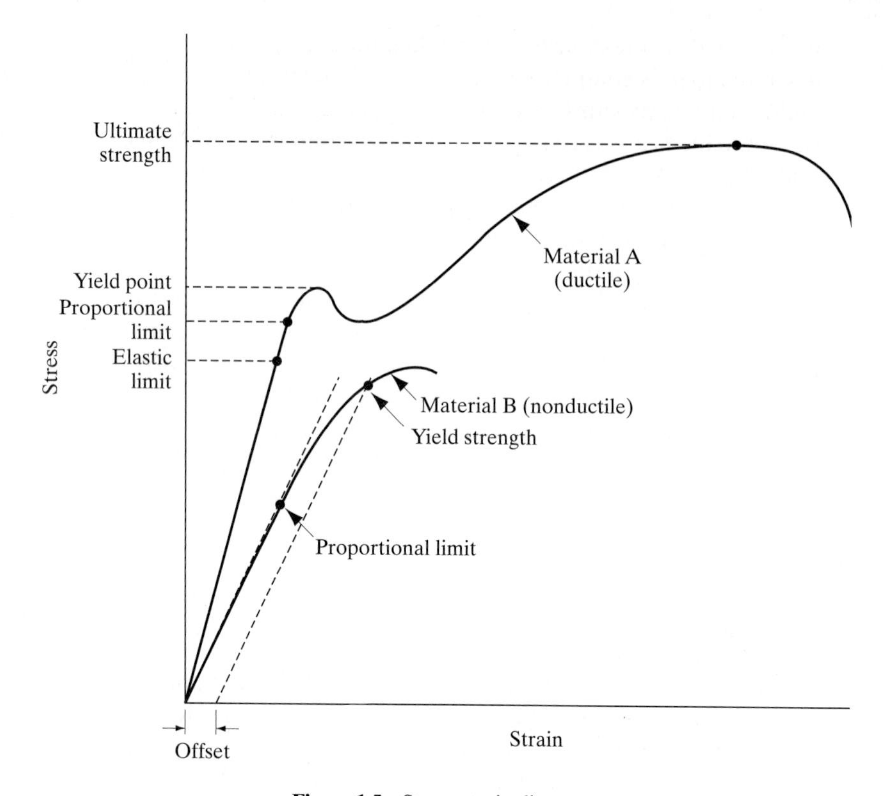

Figure 1.5 Stress-strain diagram.

point of departure of the curve from the straight line is difficult to determine; for this reason, for most materials, the elastic limit (proportional limit) is taken as the yield point of the material.

Some materials, such as material B in Fig. 1.5, do not have a yield point. Metals such as copper, aluminum, and high-strength steel exhibit this behavior. With increasing load, the stress-strain diagram deviates from the initial straight line, and when the material is unloaded from any stress in this nonlinear part of the diagram, a permanent—nonrecoverable—deformation exists in the material. This strain or deformation that remains in a previously stressed material after it is unloaded is called *set* or *permanent set.*

In materials that do not exhibit yielding, the yield point and elastic limit are predicted using the permanent set or offset approach. The stress at which the material displays a specified amount of offset upon removal of the load is the *yield strength* of the material. It should be noted, however, that the yield strength thus measured does not indicate yielding in the material at this stress; it is merely a convenient term for the measurement of a practical elastic limit.

In the offset method of yield strength prediction, a small value of offset is chosen—generally, based on recommended standards, between 0.1 and 0.3 percent (strain of 0.001 and 0.003). This value is measured from the origin—the point of intersection of the stress-strain curve with the strain axis—and marked on the strain axis. Then a line parallel to the initial straight-line part of the diagram is drawn from this offset point (Fig. 1.5). The stress corresponding to the point of intersection of this line with the stress-strain diagram is the yield strength.

The ratio of stress to strain below the proportional limit is called the *modulus of elasticity* (also called the *elastic modulus* or *coefficient of elasticity*). It refers to the stiffness in the elastic range and is generally identified by the capital letter "*E*."

There are three moduli of elasticity, which correspond to three types of stress: tension, compression, and shear. They are thus the modulus of elasticity in tension, the modulus of elasticity in compression, and the modulus of elasticity in shear (or *modulus of rigidity*). The modulus of elasticity in tension or compression is also called *Young's modulus,* named after Thomas Young, who is credited with defining this constant in 1802. When unqualified, the term modulus of elasticity invariably implies a normal stress, compression or tension. For most materials, the modulus of elasticity in tension is the same as that in compression—this is true for most materials of construction, such as steel, concrete, brick, and wood. The modulus of elasticity of all grades of structural steel is 30×10^6 psi (206,700 MPa), of reinforcing steel is 29×10^6 psi (199,800 MPa), and of structural aluminum is 10.5×10^6 psi (72,350 MPa). The modulus of elasticity of most construction materials, like concrete, wood, and cast iron, varies greatly. For example, the modulus of elasticity of wood depends on the species, grain direction, and defects, and that of concrete on mix proportions, type of aggregates, and age at the time of testing. Similarly, wide variations in the values of E can be expected between different compositions of cast iron or brass.

The modulus of elasticity in tension or compression, or Young's modulus, can be calculated as

$$E = \frac{f}{e'}$$

where f is the stress and e' is the strain. This observation that the strain is linearly proportional to the stress was first made by Robert Hooke in 1678, and is presently known as *Hooke's law.* Since the shape of the stress-strain diagram varies from material to material, there is more than one method to calculate Young's modulus. The slope of a line drawn tangent to the stress-strain diagram at the origin is the *initial tangent modulus* (Fig. 1.6). When the modulus is measured as the slope of a tangent drawn at a specified stress f_s, it is called the *tangent modulus at stress* f_s. The *secant modulus* is the slope of a line drawn connecting two points on the stress-strain curve.

The ratio between shear stress and shear strain is the modulus of elasticity in shear, the modulus of rigidity. By comparing the shear deformations under a load to the corresponding axial strain, it can be shown that the shear modulus (E_s) and the Young's modulus have the following relationship:

$$E_s = \frac{E}{2 \times (1 + p_r)}$$

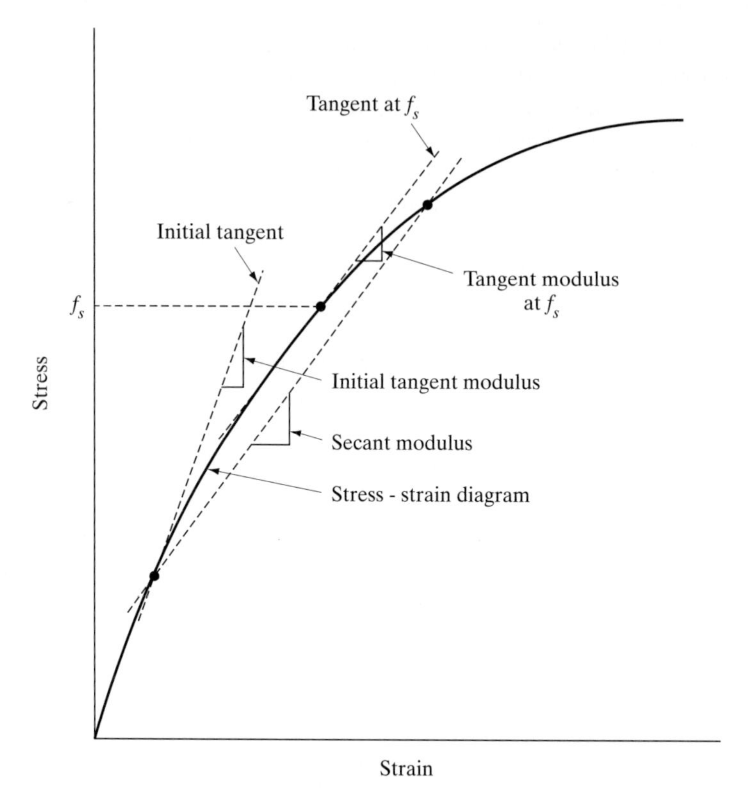

Figure 1.6 Modulus of elasticity.

where p_r is Poisson's ratio. It is difficult to determine the shear modulus experimentally, and therefore the above relationship provides a convenient path to establish it.

Ultimate strength is the maximum stress that can be applied to a material before it fractures (Fig. 1.5). It is the highest point on the stress-strain diagram. In some materials, such as steel and wrought iron, the area of cross-section at fracture is much smaller than the initial area.

1.4.4 Ductile and Brittle Materials

Plasticity is the property of a material that enables it to retain permanent set or deformation without fracture; it is thus the opposite of elasticity. (A common meaning of plasticity, which applies to soil, is the capability of being made to assume a desired form. Plastic clay is a material that has this property). Plasticity is an important requirement in the forming, shaping, and extruding of polymeric materials. In addition, plasticity shows a characteristic of materials that is important in predicting the type of failure under the action of loads. Construction materials are generally divided into two classes:

- Plastic or ductile materials
- Brittle materials

When subjected to tensile loads, a *ductile material* is capable of undergoing a high level of plastic deformation before failure. *Ductility* is the property that allows a material to undergo change of form without breaking. Wrought iron, steel, and copper are examples of ductile materials, which can exhibit large amounts of deformation as an indication of eventual failure. Concrete, mortar, brick and glass fail at small deformations or strains, and are brittle. Strong and ductile materials show higher levels of toughness than brittle or soft materials, as shown in Fig. 1.7.

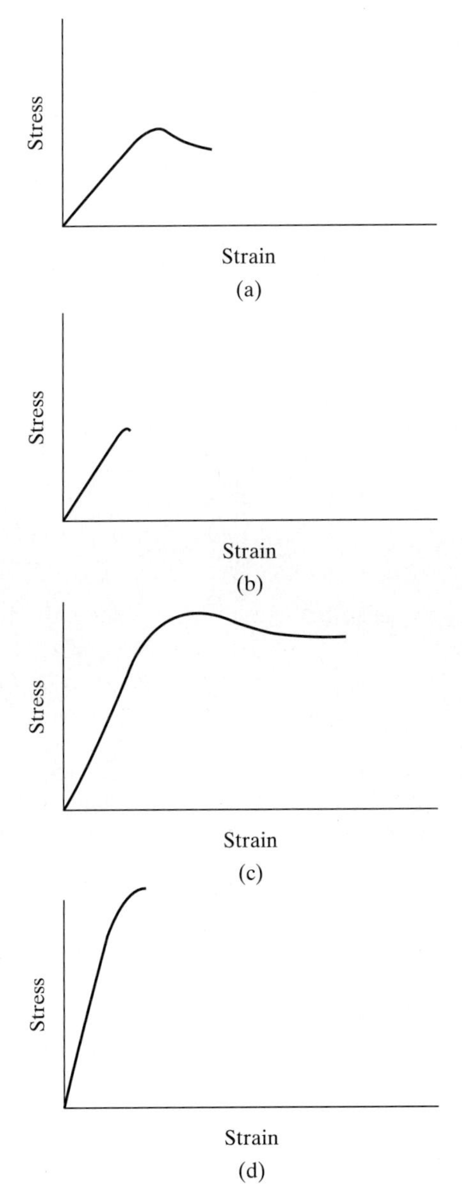

Figure 1.7 Stress-strain behavior of various materials: (a) soft and weak; (b) weak and brittle; (c) strong and tough; (d) strong and hard.

Toughness represents the ability of a material to support loads even after yielding or crack formation. It enables a material to endure shock or blows, and an increase in toughness signifies an increase in the amount of energy needed to produce a specified damage condition. The energy generates work on the body, which is the product of the deformation and the load or average stress. For most materials toughness is measured as the area under the load-deformation diagram.

Brittle materials have little or no plasticity, or they show little deformation beyond the elastic limit. Brittleness also signifies breakage with a comparatively smooth fracture (as peanut brittle), and failure in brittle materials is sudden and catastrophic. Some common brittle materials are cast iron, granite, brick, masonry, and glass, which at failure either crush or crumble with very little warning (Fig. 1.8).

Figure 1.8 Catastrophic failure of a concrete parking structure from earthquake.

When subjected to a compressive force, ductile materials simply spread, enlarging the area of cross-section. Brittle materials, however, fail by shearing along certain angles or planes (Fig. 1.9). The rupture in brittle materials can be along a diagonal plane, or in the shape of a cone or pyramid. Cylindrical specimens generally fail along a diagonal plane, for example in cast iron, or as a pyramid with splitting above, for example in concrete. The angle of rupture, called the *shear plane* or *plane of rupture,* measured from the horizontal, depends on the

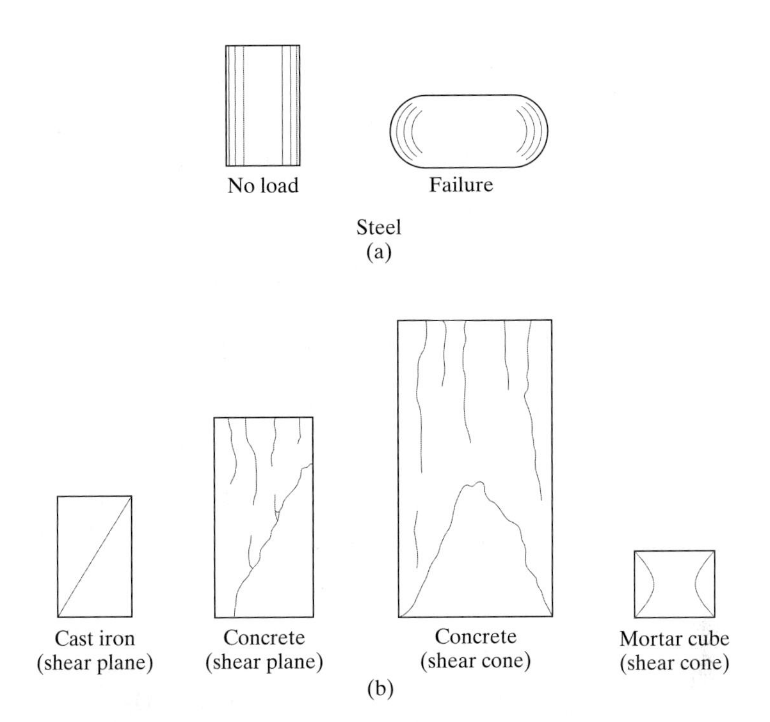

Figure 1.9 (a) Ductile and (b) brittle materials in compression.

internal friction and cohesion in the material, and generally lies between 45 and 60 degrees.

1.5 SELECTION OF MATERIALS

All primary materials of construction, or structural materials, must perform the following functions:

- Carry prescribed loads
- Satisfy serviceability and durability requirements
- Be aesthetically pleasing
- Be economically practical
- Be environmentally friendly

The most significant requirement of a material used in civil engineering projects (such as roads, buildings, dams, foundations, bridges, and power plants) is that it be able to carry the design loads. In other words, the material should have adequate strength. Concrete used as a foundation slab should be able to carry the load from the superstructure and pass it on to the ground below safely and without causing settlement; a wood beam supporting a timber floor should be strong enough to transfer the floor loads to the supporting walls; the masonry of a gravity dam should possess satisfactory compressive strength; and so on.

As explained, however, strength is a generic term that means different things in different applications. Lumber used as a floor joist or signpost must command high bending strength, but when used as a pile must have significant compressive strength. However, no matter how the loads are applied, all materials of construction are required to carry them safely.

In addition to strength, materials are required to satisfy serviceability requirements, such as deformation limits, durability aspects, constraints on performance, and adaptability. Typically, serviceability implies satisfactory performance at all times. Recorded data on past performance, laboratory test results, and established practices will help to assess the serviceability aspects of a material.

In general, a material should be able to satisfy nearly all functional aspects pertaining to a specific job. For example, a material whose surface is slick and slippery when moist is a poor choice for paving roads; a mortar that loses its binding property and crumbles with time is not a good choice for the construction of masonry walls; lumber that deteriorates from moisture or is susceptible to termite attack cannot be used in an exposed setting. When using varieties of materials in the same project the effects of the combination of two or materials on the durability of the structure cannot be ignored, especially in assessing the long-term performance.

A material is sometimes selected based solely on its appearance. Although most nonstructural materials—such as floor coverings, paints, and doors and windows—are chosen based on aesthetic considerations, the same demand also exists for some structural materials. The type of masonry suitable for public building may depend on how the finished surface will look or how well it will match with the ambient architecture. A concrete wall surface may be given different finishes depending on aesthetic appeal. Materials for the exterior finish of a residential building, such as stucco and siding, may be selected based on appearance.

In addition to the aspects just discussed, one of the key factors governing the choice of material is the cost or the economic constraints. A typical residential building in the United States has wood or wood products for all its elements (walls, floors, and roof), but the same structure in Mexico, India, or Saudi Arabia will have masonry for its walls and concrete for roof and floor. These choices are based, in addition to environmental factors, on the cost of construction; timber construction is cheaper in the United States but is prohibitively expensive in many other countries.

In every location around the world, the selection of appropriate building materials is driven primarily by economic considerations, which depend on the local availability of a material, training of the labor force, duration of construction, and usage. A typical one-story shopping mall in the suburban United States may be designed using perimeter masonry walls, tubular steel columns, and timber roof. This type of construction, using various types of materials, has several advantages, for it combines the lightness of wood, the ductility and ease of erection of steel, and the fire-resistance of masonry, and it stays economical. But within the downtown area of a city, the shopping center may be located inside a high-rise building of structural steel, again to meet economic considerations.

Environmental compatibility of a material is gaining priority in the materials selection process. A construction material should satisfy all strength, serviceability, and architectural requirements, and at the same time, must not cause environmental problems. For example, a concrete pavement for a large parking space may induce flooding of adjacent land by forcing rainwater runoff or preventing water from soaking into the ground. Cement dust charging out of cement plants or filtering through batch plants may be carried through wind into the atmosphere, causing hazardous living conditions in the neighboring communities. Finish materials in a building, such as carpets, floor coverings, and wall panels, may give out dangerous fumes in case of fire, creating life-threatening conditions.

Consequently, aspects pertaining to the pollution and environmental threats resulting from the use of materials must be investigated, albeit the most common or conventional materials of construction are known to be environmentally compatible. A flowchart describing the materials selection procedure is shown in Fig. 1.10.

1.6 STANDARDS

When we use a construction material in any civil engineering project, a number of questions come up. For example, if we are designing a concrete footing to anchor a fence post, the series of questions may include the following: What is "concrete"? What are the proportions of concrete ingredients? Is there more than one type of concrete? What type of aggregate should be used? How much water should be added while mixing? Should the sand remain dry before adding to the mixer? What procedures should be followed after placing the concrete? What changes in the mixture are necessary when the ground temperature is low? Can the same concrete be used for other types of work, such as a sidewalk?

Knowing the answers to all these questions requires that one be proficient in all aspects of concrete construction, including a thorough understanding of cement chemistry, hydration processes, geology, concrete behavior, and thermal effects. Even then, there is no guarantee that the concrete thus manufactured will be satisfactory; it can fail because of bad quality cement, weak aggregates, contaminated sand, and many other practical reasons. To guarantee satisfactory performance from the materials used in construction projects, they should be manufactured to satisfy minimum quality and product standards. In addition, the methods of use or construction should follow standard procedures.

To establish uniformity in materials and products, and methods of production, application, inspection, and testing, many organizations have developed standards for materials, testing, and inspection. Over the years, these guidelines and mandatory recommendations have gone through series of revisions and been updated. In addition to material standards, several organizations also present specifications for the method of application.

Standards represent efforts by organizations—private (like the American Iron and Steel Institute), government (such as the Arizona Department of Transportation), or voluntary (for example, the American Concrete Institute)—for agreement on common procedures or goals. In the United States, there are thousands of

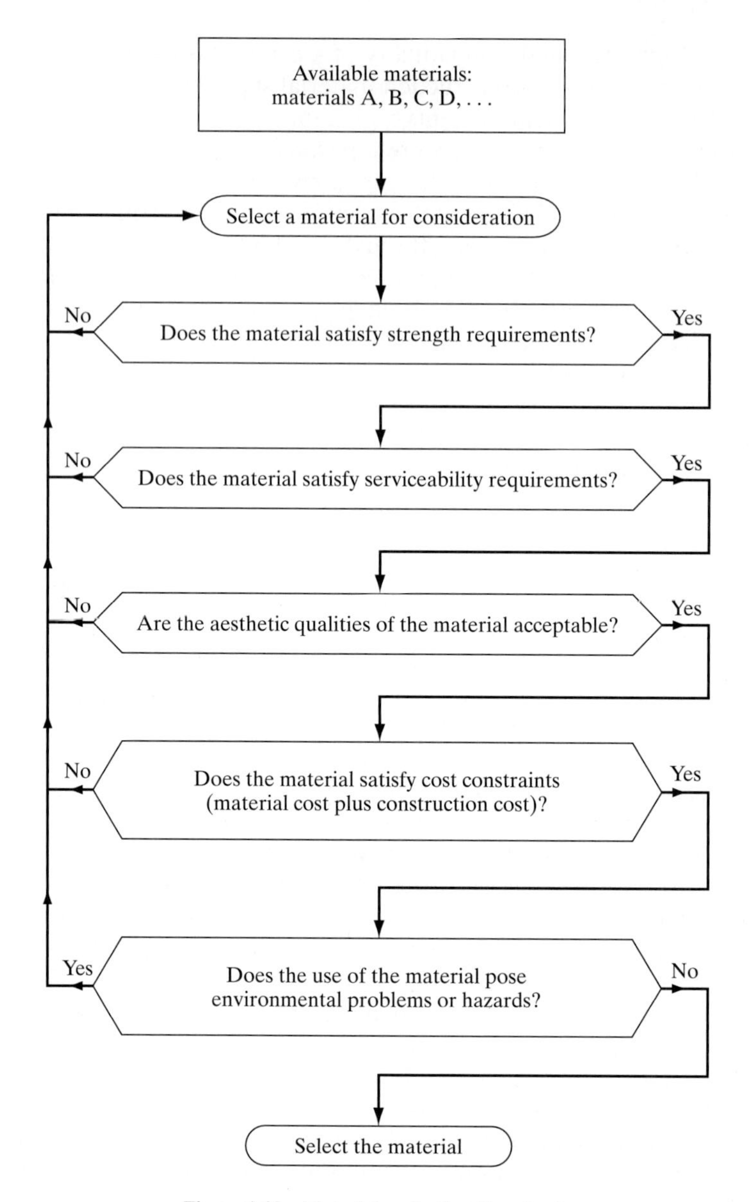

Figure 1.10 Materials selection flowchart.

standards developed by various groups and recognized by many user agencies. The American National Standards Institute (ANSI, 1430 Broadway, New York, NY 10018), National Forest Products Association (NFPA, 1619 Massachusetts Avenue, N.W., Washington, DC 20036), Portland Cement Association (PCA, Old Orchard Road, Skokie, IL 60076), American Concrete Institute (ACI, P.O. Box 9094,

Farmington Hills, MI 48333), American Institute of Steel Construction (AISC, 400 N. Michigan Avenue, Chicago, IL 60611), American Iron and Steel Institute (AISI, 1000 16th Street, N.W., Washington, DC 20036), and Brick Institute of America (BIA, 1750 Old Meadow Road, McLean, VA 22102) are some of the prominent organizations involved in the development of product and use standards for common civil engineering materials. In addition, many local, county, city, and state agencies, defense organizations, and private user groups have published independent standards and specifications.

The standards adopted by the American Society for Testing and Materials (ASTM, 100 Barr Harbor Drive, West Conshohocken, PA 19428, est. 1898) are of relevance and great help to anyone interested in the manufacture, use, and testing of materials. A number of documents published by this society (referred to as ASTM standards or specifications in this text) deliver commonly adopted standards for materials and laboratory testing. The materials standards include definitions, formulation of materials specifications, and recommendations on construction practices, and the testing standards show the development and description of laboratory procedures.

Where feasible, relevant ASTM standards are referenced in this book. Even though the laboratory procedures described at the end of each chapter are based on the recommendations of the ASTM, they may not represent them precisely. The objective pursued in this text is neither to represent the ASTM (thus duplicating the techniques and specifications presented by them), nor to advocate new, simplified, or different procedures and standards. The aim of this text is to introduce varied aspects of material specifications and material testing to the student of civil or construction engineering, thereby to develop an appreciation of material characteristics and an overview of the physical, mechanical, and chemical properties of construction materials—in other words, to get a feel for the importance of knowing these properties for the design and construction of safe and durable engineering structures. The reader is cautioned against using the procedures described in this text, which are conveniently simplified for use in academic laboratories, to satisfy agency specifications; the book is not a substitute for ASTM or any other standards.

Many of the products available in the construction industry or in supply stores are manufactured to satisfy one or more pertinent ASTM standards. In other words, by knowing the names or numbers of the standards or specifications, it is possible to establish the chemical, physical, or other properties of these materials or products. For example, a clay brick unit of grade SW (ASTM Standard Specifications C62) is expected to have been manufactured to satisfy the physical requirements, such as minimum compressive strength, described in C62. Consequently, this masonry unit can be used in any project that stipulates requirements matching with those of C62, without actually verifying the properties of the brick or testing it. Thus it can be seen that the process of standardization has helped us in choosing the right material, from a list of available materials, for a given situation, without the need for individual testing of available products.

Though the text makes frequent reference to a number of product standards and testing procedures recommended in ASTM standards, there are a good number of other private and public agencies that write their own specifications. The use of a material in any project should be based on the specifications of the agency that runs and manages the project. Its rules, procedures, and standards should be reviewed and followed before recommending a material for the project.

PROBLEMS

1. Name three materials used in wall construction.
2. Name three materials used for slabs.
3. Give two examples each of ductile and brittle materials.
4. Name two composite materials used in building construction.
5. Name three important physical properties of materials.
6. A 25-in. long member deforms to a length of 26 in. when subjected to an axial force. What is the strain?
7. Define Poisson's ratio.
8. Define yield point, proportional limit, and offset.
9. Explain the difference between tangent modulus and secant modulus.
10. Draw typical stress-strain diagrams for (a) aluminum and (b) rock.
11. Name two ASTM standards.
12. What are the assumptions made in the bending theory?

2

Aggregates

Aggregate is a rocklike material of various sizes and shapes, used in the manufacture of portland cement concrete, bituminous (asphalt) concrete, plaster, grout, filter beds, railroad ballast, base course, foundation fill, subgrade, and so on. The ASTM standards (C125 and D8) define aggregate as a granular material such as sand, gravel, crushed stone, or iron-blast-furnace slag used with a cementing medium to form mortar or concrete, or alone as in base course or railroad ballast.

Aggregates are derived primarily from rocks or stones of various types, which will be discussed in the following section. Subsequent sections discuss different types of aggregates and their properties, and examine how they affect the properties of common construction materials such as portland cement concrete and asphalt concrete.

2.1 ROCKS

Most aggregates are obtained by crushing and breaking stones or rocks, and are thus employed without alteration of their natural state. Rocks, which are aggregations of minerals, are divided—in the usual geological classification—into three family categories: igneous, sedimentary, and metamorphic.

Igneous rocks make up about 95 percent of the earth's crust. Rocks such as granite, basalt (trap-rock), and lava (or magma, which is the hot, molten material from deep underground that reaches the surface of the earth), are formed by consolidation from a fused or semifused condition, and are usually nonlaminated and nearly crystalline in structure. The dark-colored rocks formed by solidification at or near the earth's surface and composed of microscopic crystals are frequently referred to as trap-rock. Igneous rocks are classified based on texture (as coarse-grained, fine-grained, or glassy), and on mineral composition. Granite is coarse-grained, basalt is fine-grained, and pumice is glassy.

Sedimentary rocks, such as sandstone, limestone, dolomite, and shale, are formed by the solidification of materials transported and deposited, primarily by water and ice. Covering nearly 75 percent of the land surface, these rocks are the hardened layers of sediments of mainly igneous rocks that were physically, chemically, or biochemically weathered and eroded. They are distinctly stratified, featuring the original cleavage planes, and divided into two types: detrital (formed of particles that are cemented together) and chemical (formed from the precipitation of mineral matter). Sandstone and shale belong to the detrital category, and limestone and gypsum to the chemical category.

Types of rock that are formed by the gradual changes in the structure and character of either igneous or sedimentary rocks through heat, water, pressure, and other agents are called metamorphic rocks. Examples of metamorphic rocks, which have crystalline structure, are gneiss, quartzite, marble, and slate. In addition, slate, gneiss, and schist possess foliation or foliated texture (rearrangement of mineral particles into a parallel alignment), whereas marble, serpentine, and quartzite are nonfoliated, or possess no cleavage. A metamorphic rock is generally coarser, denser, and less porous than the rock from which it was formed.

A blanket of loose, noncemented, disintegrated rock fragments and mineral grains formed from weathering of bedrock is called *regolith.* The part of regolith that can support rooted plants is called *soil.*

2.1.1 Some Common Rocks and Rock Minerals

Atoms are the building blocks of all materials. When the ordered arrays of atoms, called *lattices* or *space lattices,* form a group, it is called a *crystal.* And many crystals form a pattern, called *solid material.* A *grain* or *crystalline grain* is a crystal without smooth faces. *Crystallization,* similarly, is the process of arranging of the atoms in a space lattice to form small solids of regular geometric outlines such as cubes, hexagons, or tetragons. When a material retains a random disorder of atoms or does not have a geometric arrangement of atoms, it is called *amorphous material.* The most common amorphous materials are glass, polymers, and asphalt.

Silicon (Si), which is the basis of all electronics, comprises about 25 percent of the earth's surface. An opaque element with a shining metallic luster, silicon is commonly found in the form of sand (silicon oxide, SiO_2) and has a large number of crystal structures, often distorted. Quartz, which is also SiO_2, and is used to simulate diamonds and in watches, is a clear, colorless crystal. Silicon, iron (Fe), calcium (Ca), and sulfur (S) are some of the many elements found in rocks, and soils, as minerals.

Minerals—formed naturally and usually inorganic—are crystalline substances having characteristic physical and chemical properties that depend on their composition and internal structure. They constitute the elementary building blocks from which rocks are made. All samples of a particular mineral have identical atomic arrangements of component elements. A few minerals (such as carbon, iron, gold, and silver) are elements, but most are chemical compounds. They originate by precipitation from solutions, the cooling and hardening of magma, the condensation of gases, or metamorphism.

Minerals can be identified by a number of physical properties, such as cleavage (which is the tendency to break off in certain preferred directions), fracture, color, crystal shape, hardness, specific gravity, streak, and striations (bands that run across surfaces of certain minerals).

Hardness—determined by scratching the smooth surface of a mineral with the surface of another material of known hardness, also called scratch hardness—is a common physical characteristic used in the identification of minerals. The hardness defines the material's resistance to penetration, and the standard laboratory hardness tests depend on resistance to plastic deformation. A mineral with a high hardness (or hardness number, or Mohs number—named after the German mineralogist Friedrich Mohs who defined it in 1812) will scratch a mineral with a lower hardness number. Diamond, which is the hardest natural mineral known, with a hardness of 10, is about forty times as hard as talc, which has a hardness of 1. In comparison, asphalt has a hardness of 1.3, mineral gypsum 2, fingernail 2.5, mineral calcite 3, and mineral quartz 7.

As mentioned, under proper conditions most minerals form crystals, which have three-dimensional shapes formed from atoms and occur in 32 classes, which in turn are divided into 7 possible systems based on the relationship of their axes: cubic (also called isometric), hexagonal, monoclinic, orthorhombic, tetragonal, triclinic, and trigonal. When a crystalline material is heated, the atomic structure—the movement of atoms around their mean position—becomes more energetic and is called vibration. Eventually the movement becomes so energetic that the atoms break the bonds that hold them together and the material melts. In this liquid form the atoms move freely and have a random disorder—as in an amorphous material. As the liquid is cooled, the atoms start to move back into their crystal positions. However, when the liquid is cooled rapidly—a process is called *quenching*—the result is an amorphous material.

Unlike crystals, glass has no ordered arrangement of atoms (that is, no crystal form) and is amorphous. It softens easily on heating and has no definite melting point as do true solids. Mineral quartz, which is an impure variety of sand, possesses glasslike characteristics and the same molecular composition as glass. When heated above its melting point and rapidly cooled, the silicate chain in quartz forms glass—a solid with disordered structure. This illustrates that the same mineral can exist in both crystalline and noncrystalline form depending on the process of manufacture.

Common silicate minerals are quartz (silica, SiO_2), opal ($SiO_2 \cdot nH_2O$), glass (SiO_2), feldspars (potassium aluminum silicate, sodium aluminum silicate, or calcium aluminum silicate), and hornblende (complex silicates of calcium, sodium, aluminum, iron, and magnesium). The minerals of the feldspar group are the most abundant rock-forming minerals in the earth's crust, and are the important constituents of all three major rock categories. Some iron oxide minerals are magnetite (Fe_3O_4) and hematite (Fe_2O_3). Two common carbonate minerals are calcite ($CaCO_3$) and dolomite ($CaCO_3$, $MgCO_3$). In addition, gypsum ($CaSO_4 \cdot 2H_2O$ or $CaSO_4$) and sulfate alkalis (Na_2SO_4) are also found in rocks. A large group of hydrous aluminum and potassium silicate minerals goes by the name mica. It is often found as sheet mica and also occurs in such rocks as crystalline limestone, dolomite, and serpentine. Nearly 55 percent of the mineral composition of slate is mica.

Common Rocks. The principal mineral constituents of granite are quartz and feldspar, with varying amounts of other minerals such as mica and hornblende. These silicate minerals are randomly oriented and distributed. Granite exists in a variety of colors—ranging from green to gray and from pink to red—making it highly attractive as a veneer in such uses as countertops, fronts of buildings, and floors. High in strength, hardness, and roughness, its quarrying is facilitated by the existence of planes of weakness, making it possible to remove large rectangular blocks from the parent rock. A great number of ancient structures around the world were carved out of or built using granite. The Buddhist temples of Ajanta, in the state of Maharashtra in India, were hollowed out of granite cliffs nearly two thousand years ago.

Volcanic (lava) rocks like basalt may contain a large number of holes, but are fine-grained and strong like granite. Shale, a soft, soil-like rock, is primarily composed of clay and silt. Some types of shale may be hard, strong, and durable, but many others, without a durable binder, are more like hard compacted soils.

Limestone is the name applied to all stones that contain either carbonate of lime (calcite), or carbonate of lime-magnesium (calcium magnesium carbonate, dolomite) along with calcite, as an essential constituent. These rocks, which come in many different colors, structures, and textures, contain other constituents (impurities) such as oxides of iron and silica, clay, and bitumen. Common types of limestone have very fine-grained structure and are gray or buff in color. Iron oxide gives the brown, yellow, or red color to limestone, and carbon impurities the blue, black, or gray color. Most deposits of limestone were formed from the skeletons of marine animals, but a few were chemically precipitated from solution. Limestone is used in many applications, such as for making iron, as a raw material in the manufacture of cement and quicklime, as aggregate filler in portland cement concrete and asphalt concrete, and for load-bearing masonry. Quicklime (CaO), also called caustic lime, which is derived from limestone (and discussed in detail in Chapter 4), is a colorless crystalline or white amorphous substance employed in the manufacture of glass, porcelain, and portland cement, in masonry construction, and for treating soil. The Sphinx of Giza—a lion's body 66 feet tall and 242 feet long, topped with the head of a man—was carved from layers of fifty-million-year-old red limestone, the color preferred by the ancient Egyptians. The pyramids were built using limestone and lime or gypsum mortar.

Marble, the term commonly applied to any limestone or dolomite that can take good polishing, is limestone that has been exposed to metamorphic action, rendering it more or partly crystalline. Principally used as ornamental stone for its aesthetic qualities, both indoors and outdoors, marble is found in numerous colors depending on the type of impurities present. Like all types of limestone, marble is corroded by water and acid fumes, and may prove unsuitable for use in some exposed places. The Taj Mahal—the 245-foot-tall monument built around 1630 by Emperor Shah Jahan to mark his monumental love toward his beloved wife—was constructed using white marble.

Quartzite, a rock composed of firmly cemented quartz grains, results from the metamorphism of sandstone. The fracture zone in quartzite, as opposed to that of sandstone, is across the cemented sand grains. Chalk is a soft calcium carbonate mineral similar in composition to limestone. The chief constituents of this rock are the shells of minute animals called foraminifera. The white cliffs of Dover, England are deposits of chalk formed from algae. Chalk is used in the manufacture of cement and quicklime, and as a writing and marking material. The harder variety is also used as a building stone.

Sandstone—rounded and angular grains of sand (nearly 80 percent silica or quartz) cemented together and compacted to form a solid rock—is found in many colors and varying hardness. The character of the cementing material—which is silica, lime, iron oxide, or clayey matter—determines the quality of sandstone. A siliceous bond results in hard and durable rock, whereas iron oxide, which gives the red or brown tone to the rock, makes it durable but weak. Lime as a cementing material, on the other hand, makes sandstone soft and less durable. The types of sandstone containing varying amounts of feldspar or mica grains are not as strong as those that contain only quartz. Friable sandstone—from which grains can be liberated easily by rubbing—has a weak, incomplete binder. Sandstone is widely used in construction and industry, and after crushing can also be used for ordinary sand.

Properties. Rocks and crushed stones, which are used as aggregates in various construction projects, are generally very durable, a property that is a measure of the ability of a material to withstand deterioration from weathering elements—primarily water. Rocks that are not durable have a tendency to cause the breakdown of components or structure. Structural imperfections such as fissures and cracks provide access for water, leading to frost action; the extent of destruction from the resulting expansion depends on the grain texture, mineral composition, and structure. Fine-grained textures withstand the volume changes from temperature fluctuations better than coarse-grained textures. Dense rocks are less pervious than porous rocks. Mineral constituents such as sulfides and iron oxide are least resistant to weathering action, whereas aluminates and silicates can resist decay agents very well. Carbonates (calcium and magnesium) weather rather rapidly.

Absorption in rocks, which is the ability to retain water, is directly dependent on porosity; rocks with large and straight pores absorb rapidly, whereas those with small and tortuous pores absorb slowly. Porosity, which is defined in Section 2.3, indicates the volumes of pores relative to the total volume of the rock (the volume of solids plus the volume of pores). Porosity can be as high as 20 percent in some types of sandstone and as low as 0.3 percent in marble. In addition to pores that exist within the structure of a rock, microfissures—intracrystalline and along the crystal boundary—which are planar cracks found commonly in hard rocks, also contribute to increase in absorption. Granite (an igneous rock) and slate (a metamorphic rock) absorb very little—less than 1 percent of the dry weight—and have been used as durable roof covering. Sandstone, which contains a large proportion of interconnected pores, and some types of limestone, may absorb as much as 10 percent. In addition to controlling the transmission of fluids through the interior, fissures and pores contribute to inelastic response to loads, reducing the tensile strength of the rock.

Density or unit weight of a rock depends on the porosity, type of minerals, and crystalline structure. The values of specific gravity—which measures the density of a material in relation to the density of water—of common rock-forming minerals range from 2.1 to 7.6. Gypsum has a specific gravity around 2.3; quartz, 2.65; calcite, 2.72; dolomite, 2.81; and pyrite, about 5.0. The density of fine-grained rocks like granite is about 165 pcf (pounds per cubic foot) or 2650 kg/m^3, and that of mica is 175 pcf (2810 kg/m^3). Typical physical and mechanical properties of common rocks are listed in Table 2.1.

TABLE 2.1 SOME PROPERTIES OF COMMON ROCKS

Type of rock	Specific gravity	Porosity (%)	Density pcf (kg/m^3)	Compressive strength ksi (MPa)	Modulus of rupture ksi (MPa)	Modulus of elasticity ksi (MPa) $\times 10^{-3}$
Granite	2.65	0–2	165 (2650)	15–35 (103–241)	1.2–2.2 (8.3–15.2)	6–10 (41.3–68.9)
Limestone	2.66	0.5–30	168 (2700)	5–35 (34.4–241)	0.25–2.7 (1.7–18.6)	4–14 (27.6–96.5)
Marble	2.63	0–1.5	175 (2750)	10–30 (68.9–206.7)	0.85–2.3 (3.4–15.8)	4–14 (27.6–96.5)
Sandstone	2.54	1–20	160 (2580)	7–30 (48.2–206.7)	0.5–2.0 (3.4–13.8)	1–7.5 (6.9–51.7)
Slate	2.74	—	170 (2740)	—	—	—
Shale	2.00	2–30	140 (2255)	—	—	—

2.2 TYPES OF AGGREGATES

Crushed stone, sand, and gravel are the three main types of aggregate commonly used in the manufacture of portland cement concrete and asphalt concrete, used in buildings, bridges, highways, dams, and airports (Fig. 2.1). Although aggregates are available at relatively low costs in most locales around the globe, availability is not universal. Some areas may be devoid of good quality sand and gravel, and a few others lack sources of crushed stone or rocks that can be mined easily and economically. Furthermore, aggregates available locally may be unsuitable for use in construction due to poor quality, potential for chemical reaction, and low strength.

Open-pit mining and quarrying are the two practices commonly used to extract aggregates from the parent rock. After drilling and blasting, the broken rocks are extracted with power shovels, bulldozers, and draglines, and are then transported

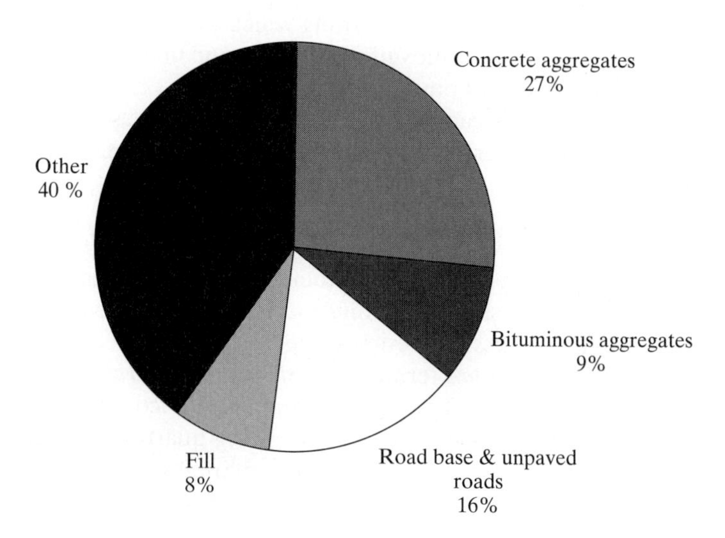

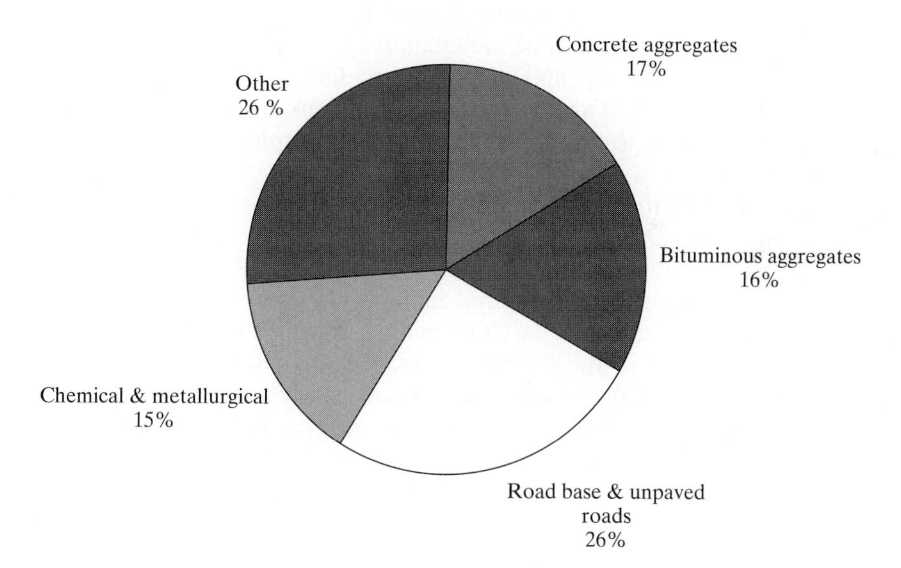

Figure 2.1 Uses of Aggregates.

to processing facilities on trucks or conveyors. The processing may involve primary and secondary crushing, following which the particles will be sorted to size. Silt and clay impurities are removed by washing, and finally the aggregate is blended to specified proportions of particles of various sizes.

Aggregates can be divided into several types based on the type of rock from which they are derived, the method of manufacture, the size, or the density.

2.2.1 Based on Source or Method of Manufacture

Rock fragments used either in their natural state or after some mechanical processing (such as crushing, washing, and sizing) are called *natural aggregates* or *natural mineral aggregates*. Sometimes, aggregates produced from crushing and processing of quarried stone are termed *manufactured mineral aggregates*.

Sand is formed from weathering and decomposition of all types of rock, the most abundant mineral constituent being quartz. It is used in a variety of products from brick to glass to concrete to explosives. *Manufactured sand* is produced by crushing stone, gravel, or air-cooled blast-furnace slag, and is characterized by sharp and angular particles.

Gravel or *river gravel* consists of materials with round and smooth edges and of size varying between 3/16 in. (4.75 mm) and 3 in. (76 mm) in diameter. In natural deposits around lakebeds, glaciated areas, and adjacent to streams, gravel can be seen as the predominant deposit, typically layered with sand. These deposits are produced by erosion of mountainous bedrock and surficial materials, and following transportation and abrasion, are finally deposited around the banks of rivers. The principal agent responsible for transport, obviously, is water or ice, and the deposition is the result of change in topography and gradient along the course of voyage.

Crushed stone is the term used to identify aggregates that are obtained by mechanically crushing rocks, boulders, or cobbles, resulting in particles that are angular in shape and have rough surface texture. Rocks most commonly used in the production of manufactured mineral aggregates are granite, sandstone, limestone, and dolomite, the last two making up for about 70 percent of the crushed stone produced in the United States. All of these have fairly high compressive strength, the average of which ranges between 13 and 34 ksi (90–235 MPa). Granite is hard, tough, and dense, with an average specific gravity of 2.6 to 2.7, whereas limestone and sandstone, with a specific gravity ranging between 2.0 to 2.6, can be hard to soft, heavy to light, and dense to porous. Some typical physical and mechanical properties of aggregates obtained from common types of rocks are given in Table 2.2. Both sandstone and limestone may contain impurities such as clay, rendering the aggregate particles soft, friable (grains can be rubbed off), and absorptive. Generally, aggregates with a crushing strength lower than about 14 ksi (100 MPa) may be unsuitable for the manufacture of normal weight portland cement concrete.

TABLE 2.2 AGGREGATES: TYPES AND SPECIFIC GRAVITY

Aggregate	Type	Average specific gravity	Average bulk density (dry) pcf (kg/m^3)
Granite	Normal weight, mineral	2.65	100–110 (1600–1760)
Gravel	Normal weight	2.70	100–105 (1600–1680)
Limestone	Normal weight, mineral	2.65	95–110 (1520–1760)
Perlite	Lightweight, natural	2.50	—
Pumice	Lightweight, natural	0.75	—
Quartzite	Normal weight, mineral	2.69	—
Scoria	Lightweight, natural	0.75	—
Expanded shale, clay, and slate	Lightweight, manufactured	1.00	50 (800)
Sawdust	Lightweight, organic	0.50	—
Expanded vermiculite	Lightweight, manufactured	0.20	6.5 (105)
Expanded perlite	Lightweight, manufactured	0.75	12.5 (200)
Sand	Normal weight	2.60	103–110 (1650–1760)
Barite	Heavyweight	4.50	150 (2400)
Sandstone	Normal weight, mineral	2.55	90–100 (1440–1600)
Basalt	Normal weight, mineral	2.85	—
Magnetite	Heavyweight	4.50	150 (2400)
Hematite	Heavyweight	5.00	190 (3040)

Aggregates can also be manufactured by crushing waste concrete and clay bricks, and these products can be an economical alternative when good quality mineral aggregates are scarce or expensive. Aggregates that are porous and lighter are manufactured by firing natural raw materials such as clay or shale, producing volumetric expansion, and are called manufactured lightweight aggregates. However, natural sand, crushed stone, and gravel are by far the most common types of aggregates in construction and are readily selected whenever they satisfy minimum quality specifications and can be economically delivered in sufficient quantity to the construction site.

2.2.2 Based on Size

The size of aggregate particles varies, ranging between that of dust and 2 in. (50.8 mm)—and may be higher in special cases. Based on the maximum size of particles in a sample, aggregates are divided into two groups:

- Fine aggregate
- Coarse aggregate

Fine aggregate (also called *sand*) consists of natural or manufactured particles ranging in size from 0.006 in. (150 μm) to 3/16 in. (4.75 mm). In concrete construction, fine aggregate is defined as an aggregate with particles predominantly of size smaller than 3/16 in. (4.75 mm) and equal to or larger than 0.0029 in. (75 μm).

Coarse aggregate consists of rounded river gravel, crushed stone, or manufactured aggregate with particles of size equal to or larger than 3/16 in. (4.75 mm). G*ravel* is naturally rounded aggregate of particles larger than 1/4 in. (6.35 mm), and *crushed gravel* is fine aggregate made by crushing gravel or cobbles. The size of a particle that is the upper limit for fine aggregate or the lower limit for coarse aggregate may depend, however, on the type of construction and the agency specifications. For example, an aggregate particle size larger than 3/32 in. (2.36 mm) is sometimes classified as coarse aggregate for use in asphalt concrete construction.

2.2.3 Based on Density

Based on specific gravity or density measured in bulk, aggregate is divided into three types:

- Lightweight
- Normal-weight
- Heavyweight

Specific gravity is defined as the ratio of the mass of a unit volume of a material—at a specific temperature—to the mass of the same volume of gas-free distilled water at that temperature. It represents the ratio between the density of the material and the unit weight of water, which is 62.4 pcf, 0.01 MN/m^3, or 1 g/l (1 g/cm^3).

Unlike that of metals, the specific gravity of porous materials such as aggregates and wood can be computed in several ways or under different moisture conditions, as discussed in Section 2.3.1. Specific gravity values of most natural mineral aggregates lie in the range of 2.4 to 2.9, meaning that they are 2.4 to 2.9 times as heavy as water.

Densities of normal-weight aggregates, measured in bulk—called the bulk density or specific weight—range between around 95 pcf (1520 kg/m^3) and 105 pcf (1680 kg/m^3). Bulk density, which is explained in Section 2.3.2, is the mass of a unit volume of an aggregate at a specific temperature, and has the units of g/l, g/m^3, pcf (lb/ft^3), and so on. The measurement of the unit volume is based on the volume occupied by the particles (solid volume) plus the volume of the voids between them.

Since the weight of particles varies with moisture content, the oven-dry condition is normally used as the control in the calculation of bulk density.

Normal-Weight Aggregate. Crushed stone, gravel, and ordinary sand are examples of normal weight aggregates. They are commonly used in the manufacture of normal-weight concrete, asphalt concrete, and roadway sub-base. The average values of specific gravity for sand and granite are 2.6 and 2.65, respectively. As indicated, the bulk density of normal-weight aggregates is around 95 to 105 pcf (1520 to 1680 kg/m^3).

Lightweight Aggregate. Lightweight fine aggregate is any aggregate with bulk density less than 70 pcf (1120 kg/m^3) and lightweight coarse aggregate is any aggregate with bulk density less than 55 pcf (880 kg/m^3).

Lightweight aggregates are used as ingredients in the manufacture of lightweight concrete, for making lightweight masonry blocks (to improve their thermal and insulating properties and nailing characteristics), and lightweight floor or roof slabs.

There are two types of lightweight aggregates:

- Natural lightweight aggregates
- Manufactured (also called synthetic) lightweight aggregates

Natural lightweight aggregate consists of particles derived from natural rocks, primarily those of volcanic origin, including pumice, scoria, and tuff. *Manufactured lightweight aggregate* is produced by expanding under heat raw materials such as blast-furnace slag, shale, slate, and fly ash, plus a few lightweight rocks and minerals like perlite and vermiculite. The manufacturing process is fairly simple: The raw materials, which are crushed or ground and pelletized, are heated to more than 1830 °F (1000 °C). At this high temperature the material bloats—like popped corn—from the rapid generation of gases. When cooled, the particles produced are very porous, highly absorptive, and have low specific gravity. The most common natural lightweight aggregates are pumice, scoria, and vermiculite; the most common manufactured lightweight aggregates are expanded shale, expanded slate, expanded clay, expanded perlite, and expanded slag.

Pumice (volcanic glass), the most widely used natural lightweight aggregate, is usually whitish gray to yellow in color, but may also be brown, red, and black. Large quantities of pumice are found in volcanic areas of the western United States. It has a porous structure and can be identified as a finely vesicular glassy rock containing tubular bubbles.

Scoria, which is also of volcanic origin, resembles industrial cinders, and is usually red to black in color. (Cinders are residues from high-temperature combustion of coal in industrial furnaces.) The pores in scoria are larger than those found in pumice and are more or less spherical in shape.

Perlite is a siliceous volcanic rock commonly mined in the western United States *Expanded perlite* is an aggregate derived from crushing perlite rocks and then heating quickly to above 1500 °F (815 °C), driving out the water. The particles expand to produce a lightweight, noncombustible, porous, and glasslike material that can be used as lightweight coarse aggregate in the production of lightweight concrete. Ground expanded perlite can be used in place of natural sand in lightweight concrete, with very good insulating properties. Concrete made with expanded perlite has a density ranging between 20 pcf (320 kg/m^3) and 40 pcf (640 kg/m^3), limited compressive strength, and high shrinkage. Perlite in its natural state or after crushing can also be used in the manufacture of cement mortar.

Vermiculite—a type of mica—is also used in the manufacture of lightweight concrete. It is produced by heating crushed raw material at about 2000 °F (1090 °C) until it expands to about 20 times its original volume. It is too soft and weak a material for use in structural concrete—concrete that requires sufficient strength—but instead it can be used as a replacement for ordinary sand in cement mortar. The bulk density of vermiculite aggregate is in the range of 4 to 12 pcf (64 to 192 kg/m^3), which is nearly the same as that of expanded perlite. Concrete made with expanded vermiculite or expanded perlite has low compressive strength, low density (15 to 50 pcf, or 240 to 800 kg/m^3), high shrinkage, but excellent insulating properties. It is used primarily for indoor applications where the prime consideration is the thermal insulating property of the resulting concrete.

Blast-furnace slag is a nonmetallic product, consisting essentially of silicates and aluminates of calcium (or lime) and other bases, developed in a molten condition simultaneously with iron in a blast furnace. *Expanded slag* is produced by expanding blast-furnace slag by mixing it with water while it is still molten. The violent reaction between the molten slag and the water creates aggregate particles that are porous in structure. The particles are hard and possess considerable strength, but their use in structural concrete is limited because of their high sulfur content.

Expanded shale, expanded clay, and *expanded slate* are aggregates also belonging to the manufactured lightweight aggregate category. They are produced by crushing the raw materials—shale, clay, or slate—and heating them to the fusion point—approximately 2460 °F (1350 °C), when they become soft and expand up to 600 to 700 percent of the original volume because of the entrapped gases. They can also be produced by sintering, a process in which a mixture of raw materials in moving grates, including coal or ashes, is exposed to flames under forced draft. The resulting product is then crushed to the sizes desired.

Some lightweight aggregates are also used to manufacture *nailable concrete.* Sawdust, expanded slag, pumice, and scoria are some of the most commonly used aggregates in the production of this type of concrete, made by mixing cement, sand, and sawdust or other lightweight aggregates with sufficient water. When sawdust is used it should have particles of size 3/32 to 3/4 in. (2.5 to 19 mm), be free from tannic acid, and contain very low amounts of bark. (Tannic acid, or tannin, and sugar will retard the setting of cement.) Sawdust from pine and fir, containing little or no tannin, can be a good aggregate. Various methods of processing sawdust include pretreatment with lime and calcium chloride, aging for periods up to one year, and presoaking for a period from 5 minutes to 24 hours, and washing. A drying period usually follows this processing. The sawdust from most of the softwood species can be rendered compatible with portland cement if a mixture of lime and cement is used as the binder. *Sawdust concrete* is used for floor finishes and in the manufacture of precast floor tiles. Concrete made with a mix proportions of 1:3—cement to sawdust by volume—and enough water to produce good workability can have a density of 55 pcf (880 kg/m^3) and a compressive strength of some 700 psi (4.8 MPa), which can be increased by increasing the cement content.

Bulk density of various types of lightweight aggregates range from 10 to 70 pcf (160 to 1120 kg/m^3), but most have values less than 50 pcf (800 kg/m^3). Expanded perlite (commonly used in the manufacture of insulating concrete) has an average bulk density of 15 pcf (240 kg/m^3); pumice, about 30 pcf (480 kg/m^3); and expanded clay, expanded shale, and expanded slate, about 50 pcf (800 kg/m^3).

Heavyweight Aggregate. *Heavyweight aggregate* is an aggregate of high density and is used primarily in the manufacture of heavyweight concrete, employed for protection against nuclear radiation and as bomb shelters. The unit weight of heavyweight concrete varies from 150 pcf (2400 kg/m^3)—using normal-weight aggregates—to 400 pcf (6400 kg/m^3)—using steel punchings as coarse and fine aggregates.

A number of naturally occurring materials such as mineral ores and barite ($BaSO_4$, also called heavyspar) have been used for the manufacture of heavyweight concrete. Barite is a quarry rock with a high barium sulfate content (90 to 95 percent), a small percentage of iron oxide, and some crystalline minerals. It has a specific gravity in the range of 4.0 to 4.6, hardness 5 to 6, and an average bulk density of 150 pcf (2400 kg/m^3). Some iron ores such as limonite, Hematite (Fe_2O_3), and magnetite ($FeFe_2O_4$) are used as fine and coarse aggregates after processing. They have specific gravities in the range of 4.1 to 5.2, hardness of 5 to 6.5, and an average bulk density of 180 pcf (2880 kg/m^3), and are used to produce concrete of unit weight in the range of 210 to 260 pcf (3360 to 4160 kg/m^3). Manufactured iron products are used to make extra-heavyweight concrete—concrete with density around 300 pcf (4800 kg/m^3). Steel punchings, steel bars, and steel shots have specific gravity between 6.2 and 7.7, but may be more expensive compared to natural heavyweight aggregates.

Specific gravity and other properties of some of the commonly used aggregates are listed in Table 2.2. Due to variations in the mineral constituents between rocks from different sources, an individual aggregate when tested may present values not matching the values listed in the table.

2.3 PROPERTIES OF AGGREGATES

A number of physical and mechanical properties affect durability, strength, and performance of construction products made with aggregates. Although aggregates themselves can be put to use as a construction material—as in railroad ballast—it is more common to use them along with a binder such as portland cement or asphalt. In this regard, the aggregate particles should bond well with the bonding agent and also retain their strength, shape, and texture throughout the service life.

Although mechanical and chemical properties of an aggregate are of relevance in its selection for use in portland cement concrete or asphalt concrete, it is the spectrum of physical properties that is of more significance and importance. For example, the moisture retained in sand affects the workability of fresh concrete, as well as the strength and durability of hardened concrete, much more than the hardness of the sand. However, this should not be construed to mean that the mechanical and chemical properties are of no consequence. In fact, hardness and strength of aggregates are taken into consideration in road design and construction. But a minor variation in a mechanical property may not be as significant as a small difference in a physical property.

The most important properties to consider in selecting an aggregate for a particular application are as follows:

- Specific gravity
- Bulk density
- Porosity
- Voids
- Absorption
- Moisture content
- Shrinkage
- Gradation and fineness modulus
- Modulus of elasticity
- Compressive strength
- Chemical reactivity

All but the last three of these fall into the category of physical properties. The following sections discuss these properties in detail.

2.3.1 Specific Gravity and Moisture Content

Specific gravity is the ratio of the mass (or weight) of a unit volume of a material—at a specific temperature—to the mass of the same volume of gas-free distilled water at that temperature. (Note that the term weight is used in this text to represent a measure of mass and, accordingly, the two terms, weight and mass, are interchangeable.)

$$\text{Specific gravity} = \frac{\text{weight of material}}{\text{volume of material} \times \text{density of water}}$$

The aggregate particles contain pores, and due to this, the volume of a particle depends on how it was assessed: with or without the pore volume. In other words, the quantity in the denominator of the preceding equation is not the same when the pore volume has been included as when it has been excluded from the volume calculation. In addition, the numerator, which is the weight of material, will change with the amount of moisture in the material. Because of these different possibilities that exist in its measurement, the specific gravity of an aggregate should be associated with or qualified by the moisture level and the method of measurement of the volume.

Moisture Content and Moisture Levels. It was previously indicated that aggregates are generally porous and thus absorb moisture. The moisture content (MC) of an aggregate, which is generally expressed as a percentage, is calculated as follows:

$$\text{Moisture content} = \frac{\text{weight of moisture}}{\text{oven-dry weight}} \times 100$$

If a 110-g sample of sand shows a weight of 100 g after it is dried in an oven, the weight of moisture is equal to 10 g, and the moisture content is [10/100], or 0.1, or 10 percent.

The moisture content represents the amount of moisture existing at the time of measurement. Two types of moisture measurement are recognized in aggregate particles:

- Absorbed moisture
- Surface moisture

Moisture retained within the pores is the absorbed moisture, and that which is held on the surface is surface moisture. Based on the moisture level, aggregate particles are divided into four states (Fig. 2.2):

- Oven dry
- Air dry
- Saturated surface dry
- Wet

These moisture conditions are explained in the following sections.

Water absorption or *absorbed moisture* (also called *absorption capacity*) is defined as the weight of water absorbed by dry aggregate particles in reaching a moisture level or condition called saturated surface dry condition (SSD), and is expressed as a percentage of the oven-dry weight.

The SSD condition represents the state of moisture when all of the pores within a particle, or within all particles in a sample, are filled with water and the

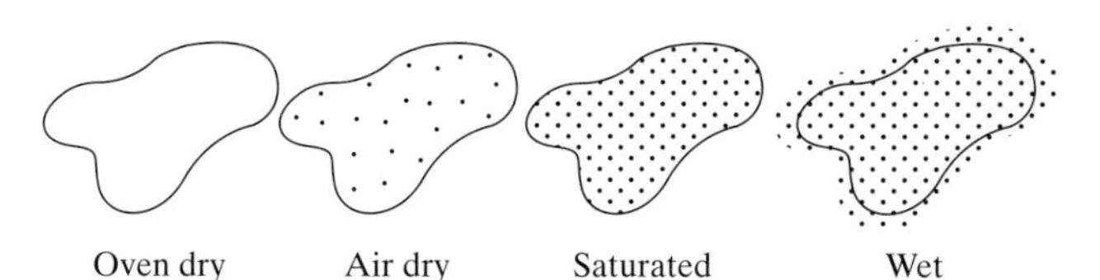

Figure 2.2 Moisture conditions of aggregates.

surface of the particle stays dry. In this condition, it is assumed that the particle will neither absorb moisture from nor contribute moisture to the surroundings. Absorption capacity is also defined as the increase in the weight of aggregate from water retained within the pores.

Absorption capacity, expressed as a percentage of the dry weight of the aggregate (which is the weight established after drying the particles in an oven to a constant weight), varies from one type of aggregate to another. Granite has a very low absorption capacity, less than 1 percent, whereas porous sandstone has very high absorption, as high as 13 percent. Most lightweight aggregates have high absorption values, generally in the range of 5 to 20 percent. The following equation is used to calculate the absorption capacity.

$$\text{Absorption capacity} = \frac{\text{saturated-surface-dry weight} - \text{oven-dry weight}}{\text{oven-dry weight}} \times 100$$

In a laboratory, the *moisture content* (MC, also called *total moisture content*) is found by determining the loss in weight of a representative sample of aggregate—typically 2000 g of coarse aggregate or 1000 g of fine aggregate—after it has been dried in an oven at a temperature of 110 °C (230 °F) to a constant weight. The following equation is used for calculating the MC of the test sample.

$$\text{Moisture content} = \frac{\text{original sample weight} - \text{oven-dry weight}}{\text{oven-dry weight}} \times 100$$

Moisture is retained within the pores of particles and also on their surfaces. If the moisture content is more than the absorption capacity, the aggregate is termed *wet*—meaning that there is free moisture on the aggregate surface—and if it is less, the aggregate is termed *dry*. Dry particles are absorbent, and the amount of moisture they will absorb depends on the existing moisture level. A wet condition generally exists after rains. The *free moisture* or *surface moisture,* also expressed in percentage, is thus obtained by subtracting the absorption capacity from the moisture content:

$$\text{Surface moisture} = \text{MC} - \text{absorption capacity}$$

Crushed stone and gravel—the coarse aggregates—absorb very little and also hold little water on the surface. Fine aggregates, however, have high absorption capacity and even higher surface moisture. If the moisture content of an aggregate sample is 6 percent and its absorption capacity is estimated as 5 percent, the surface moisture is $(6 - 5)$, or 1 percent. Thus, 10 m^3 of this aggregate, when used in the manufacture of concrete, will contribute an equivalent of 1 percent of its weight to the mixing water. If the bulk density of the sample is 1500 kg/m^3, the total amount of water contributed by the aggregate to the mix = $(10 \times 1500 \times 1/1000) = 15$ kg. Similarly, if the moisture content of this aggregate is only 3 percent, it will absorb water up to $(5 - 3)$ or 2 percent of its weight from the mixing water. On the job, fine aggregates usually contain about 4 to 5 percent surface moisture. After rain or washing, this may reach as high as 10 percent.

The surface moisture in fine aggregates can cause an increase in the volume—compared to the dry volume—called bulking. *Bulking* is defined as the percent increase in volume of an aggregate over its dry-rodded volume, and is caused by the pulling apart of the particles due to the surface tension. The increase in volume depends on the moisture content and the size of particles. Typical variations in the amount of bulking with changes in moisture content for fine and coarse sand are illustrated in Fig. 2.3. The finer the sand, the higher the bulking. Maximum bulking in coarse sand may be about 15 percent, and in fine sand, more than 35 percent. This means that 1 ft^3 of fine moist sand with a moisture content of 8 percent has a dry volume of only (100/135) or 0.74 ft^3. The bulking is around maximum when the moisture content is in the range of 5 to 8 percent; samples with lower or higher moisture content will bulk less. Bulking should be taken into consideration when concrete ingredients are measured by volume—called volume batching.

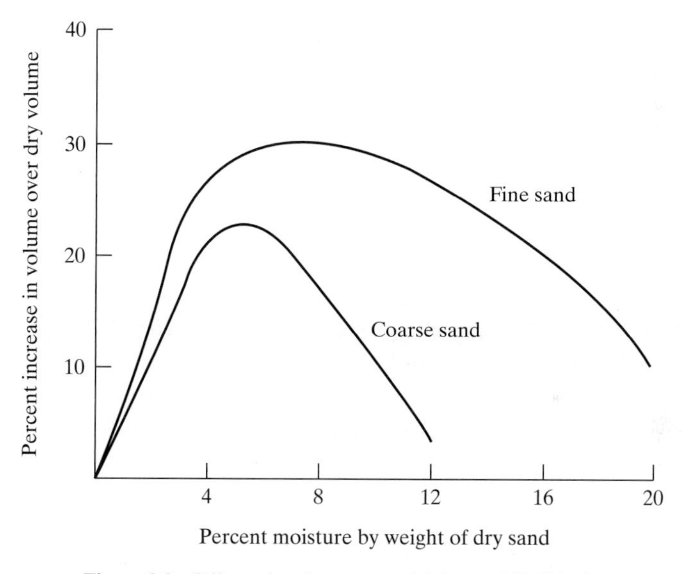

Figure 2.3 Effect of moisture on voids in sand (bulking).

Bulk Specific Gravity and Apparent Specific Gravity. As discussed, the specific gravity of porous materials such as aggregates can be measured in different ways due to changes in the weight caused by different moisture levels and differences in the manner in which the volume is measured. When the volume of pores is included in the measurement of the volume of material, the resulting specific gravity is said to be bulk. *Bulk specific gravity* (bulk SG) is computed as the ratio of the weight in air of a unit volume of a permeable material—which includes both permeable and impermeable pores, at a specific temperature—to the weight in air of an equal volume of gas-free distilled water. When the weight in air includes the weight of water within all pores, or when the density computation is based on saturated particles, the bulk specific gravity is for SSD condition. When the moisture content is less than that for SSD condition, the bulk specific gravity is for dry condition. And when the density assessment is based on the weight of oven-dry particles, the bulk specific gravity is for oven-dry condition.

$$\text{Bulk } SG = \frac{\text{density of particles based on solid volume plus pore volume}}{\text{density of water}}$$

Apparent specific gravity (apparent SG) is the ratio of the weight in air of a unit volume of the impermeable portion of the material—at a specific temperature—to the weight in air of an equal volume of gas-free distilled water at that temperature. This measurement pertains to the density of the solid material or of the constituent particles, excluding the pore space that is accessible to water (or permeable pores).

$$\text{Apparent } SG = \frac{\text{density of particles based on solid volume}}{\text{density of water}}$$

In the manufacture of portland cement concrete, cement mortar, and asphalt concrete, the aggregate particles, which occupy a major portion of the volume, do not undergo any physical or chemical changes. In other words, a piece of crushed stone used in concrete retains its shape, size, and porous structure throughout the structure's service life. Because of this, it is the bulk specific gravity that is of primary interest in construction technology. Consequently, the term specific gravity most often stands for bulk specific gravity, generally in SSD condition.

It was previously indicated that the specific gravity (or bulk SG) of normal-weight aggregates is anywhere from 2.5 to 2.75, with 2.65 being the average value for crushed stone, gravel, or sand.

Porosity. The porosity of stone and its effects on the durability of structures have been briefly discussed. All aggregates are porous; some are more porous, and some are less. Granite has very few pores, whereas expanded perlite or expanded clay is largely porous. *Porosity* is defined as the ratio of the volume of pores in a particle to its total volume (solid volume plus the volume of pores).

$$\text{Porosity} = \frac{\text{volume of pores}}{\text{total volume of particles}}$$

Most varieties of granite and limestone have very low porosity (about 1 percent), whereas a large majority of sandstone rocks generally have high porosity (about 3 percent). However, tests have shown that porosity of a few types of limestone and sandstone can be as high as 13 and 30 percent, respectively, which are also the typical ranges in lightweight aggregates.

The porosity and the character of the pores affect not only the absorption of particles but the freezing characteristics as well. Freezing occurs when the pores are filled with water and the ambient temperature drops, resulting in an increase in the water volume. When the aggregate particles are not strong enough to withstand the stresses resulting from the internal expansion, they crack or fail.

A porous stone is less resistant to cycles of freezing and thawing than a stone of dense crystalline structure. Particles in which the pores are connected by minute tubes are more greatly weakened by freezing than are stones with larger pores and larger connecting tubes. If a porous aggregate particle absorbs so much water that the remaining pore space is insufficient to accept the expansion expected during freezing, the particle will crack and fail. The tendency for failure, as tests have shown, depends on the size and tensile strength of the particle. Granite possesses excellent resistance to freeze-thaw cycles. Sandstone and limestone also have good resistance but typically not as high as granite.

Porosity also affects the strength, permeability, and water absorption of aggregates, in turn affecting the behavior of both freshly mixed and hardened concrete. When concrete will be exposed to cold temperatures and moisture, resistance to cycles of freeze-thaw is important to ensure long service life, for without adequate capacity to endure the internal movement the concrete will disintegrate. The freeze-thaw resistance of concrete depends on a number of factors, but mostly on the aggregate porosity, absorption, and pore structure.

2.3.2 Bulk Density and Voids

Unit weight or *bulk density* (also called *dry-rodded weight, specific weight,* and *dry-loose weight*) is the mass of aggregate per unit bulk volume. The mass of the sample is determined in oven-dry condition (after driving away all moisture), and the volume is the solid volume of particles plus the volume of voids—the space between the particles.

The unit weight is established by filling a standard metal container with a representative sample of the aggregate, oven dried to a constant weight, and recording the weight of the material in the container. The volume occupied by the material, or the volume of the container, includes the solid volume of all the particles and the void space between the particles. The bulk density of normal-weight aggregates generally ranges between 95 and 105 pcf (1520 and 1680 kg/m^3), and of lightweight aggregates, between 10 and 70 pcf (160 and 1120 kg/m^3). Heavyweight aggregates have bulk density exceeding 150 pcf (2400 kg/m^3). If the moisture stored within the particles is subsumed, the bulk density increases.

The unit weight of a material provides an estimate of the weight of the material needed to fill a certain volume. For example, the amount of sand with a unit weight of 100 pcf required to fill an area of size 5 ft $\times$ 10 ft to a depth of 2 ft—a volume of 100 ft^3—is equal to (100 $\times$ 100) or 10,000 lb. Additionally, by knowing the unit weight and the specific gravity, the solid volume of the material and the void content can be calculated.

Voids represent the amount of air space between the aggregate particles, and the *void content* is the total volume minus the solid volume of particles. The void content is generally expressed as a percentage of total or gross volume, which is the solid volume plus the volume of voids. For example, a forty percent void content means the solids (including the pores within the solids) occupy 60 percent of the total volume (container volume), and the rest, 40 percent, is filled with air—assuming that the aggregate particles are dry.

The void content can be calculated using the specific gravity and the bulk density:

$$\text{Void content (\%)} = \frac{SG \times W - B}{SG \times W} \times 100$$

where SG is the specific gravity, W is the density of water, and B is the bulk density.

The void content in normal aggregates varies from 30 to 50 percent depending on the size, shape, and texture. Fine aggregates have void content typically around 35 to 40 percent, and coarse aggregates, from 30 to 50 percent, depending on the maximum size. Gravel, generally, has lower void content than crushed stone, and thus requires less binder when made into concrete. But crushed stone, with a larger surface area relative to the rounded particles of gravel, can better interlock and bond with the cement paste. In addition, the rough texture of crushed stone, compared to the smooth surface of gravel, provides better adhesion.

The larger the maximum size, the lower the void content. In concrete mixtures, the aggregate used is a combination of fine and coarse, called mixed aggregate, for which the void content varies with the composition of the mix, as shown in Fig. 2.4.

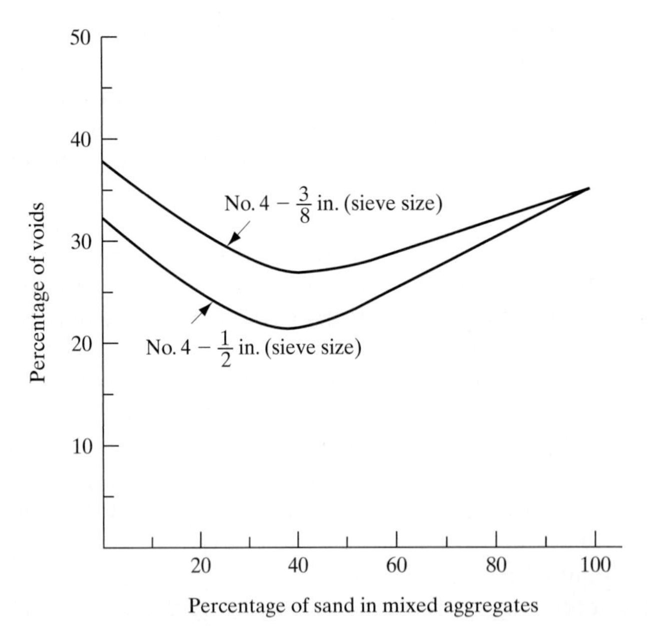

Figure 2.4 Variation in the volume of voids for various sand–gravel mixtures.

In a mixed aggregate containing about 40 percent sand and 60 percent gravel, the void content can be as low as around 25 percent. When used in the manufacture of concrete, this aggregate occupies about 75 percent of the total volume, and the balance is made up of cement, water, and air voids. A large void content in mixed aggregates thus calls for more cement paste (cement plus water).

2.3.3 Modulus of Elasticity and Strength

The modulus of elasticity (also called Young's modulus) is the ratio of stress to the corresponding strain below the proportional limit. It is obtained as the slope of the elastic region of the stress-strain diagram, and varies with the material. At a given stress level, a material with a higher modulus deforms less than a material with a lower modulus.

The modulus of elasticity of granite and limestone (in compression) is in the range of 2,000 to 7,000 ksi (13.8 to 48.3 GPa). Sandstone generally has a lower modulus, which may range from 1,000 to 5,000 ksi (6.9 to 34.5 GPa). The modulus of elasticity of portland cement concrete increases with increase in the modulus of elasticity of the aggregates used in the mix. The creep and shrinkage characteristics of concrete are also affected by the modulus of the aggregates. The stiffness of aggregates restrains the movement of paste originating from creep, shrinkage, and thermal variations, provided that the aggregate is stiffer than the paste. The compressive strength of concrete also depends, to some degree, on the type of aggregate used in the mix. But the influence of the aggregate type is more pronounced on the elastic modulus or deformation properties of concrete, through a combination of characteristics such as water demand and workability.

However, the modulus of elasticity or the deformation characteristics of aggregates are seldom considered in the selection of aggregate for a particular job. Rocks typically do not obey Hooke's law, although granite, limestone, and a few other stones exhibit a stress-strain relationship that is nearly elastic. Sandstone, which has a porous texture, exhibits a nonlinear stress-strain relationship even at very small stresses.

The compressive strength of aggregates depends on the compressive strength of the original rocks from which the aggregates are quarried. Aggregates processed from granites and limestone have high compressive strengths, and those from sandstone and lightweight aggregates, low compressive strengths. The average low compressive strength of limestone and granite is typically about 13 ksi (90 MPa), whereas that of sandstone is 6 ksi (40 MPa). Basalt has a very high compressive strength, generally around 23 ksi (160 MPa). The flexural strength—strength measured in bending—of basalt is also very high, about 5 ksi (35 MPa). Granite and limestone have a flexural strength close to 2.2 ksi (15 MPa).

The strength of an aggregate may affect the strength of the building material made with it, when the two strength values are close to each other. For example, the strength of portland cement concrete is not significantly affected by the strength of the aggregates used in the mix, if the desired strength in concrete is much lower than the aggregate strength. The compressive strength of common normal-weight aggregates is generally much higher than the strength of products made with them such as the masonry unit, portland cement concrete, and asphalt concrete. The compressive strength of granite may range from 15 to 30 ksi (100 to 200 MPa), and when it is used to make concrete of, say, 4 ksi (27.5 MPa), which is substantially lower than the aggregate strength, it does not really matter what the value of the aggregate strength is. It can thus be concluded that the strength of this concrete remains independent of the strength of the aggregate. However, when the strength of the aggregate is low, or if it nearly matches the required strength of the building product, as may be the case with many lightweight aggregates, the product strength may vary significantly with the type of aggregate used.

2.3.4 Gradation and Fineness Modulus

Fine and coarse aggregates used in the manufacture of portland cement concrete and asphalt concrete are made up of particles of various sizes, ranging from a nominal maximum to the smallest specified. Concrete is not generally produced using particles of identical size. Instead, the aggregates we use consist of particles that are big, medium, and fine—like a family. *Gradation* (also called *particle-size distribution* or *grain-size distribution*) refers to the proportions—by mass or weight—of aggregate particles distributed in special particle-size ranges. It is an important property, affecting several characteristics of concrete such as mix proportioning, workability, economy, porosity, durability, and shrinkage. Aggregates conforming to the specified grading limits—the range of proportions in each size—produce a mix with low void content, thus requiring a minimum quantity of binder (cement paste). In other words, an aggregate in which all the particles are of the same size will have more voids than an aggregate in which the particle sizes vary. Gradation refers not only to individual aggregates—fine or coarse—but also to the combined or mixed aggregates—fine and coarse mixed together. Much of the data on the beneficial effects on concrete strength and density of utilizing graded particles is credited to the pioneering studies conducted around 1902 by W. B. Fuller and S. E. Thompson of New York.

Grading refers to the process that determines the particle-size distribution of a representative sample of an aggregate. The practice of combining aggregate particles of various sizes to obtain a specified gradation is called *blending*.

A *normally graded aggregate* is the one that conforms to the grading limits specified by an agency or organization. Various agencies and standard-writing organizations prescribe limits on the percentages of particles that should be larger (or smaller) than a particular sieve. Recommendations may differ from one agency to another, and for each class of work. The grading limits recommended by some national organizations such as ASTM will be discussed subsequently.

A *dense-graded aggregate* is an aggregate that has a particle-size distribution that results in the least voids or lowest void content. An *open-graded aggregate* is an aggregate that has a particle-size distribution that results in relatively large voids or void content. The void content, as explained, represents the space between the particles, and is expressed as a percentage of the gross volume, which is the volume of the dry solids plus the volume of the voids. A *gap-graded aggregate* is an aggregate in which one or more intermediate sizes are missing.

When designing the mix proportions of portland cement concrete, a major goal is to achieve a maximum density with a minimum amount of paste, while attaining the desired quality and strength. A properly graded aggregate is important in accomplishing this. The particle-size distribution of both coarse and fine aggregates has an impact on how easy it is to work with fresh concrete—a characteristic referred to as *workability.* It is an aspect that is a composite measure of several tasks such as consolidation, finishability, and pumpability. Gradation also affects some properties of hardened concrete such as strength and density, as well as the cost of concrete.

Sieve Analysis. Gradation of aggregates is determined from sieve analysis, in which a representative sample of the aggregate is passed through a series of sieves and the weight retained in each sieve—expressed as a percentage of the sample weight—is compared with the grading limits specified.

A *sieve* is an apparatus—round or square in shape—with square openings, identified either by the size (clear) of the opening or by a number. The higher the number, the smaller the opening. The standard sieves for coarse aggregate are No. 4, 3/8 in., 1/2 in., 3/4 in., 1 in., $1\frac{1}{2}$ in., 2 in., and $2\frac{1}{2}$ in. The standard sieves for fine aggregate are Nos. 100, 50, 30, 16, 8, and 4. The sizes of openings for these sieves are shown in Table 2.3. It can be seen that most of the sieves are arranged so that each has an

TABLE 2.3 SIEVE SIZES

Sieve designation		Normal opening		
in	mm	in.	mm	Type[a]
2	50	2	50	H
$1\frac{1}{2}$	37.5	1.5	37.5	F
1	25	1	25	H
3/4	19	0.75	19	F
1/2	12.5	0.5	12.5	H
3/8	9.5	0.375	9.5	F
No. 4	4.75	0.187	4.75	F
No. 8	2.36	0.0937	2.36	F
No. 16	1.18	0.0469	1.18	F
No. 30	600 μm	0.0234	0.60	F
No. 50	300 μm	0.0117	0.30	F
No. 100	150 μm	0.0059	0.15	F
No. 200	75 μm	0.0030	0.075	F

[a]H: half sieve; F: full sieve.

opening size that is one-half that of the upper or twice that of the lower sieve. For example, the No. 50 sieve has an opening of 0.3 mm, which is twice the size of the opening of the No. 100 sieve (equal to 0.15 mm), and half that of the No. 30 sieve (equal to 0.6 mm). These sieves are called *full sieves*. Sieves that do not fall into this arrangement, such as 1 in. and 1/2 in. sieves, are called *half sieves.*

Sieve (or *mechanical*) analysis consists of determining the proportionate amounts of particles retained on or passing through each of a set of sieves arranged in decreasing sizes. Using the percentages of weights retained in each sieve, a graph—called a *particle-size distribution curve*—is drawn by plotting the log of the opening size on the x-axis, and the percentage of particles, by weight, coarser than or finer than the particular sieve on the y-axis (Fig. 2.5). A uniform shape, like an escalator, shows uniform gradation (graph [a]). A horizontal or near-horizontal shift, like a landing (graph [b]), shows that some intermediate particle sizes, corresponding to

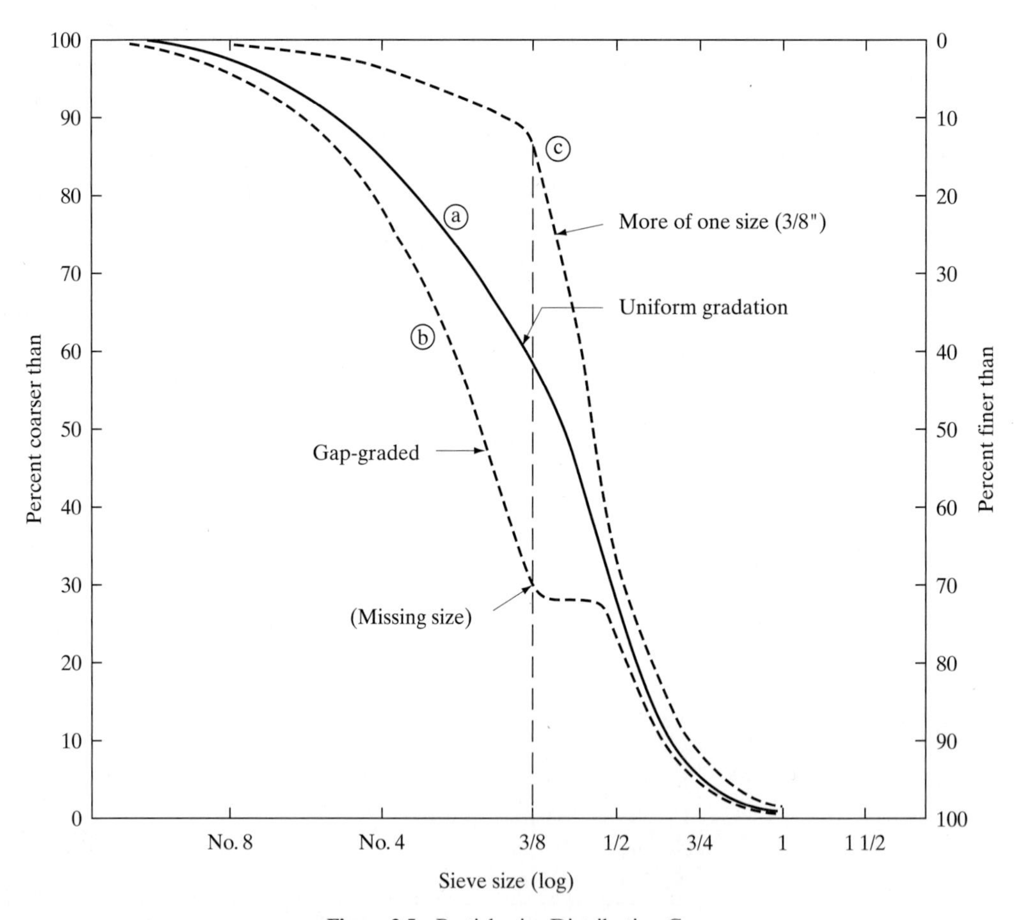

Figure 2.5 Particle-size Distribution Curves.

the top of the landing, are missing (gap-graded aggregate). When a large part of the aggregate is made up of particles of one size, it is reflected in the graph as a near vertical drop (graph [c]). A gradation chart that remains closer to the origin has a large percentage of finer particles—a, finer gradation—and similarly, when it shifts away from the origin, a large percentage of coarser particles—a coarser gradation.

The sieve analysis of fine aggregate is commonly carried out in a mechanical shaker that is designed for 8 in. (203 mm) diameter round sieves; that of coarse aggregate may be carried out in the same shaker designed for 8 in. round sieves, or by using larger rectangular sieves mounted in frames designed to accommodate them. The mechanical shaker imparts lateral (or lateral and vertical) motion to the set of sieves, causing the particles thereon to bounce and turn so as to present different orientations to the sieving surface. This aspect is important, for the sieve openings are square whereas the particles are neither square nor round.

The test sample is placed on the top sieve, the lid is put on, and the set of sieves is agitated in the mechanical shaker. By measuring the weights of particles retained in each sieve—after the agitation—the percentage of weight passing through (finer than) or retained (coarser than) in each sieve is calculated (Table 2.4). These values are plotted in a chart, as previously described, and then compared with the specified or standard limits to establish the suitability of the aggregates.

TABLE 2.4 SIEVE ANALYSIS TEST RESULTS

Sieve no.	Opening (mm)	Weight of sieve (g)	Sieve plus material (g)	Weight retained (g)	Percent retained	Percent coarser than	Percent finer than
3/4	19	780	1080	300	7.0	7.0	93.0
1/2	12.5	760	1770	1010	23.5	30.5	69.5
3/8	9.5	750	2070	1320	30.7	61.2	38.8
4	4.75	760	1980	1220	28.4	89.6	10.4
8	2.36	720	1090	370	8.6	98.2	1.8
Pan		350	425	75	1.7	99.9	0.1
			Cumulative:			256.0	

Sample weight = 4300 g

Note: Half sieves and pan are not included in the cumulative.

Maximum Aggregate Size and Grading Limits. *The maximum size* of coarse aggregate is defined as the smallest sieve opening through which the entire sample passes. In practice it is common to use a *nominal maximum size,* which is the smallest sieve opening through which most (not all) of the particles pass. About 5 percent of the sample by weight may still be retained on this sieve. For example, 3/4 in. (nominal) maximum size aggregate has about 95 percent of the particles (by weight) passing through the 3/4 in. sieve. The choice of a maximum size depends on the type of job. An overly large aggregate may produce nonuniform concrete. The ACI code limits the maximum size of coarse aggregate to one-fifth of the narrowest dimension of a concrete member. For example, a 6-in. thick retaining wall may not use an aggregate larger than (6/5) in., or 1-in. aggregate. In the case of slab-on-grade, this limit is increased to one-third the thickness of the slab, although in most cases a maximum size smaller than this is used. In reinforced concrete construction, the maximum aggregate size should not be larger than three-fourths of the minimum clear distance between the bars. The maximum aggregate size affects the paste requirement and placement of concrete.

Typical limits of particle sizes for coarse aggregates used in road construction and in other forms of concrete are shown in Tables 2.5 and 2.6. The grading requirements of fine aggregates (natural sand and crushed stone) used in concrete masonry construction are given in Table 2.7. It can be seen from these tables that for each type of aggregate, two values (percent finer than) are specified for each sieve in the set, representing the lower and upper limits of the percentage of the sample that may be finer than that sieve. For example, for 1-in. nominal maximum size aggregate in concrete work, the percentage finer than the 1/2 in. sieve has the upper limit of 60 and lower limit of 25. This means that the amount of particles finer than 1/2 in. can be

TABLE 2.5 GRADING REQUIREMENTS OF AGGREGATES FOR CONCRETE CONSTRUCTION

Nominal size	Amounts finer than each sieve, in percent of weight									
	1 in.	3/4 in.	1/2 in.	3/8 in.	No. 4	No. 8	No. 16	No. 30	No. 50	No. 100
1 in. to No. 4	95–100	—	25–60	—	0–10	0–5	—	—	—	—
3/4 in. to No. 4	100	90–100	—	20–55	0–10	0–5	—	—	—	—
1/2 in. to No. 4	—	100	90–100	40–70	0–15	0–5	—	—	—	—
3/8 in. to No. 8	—	—	100	85–100	10–30	0–10	0–5	—	—	—
Fine aggregate	—	—	—	—	95–100	80–100	50–85	25–60	10–30	2–10

Source: ASTM C33 Copyright ASTM. Reprinted with permission.

TABLE 2.6 GRADING REQUIREMENTS OF AGGREGATES FOR ROAD AND BRIDGE CONSTRUCTION

Nominal size	Amounts finer than each sieve, in percent of weight									
	1 in.	3/4 in.	1/2 in.	3/8 in.	No. 4	No. 8	No. 16	No. 30	No. 50	No. 100
1 in. to No. 4	95–100	—	25–60	—	0–10	0–5	—	—	—	
3/4 in. to No. 4	—	—	—	—	—	—	—	—	—	—
1/2 in. to No. 4	—	—	90–100	40–70	0–5	0–5	—	—	—	—
No. 4 to 0	—	—	—	—	85–100	—	—	—	—	10–30

Source: ASTM D448. Copyright ASTM. Reprinted with permission.

TABLE 2.7 GRADING REQUIREMENTS OF VARIOUS TYPES OF SAND

Type of sand	Amounts finer than each sieve, in percent of weight									
	3/8 in.	No. 4	No. 8	No. 16	No. 20	No. 30	No. 40	No. 50	No. 100	No. 200
20–30 sand	—	—	—	100	85–100	0–5	—	—	—	—
Graded sand	—	—	—	100	—	96–100	65–75	20–30	0–4	—
Sand for concrete	—	95–100	80–100	50–85	—	25–60	—	10–30	2–10	—
Natural sand for masonry mortar	—	100	95–100	70–100	—	40–75	—	10–35	2–15	—
Manufactured sand for masonry mortar	—	100	95–100	70–100	—	40–75	—	20–40	10–25	0–10

Source: ASTM C144 Copyright ASTM. Reprinted with permission.

equal to or less than 60 percent by weight and equal to or more than 25 percent by weight. Or, the amount of particles coarser than 1/2 in. can be equal to or more than 40 percent by weight (100 − 60) and equal to or less than 75 percent by weight (100 − 25). For the same sample, the limits of the percentages finer than the 1 in. sieve are 95 and 100, which also means that the limits of the percentages coarser than 1 in. are 0 and 5. From the tables it can be seen that the percentages of the amounts of the maximum and the minimum sizes in all types of aggregates are less than 10 percent.

If, in a particular sample of 1-in. nominal maximum size aggregate, the percentage coarser than 1 in. in size is equal to 5, the percentage of particles of size between 1 and 1/2 in. can be anywhere between (40 − 5) and (75 − 5), or 35 to 70 percent. This means that an aggregate gathered by mixing together 5 percent of 1-in. particles and about 50 percent of 1/2-in. particles, with appropriate quantities of No. 4 and No. 8 sieves, would satisfy the specified limits for 1-in. aggregate. Similarly, 1/2-in. maximum aggregate (1/2 in. to No. 4) may be obtained by mixing together about 5 percent of 1/2 in., 45 percent of 3/8 in., 40 percent of No. 4, and 10 percent of No. 8 particles.

Figures 2.6–2.8 show typical standard gradation charts—graphs drawn from the upper and lower limits specified in Tables 2.5–2.7. In these charts, the percentages finer than or coarser than each sieve are plotted as ordinates and the sizes of sieve openings as abscissas. As previously indicated, the latter is normally represented as the log of the sieve opening, so that each full sieve will get the same share of the x scale. In the usual procedure, the gradation chart for a representative sample is superimposed over these curves representing the specified upper and lower limits. If the curve drawn for the test sample falls within the limiting curves, the sample is said to satisfy the approved specifications or standards. If, instead, the test curve deviates from the limits or slips outside of the curves, the sample should be rejected or modified so as to fall within the limits.

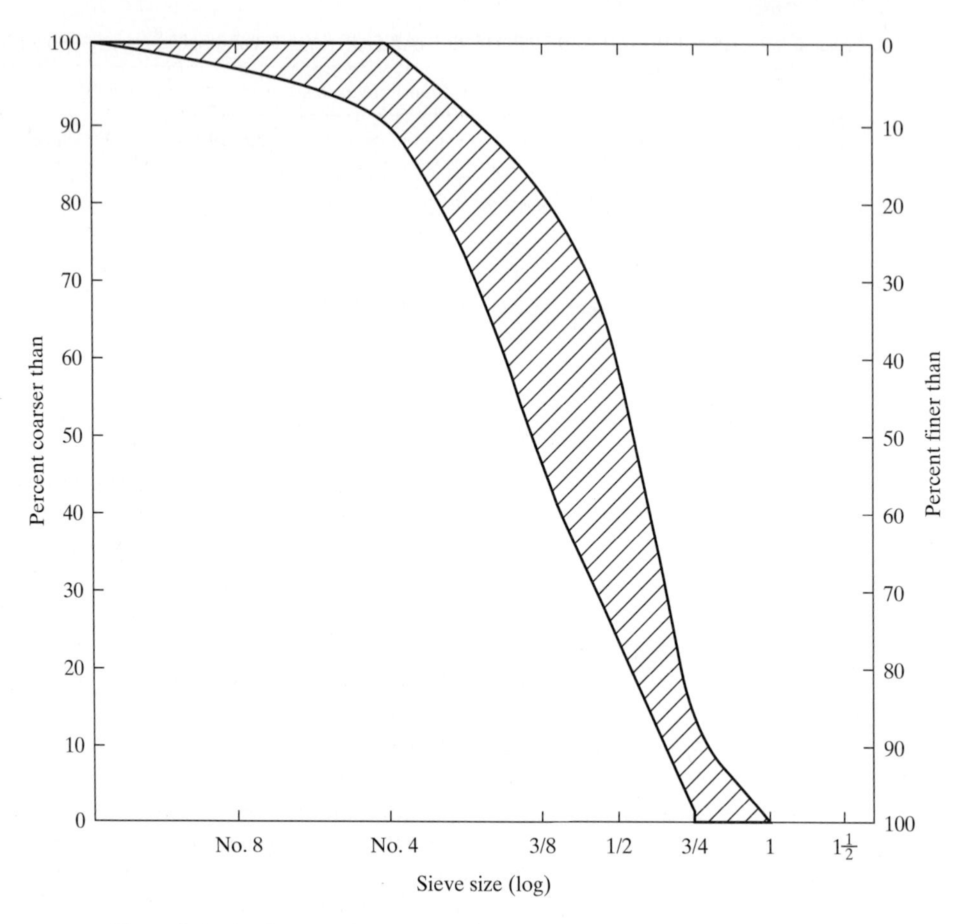

Figure 2.6 Grading Requirements of 3/4 in. to No. 4 Aggregate for Concrete Construction (from ASTM C33).

Although every sample with a gradation curve that falls within the lower and upper bounds does satisfy the specifications, the gradation characteristics will vary from one sample to another. A sample with a curve lying next to the lower bound has more of the finer particles, and similarly a sample with a curve staying close to the upper bound has more of the coarser materials. For example, it was illustrated that a mixture of 5 percent of 1/2 in., 45 percent of 3/8 in., 40 percent of No. 4, and 10 percent of No. 8 will satisfy the limits specified for 1/2 in. to No. 4 aggregate. Instead, if we combine 5 percent of 1/2 in., 55 percent of 3/8 in., 30 percent of No. 4, and 10 percent of No. 8, the percent finer than each sieve will be as follows:

1/2 in.: $(100 - 5) = 95$

3/8 in.: $(100 - 5 - 55) = 40$

No. 4: $(100 - 5 - 55 - 30) = 10$

No. 8: $(100 - 5 - 55 - 30 - 10) = 0$

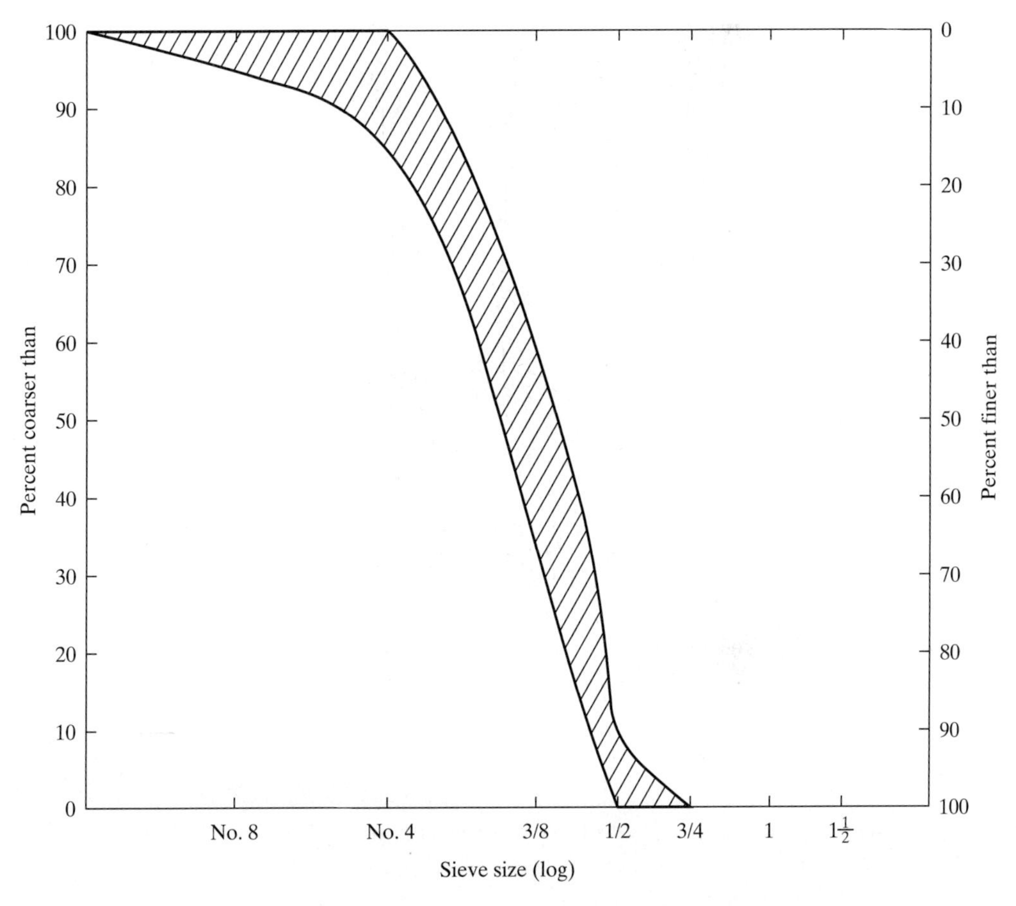

Figure 2.7 Grading Requirements of 1/2 in. to No. 4 Aggregate for Concrete Construction (from ASTM C33).

This sample still satisfies the specified limits, but as can be seen from the proportions of ingredients, it has a higher percentage of the 3/8 in. and coarser particles, than the previous sample. In other words, although both samples satisfy the prescribed grading limits, the former is finer than the latter. This difference in the size distribution—coarser versus finer gradation—does affect the properties of concrete.

Fineness Modulus. In addition to the graphical representation of particle-size distribution, sieve analysis is also used to calculate a factor called *fineness modulus.* Fineness *modulus* is a number obtained by adding the values of the percentages coarser than each full sieve in the set, and dividing the sum by 100. The sieves used in the calculation—full sieves—are some or all of No. 100, No. 50, No. 30, No. 16, No. 8, No. 4, 3/8 in., 3/4 in., $1\frac{1}{2}$ in., 3 in., and 6 in. For example, the sum of percentages coarser than all of the full sieves of sample shown in Table 2.4 is equal to

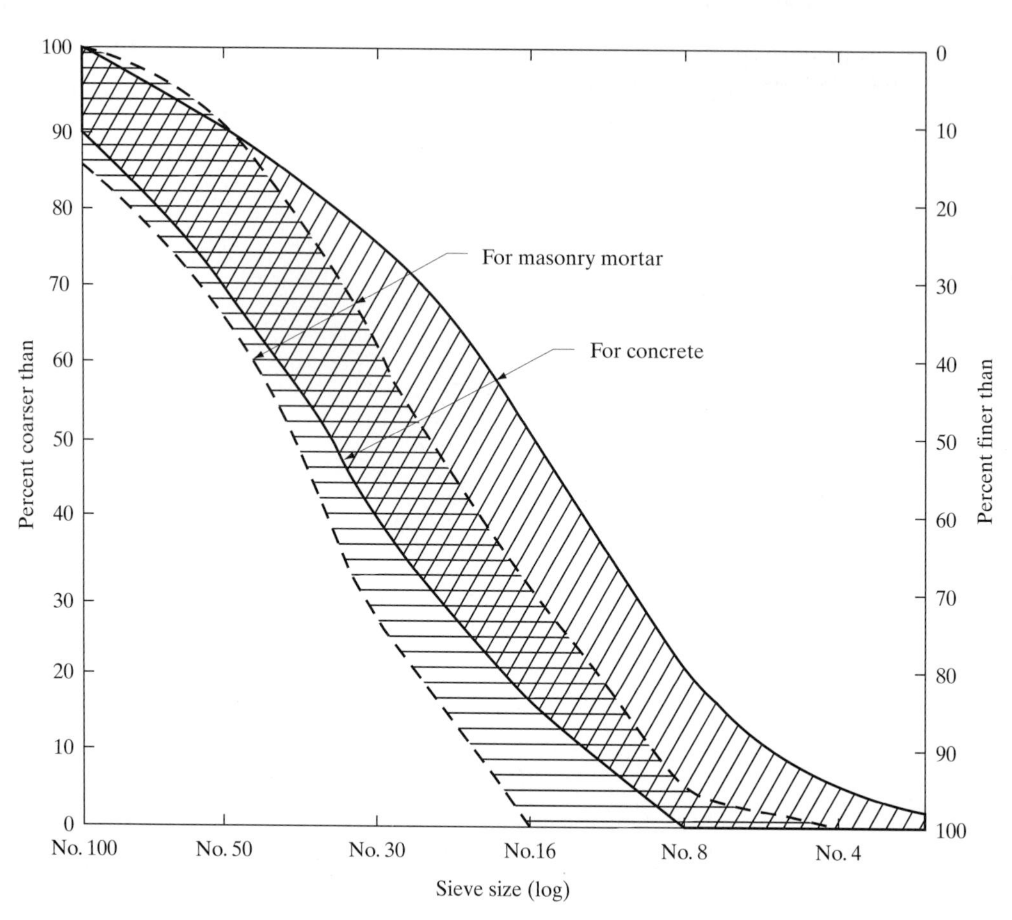

Figure 2.8 Grading Requirements of Fine Aggregate for Concrete and Masonry Mortar (from ASTM C33 and C144).

256. Dividing this by 100, we get 2.56 (or 2.6, after rounding the second digit), which is the fineness modulus of the test sample.

The fineness modulus of aggregate is generally between 2.0 and 4.0, rounded off to a single digit after the decimal point. When correlated with the gradation chart, fineness modulus proves to be less informative, however it is useful as an indicator of the average size of a sample—and the fineness modulus of fine aggregate is a property essential for the mix design of concrete. The average size of the sample is the size of the *n*th sieve counted upward from the last (smallest) sieve in the set, where *n* is equal to the fineness modulus. For example, the fineness modulus of the sample shown in Table 2.4 is 2.6, and counting 2.6 sieves from the bottommost sieve, the second sieve is No. 4, the third is 3/8, and the 2.6th sieve is somewhere midway between the No. 4 and 3/8 sieves, nearer to the latter. A fineness modulus of 3.0—for a sample using a set of sieves that includes all standard sieves between Nos. 4 and 100—indicates that the average size of the sample is the third sieve from the bottom (No. 100), and is thus No. 30. A lower fineness modulus of fine aggregate, denoted by a smaller number, indicates a larger percentage of finer materials (generally, fine sand); similarly, a higher number means fewer finer particles or plenty of coarser particles (generally coarse sand). If, in a construction project, the fineness modulus of sand varies slightly between shipments, it may not affect the mix proportioning and quality of concrete. But when the fineness modulus fluctuates drastically—such as from 2.2 to 2.6—the consistency, placement, and quality of concrete will be affected. Very fine aggregate, with a fineness modulus ranging between 1.0 and 1.7, is called *blend sand*. It is sometimes used in concrete to improve workability—primarily pumpability and finishability.

The fineness modulus of sand in concrete manufacture may lie between 2.2 and 3.2; the lower number is for fine sand, and the higher number for coarse sand. Illustrative examples presented in Section 2.4 show the method of plotting particle-size distribution and the calculation of fineness modulus.

Other Standard Aggregates. The ASTM standard C778 identifies a special type of sand called *standard sand* which comes in two grades: 20–30 sand and graded sand. Standard sand is defined as a type of silica sand composed entirely of naturally rounded grains of nearly pure quartz quarried from Ottawa, Illinois, or LeSueus, Minnesota.

While following the procedure for establishing the strength of hydraulic cements (ASTM C778) the term *graded sand* applies to the type of sand predominantly graded between No. 30 and No. 100 sieves. 20–30 sand is sand predominantly graded to pass No. 20 sieve and be retained on No. 30 sieve. *Slurry-seal sand* is a type of fine crushed sand for mixing with liquid asphalt, to be applied in a thickness of 1/8 in. (3.2 mm) to fill cracks and irregularities in old asphalt concrete pavements. *Seal-coat aggregate* is a type of aggregate featuring very small particles of crushed rock, and is used for mixing with liquid asphalt for application over aggregate base, and also to resurface asphalt concrete or portland cement concrete pavements.

2.3.5 Other Properties

Although aggregates are commonly supplied to concrete manufacturers and batch plants as graded, the method of stockpiling may change the gradation. Stockpiling forces the particles to gravitate down long slopes, contributing to poor blending of the variously sized particles. This aspect should be taken into account in assessing the quality of aggregates for use in concrete.

Several properties of aggregates have been described, including specific gravity, moisture content, bulk density, voids, porosity, modulus of elasticity, compressive strength, gradation, and fineness modulus. Most of these properties are routinely measured and made use of in concrete mix design calculations. A few other properties that may be useful in concrete and highway construction are discussed in the following paragraphs.

Texture and shape of aggregates may also affect their binding properties. Gravel has a smooth texture, whereas basalt and limestone aggregates have a rough texture. The surface of granite aggregates is crystalline, and that of sandstone is granular. Many lightweight aggregates such as pumice and expanded slag have honeycombed and porous makeup. Gravel is round, whereas most crushed stone aggregates are either angular or subangular.

The *abrasion resistance* (also called *wear resistance* or *hardness*) is an important property in determining the suitability of an aggregate for use in roads and pavements. Abrasion is the mechanical wearing and scraping of rock surfaces by friction or impact or both. Hardness is an index or a measure of the resistance of a material to deformation, indentation, cutting, or scratching. One material harder than another can permanently deform the second one, as by scratching or cutting.

The abrasion resistance of an aggregate may be an important factor to consider when it is used in concrete products that will be subjected to abrasion—as in heavy-duty or industrial floors and pavements. But the abrasion resistance of a construction material—such as a floor in a manufacturing facility—is not directly related to that of the aggregates that went into the making of the floor. A portland cement concrete floor is a composite material, the properties of which depend on several factors, such as the amount of cement, the amount of aggregates, the amount of water, finishing and maintenance, and type of aggregates.

Abrasion resistance of aggregates is measured by the LA Abrasion Test (ASTM C131)—the Test Method for Resistance to Degradation of Small-size Coarse Aggregates by Abrasion and Impact in the Los Angeles Machine. This test, developed in 1916, is both an impact test and an abrasion test. When more than 50 percent abrasion losses of the sample are encountered in the test, the aggregates may not perform well in construction. The abrasion resistance values of sandstone, limestone, and granite are nearly equivalent.

Toughness is a measure of the resistance of aggregate particles to fracture under an applied load. No satisfactory test is available for measuring the effective strength or toughness of aggregates. The resistance of aggregates to failure from impact is commonly taken as a measure of toughness. The impact resistance is taken as a measure of the strength of the aggregates in response to vibratory forces or similar shock loads. Tests have shown that the toughness of aggregates is linearly related to their compressive strength. Aggregate particles for road and pavement construction must be tough enough to support the weight of the rollers during construction, and to withstand the impact as well as the crushing action of traffic. Limestone exhibits somewhat lower toughness characteristics than sandstone or granite.

As a rule, aggregates for use in concrete and highway construction must be clean, hard, strong, and durable. They should be free of chemicals and fine material coatings that may affect their performance and bonding characteristics. Silt and clay on the surface of particles lower their bond characteristics. Extremely fine fractions—particles passing No. 200 sieve (75 μm)—are commonly called *silt* and should not be permitted in large quantities. Silt is non-plastic; it has little or no strength when air-dried. When present in considerable amounts, silt can increase the water requirement of concrete. This in turn contributes to a decrease in strength and durability.

Generally, very fine particles in sand may not present many problems in concrete unless its proportions exceeds about 10 percent. During the finishing of concrete surfaces, fine clay and silt rise to the surface, where they can be scaled off. If an excess of clay is suspected, it can be removed by washing the sand. However, too much washing may in addition remove the finer material, resulting in a harsh mix. Purity of sand can be tested at the jobsite by rubbing a small sample in the palm of the hand, and gauging the amount of discoloration produced. Very fine-grained siliceous materials (such as ground pumice) are sometimes added with normal aggregates to a portland cement concrete mix to increase the workability, strength, and durability, through pozzolanic action. If particles are porous and smooth, asphalt cement will not bond well with them. Porous aggregates will also absorb more asphalt, requiring an increased quantity.

Aggregate particles that are prone to splitting and cracking during handling or transportation should be avoided. Flat, thin, or long particles break more easily than particles of round or cubical shape. Some aggregates may contain appreciable amounts of shale, organic matter, expansive material, and soft or porous particles, which make them unfit for use. These harmful materials are called *deleterious materials,* and they act to slow down the hydration process and result in harmful by-products. A visual inspection of coarse aggregate will often reveal the presence of such deleterious materials. Sand can be inspected by rubbing a sample, as just described. When visual inspection is not reliable, standard tests may be carried out to determine the amount of impurities. ASTM C40-84 describes a standard test procedure for the approximate determination of the presence of injurious organic compounds in fine aggregate. ASTM C142 provides a test to measure the amount of clay lumps and friable particles, materials that can generally be removed by washing. Most deleterious materials, in fact, can be removed by washing or flushing with water.

Some aggregates (or compounds in them) are known to react chemically with other materials of construction, and thus are deleterious. For example, certain substances in limestone and dolomite aggregates, called *reactive silica,* are known to react with portland cement, particularly in a warm and moist environment, causing deterioration of concrete. This type of reaction is referred to as *alkali-aggregate reaction (alkali-silica reaction)* because of the chemical reaction between the silica in the aggregates and the alkalis in the cement (primarily sodium hydroxide and potassium hydroxide). The reaction is characterized by map-like cracks in the concrete, which will weaken it and allow for moisture penetration. Alkali-silica reaction is explained further in Section 3.9.1. Most ordinary aggregates possess chemical stability in concrete, which means that the particles neither undergo chemical changes nor react with cement or water.

2.4 EXAMPLES

Example 2.1

Sieve analysis of a 1000-g sample of fine aggregate resulted in the following data. Find the fineness modulus.

Sieve size	4	8	16	30	50	100
Weight retained (g)	26	130	240	252	210	138

Solution

Sieve no.	Weight retained (g)	Percent retained	Percent coarser	Percent finer
4	26	2.6	2.6	97.4
8	130	13.0	15.6	84.4
16	240	24.0	39.6	60.4
30	252	25.2	64.8	35.2
50	210	21.0	85.8	14.2
100	138	13.8	99.6	0.4
		Cumulative:	308.0	

$$\text{Fineness modulus} = \frac{308}{100} = 3.08$$

$$\begin{aligned}\text{Average sieve size} &= \text{third sieve from the bottom}\\ &= \text{No. 30} = 0.0234 \text{ in.}\end{aligned}$$

Example 2.2

Find the volume of voids in a 3-yd^3 coarse aggregate of bulk density equal to 102 pcf. The specific gravity of the particles is 2.65.

Solution

$$\text{Void} = \frac{SG \times W - B}{SG \times W} \times 100$$

$$\begin{aligned}\text{Specific gravity, } SG &= 2.65\\ \text{Density of water, } W &= 62.4 \text{ pcf}\\ \text{Bulk density, } B &= 102 \text{ pcf}\end{aligned}$$

$$\text{Void} = \frac{2.65(62.4) - 102}{2.65(62.4)} \times 100$$
$$= 38.3\%$$
$$\text{Volume of voids} = 38.3(3) \times \frac{27}{100}$$
$$= 31 \text{ ft}^3$$

Example 2.3

Using the following data, determine the moisture content, absorption, and free moisture in a sample of fine aggregate. If 75 lb of the sample is used in the mix design of concrete, what is the amount of water contributed to the mix by the fine aggregate (FA)?

$$\text{Initial sample weight} = 1.2 \text{ lb}$$
$$\text{Saturated surface dry weight} = 1.14 \text{ lb}$$
$$\text{Oven-dry weight} = 1.06 \text{ lb}$$

Solution

$$\text{Moisture content} = \frac{1.2 - 1.06}{1.06} \times 100$$
$$= 13.2\%$$
$$\text{Absorption} = \frac{1.14 - 1.06}{1.06} \times 100$$
$$= 7.5\%$$
$$\text{Free moisture} = 13.2 - 7.5$$
$$= 5.7\%$$

Amount of water contributed

$$\text{to the mix by FA} = 5.7 \times \frac{75}{100}$$
$$= 4.28 \text{ lb}$$

Example 2.4

Estimate the volume of damp fine sand with a moisture content of 8 percent if its volume on dry basis is 3 ft^3.

Solution From Figure 2.3, the expected bulking of fine sand at 8 percent moisture content is about 30 percent.

$$\text{Dry volume} = 3 \text{ ft}^3$$
$$\text{Damp volume} = 3 \times 1.3 = 3.9 \text{ ft}^3$$

2.5 TESTING

A number of laboratory tests are described in this section. Following is a list of aggregate properties and the corresponding test numbers.

Property	Test no.
Specific gravity of coarse aggregate	AGG-1
Specific gravity of fine aggregate	AGG-2
Absorption of coarse aggregate	AGG-1
Absorption of fine aggregate	AGG-2
Bulk density of aggregates	AGG-3
Voids in aggregates	AGG-3
Moisture content of aggregates	AGG-4
Sampling of aggregates	AGG-5
Gradation of aggregates	AGG-6, AGG-7

Test AGG-1: Specific Gravity and Absorption of Coarse Aggregate

PURPOSE: To determine the specific gravity (bulk and apparent) and absorption capacity of coarse aggregate (AGG-5).

RELATED STANDARD: ASTM C127.

DEFINITIONS:

- *Specific gravity* is the ratio of weight in air of a unit volume of a material to the weight of an equal volume of water.
- *Bulk specific gravity* is the ratio of the weight in air of a unit volume of aggregate (including the permeable and impermeable voids in the particles, but not including the voids between the particles) to the weight of an equal volume of water.
- *Apparent specific gravity* is the ratio of the weight in air of a unit volume of the impermeable portion of aggregate to the weight of an equal volume of water.
- *Absorption* is the increase in weight of aggregate due to water in the pores, but not including water adhering to the outside surface of the particles, expressed as a percentage of the dry weight.

EQUIPMENT: Balance, wire basket (of 3.35 mm or finer wire mesh), water tank, oven.

SAMPLE: A minimum of 4000 gm (8.8 lb) test sample for aggregate of maximum nominal size 1 in. (25 mm). The sample should not have particles of size less than 0.187 in. (4.75 mm).

PROCEDURE:

1. Weigh the test sample: A (g).
2. Immerse the aggregate in water at room temperature for a period of 24 ± 4 h.
3. Remove the sample from the water. Roll it in a large absorbent cloth until all visible films of water are removed. The sample is now in saturated surface dry (SSD) condition.
4. Weigh the sample and obtain its saturated surface dry weight: B (g).
5. Place the SSD sample in the wire basket and determine its weight in water: C (g). Note that the wire basket should be immersed to a depth sufficient to cover it and the test sample during weighing.
6. Remove the sample from the wire basket.
7. Dry the sample to constant weight at a temperature of 110 ± 5 °C (approximately 24 h), and weigh: D (g).
8. Calculate specific gravity and absorption.

$$\text{Bulk specific gravity, dry} = \frac{A}{B - C}$$

$$\text{Bulk specific gravity, SSD} = \frac{B}{B - C}$$

$$\text{Apparent specific gravity} = \frac{D}{D - C}$$

$$\text{Absorption} = \frac{B - D}{D} \times 100$$

REPORT: Report the specific gravity and absorption values for 3/4-in. gravel, 3/4-in. crushed stone, and 1/2-in. crushed stone in tabular form.

Test AGG-2: Specific Gravity and Absorption of Fine Aggregate

PURPOSE: To determine the bulk specific gravity, apparent specific gravity, and absorption of fine aggregate (AGG-5).

RELATED STANDARD: ASTM C128.

DEFINITIONS: Refer to Test AGG-1.

EQUIPMENT: Balance, pycnometer (or a volumetric flask of 500 cm^3 capacity, or a fruit jar fitted with a pycnometer top), oven.

SAMPLE: About 500 g of fine aggregate sample.

PROCEDURE:

1. Weigh the test sample: A (g).
2. Cover the test sample with water, either by immersion or by the addition of at least 6 percent moisture to the sample, and permit to stand for 24 ± 4 h.
3. Decant excess water with care to avoid loss of fines. Spread the sample on a flat nonabsorbent surface exposed to a gently moving current of warm air, and stir frequently to secure homogeneous drying. Continue until the sample approaches a free-flowing condition. When the specimen has reached a surface dry condition, it is called saturated surface dry (SSD).
4. Weigh the SSD sample: B (g).
5. Fill the pycnometer with water to the top, and weigh: C (g).
6. Remove part of the water, and introduce the SSD sample into the pycnometer.
7. Fill with additional water to approximately 90 percent of its capacity.
8. Roll, invert, and agitate the pycnometer to eliminate all air bubbles.
9. Bring the water level in the pycnometer to its calibrated capacity.
10. Determine the total weight of the pycnometer, specimen, and water: D (g).
11. Remove the sample from the pycnometer, dry to constant weight at a temperature of 110 ± 5 °C, cool, and weigh: E (g).
12. Calculate the specific gravity and absorption.

$$\text{Bulk specific gravity, dry} = \frac{A}{C + B - D}$$

$$\text{Bulk specific gravity, SSD} = \frac{B}{C + B - D}$$

$$\text{Apparent specific gravity} = \frac{E}{C + E - D}$$

$$\text{Absorption} = \frac{B - E}{E} \times 100$$

REPORT: Report the specific gravity values for all the types of fine aggregate.

Test AGG-3: Unit Weight and Voids in Aggregate

PURPOSE: To determine unit weights of and voids in a sample of coarse, fine, or mixed aggregates (AGG-5).

RELATED STANDARDS: ASTM C29, C127, C128.

DEFINITIONS:

- *Air void* is a space filled with air.
- *Unit weight* or *bulk density* is the weight in air of a unit volume of a permeable material (including both permeable and impermeable voids).

EQUIPMENT: Balance, 5/8-in.-diameter tamping rod (24 in. long), cylindrical metal measure (minimum capacity of 1/2 ft^3 for coarse aggregate of size not larger than $1\frac{1}{2}$ in. and of 1/10 ft^3 for fine aggregate).

SAMPLE: Aggregate dried to constant weight, preferably in an oven at 110 ± 5 °C.

PROCEDURE:

1. Find the empty weight of the metal measure.
2. Fill the measure one-third full with the dry sample.
3. Rod the layer of aggregate with 25 strokes. (Do not allow the rod to strike the bottom of the measure.)
4. Fill the measure two-thirds full, level, and rod as in step 3.
5. Fill the measure overflowing and rod as in step 3.
6. Level the surface of the aggregate with a finger and tamping rod such that any slight projection of the larger pieces of the coarse aggregate approximately balances the larger voids in the surface below the top of the measure.
7. Weigh the measure with the aggregate and find the net weight of the aggregate: A.
8. Calculate the unit weight: B.

$$\text{Unit weight or bulk density, } B = (A/V) \text{ lb/ft}^3 \text{ or kg/m}^3$$

   where V is the volume of the measure.

9. Calculate the void content or percent void.

$$\text{Void (\%)} = \frac{S \times W - B}{S \times W} \times 100$$

   where S is the bulk specific gravity (dry basis) from Test AGG-1 or AGG-2, and W is the unit weight of water (62.4 pcf or 999 kg/m^3).

REPORT: Calculate unit weight and void for gravel, 3/4-in. crushed stone, 1/2-in. crushed stone, and sand. Comment on the results.

Test AGG-4: Total Moisture Content and Surface Moisture Content of Aggregate

PURPOSE: To determine the percentages of total moisture and surface moisture in a sample of aggregate (AGG-5).

RELATED STANDARDS: ASTM C566, C127, C128.

DEFINITIONS:

- *Moisture content* (total) is the weight of water in the particles expressed as a percentage of the dry weight of the particles.
- *Absorption* is the increase in weight of aggregate due to water in the pores, but not including water adhering to the outside surface of the particles, expressed as a percentage of the dry weight.
- *Surface moisture* is equal to the difference between the total moisture content and the absorption.

EQUIPMENT: Balance, oven.

SAMPLE: A minimum of 4000 g of coarse aggregate (1 in. maximum size) or 500 g of fine aggregate.

PROCEDURE:

1. Weigh the sample: A (g).
2. Dry the sample to constant weight in an oven at 110 ± 5 °C for approximately 24 h and cool.
3. Weigh the dried sample: B (g).
4. Calculate the moisture content.

$$\text{Total moisture content} = \frac{A - B}{B} \times 100$$

$$\text{Surface moisture content} = \text{total moisture content} - \text{absorption}$$

Note: Absorption is calculated in Test AGG-1 or AGG-2.

Test AGG-5: Reducing Field Sample of Aggregate to Test Sample

PURPOSE: To obtain laboratory samples of aggregates from stockpiles.

RELATED STANDARDS: ASTM C702, D75.

EQUIPMENT: Shovel, scoop, broom.

PROCEDURE:

1. Obtain a sample of aggregate [about 110 lb (50 kg)] from three places in the stockpile: from the top third, at the midpoint, and from the bottom third of the volume of the pile.
2. Place the field sample on a hard, clean level surface.
3. Mix the material thoroughly by turning the entire sample three times.
4. Shovel the entire sample into a conical pile.
5. Carefully flatten the conical pile to a uniform thickness and diameter by pressing down the apex with a shovel. (The diameter should be approximately four to eight times the thickness.)
6. Divide the flattened mass into four equal quarters with a shovel.
7. Remove two diagonally opposite quarters. Brush the cleared spaces clean.
8. Mix and quarter the remaining material until the sample is reduced to the desired size.

Test AGG-6: Sieve Analysis of Coarse Aggregate

PURPOSE: To determine particle-size distribution and fineness modulus of coarse aggregate (CA) by sieving.

RELATED STANDARD: ASTM C136.

DEFINITION:

- *Fineness modulus* is the sum of the total percentages of material in the sample that is coarser than (cumulative percentages retained) each of the following sieves and divided by 100: No. 8, No. 4, 3/8 in., 3/4 in., $1\frac{1}{2}$ in., and larger, increasing in the ratio 2:1.

EQUIPMENT: Balance, sieves, mechanical shaker, oven.

SAMPLE: Dry coarse aggregate of weight equal to the following.

Weight of sample [lb (kg)]	Maximum size of CA [in. (mm)]
33 (15)	$1\frac{1}{2}$ (37.5)
22 (10)	1 (25)
11 (5)	3/4 (19)

Use the same weight (same as that for 3/4 in.) when the maximum size is less than 3/4 in. (19 mm).

PROCEDURE:

1. Dry the sample to constant weight at a temperature of 110 °C (230 °F) if the sample is lightweight or is suspected of containing appreciable amounts of material finer than a No. 4 sieve.
2. Weigh the dry sample accurately.
3. Weigh each empty sieve and the pan.
4. Nest the suitable sieves in order of decreasing size of opening from top to bottom. Place the pan at the bottom of the set. *Sieves:* No. 8, No. 4, 3/8 in., 1/2 in., 3/4 in., 1 in., $1\frac{1}{2}$ in. (and higher if needed).
5. Place the sample on the top sieve.
6. Place the lid, and agitate the sieves in the mechanical shaker for about 10 min.
7. Weigh the sieves with the material retained.
8. Determine the weight retained in each sieve. The total weight of the material after sieving should check closely with the original weight of the sample. If the amount differs by more than 0.3 percent (based on the original weight), the results should not be used.
9. Calculate the percentage coarser than and the percentage passing (refer to Example 2.1).
10. Draw the particle-distribution curve and calculate the fineness modulus and the average size of the sample.

REPORT: Show the particle-size distribution. Report the fineness modulus and average sample size.

Test AGG-7: Sieve Analysis of Fine Aggregate

PURPOSE: To determine particle-size distribution and fineness modulus of fine aggregate by sieving.

RELATED STANDARD: ASTM C136.

EQUIPMENT: Balance, sieves, mechanical shaker, oven.

SAMPLE: 500 g of dry fine aggregate.

PROCEDURE:

1. Dry the sample to constant weight at a temperature of about 110 °C (230 °F).
2. Weigh the dry sample, the empty sieves, and the pan.
3. Nest the following sieves in order of decreasing size of opening from top to bottom, and place the pan at the bottom of the set. *Sieves:* No. 100, No. 50, No. 30, No. 16, No. 8, and No. 4.

4. Place the sample on the top sieve.
5. Place the lid and agitate the sieves in the mechanical shaker for about 10 min.
6. Weigh the sieves with the material retained. Determine the weight retained in each sieve. The total weight of the material after sieving should check closely with the original weight of the sample. If the amount differs by more than 0.3 percent (based on the original weight), the results should not be used.
7. Calculate the percentages coarser than and the percentages passing (as in Test AGG-6).
8. Draw the gradation chart. Calculate the fineness modulus and the average size of the sample.

REPORT: Draw the particle-size distribution. Report the average size of the sample and the fineness modulus.

PROBLEMS

1. 100 lb of loose gravel can fill a 1-ft^3 box. Its specific gravity is 2.68. What is the dry rodded weight?
2. Find the volume of voids in the preceding problem.
3. A sample of aggregate has a moisture content of 10 percent. Its absorption capacity is 4 percent. What is the weight of free water in a 100-lb sample of this aggregate?
4. Explain the difference between apparent and bulk specific gravity.
5. What is gap-graded aggregate?
6. How is fineness modulus measured, and what does it represent?
7. In a sieve analysis using sieves No. 100, No. 50, No. 30, No. 16, No. 8, and No. 4, the fineness modulus is found to be equal to 3. What is the average size of the sample?
8. Indicate a half-sieve.
9. What is the maximum size of fine aggregate?
10. The specific gravity and bulk density of an aggregate are 2.6 and 95 pcf respectively. Find the loose (bulk) volume of a 190-lb sample.
11. Give two examples of lightweight aggregate.
12. What is the name of the gradation when the void content is the lowest?
13. What causes bulking?
14. What is the shape of a sieve opening?
15. Name a heavyweight aggregate.
16. Plot typical gradation curves for two samples, one with a fineness modulus of 2.2 and the other of 3.2.
17. What is a mixed aggregate?

18. What are the four conditions of moisture in aggregate?
19. Name three common minerals.
20. Name three common types of rocks used as aggregates.
21. A sample of gravel weighs 420 g. Its oven-dry weight is 400 g. What is the moisture content?
22. A sample of fine aggregate with a specific gravity of 2.6 has a bulk density of 105 pcf. Find the void content.
23. What is the maximum size of aggregate that will be used in a 5 in.-thick retaining wall?

3

Concrete and Other Cementitious Materials

Concrete, as it is known today, is a common construction material, the properties of which may be predetermined by design, selection of constituent materials, and quality control. The constituent materials of concrete are:

- Cement
- Aggregates
- Water
- Admixtures

These ingredients are mixed together and molded into the desired size and shape while the mixture is still wet. Within a few minutes of mixing, the cement and water begin to undergo a chemical reaction referred to as *hydration.* This reaction, which continues with time, produces a hard, strong, and durable material called *hardened concrete* or merely *concrete.*

3.1 TYPES OF CEMENT

Cement is any material that binds or unites—essentially like glue. Rubber cement, for instance, is a sticky liquid that can be used for "cementing" pictures or photographs on a board. When a broken tool needs repair, we say that we "cement" the parts together. However, in civil engineering or construction, the words "cement" or "cementitious material" nearly always refer to an ingredient in concrete, mortar, or grout.

Two types of cement are used in the building industry: *hydraulic* and *nonhydraulic*. Hydraulic cement is any cement that turns into a solid product in the presence of water (as well as air), resulting in a material that does not disintegrate in water. In other words, hydraulic cement sets and hardens in water. Nonhydraulic cement, on the other hand, requires no water to transform it into a solid product. Some common varieties of both types of cement are listed in Table 3.1.

TABLE 3.1 HYDRAULIC AND NONHYDRAULIC CEMENTS

Material	Cementitious nature
Portland cement	Hydraulic
Lime	Nonhydraulic
Gypsum	Nonhydraulic
Natural pozzolan	Pozzolanic or latent hydraulic
Fly ash	Pozzolanic or latent hydraulic
Silica fume	Latent hydraulic
Ground blast-furnace slag	Hydraulic or latent hydraulic

3.1.1 Hydraulic Cement

The most common hydraulic cement is portland cement, a finely pulverized material that develops its binding property thanks to its reaction with water. It is manufactured by heating a mixture of limestone and clay until it almost fuses and then grinding the clinker to a fine powder. Materials that exhibit hydraulic activity only in consequence of chemical reaction with other compounds (such as lime produced during the hydration of portland cement) are said to have *latent* hydraulic properties. Many varieties of hydraulic cement can be produced by mixing together latent hydraulic materials and portland cement, and some of the commonly available cements are described in the following paragraphs.

Natural Cement. Natural cement is also a variety of hydraulic cement. It is obtained by burning argillaceous (clayey) or magnesium limestone (called cement rock) having the proper chemical composition and without the addition of any other material, and grinding the resulting clinker. In 1818, during the geological surveys for the Erie Canal, natural cement rock—rock containing lime, silica, and alumina—was discovered near Chittenango, New York. A number of historic structures in the United States—including the Brooklyn Bridge, the Statue of Liberty, and the American Museum of Natural History—were built with natural cement, produced from burning cement rock. This slow-hardening cement used to be manufactured in Rosendale, New York, and Louisville, Kentucky.

Blended Cement. Blended cement consists of interground blends of portland cement clinkers and fly ash, natural or calcined pozzolan, or slag, within the constituent limits specified. They may also consist of blends of blast-furnace slag or pozzolan with lime. These cements generally have higher resistance against alkalis, reactive aggregates, and seawater. They hydrate slowly and the rate of gain in strength is lower than that of ordinary portland cement. The common blended cements are portland pozzolan cement, pozzolan-modified portland cement, portland blast-furnace slag cement, slag-modified portland cement, and masonry cement.

Portland Pozzolan and Pozzolan Cement. Pozzolan is a natural or artificial product composed chiefly of lime, silica, or alumina. In a finely divided form it reacts with lime in the cement paste to form a hydraulic product. Portland pozzolan cement is produced by intimately blending portland cement with a pozzolan. ASTM C595 identifies two types of portland pozzolan cement: Type IP, for use in general concrete construction, and Type P, for use where high strength at early age is not required. The amount of pozzolan is about 15–40 percent by weight of the cement. Pozzolan-modified portland cement is manufactured in the same manner as portland pozzolan cement, but the pozzolan component is less than 15 percent by weight.

Portland-pozzolan cement has a lower heat of hydration and lower thermal shrinkage than ordinary cement. It is better suited for massive structures like dams, piers, and footings. Concrete made with this cement has better watertightness and improved resistance to sulfate attack.

Portland Blast-furnace Slag Cement. Portland blast-furnace slag cement consists of an intimate and uniform blend of portland cement and fine granulated blast-furnace slag. Blast-furnace slag is a nonmetallic product, consisting essentially of silicates and aluminosilicates of calcium and other bases, which is developed, in a molten condition, simultaneously with iron in a blast furnace. When the molten slag is rapidly chilled (as by immersion in water) the resulting glassy granular product is granulated blast-furnace slag, and the process is known as granulating.

Portland blast-furnace slag cement is produced by intergrinding portland cement clinker and granulated slag, or blending portland cement and finely ground slag, or a combination of intergrinding and blending. The amount of slag is 25–70 percent of the total weight of the cement. Slag-modified portland cement is similar to portland blast-furnace slag cement, but the slag constituent is less than 25 percent of the weight of the cement.

Slag Cement. Slag cement is a hydraulic cement consisting mostly of an intimate and uniform blend of granulated blast-furnace slag with portland cement or hydraulic lime or both. The amount of slag is at least 70 percent of the weight of the slag cement.

White Cement. White cement is a cement suitable for exposed aggregate finishes and for making colored cements with pigment addition. The white color is achieved by nearly eliminating the iron content from ordinary portland cement. Iron-free clay (kaolinite or china clay) must be used in its manufacture, and bauxite (aluminum oxide) is often needed to achieve the required alumina content.

Masonry Cement. Masonry cement contains a finely divided material such as lime or ground limestone and an air-entraining agent, mixed with portland cement. Its chemical makeup is not regulated and varies between brands. The admixture and limestone in the cement provide good workability and water retention, the two most important properties of fresh mortar.

Plastic Cement. Plastic cement is a mixture of approximately 96 percent portland cement and plasticizing and air-entraining agents. It is used primarily for exterior plaster and satisfies the durability and hardness requirements for exposed surfaces.

High-alumina Cement. High-alumina cement (also called *calcium aluminate cement*) is produced by burning limestone and bauxite, and contains sintered calcium aluminate (monocalcium aluminate, CA) instead of the calcium silicates of portland cement. The amount of alumina (Al_2O_3) in this cement is about 40 percent, compared to around 4 percent in portland cement. High-alumina cement gains strength very quickly and has refractory properties. Concrete made with this cement should be kept cool during the initial period after mixing to limit the heat of hydration. If the concrete temperature rises above 85 °F (30 °C), such as from exposure to heat, it leads to loss in strength due to an increase in porosity and decrease in solid volume. Strength loss can be decreased by using low water to cement ratio. It is well known that high-alumina cement concrete may experience strength retrogression upon drying. A number of structural failures around the world have more or less resulted in a ban on the use of this cement in structural applications, but it is still used in the production of refractory concrete.

3.1.2 Nonhydraulic Cement

The two common nonhydraulic cements are gypsum and lime. Gypsum is obtained by calcining natural gypsum; lime, by calcining limestone. A mixture of calcined gypsum and water was used by Egyptians around 3000 B.C. for the construction of pyramids. Lime mortar has been used by many civilizations around the world for its bonding ability and low permeability (for the prevention of water penetration).

Lime. Lime, which generally refers to quicklime or calcium oxide, is obtained by burning limestone, or calcium carbonate, at a temperature in excess of 900 °C (1650 °F). This drives off water and carbon dioxide, leaving behind calcium oxide—a white porous solid, called quicklime or unslaked lime. Strongly alkaline, it is entirely soluble in water, thus forming one of the strong bases, increasing in bulk—an expansion of about three times the original volume—and evolving much heat, and changing into calcium hydroxide, or hydrated lime.

The setting of lime is a chemical process in which the evaporation of excess water from the lime paste is followed by the gradual replacement of the remaining water of the hydroxide by calcium dioxide from the atmosphere. This reaction, which occurs in the presence of moisture or humidity, reverts the hydrated lime back to the original calcium carbonate—limestone.

Lime is one of the oldest building materials, used in the construction of masonry, roads, and dams. The Great Wall of China was largely laid in lime mortar. Independence Hall in Philadelphia (1734) was also built using lime mortar. Today, lime is mixed in combination with portland cement for mortars in masonry construction and in plastering. It is also put to use to stabilize soils and neutralize acidic soils. Lime-cement columns have been used to strengthen soft compressive soils at several modern construction sites, such as the I-15 project in Utah and the BART extension to the San Francisco airport. Hydrated cement-lime in these underground piles releases a high amount of heat, which draws water from the surrounding clays, speeding the consolidation of soil.

Lime is available in both forms: quicklime and hydrated (or slaked) lime, $Ca(OH)_2$. Because it is faster and safer for handling purposes, hydrated lime is more widely used than quicklime. Quicklime, which can be purchased crushed, ground, or pulverized, must be slaked prior to use. Care is necessary while adding water, to prevent splattering that can cause severe burns. Following the addition of water, the lime putty (paste) should be screened and left to age—from a few hours to several days—to allow for the slaking of all particles. Bagged quicklime must be stored in a dry place to prevent it from coming into contact with moisture.

Hydrated lime is commonly sold in 50-lb bags. Type S hydrated lime (also called pressure-hydrated lime) contains up to 8 percent combined unhydrated calcium oxide and magnesium oxide. This highly hydrated product renders superior plasticity to mortar.

Type N hydrated lime (also called normal hydrated lime) should be soaked in water overnight prior to use to improve its plasticity and workability. The soaking is accomplished by sifting the powdered lime evenly in a pan half filled with clean water, and allowing the lime to settle into the layer of water without stirring or mixing. The spreading of lime should be continued until a thick paste is formed, and then the mixture is left to soak until ready for use—normally, the next day. The film of water on top of the paste prevents lime from coming into contact with the atmospheric carbon dioxide.

When hydrated lime is used it is assumed that a unit volume of dry hydrated lime produces a unit volume of lime putty. The specific gravity of lime ranges between 2.3–2.6, and the bulk density between 35–40 pcf (560–640 kg/m^3). A bag of lime is assumed to produce approximately 1.15 ft^3 of lime putty.

Gypsum. Pure gypsum is fully hydrated lime sulfate, ($CaSO_4 + 2H_2O$), also called calcium sulfate dihydrate. It is the most common sulfate mineral, occurring in many places and forms. Alabaster is a fine-grained, translucent variety of gypsum; quarried in England and Italy, this white-colored stone is easy to carve and is used to make decorative objects. Selenite is a white semitranslucent crystalline gypsum rock, occurring in small deposits. (Gypsum rock formations may contain impurities of up to about 6 percent.) Plaster of paris, which is used to make molds and wallboards (gypsum boards), is made from gypsum.

Calcining gypsum, which is done at relatively low temperatures, drives away water of crystallization, forming either calcium sulfate hemihydrate (from incomplete dehydration of gypsum), $CaSO_4 + \frac{1}{2}H_2O$ or calcium sulfate, also called anhydrite or anhydrous calcium sulfate, from complete dehydration. Plaster of paris is the result of incomplete dehydration; calcium sulfate hemihydrate is also used as the cementitious base for most gypsum plasters.

Gypsum plaster is made by combining calcined gypsum, which is soluble in water, with sand, water, and other ingredients. It sets very quickly and develops strength rapidly; this process involves recombination of the partially or totally dehydrated calcium sulfate with water to reform hydrated calcium sulfate, or gypsum mineral. But as gypsum is quite soluble in water, gypsum plaster lacks resistance to water. *Keene's cement,* which is used for plastering, is made by combining anhydrous gypsum with portland cement.

3.2 VARIOUS CEMENTITIOUS MATERIALS

The most common cementitious material is concrete, or portland cement concrete, which is a composite of cement, fine aggregate, coarse aggregate, and water, with or without admixture. Others are mortar, grout, shotcrete, and plaster (Fig. 3.1). Mortar

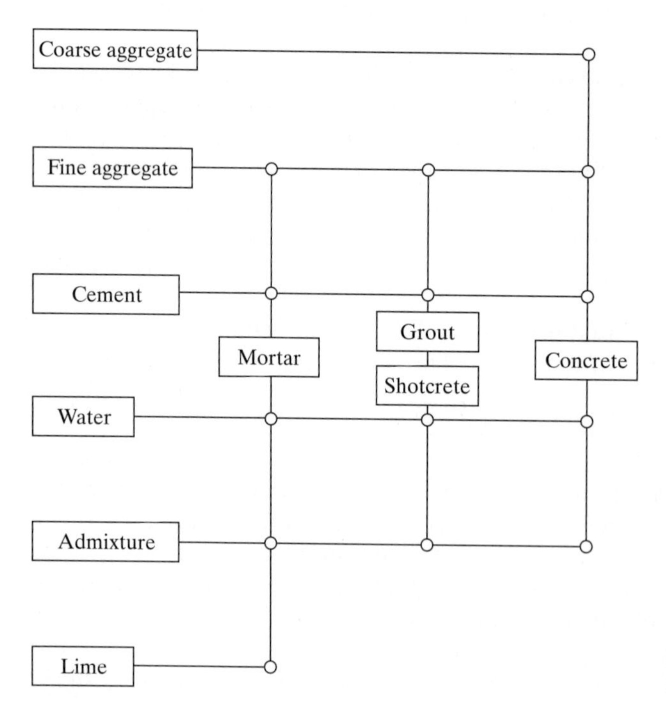

Figure 3.1 Concrete and other cementitious materials.

(as well as plaster) is a mixture of sand, cement, lime or gypsum, and water. Mortar is used to bond individual masonry units, such as bricks and blocks, in a masonry wall, and plaster is used for plastering. Grout is a mixture of cement, sand, and fine gravel, to which sufficient water is added that it can be poured without causing segregation—which is the tendency for the large and fine particles in fresh concrete to become separated, resulting in a porous and honeycomb structure. Stucco is a type of plaster made with cement, water, and fine sand, along with lime or gypsum. It can be applied—generally in three coats—directly to masonry walls, but over wood sheathing, some type of wire mesh must be utilized to tie the stucco and the sheathing together. It is a desirable wall covering that is strong, fire resistant, and fungus resistant. Shotcrete, also called gunite or spray concrete, refers to pneumatically placed concrete, in which a dry mixture of cement and sand is blown through hoses and water is injected at the nozzle. The maximum size of aggregate for shotcrete is generally 3/4 in. (20 mm). This material and technique are employed in the repair of damaged structures and in the construction of domes, shells, tunnel linings, culverts, and swimming pools. Shotcrete is also used to rehabilitate structures (such as bridges, culverts, sewers, and dams) damaged by freeze-thaw action, aggressive chemicals, or fire.

3.3 USES OF CONCRETE

Concrete is one of the most common construction materials and is employed in a wide variety of applications, ranging from piles to multistory buildings and from railroad ties to dams (Figs. 3.2–3.3). It is also used in foundations, pavements, walkways, storage tanks, and many other structures. In fact, it is hard to find a structural application in which concrete is not put to use in one form or another. It is one of the most economical materials of construction, being very versatile by nature and flexible in application.

Important properties of concrete include:

- Compressive strength
- Durability and freeze-thaw resistance
- Wear resistance
- Impermeability
- Abrasion resistance
- Resistance to environmental attacks (from seawater, sulfates in soil, and so on)

Not all of these properties are important for every application, but most are. For example, in building a liquid-retaining structure such as a storage tank or dam, the primary requirements are impermeability, resistance to chemical attacks from liquids, and weather resistance. In nonstructural applications such as building facades and sign walls, concrete may also be required to possess adequate thermal resistance, light weight, and pleasing appearance.

Unlike wood and steel, which deteriorate through contact with moisture, concrete possesses excellent resistance to water. This property is of prime importance in

Figure 3.2 Concrete is a major material in building construction.

the building of concrete dams, aqueducts, pipes, canals, storage tanks, and foundations. Freshly made concrete is plastic in consistency and can easily be molded into any shape and size, and is thus adopted for designing shells, folded plates, circular pipes, and arches. The materials for the manufacture of concrete are readily available and relatively inexpensive, and large quantities of industrial waste products can be recycled and used as ingredients in concrete preparation.

Concrete derives its strength and other properties from the key physical properties of the aggregates; the type, quantity and quality of the cement; and the mix proportions. The aggregates used in concrete are not physically transformed or chemically altered during the hardening process. But the reaction between the cement and water produces compounds that harden with time, binding—or cementing—the individual particles. Thus, to understand the properties of concrete,

Figure 3.3 Hoover Dam is one of the oldest concrete structures in the United States.

it is important to know the composition of the cement and the chemical process that follows when it comes into contact with water. These will be reviewed in the following sections.

3.4 PORTLAND CEMENT

The term "portland cement" describes a hydraulic cement that is produced by pulverizing clinkers removed from cement kilns, and that consists essentially of hydraulic calcium silicates. Joseph Aspdin, a bricklayer in Leeds, England, first used the term in 1824 to describe a patented cementitious product formed by heating a mixture of clay and limestone or chalk in a furnace to a temperature sufficiently high to drive off carbon dioxide. The name was derived from the fact that the color of the concrete mixture, made with this hydraulic cement, resembled that of some natural stones on the Isle of Portland, Dorset County, England. However, I. C. Johnson, an English cement manufacturer, has also been claimed to be the real inventor of portland cement. In 1845, his company began the production of portland cement, essentially as we have it today.

David Saylor of Allentown, Pennsylvania, who was granted a patent in 1871, is credited with the earliest manufacture of portland cement in the United States. The first rotary kiln for the manufacture of portland cement was erected in 1886 at Roundout, New York. The Coplay Cement Company, whose first president was Saylor, shipped their first cement in 1874. Cement consumption in the United States began to increase dramatically after around 1910.

3.4.1 Manufacture

Portland cement is manufactured by heating to incipient fusion a carefully controlled mixture of limestone and clay, with or without secondary raw materials (such as bauxite and iron ore), and then grinding the resulting product, called "clinker," into a fine powder. Most of the ingredients for the manufacture of portland cement are found in nature (Table 3.2). Limestone, shale, slate, clay, chalk, marl (which is a

TABLE 3.2 COMMON RAW MATERIALS AND COMPOUNDS USED IN THE MANUFACTURE OF PORTLAND CEMENT

Raw material	Compound
Cement rock	Lime, silica, alumina
Limestone	Lime
Chalk	Lime
Slag	Lime, silica, alumina
Oyster shell	Lime
Marble	Lime
Clay	Silica, alumina
Shale	Alumina
Fly ash	Alumina, silica
Kaolin	Alumina
Bauxite	Alumina
Sand	Silica
Quartzite	Silica
Traprock	Silica
Iron ore	Iron oxide
Iron dust	Iron oxide
Pyrite	Iron oxide, sulfate
Gypsum	Sulfate

natural soil deposit found in lake bottoms and swamps, consisting of clay and calcium carbonate and used especially to fertilize soil deficient in lime), silica sand, and iron ore are plentiful throughout the world. Each manufacturing facility may use a different combination of raw materials, depending on their chemical compositions, but limestone (containing less than 3 percent $MgCO_3$) and clay are the most common.

The three primary constituents of the raw materials used in the manufacture of portland cement are:

- Lime (CaO)
- Silica (SiO_2)
- Alumina (Al_2O_3)

Lime is derived from limestone or chalk; silica and alumina from clay, shale, or bauxite. In addition to these compounds, most raw materials contain small amounts of iron oxide (from iron ore or clay), magnesia, sulfur trioxide, alkalis, and carbon dioxide. But lime and silica make up for about 60 percent and 20 percent of the ingredients in raw materials respectively, and iron oxide and alumina account for about 10 percent. Iron ore also functions as a flux, lowering the clinkering temperature.

Portland cement is manufactured by one of two basic processes: *wet* and *dry process*. In both, the raw materials are homogenized by crushing, grinding, and blending so that approximately 80 percent of the raw materials pass a No. 200 sieve. In the wet process, the mix (in the form of slurry containing about 30–40 percent water) is heated to about 2750 °F (1510 °C) in horizontal revolving kilns 250–500 ft (76–153 m) in length and 12–16 ft. (3.6–4.8 m) in diameter. Natural gas, petroleum, or coal (along with supplemental fuels such as coke, tires, rice hull, and waste products) are used for burning. At this high temperature, the oxides of calcium, silica, aluminum, and iron are chemically combined to form cement clinkers. The rotation and shape of the kiln allow the blend to flow down the kiln, submitting it to gradually increasing temperature. In the dry process, the mixture is fed into the kiln and burned in a dry state, which provides considerable savings in fuel consumption and water usage, but the process is dustier. While the wet system is usually more efficient than dry grinding, the high fuel requirement may make it uneconomical.

In the kiln, initially, water from the raw materials is driven off and limestone is decomposed into lime (CaO) and carbon dioxide. In the burning zone portion of the kiln, silica and alumina from the clay undergo a solid-state chemical reaction with lime to produce calcium aluminate. As the material moves through hotter regions in the kiln, calcium silicates are formed. These products, which are black or greenish-black in color, are in the form of small pellets, called cement clinkers. In a cement production facility the raw materials are continuously sampled and analyzed to ensure that the clinkers contain approximately 65 percent lime, 21 percent silica, 5 percent alumina, and 3 percent iron oxide.

The cement clinkers are hard, irregular, ball-shaped particles—about 3/4 in. (18 mm) in diameter—and are cooled to about 150 °F (51 °C) and stored in clinker silos. When needed, the clinkers are mixed with about 2 to 5 percent gypsum—to retard the setting time of the cement—and then ground to a fine powder—particles less than 45 μm in diameter—in ball mills or a roller press. The cement thus obtained is then stored in storage bins or cement silos, or bagged for shipment.

In the United States, a standard bag of cement weighs 94 lb (42.6 kg) and has a bulk volume of 1 ft^3 (0.028 m^3), which means the bulk density of cement is 94 pcf (1500 kg/m^3). In many other countries, cement is sold in 50-kg bags. Cement bags should be stored on pallets in a dry place, preferably covered with tarpaulins or similar waterproof coverings. Cement that comes into contact with moisture forms lumps and sets more slowly than does cement that is kept dry. Most bagged cements in moist climates have a shelf life of 6 to 8 weeks.

3.4.2 Cement Chemistry

Portland cement typically consists of about 65 percent CaO, 21 percent SiO_2, 4.5 percent Al_2O_3, and 3 percent Fe_2O_3. In addition, small quantities—less than 2.5 percent—of SO_3, MgO, Na_2O, and K_2O are found. In the clinker, these oxides exist as special compounds in the form of extremely fine interlocking crystals (although some amorphous materials do exist). The majority of compounds are made up of silicates and aluminates.

About 75 percent of portland cement is composed of calcium silicates; compounds of aluminum, iron, and gypsum make up the balance. The four major compounds of portland cement are:

- Tricalcium silicate ($3CaO \cdot SiO_2$)
- Dicalcium silicate ($2CaO \cdot SiO_2$)
- Tricalcium aluminate ($3CaO \cdot Al_2O_3$)
- Tetracalcium aluminoferrite ($4CaO \cdot Al_2O_3 \cdot Fe_2O_3$)

These compounds and their abbreviated symbols are shown in Table 3.3. It is customary in cement chemistry to represent these compounds in shorthand notation: Each oxide formula is abbreviated to a single letter and the number of oxide compounds is designated by a subscript placed after the letter. For example, "C" stands for lime (CaO), and "S" for silica (SiO_2). Thus the compound $3CaO \cdot SiO_2$, or tricalcium silicate, is written as C_3S.

TABLE 3.3 PRINCIPAL COMPOUNDS OF CEMENT

Compound	Chemical formula	Industry code (abbreviation)	Percentage amount (range)	Rate of reaction with water
Tricalcium silicate	$3CaO \cdot SiO_2$	C_3S	35–65	Medium
Dicalcium silicate	$2CaO \cdot SiO_2$	C_2S	15–40	Slow
Tricalcium aluminate	$3CaO \cdot Al_2O_3$	C_3A	0–15	Fast
Tetracalcium aluminoferrite	$4CaO \cdot Al_2O_3 \cdot Fe_2O_3$	C_4AF	6–20	Medium

The proportions of these compounds (also called phases) are computed from the chemical analysis of cement following a procedure outlined in ASTM C150. For example, the percentage of tricalcium silicate is calculated from the amounts of lime, silica, alumina, iron oxide, and sulfur oxide. Similarly, the percentage of dicalcium silicate is computed from the amounts of tricalcium silicate and silica, as follows:

$$\%\ \text{dicalcium silicate} = 2.867 \times (\%\ \text{silica}) - 0.7544 \times (\%\ \text{tricalcium silicate})$$

In addition to the four major compounds, listed in Table 3.3, minute quantities of other compounds are also found in cement. Depending on the relative quantities of these major compounds, the properties of hardened cements vary. When the cement comes in contact with water, hydrates of various compounds are formed, which give hardness and strength to the mix. This process, called hydration, is explained in Section 3.5.

The strength characteristics of portland cement depend on the relative proportions of the silicates C_3S and C_2S, which constitute about 75 percent of the cement. A bigger percentage of C_3S produces higher heat of hydration and accounts for faster gain in strength. A higher amount of C_2S improves the later-age strength of cement. Over the years, the proportion of C_3S in cements has steadily increased while that of C_2S has decreased. For example, portland cement manufactured around 1935 had 55 percent C_3S (compared to 30 percent in 1890) and 23 percent C_2S (compared to 36 percent in 1890). Type I portland cement manufactured today typically has around 60 percent C_3S and 16 percent C_2S. C_3A has a poor cementing property, turns out higher heat of hydration, and contributes to faster gain in strength; in addition, it results in poor sulfate resistance and increases the volumetric shrinkage upon drying. C_4AF acts as filler with little contribution to the strength of cement.

3.4.3 Types of Portland Cement

Through modifications in the relative amounts of the main ingredients and in the fineness, portland cement is manufactured in eight different types: Type I, Type IA, Type II, Type IIA, Type III, Type IIIA, Type IV, and Type V (Table 3.4).

Type I portland cement (standard portland cement, or ordinary portland cement) is a general purpose cement, and is the most commonly used portland cement. It is recommended when special properties of other cements are not required. Type IA is ordinary portland cement interground with an air-entraining admixture. Every cement manufacturer produces Type I cement; chemical compositions vary among the manufacturers, however, as they depend on the raw materials.

Type II portland cement (modified portland cement) is a general-purpose cement for use when moderate sulfate resistance or moderate heat of hydration is desired. Type IIA is obtained by combining Type II cement with an air-entraining admixture. Types II and IIA cements have better resistance to deterioration from sulfate attack than Types I or IA, and are more suited for places where sulfate concentration in groundwater is higher than normal but not severe. Type II cement has

TABLE 3.4 PORTLAND CEMENT TYPES AND USES

	Standard chemical requirements				
Cement types	C_3S (max.)	C_2S (min.)	C_3A (max.)	$C_4AF + 2C_3A$ (max.)	Uses
I and IA					General use; when special properties are not required
II and IIA			8		General use; has moderate sulfate resistance and heat of hydration
III and IIIA			15		When high early strength is required
IV	35	40	7		When low heat of hydration is required
V			5	25	When high sulfate resistance is required

relatively low C_3A content, typically around 6 percent. The heat of hydration generated during the hydration of this cement is lower than that of Type I cement, and thus is apt for use in mass concrete works such as piers and heavy footings. The lower heat of hydration is also beneficial when placing concrete in hot weather. Type II cement is used, for example, in highway pavements, reservoir linings, foundations, high-rise buildings, piers, and massive structures.

Type III portland cement (high early strength portland cement) is for use when early strength is desired. Type IIIA is Type III cement with an air-entraining admixture. These cements have more C_3A than any other type of portland cement (as high as 15 percent) and are finer, making the cement set and harden rapidly. It is used when formwork is to be removed as early as possible and the structure is to be brought into service quickly. Under favorable conditions, the forms of slabs can be stripped in three days, compared to the normal seven-day waiting period with Type I cement—this generates substantial savings in the cost of lumber required for forming. In precast construction, the forms are stripped even earlier, typically within 24 hours. Type III cement is also used in cold weather operations to decrease the time needed to protect the concrete from freezing. Concrete made with this cement sets and hardens rapidly; the one-day strength of concrete using Type III cement is about the same as the three-day strength of concrete made with Type I cement, and the three-day strength is nearly equal to the seven-day strength using Type I cement.

Type IV portland cement (low-heat cement) is for use when the heat of hydration must be lowered. It contains the smallest amounts of C_3S and C_3A (the compounds that produces the most heat during hydration) of all types of cement. In large structures such as concrete dams and massive foundations, where the temperature gradient within the cross-section must be carefully controlled, the use of Type I cement may result in cracking of the concrete. The use of low-heat cement instead produces lower heat of hydration and slower hardening, resulting in better control of the rise in concrete temperature. However, Type IV cement is no longer manufactured in the United States as more economical and practical methods of temperature control, such as the use of pozzolans, have rendered this cement unnecessary.

Type V portland cement (sulfate-resisting cement) is chosen when higher sulfate resistance is required. Sulfates of sodium, calcium, and magnesium are often found in soils, lakebeds, and seawater. These compounds react with calcium hydroxide, or $Ca(OH)_2$, formed during the hydration of cement to produce gypsum, calcium sulfate, and other compounds. Calcium sulfate reacts primarily with C_3A to produce calcium hydroxide and an expansive compound, tricalcium sulfoaluminate. This, along with sodium-magnesium sulfoaluminate, causes volume expansion in concrete. Cements containing low amounts of C_3A—generally less than 3 percent—have higher resistance to sulfate attack. Thus, Type V cement, which has a maximum limit of 5 percent C_3A, is best suited in applications requiring high sulfate resistance. Factors other than the amount of C_3A, such as the ratio between C_3S and C_2S, are also found to affect the sulfate resistance, however, and the use of Type V cement may increase the corrosion danger in reinforced concrete construction. As shown by a number of deterioration problems in fairly new concrete structures in the Middle East, the primary mechanism of deterioration of structures at or near seawater (which contains a large amount of chlorides and less of sulfates) is corrosion of reinforcement. Type V cement may impair the corrosion resistance offered by concrete.

Type V cement is commonly chosen for concrete construction below ground, in inland lake water and seawater, and around sewage disposal sites. This cement is found to be somewhat resistant to the destructive action of organic acids.

Sulfate resistance can also be improved by adding pozzolans such as fly ash, along with ordinary portland cement, to concrete. The addition of a pozzolanic material while mixing the ingredients removes the excess calcium hydroxide from the hydrated cement paste and thus supports a more stable environment against deterioration.

Air-entraining cement (Types IA, IIA, and IIIA) can be used to obtain air-entrained concrete. As explained earlier, each of these three types of portland cement comes premixed with an air-entraining admixture. However, it is more common to add an air-entraining admixture to ordinary (nonair-entrained) concrete at the job site, since the quantity of the admixture can be varied to satisfy the design or durability requirements. Additional details on air-entrained concrete are furnished in Section 3.11.1, "Chemical Admixtures."

3.4.4 Fineness

Fineness of cement relates to the size of the cement grains. It has a considerable influence on the behavior of cement during its hydration, and can be measured from the percentage of particles that pass a No. 200 sieve (75 μm). For instance, about 80% or more of the particles in Type I cement pass through this sieve. But this method of measuring fineness is not generally followed, as it does not provide information on the size distribution of particles.

Fineness is commonly established by measuring the specific surface, or surface area per unit mass (m^2/kg), following a procedure described in either ASTM C115 (the Wagner turbidimeter method) or ASTM C204 (the Blaine air-permeability test). In the Blaine air-permeability method, the rate of flow of air—of known volume—through a prepared bed of cement placed in the cell of the apparatus is measured. From the time required to pass the bed, specific surface is calculated empirically.

Fineness affects primarily the hydration of cement. The rate of hydration increases with increasing fineness, which increases the rate of strength development and evolution of heat. For example, an increase in fineness from 200 m^2/kg to 300 m^2/kg, an increase of 100 percent, increases the 7-day strength of a typical mortar from 3700 psi (25.5 MPa) to 4300 psi (29.7 MPa), an increase in strength of about 16 percent. However, after about 90 days, this increase in strength—accompanying the finer cement—is negligible. Even though finer cements show higher early-age strength and increased workability (in low cement-content mixtures), they contribute to excessive cracking and lower the resistance to freezing and thawing. Increase in fineness decreases the amount of bleeding, but increases the amount of water required to maintain a level of workability.

Type III cement is ground finer than other types of cement. Generally, most manufacturers grind their cements finer than the ASTM recommended minimum, which is 280 m^2/kg for all types of cement. Over the years, cements have been getting finer; the minimum fineness around 1935 was 150 m^2/kg, whereas Type I cements sold today have fineness ranging between 330–420 m^2/kg.

3.4.5 Strength of Cement

The term "strength of cement" is something of a misnomer, for it is established by measuring the compressive strength of the mortar, not the cement. Size 2 in. (50 mm) cement mortar cubes of mix proportions (by weight) 1 part cement, 2.75 parts standard graded sand, and 0.485 part water (0.46 for air-entrained cement), for all types of cement, are tested at 3, 7, and 28 days. The compressive strength thus established does not represent a fundamental property of either the cement or the mortar, for the strength of either material changes when the mix proportions are modified or the size and shape of the specimen are varied. But the test, described in ASTM C109, provides a relative measure of strength, in that when the amount of cement remains unchanged the use of higher-strength cement enhances the overall strength of the concrete.

The ASTM prescribes a minimum strength of 1800 psi (12.4 MPa) at 3 days and 2800 psi (19.3 MPa) at 7 days for Type I cement. For Type II cement these values are 1500 psi (10.3 MPa) and 2500 psi (17.2 MPa) respectively, and for Type III cement, the minimum strength at 3 days is 3500 psi (24.1 MPa). Typically, Type I portland cement manufactured in the United States exhibits a strength close to 3500 psi (24.1 MPa) at 3 days, 4500 psi (31.1 MPa) at 7 days, and 5300 psi (36.6 MPa) at 28 days. These numbers show that the cements manufactured nowadays have strengths much higher than the recommended minimum, and that they reach a strength value at 7 days that is nearly the same as the 28-day strength of cement manufactured around 1935.

3.4.6 Consistency of Cement

Normal consistency, measured using ASTM C187, is a property used to establish the setting properties of cement. The testing is done in a Vicat apparatus using a 10-mm-diameter needle under a load of 300 g. A paste is said to have normal consistency when the plunger penetrates 10 ± 1 mm below the original surface in 30 s. The amount of water required, expressed as a percent of the dry weight of cement, is the water to cement ratio of normal consistency, and ranges between 24–33 percent for most brands of cement.

3.5 HYDRATION

Dry portland cement does not possess the cementing or binding property, for it is a hydraulic material. The chemical reaction between the compounds of cement and water yields products that achieve the binding property after hardening. This process, the reaction between cement and water, is called *hydration.*

Hydration, although it is a continuous phenomenon under favorable conditions, can be regarded as made up of two stages: setting and hardening. When mixed with water, cement first sets and then rather slowly hardens. These stages will be discussed in the following sections.

3.5.1 Setting

When cement is mixed with sufficient water the resulting paste loses its plasticity and then slowly forms into a hard rock. In a favorable environment, within one or two hours after the mixing of cement and water the sticky paste loses its fluidity; within a few hours after mixing, noticeable stiffening commences. This mechanism is called *setting,* and is divided into two stages: initial and final set. A precise definition of either initial set or final set is not possible; generally speaking, however, the former is when the paste is beginning to stiffen, and the latter is when it is beginning to harden and able to sustain some loads.

The time lapse from the addition of water to the mix to the initial set is called initial setting time, and similarly, to the final set, final setting time. They are measured, in a laboratory, as the time required for the cement paste to withstand a certain arbitrary pressure. Accordingly, the two readings pertain to the rate of stiffening of the cement only and not to the concrete made with it.

The time taken for a 1-mm-diameter needle in the Vicat apparatus to penetrate a depth of 25 mm into the cement-paste sample is the initial setting time. The final setting time is reached when in the modified Vicat apparatus only the needle penetrates the surface, while the attachment fails to do so.

For Type I cement, the initial and final setting times are about 2–4 h and 5–8 h, respectively. Fineness of cement, chemical composition of cement, and storage conditions affect the setting time. In addition, the amount of water and the ambient temperature also affect the rate of setting. The finer the cement, the faster the rate of setting and also the hydration. Increases in the amounts of tricalcium aluminate and tricalcium silicate decrease the setting times. In cold weather applications, use of hot water for mixing accelerates the setting, but may contribute to flash setting. This can be avoided by mixing aggregate and water together (both may be hot) for about a minute before adding the cement.

The rate of setting is also a measure of the rate of release of heat of hydration. The compound gypsum, which is added to clinker when it is ground, retards setting and prevents flash set, which is accompanied by an appreciable liberation of heat. *Flash set* is defined as the rapid development of permanent rigidity of the cement paste—along with high heat. The rapid development of rigidity without the evolution of heat is called *false set*. It is a premature stiffening of the paste, and is attributed to unstable gypsum in the cement. Fresh concrete can be reworked after false set without the addition of extra water.

When a concrete mixture reaches a state in which its form cannot be changed without producing rupture, it is said to have set. The setting time of concrete depends on the setting properties of cement and many other variables. Mixtures with higher water content take a longer time to set. The higher the ambient temperature, the more rapid the setting. A concrete mix may begin to set in 2 h at 70 °F (21 °C) ambient temperature, but if the temperature rises to 95 °F (35 °C) it sets in 1 h. If the ambient temperature drops to 50 °F (10 °C) the setting may take about 3 h. Elevated temperature of mixing water increases the rate of setting; at temperatures close to freezing, concrete does not set.

3.5.2 Hardening

In ordinary cement paste and concrete, setting occurs within a few hours, but hardening—which is the development of strength over an extended period of time—is not completed for months or years. Hardening of concrete is the net outcome of hydration.

Portland cement is a mixture of several compounds (oxides) all of which can hydrate, or react with water. Hydration is the key for strength development in concrete. But not all compounds hydrate at the same rate, and as a consequence, the strength development of cement is a function of time and temperature, in addition to the composition of cement. The rate of hydration of cement depends on the relative proportions of silicates and aluminates, the fineness, and the ambient conditions—humidity and temperature.

Tricalcium aluminate is the most reactive compound in cement, and it hydrates at a much faster rate than do the silicates. The stiffening characteristics and setting times of cement are due largely to the hydration products involving aluminates. In fact, gypsum is added to clinkers to slow down the hydration of tricalcium aluminate and prevent flash set. The silicates, however, play a dominant role in the hardening process, which is responsible for strength development.

Tricalcium silicate hardens rapidly and is largely responsible for early strength development. Dicalcium silicate hardens slowly and is accountable for later-age strength, the strength increase beyond about two weeks.

The hydration of silicates produces crystalline calcium hydroxide and calcium silicate hydrate. The latter is a poorly crystalline (or amorphous) material made up of extremely small particles. The physical properties of the paste and the mechanical properties of hardened cement depend primarily on calcium silicate hydrate. In a completely hydrated paste, this product occupies about 50–60 percent of the volume of solids.

Reaction between tricalcium aluminate, gypsum, and water results primarily in calcium sulfoaluminate hydrate, also known as ettringite. Reaction between tricalcium aluminate and water causes the formation of calcium aluminate hydrate, which is accompanied by considerable evolution of heat, and the mixture sets almost instantly. In the presence of gypsum, the heat evolution is lessened. Tricalcium aluminate again combines with ettringite and gypsum to produce another compound, monosulfoaluminate, which may once more form ettringite, followed by a large increase in volume. Hydration of tetracalcium aluminoferrite produces calcium sulfoaluminate hydrate and monosulfoaluminate.

Calcium hydroxide, which can occupy as much as one-quarter of the volume of hydration products, does not contribute as a binder. It is prone to combine with carbon dioxide from the environment to form a soluble salt that can cause efflorescence.

It is estimated that one cm^3 of cement on complete hydration occupies a volume of about two cm^3, which means that one cm^3 dry cement grains (solid volume) produces hydration products that occupy roughly 2 cm^3 volume. In other words, 1 g (or, 1 lb) of cement requires about 0.32 g (or, 0.32 lb) of water to complete the hydration, for the specific gravity of cement is 3.15. Hydration can also be conceptualized as a process in which the space occupied by cement and water is being replaced more and more by the products of the reaction between the two. The space not filled by either the cement or the hydration products makes up the voids or pores within concrete, filled with water or air. These voids do not possess any strength, and weaken the paste. Under favorable conditions of moisture and temperature, hydration can continue as long as space for the products of hydration is available.

Heat of Hydration. Hydration is always accompanied by the release of heat. In other words, hydration of portland cement is an *exothermic reaction*. The amount of heat generated, or the heat of hydration, depends on the chemical composition of the cement, the fineness, and the ambient temperature. The heat of hydration brings about a rise in concrete temperature, which reaches a maximum at 2 to 4 days after casting and stays relatively low beyond some 14 days. A temperature rise as high as 45 °F (23 °C) has been observed in concrete dams. Of the four major compounds in cement, tricalcium aluminate liberates the most heat; dicalcium silicate, the least. The former may cause flash set when mixed with water, accompanied by the release of considerable heat. Tricalcium silicate releases twice as much heat as does dicalcium silicate. Type IV cement, which has low levels of C_3A and C_3S, generates less heat compared to other types of cement. The rate of heat generation speeds up with increase in fineness. Fly ash and other pozzolanic materials cut down the rate of heat generation.

In massive concrete structures, the evolved heat is dissipated so slowly that the temperature of the inner core of the structure shows a marked increase, followed by thermal expansion. This, along with the thermal gradient between the inner core and the outer surface, can cause high tensile stresses in the interior. Structures that are not provided with appropriate measures to control the heat of hydration or the temperature rise are bound to crack. Both the use of low-heat cement and the addition of pozzolans will help lower the heat.

In cold-weather concrete construction, this heat of hydration is beneficial, helping to assure setting and maintain continuous hydration.

3.6 PROPERTIES OF CONCRETE

Portland cement concrete is a mixture of cement, aggregates, and water (Fig. 3.4). It can be envisioned as a mixture of aggregates (fine and coarse), bound together by cement paste (the slurry made up of cement and water). The aggregates take up about 60–75 percent of the volume of concrete, and the paste constitutes about 25–40 percent (Fig. 3.5). Of the cement paste, the cement occupies about 25–50 percent of the volume, and water makes up the balance. In addition to these ingredients, fresh concrete also contains air, its volume ranging from 2–8 percent of the total volume.

As discussed in Chapter 2, aggregates are divided into two types: fine (sand) and coarse (gravel). Fine aggregate—particles mostly of size equal to or less than 4.75 mm (or a No. 4 sieve)—fills up about 30–45 percent of the total volume of aggregates. Coarse aggregate—particles predominantly retained on a No. 4 sieve—makes up the balance (or nearly two-third the volume of aggregates). Both aggregate types—fine and coarse—are graded, meaning that they contain particles of various sizes, large to small, so as to produce a mixture with a minimum void content. Each particle of both coarse and fine aggregate should be coated with cement paste, which upon hydration provides the necessary bond. The strength of concrete depends both on the strength of the aggregate particles and that of the paste.

Fresh concrete does not possess any strength, even though nearly 75 percent of its volume is made up of particles that have substantial compressive strength. As the hydration proceeds, the products of this reaction (between compounds of cement and water) render the strength, which increases with the rate of hydration, and eventually concrete is able to withstand the service loads.

The properties of concrete depend primarily on the ingredients, mix proportions, and rate of hydration. A number of properties of concrete at both the mixing (or plastic) stage and the solid stage are relevant to a construction professional.

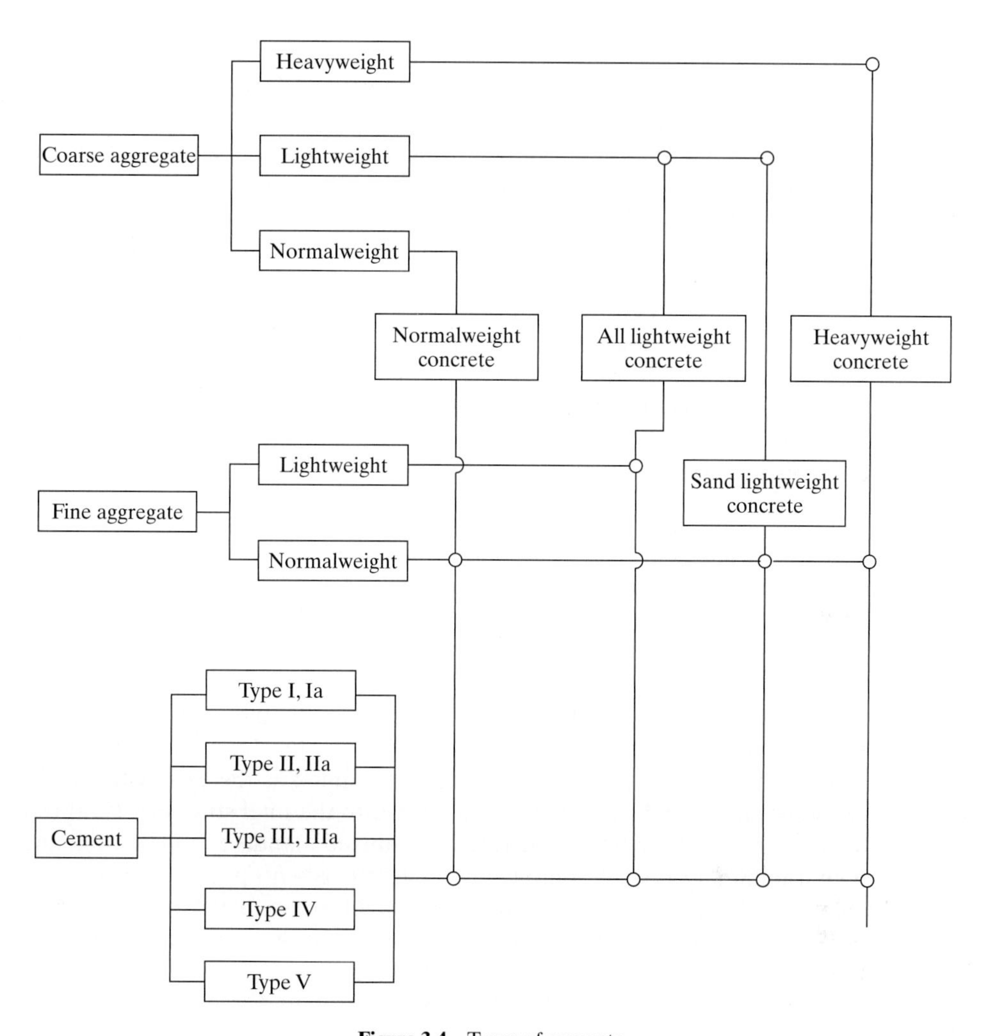

Figure 3.4 Types of concrete.

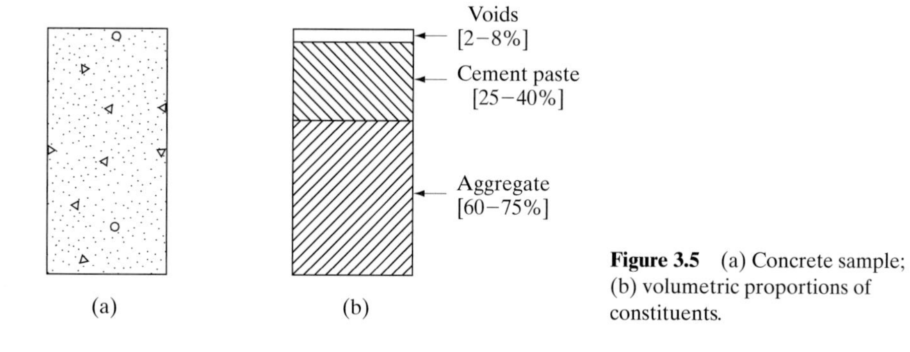

Figure 3.5 (a) Concrete sample; (b) volumetric proportions of constituents.

Several variables affect the characteristics of concrete during the mixing stage, and a number of parameters pertaining to fresh concrete affect the properties of hardened concrete.

During the mixing stage, fresh concrete is plastic and fluid, and is capable of being transported, placed, compacted, and molded into any shape. After attaining a level of hydration, it becomes a hard and brittle mass, qualified to offer structural properties such as adequate compressive strength and stiffness. Unsatisfactory performance during the mixing stage nearly always leads to inadequate strength properties at the hardened stage. Good concrete has desired attributes in the mixing stage and meets the specified strength requirements in the solid stage. The following sections elaborate on several properties that affect the performance of plastic as well as hardened concrete.

3.6.1 Properties of Fresh Concrete

Concrete at the mixing stage—fresh concrete—should be such that it can be transported, placed, compacted, and finished without harmful segregation. A proper mix should maintain its uniformity inside the forms and should not bleed excessively. It should set within a reasonable amount of time and hydrate in a manner that ensures adequate strength when the structure is put to service.

Workability. All aspects of fresh concrete—those associated with several tasks, from the selection of materials to the final finishing—are collectively called workability. Workability of concrete can be defined as the ease with which a fresh-concrete mix can be handled from the mixer to the final structure. It represents the ability (or lack thereof) of concrete to be mixed, handled, transported, and placed with a minimum loss of homogeneity. There exists no procedure to quantitatively measure workability, but subjective assessment is possible. A nonworkable mix can easily be identified from its inability to satisfy one or more of the concreting tasks: mixing, transporting, compacting, and finishing.

Workability is commonly expressed through three mutually independent characteristics:

- Consistency
- Mobility
- Compactibility

In addition, for some structures such as floor slabs, finishability can be added to the three preceding characteristics.

Workability depends on the proportions of ingredients; the physical characteristics of the cement and aggregates; the equipment for mixing, transporting, and compacting; the size and spacing of reinforcement; and the size and shape of the structure. Good workability requires a fairly high proportion of cement, adequate quantity of fine materials, low coarse aggregate content, and high water content. A fair amount of fine particles is necessary to have plasticity in the mix. Deficiency in fine aggregates results in a harsh mixture, prone to segregation and difficulty in mixing. An increase in the amount of fine aggregate and the addition of more water improve workability.

An excessive amount of any size fraction results in poor workability. Rounded grains of natural sand produce better workability than do the angular and elongated grains of crushed stone; however, coarse aggregate from crushed stone can contribute to good workability when properly graded. The workability of lean mixtures can be improved by the addition of finely divided materials such as hydrated lime, micro-silica, fly ash, and (of course) more cement. Air-entraining admixtures improve workability and decrease the tendency for bleeding and segregation, but they lower the density as well as the strength of hardened concrete. Reduction in bleeding may present problems in the finishing of slabs; air-entrained concrete should be finished earlier than normal concrete.

Increase in ambient temperature speeds up setting and hardening, requiring additional mixing water to maintain a level of workability. In other words, for the same cement content per unit volume of concrete, the water to cement ratio should be higher in warm climates than for concrete of the same slump in cold climates.

Consistency and Slump. Consistency is a measure of concrete's wetness or fluidity, and depends on the mix proportions and properties of ingredients. It is an important component of workability. Wet mix is generally more workable than dry blend, but mixtures of the same consistency may vary in workability.

Consistency is generally measured with a *slump test* (ASTM C143). The result from this test, called slump, is also used as an indirect measure of the characteristics of workability, even though the test measures only a single characteristic: consistency. It is most appropriate to use the slump test to measure changes in workability, such as those between consecutive batches, or for relative comparisons of workability between different types of concrete.

The slump test was developed in 1913 in the United States by C. M. Chapman. It utilizes a metal slump cone (Fig. 3.6) having a bottom diameter of 8 in. (200 mm)

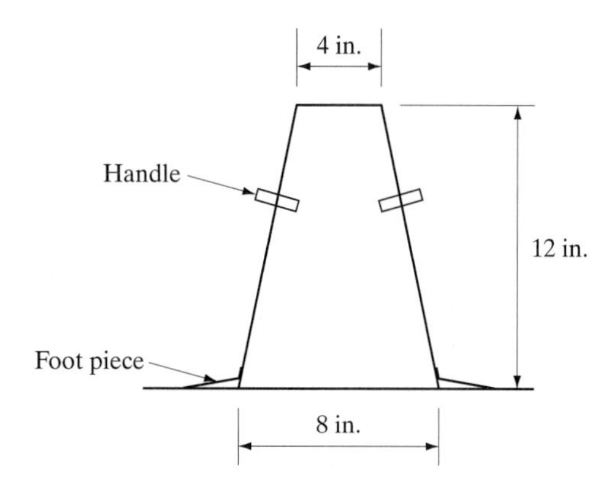

Figure 3.6 Slump cone (mold).

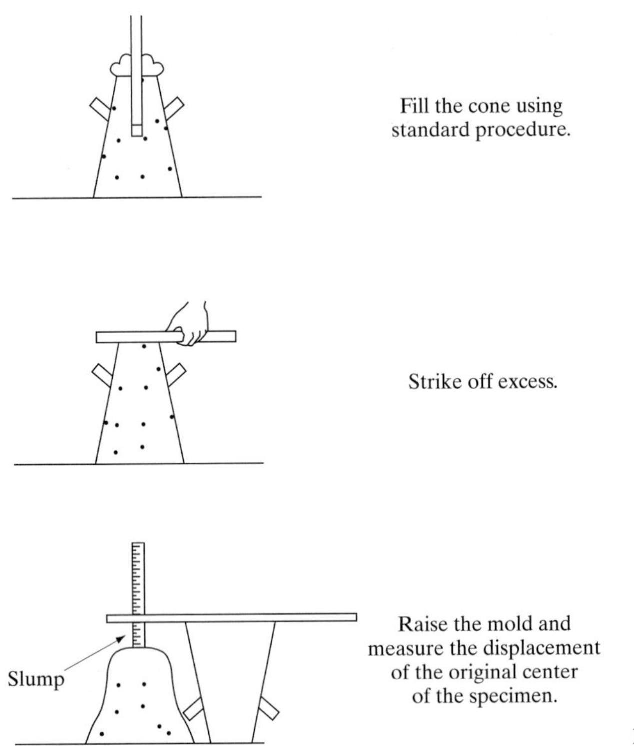

Figure 3.7 Slump test.

and a top diameter of 4 in. (100 mm). The height of the cone is 12 in. (300 mm). The cone—with inside surface dampened—is placed on a smooth, flat, nonabsorbent surface and, while it is held firmly in place by standing on the foot pieces, is filled with fresh concrete in three layers, each approximately one-third the volume of the mold. After placing, the first layer is tamped or rodded throughout its depth 25 times with a 5/8 in. (16 mm) diameter and 24 in. (610 mm) long tamping rod having one end rounded to a hemispherical tip (Fig. 3.7). The process is repeated twice more, rodding each layer with strokes that just penetrate the underlying layer. The excess protruding beyond the top of the cone is then struck off with a rolling and screeding motion of the tamping rod placed horizontally, filling the cone exactly. The mold is immediately removed by raising it vertically with a steady upward lift. Without the lateral support, the concrete will subside into one of the four shapes shown in Fig. 3.8. The number of inches (or mm) the original center of the cone sinks or settles uniformly—or *slumps*—is measured to the nearest 1/4 in. (6 mm). This measurement is called slump. The slump test is made immediately after the concrete has been discharged from the mixer or transit truck, and the slump is read promptly after the metal cone is lifted. The sample of concrete for the test should be taken from the middle portion of the batch, before any water is added to the mixer at the jobsite.

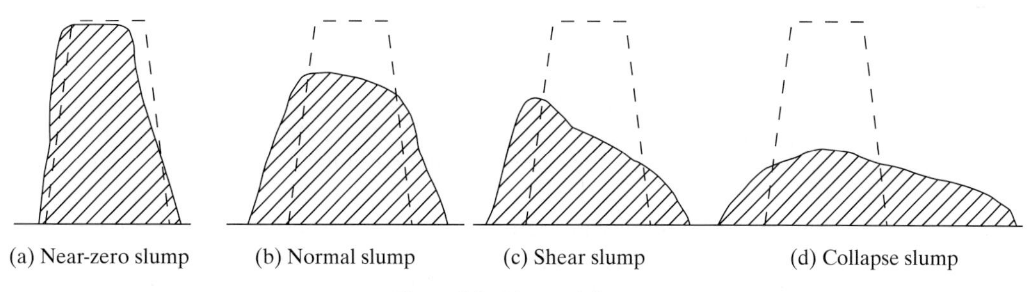

Figure 3.8 Types of slump.

A very dry mix, one with a low water content, will have a near zero slump. Lightweight aggregates absorb much water, leading to a harsh mix with low slump. Addition of air-entraining admixture, fine sand, or water may improve the consistency—and workability—of such mixtures. A very wet mix will have a slump value close to 10 in. (25 cm). The slump of concrete used in most structural applications varies between 2–7 in. (51–178 mm). Addition of water-reducing admixture or superplasticizer may increase the slump to 8–9 in. (203–230 mm).

Concrete that is properly proportioned will slump gradually and retain its original shape. A poor mix will separate, crumble, and fall apart. Concrete that lacks plasticity and cohesion produces a *shear slump* (Fig. 3.8). Lean, harsh, or extremely wet concrete produces a *collapse slump,* in which separation of finer materials (including water) from coarser particles is normally seen. If decisive falling away or shearing off of a portion of the concrete mass occurs during the test, the measurement should be disregarded. If two consecutive tests show the same finding, the concrete definitely lacks the necessary plasticity and cohesiveness. Due to the hydration process and the evaporation of water, the value of slump decreases with time after mixing, which is called *slump loss,* the rate of which increases with increase in ambient temperature.

The minimum and maximum recommended values of slump for various types of construction are listed in Table 3.5. These numbers are somewhat arbitrary

TABLE 3.5 RECOMMENDED SLUMPS FOR VARIOUS TYPES OF CONSTRUCTION

	Slump (in.)	
Types of construction	Maximum	Minimum
Footings, caissons, foundation walls, and substructure walls	3	1
Beams, columns, and walls	4	1
Pavements and slabs	3	1
Mass concrete	2	1

limitations, based primarily on the size and shape of the structure. Nowadays, one or more admixtures are commonly added to all types of concrete, and as a result the recommended limits on slump are not critically binding. For thin walls, a wet mix or a mixture with high slump may be required. Most specifications permit the use of concrete of lower than recommended slump, provided it is properly placed and consolidated.

A slump test is generally carried out at the construction site so as to compare the slump of freshly placed concrete with the limits specified in the design. Rather than looking at the value of slump in terms of its acceptability, it may be more appropriate to judge the measured readings for possible effects on the workability. Concrete delivered to the site in different trucks and at different periods should have nearly the same slump, to maintain uniformity and homogeneity. The measured value of slump may help in estimating the modifications in the amounts of ingredients—primarily water and fine aggregate—necessary to maintain consistency from batch to batch.

Changes in slump may also be due to variations in water content and alterations in the grading and proportioning of aggregates. Changes in the amount, fineness, and brand of cement, and fluctuations in ambient temperature also contribute to variations in slump. But slump alone does not indicate either the quality or workability of concrete, and should not be used for quality comparison of different batches of concrete. A lower than prescribed slump may not represent a better quality concrete; similarly, a higher value may not suggest that the concrete is bad. A variable slump—between batches—may lead to deviation in the compressive strength due to changes in the water-to-cement ratio. Low or insufficient slump results in poor consolidation and low-quality concrete. However, a mixture with lower than required slump may be used if it can be well compacted within the forms.

Unfortunately, slump is not subject to precise production or measurement, which is a gentler way of saying that the slump test is an inaccurate engineering measurement. Moreover, temperature, delay in placement, and material variations can cause a dip in slump with time. For this reason it is more practical to specify a working limit on slump, with a tolerance of 2 in (51 mm). For example, a specification of 3 ± 1 in. slump is more construction-friendly than requiring a slump of 3 in.

Another method of measuring consistency is the *ball penetration test* (ASTM C360). In this test (also called *Kelly Ball Penetration Test*), the penetration of a metal weight into freshly mixed concrete is used as a means to determine the consistency or workability. The apparatus consists of a cylinder with a hemispherically shaped bottom attached to a handle weighing 30 lb (13.6 kg). A frame is used to guide the handle and to serve as a reference for measuring the depth of penetration (Fig. 3.9). The apparatus is placed on a freshly placed smooth concrete surface, and the ball is set on the surface with the handle vertical. The weight is released, and the penetration is read—to the nearest 1/4 in. (6.4 mm)—after the weight comes to rest. A minimum of three readings is taken for each batch. The test is simple, quick, and can be compared with the slump values.

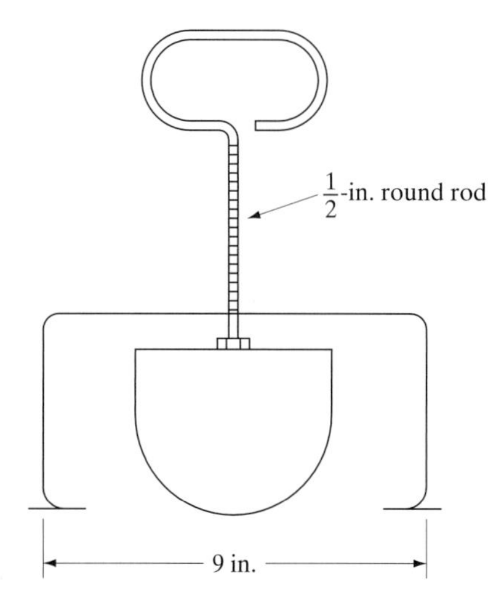

Figure 3.9 Ball penetration apparatus.

3.6.2 Factors Affecting Consistency and Workability

As mentioned, a number of factors influence the consistency and workability of fresh concrete. A summary is provided in the following paragraphs.

Workability is relatively insensitive to the proportion of cement, but is heavily dependent on the water content. Increase in the fineness of cement without corresponding increase in the water content decreases the workability. Some admixtures—such as water-reducing admixtures or superplasticizers, and air-entraining admixtures—improve workability.

Workability decreases with increase in the surface area of aggregates. The surface area is governed by the shape, grading, and maximum size of the aggregate. As the maximum size of coarse aggregate is increased and as the particles become round, the surface area is decreased and the workability is improved. In other words, as the maximum size of aggregate is decreased, the amount of coarse aggregate should be lowered to retain a level of workability. A coarser gradation of an aggregate increases its fineness modulus and in turn improves the workability. As the fineness modulus of fine aggregate is increased, the amount of coarse aggregate should be decreased to maintain the workability level. Increase in the proportion of coarse aggregate generally benefits workability.

Workability is also influenced by the rate of hydration and the loss of water through evaporation. With increase in ambient temperature the hydration rate speeds up, which increases the rate at which water is used up for hydration. In addition, the rise in temperature enhances the rate of evaporation. Delay in placing or consolidation of fresh concrete leads to greater evaporation loss, decreasing the workability. There are no laboratory tests to quantitatively measure the workability of fresh concrete. However, when a concrete mix cannot be placed, compacted, or finished effectively it is said to be unworkable. Similarly, when the compacted mix shows a tendency for segregation, it is said to be unworkable.

3.6.3 Segregation and Bleeding

Segregation and bleeding are two major properties that affect the quality and performance of concrete during its placement. Both of these result in nonhomogeneity; excessive bleeding, in addition, gives rise to a weak top layer.

Segregation. Segregation is the tendency for separation between large and fine particles of fresh concrete. Rock pockets (or honeycombing), sand streaks, crazing, and surface scaling are usually related to segregation. A honeycomb is a void left in concrete from failure of the mortar to fill the space among the coarser particles.

Generally, mixtures that are very wet or deficient in finer particles (such as finer sand) tend to segregate. High-slump concrete is more prone to segregation than low-slump concrete. Segregation and the accompanying nonhomogeneity affect the strength and durability of hardened concrete.

Air-entrainment decreases the tendency to segregate. Moving the concrete over long distances as it is being placed within the forms leads to segregation. Too much vibration during consolidation may also cause segregation. Finely divided materials (such as fly ash, powdered limestone, and silica-fume) decrease the tendency to segregate. To prevent segregation, fresh concrete must be dropped vertically and not at an angle.

Bleeding. A concrete mix that does not possess proper consistency is unable to hold the mix water, which slowly gets displaced and then rises to the top of the form. This water will eventually be lost, either through evaporation or by leakage through the joints and sides of the forms. This process of separation of water from the mix is called bleeding. Some amount of bleeding—which is a form of segregation—is normal in all concrete construction. When freshly deposited concrete sets, some of the aggregate particles and cement grains are partly suspended in water, so they tend to settle. When the solids settle, some water is displaced and appears at the surface, and this is normal bleeding.

Excess bleeding results in the movement of a high amount of water and finer particles to the top of the concrete, producing a nonhomogeneous mix and increasing the water-to-cement ratio at the top. The water lost through the joints and sides of forms carries with it some cement, contributing to weaker concrete. Bleeding is also found to be a cause for the formation of fine cracks below the larger particles of the coarse aggregate. Overvibration, overtroweling, and lean mixes increase the potential for bleeding. Air-entrainment is an effective method of controlling bleeding. Increase in the fineness of cement and decrease in the water-to-cement ratio decrease bleeding. Rich mixes—mixes with more cement—bleed less; mixes that lack fine sand bleed more. Bleed water in slabs should be allowed to evaporate, or be removed by dragging a hose across the surface before finishing.

Premature finishing may be a cause of excessive bleeding, and may lead to loss of some entrained air, making the concrete vulnerable to scaling when exposed to low temperatures. *Scaling* refers to surface cracks, and the removal of the surface layer in concrete slabs, caused by the pressure generated when the water in concrete pores freezes.

Bleeding in appropriate amounts is beneficial for the finishing of concrete, for it helps to bring the soft materials to the surface. The top surface of floors, pavements, and roofs must be finished smooth, and this is accomplished by working the concrete before it sets in such a way that the top layer consists solely of cement paste mixed with finer sand—a process called *floating*. This layer is then finished smooth with a metal trowel.

3.7 MIXING, PLACING, AND CURING

The purpose of mixing the ingredients is to ensure that each particle of aggregate in fresh concrete will be coated with the cement paste. The first stage of mixing involves weighing out or measuring out all of the ingredients for a batch of concrete—the process called *batching*. The materials that are mixed at any one time constitute a *batch*, and the size of a batch is usually designated by the number of bags of cement it contains. Each bag of cement has a bulk volume of 1 ft^3 and weight of 94 lb. A six-sack (six-bag) batch means six sacks of cement per yd^3 of concrete, or $6 \times 94 =$ 384 lb cement per cubic yard of concrete. The amount of concrete—in ft^3—produced per sack of cement is called *yield* or *return*. A six-sack batch will yield $27/6 = 4.5$ ft^3 of concrete.

Concrete can be mixed by hand or in a machine. *Hand mixing* is done in a tight wooden or metal box or in a wheelbarrow. First the sand is spread in a uniform thickness and subsequently covered by the cement. The two materials are given three turns and the mixture is formed into a wide crater. The coarse aggregate is then dumped into this crater. The mixing water is then poured over the coarse aggregate, taking great care to ensure that none of the water escapes, and the edges of the pile are gradually turned in to absorb the water. After this the entire mix is given four or more complete turns until it is of uniform consistency. Mostly, concrete is conveyed from a *ready-mixed plant* (or *batch plant*) to the job site in rotating-drum trucks or transit mixers (Fig. 3.10). Proportioning, batching, mixing, and delivery are all done by the concrete supplier or the batch plant operator. The batch plant must be located within a reasonable distance of the construction site; when the site is in a remote location, a field batch plant is necessary.

Machine mixing is accomplished using either stationary mixers (plant mixers) or mobile mixers (transit mixers). Stationary mixers come in two types: one-opening and two-opening. In a one-opening type mixer, the materials are charged and discharged through the same opening. In the two-opening type, the materials are charged from the opening at the rear and the mixed concrete is discharged from the opening at the front. Both types mix concrete in a rotating drum equipped with

Figure 3.10 Concrete batch plant.

internal blades designed to lift and tumble the ingredients. Concrete that is thoroughly mixed in a stationary mixer, called *ready-mixed concrete* or *ready mix,* can be delivered to the construction site either in a truck agitator or a nonagitating truck. A truck agitator is a special truck-mixer operating at agitating speeds, and is used to transport concrete for all uses, such as for pavements, buildings and foundations. A nonagitating truck is used for short hauls.

A common truck mixer or transit mixer is a special truck used for both mixing and transporting concrete to a job site over short or long hauls (Fig. 3.11). No central mixing plant is needed at the site; a batch plant, or proportioning plant, where exact amounts of ingredients are loaded into the truck, is all that is necessary. The truck carries a mixer and a water tank from which the driver can add the required amount of water into the mixer at the proper time. Generally, part of the water is introduced

Figure 3.11 Truck mixer and pump.

before the truck arrives at the job site or at the batch plant. The mixer is kept revolving slowly en route to prevent segregation of nearly dry particles. To preserve the quality, concrete should be conveyed to the construction site as rapidly and directly as practicable, and be deposited within 90 minutes or before the drum has revolved 300 turns after the introduction of water to the dry ingredients. To prevent delay in placement, proper attention must be given to scheduling the ready-mixed concrete deliveries and for providing easy access for the truck to the site. Men and equipment to place, finish, and cure the concrete should be present at the start of the concreting.

In a batch plant, the coarse aggregate generally is added dry, and the fine aggregate, moist (Fig. 3.12). In hot and humid climates, the coarse aggregate stockpile must be equipped with sprayers to reduce the temperature of aggregates through evaporation. In high-temperature applications, the internal temperature of concrete can be kept low by cooling the materials and using ice instead of the mixing water. Preventing excessive internal temperature rise in mass concrete is important to reduce cracking from thermal stresses.

Ready-mixed Concrete. Ready-mixed concrete is produced and delivered using one of three operations: central-mixing, shrink-mixing, and truck-mixing. Central-mixed concrete is completely mixed in a stationary mixer at the mixing plant and conveyed (agitating) in a truck mixer. The truck mixer may be used at the

Figure 3.12 Aggregate storage silos.

job site for remixing. Shrink-mixed concrete is partially mixed in a stationary mixer at the plant to reduce—or shrink—the total volume, and is completely mixed in a truck-mixer during transit. Truck-mixed concrete is completely mixed in a truck-mixer via 70–100 revolutions at the mixing speed of the drum—8 to 12 revolutions per minute. The batch plant in this case only batches the ingredients and does not mix them, and is called a *dry plant.*

If the requirements for uniformity of concrete are not met within 100 revolutions of mixing, that mixture should not be used until the conditions are corrected. The discharge of concrete should be completed within 90 min or before the drum has completed 300 revolutions, whichever comes first, after the introduction of water.

Inclined-axis transit mixers or truck-mounted mixers come in two styles: rear-discharge and front-discharge. The latter type requires a special truck chassis and is more expensive. The size of a concrete mixer is designated by the rated capacity of the volume of fresh concrete that can be mixed in a single batch. It can be as small as 2 ft^3 (1.06 m^3) and as large as 15 yd^3 (11.5 m^3). A 16-ft^3 mixer, for example, is capable of producing a maximum of 16 ft^3 (0.45 m^3) of concrete, with a 10 percent overload capacity per batch.

When the purchaser assumes the responsibility for the proportioning of ready-mixed concrete, the cement content and the maximum allowable water content are specified, including the type and amount of admixture. If the purchaser requires the manufacturer of ready-mixed concrete to assume the full responsibility for the selection of mix proportions, the former should specify the required compressive strength and the minimum cement content in number of bags per yd^3. The manufacturer is generally required to furnish the quantities of ingredients and the evidence that the proportions selected will produce concrete of desired quality.

In the absence of designated general specifications, the purchaser should specify the following:

- Designated maximum size of coarse aggregate
- Desired (or maximum) slump at the point of delivery
- Air content (for air-entrained concrete)
- Mix design requirements—such as cement content, maximum water content, and admixtures, or minimum compressive strength at 28 days
- Unit weight required (for structural lightweight concrete)

The concrete slump, measured within a period of 30 min from either arrival at the job site or after the initial slump adjustment, should be within the tolerance limit of the specified value. Slump tests should not be made before 10 percent or after 90 percent of the batch has been discharged.

Strength, slump, and air-content tests should be made with a frequency of no less than one test for each 150 yd^3 (115 m^3) of concrete. On each day the concrete is delivered, at least one strength test should be made for each class of concrete. Two standard test specimens—6-in. diameter and 12-in. height cylinders—are required for each strength test. The average of the two strength-test values is taken as the *test strength*. No more than 10 percent of the strength test results should have values less than the specified strength, and the average of any three consecutive strength tests should be equal to or greater than the specified strength.

The number of samples for the strength test of each class of concrete—placed each day—should be adequate to represent the total quantity of concrete placed. Three criteria establishes the required minimum sampling frequency for each class of work:

- Not less than once each day a given class of work is placed
- Not less than once for each 150 yd^3 (115 m^3) of each class of concrete placed each day
- Once for each 5000 ft^2 of surface area of slabs or walls placed each day

When the total quantity of a given class of concrete is less than 50 yd^3, strength tests are not required when evidence of satisfactory strength is submitted and approved.

Machine Mixing. The dry ingredients are introduced into the mixing drum either by hand or with a mechanical skip. When the mixer is equipped with a skip, the dry ingredients are placed together in the skip and then dumped into the mixer, while the water runs into the drum from the side opposite the skip. The mixing time may vary from about 2 to 5 minutes depending on the volume of concrete, and should be at least $1\frac{1}{2}$ min after all the materials are in. After mixing, the fresh concrete is discharged from the mixer through the discharge chute.

In small drum mixers—having revolving drums with fixed blades inside—the ingredients are introduced into the drum by hand. Part of the mixing water is poured into the drum first. Next, part of the coarse aggregate is added while the mixer is running. Adding coarse aggregate to water before the addition of cement prevents the latter from sticking to the drum. Then, while the mixer is still running, the measured amount of cement is added, followed by the fine aggregate and the remaining coarse aggregate in proper proportions. Lastly, the balance water is introduced into the drum, and the mixing is done for three minutes. The mixer is then stopped for two minutes, at the end of which it is made to run for an additional two minutes. Fully mixed concrete is subsequently dumped into the wheelbarrow while the mixer is still running.

When proper mixing is effectuated, there is hardly any difference between hand-mixed and machine-mixed concrete. But when the volume of concrete is large, hand mixing does not produce a uniform or homogeneous mixture.

3.7.1 Pumping and Placing

Fresh concrete should be placed within the molds as soon as it is mixed thoroughly. Proper care is necessary to ensure that the concrete flows through narrow spaces and between reinforcing bars.

Forms. Formwork for placing concrete comes in many types:

- Plywood and steel frame
- All aluminum
- Plywood attached to steel hardware
- All plywood
- All steel
- Fiberglass
- Wood or lumber (dimension category)

Forms should be strong enough to withstand the weight of fresh concrete and the loads from the construction crew and machinery.

Form coatings or sealers are applied to the form surfaces to protect the form material, facilitate the action of the form-release agents, and prevent discoloration of the concrete surface. Prior to each use, release-agents are applied to the form contact surfaces to minimize concrete adhesion and to facilitate stripping. If the hardened concrete surface will be treated by painting or tiling, the form coating or release agent should not impair the adhesion of the finish material. Stripping of forms should not cause excessive deflection or distortion of the concrete surface.

When no strength tests are performed and the temperature surrounding the units has remained more than 50 °F (10 °C), the forms and supports for walls, columns, and sides of beams may be removed after 12 hours. The forms and supports of slabs may be removed after 14 days if the temperature has remained above 50 °F. For spans larger than 20 ft, the supports must remain for 21 days at temperatures at or higher than 50 °F.

Before the placement of concrete in footings and pavements, the subsoil and the insides of forms should be moistened to prevent the dry soil from absorbing the mixing water. Oiled forms and damp soil need not be moistened. The forms should be braced and aligned before placing the concrete (Fig. 3.13). The form alignment should be checked before and after the placement of concrete to ensure that the sides are within the required tolerances.

Figure 3.13 Formwork for footing and slab.

Pumping. Fresh concrete is typically conveyed to the job site in wheelbarrows, carts, belt conveyors, cranes, or chutes. Alternately, a concrete pump can be used to push the concrete to its final position. Concrete pumping is the standard method of placement in high-rise construction.

Pumps must be of adequate capacity and be capable of moving the concrete without effecting segregation. Today's pumps have the capacity for a maximum horizontal reach of 2500 ft (760 m), and a maximum vertical reach—in a single lift—of 1400 ft (420 m) or more—and for a volume of 170 yd^3 (130 m^3) per hour. Equipment made of aluminum or aluminum alloys should not be used for pump lines, chutes, or tremies, for the reaction between cement alkalis and aluminum has been found to drastically reduce the strength of concrete.

Pumpable concrete has an average to high slump, typically 4 in. (10 cm) or more. The stability of the mix is the key to having a homogeneous material. When pumping concrete for tall buildings, the pressure in the hose may force the water out of the mix, driving the coarse particles to interlock and consequently blocking the hose. The addition of fine particles or small amounts of silica fume stabilizes the mix. To prevent the clogging of the hose at the start of pumping, a slurry mixture of fine sand, cement, and water is sometimes forced through the hose to act as a lubricant.

Placing. Fresh concrete should be placed as near as possible to its final position. In slab construction, the concrete should be placed first around the perimeter at one end, with each batch dumped against the previously placed concrete. To prevent segregation, concrete should not be placed in big piles and later moved horizontally to its final position. It should be discharged into the face of the concrete already in place and not away from it. In beams and other deep members, concrete should be placed in horizontal layers of uniform thickness of 6 to 20 in. (15 to 50 cm), and each layer should be consolidated before the placement of the next layer. In wall construction, the placement should be a continuous operation and in uniform lifts of no more than 4 ft (1.2 m). The first batch should be placed at both ends of the section and the concreting should progress toward the center. The same procedure can be followed for beams.

In the construction of walls, lower-slump concrete should be placed at the top layer to compensate for the water gain from bleeding. When placing concrete in tall forms, excessive bleeding (in general) can be prevented by placing it slowly. The free vertical drop at any point should not exceed 3 ft (0.9 m). Excessive slumps may cause separation of the coarse particles from the mortar, leading to rock pockets or honeycombing. When making slopes, concrete should be placed at the bottom of the slope and moved up. For construction under water, concrete can be placed using a *tremie,* which is a sheet-metal hopper with a pipe leading out of the bottom. It can be described as a large tube of adjustable length made of sheet metal and provided with a hopper at the top. It is supported vertically in the water, with provisions for horizontal movement over the area being concreted. The bottom of the tube rests on the ground and can be closed by a valve arrangement. The top of the tremie, which is above the water level, is filled with fresh concrete and then it is lifted a short distance, allowing the concrete to escape from the bottom. After depositing a layer of concrete this way, the procedure is repeated to form the next layer.

3.7.2 Concrete Tools

A number of tools are required to place, compact, and finish concrete. Some common tools are shown in Fig. 3.14. A wheelbarrow is commonly used to transport fresh concrete from the mixer to the forms. Ordinary square-end shovels are used for mixing and depositing. Spreading is done with a short-handled square-end shovel

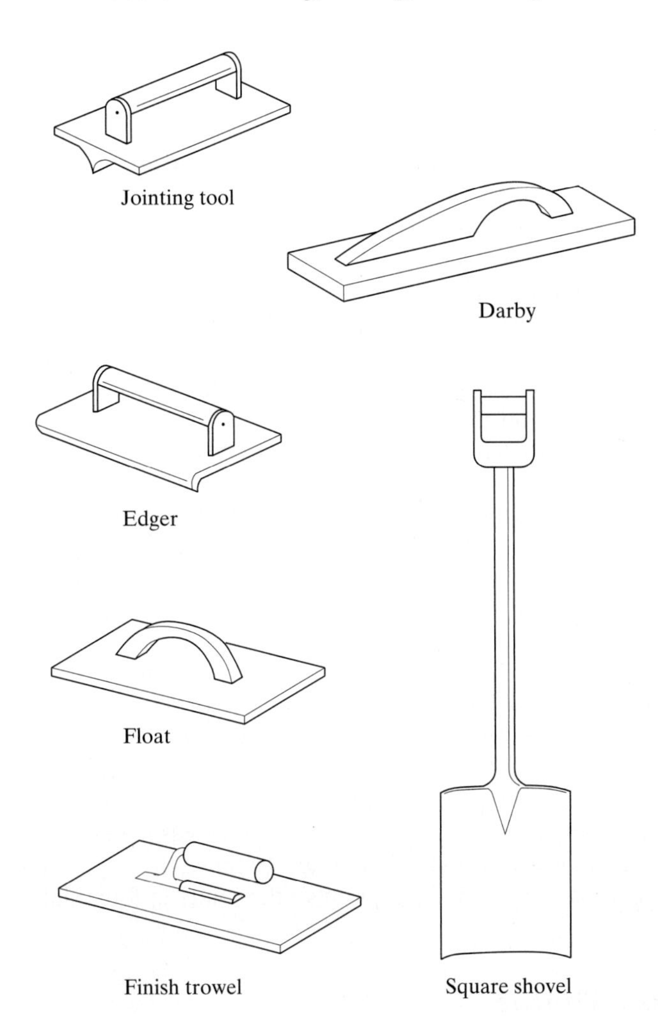

Figure 3.14 Concrete tools.

or a special hoe. Consolidation is usually accomplished using *internal vibrators,* but they should not be used either to move or to spread the concrete. *Form vibrators* (*external vibrators*), which are devices attached to the forms, are used for thin sections. *Surface vibrators,* which vibrate the concrete from its surface, are effective up to limited depths. Hand tamping and spading provides adequate compaction if the concrete has a high slump, of about 6 in. (15 cm); vibrators may be needed, however, around corners and inserts. Puddling of concrete—moving a piece of dimension lumber (2×4 in.) or steel rod vertically up and down—helps to remove the entrapped air and consolidate the mixture. During consolidation, care should be taken to avoid hitting or scraping the inside surface of forms, as that could remove the form-release agent.

Concrete should be vibrated at close enough intervals so that the field of influence, or the radius of action, from one insertion point overlaps that of the previous insertion point. All reinforcement should be supported adequately. Without support, the weight of fresh concrete and of workers can bend or displace reinforcement.

To help distribute the weight over fresh concrete a *knee board* made from lumber and plywood (24 × 12 in. [60 × 30 cm]) is used. A tamper consisting of a 4-ft (1.2 m) handle on an 8 × 8 × 12 in. (20 × 20 × 30 cm) piece of timber is used to compact concrete for sidewalks and pavements. A metal tamper can also be used. The striking-off operation, also called *screeding,* is required to even out the surface at the specified depth. A *strikeboard,* usually of 2 × 4 in. (5 × 10 cm) lumber, long enough to rest across the top of the form, is used for this operation. The procedure involves moving the straight edge, supported on the screed or form edges, back and forth lengthwise in short strokes (sawing motion) while moving it slowly along the top surface. For the finishing of large slabs and pavements, vibratory screeds are employed. Screeding is carried out immediately after placement and has the greatest effect on surface tolerance.

Leveling the surface after it is struck off to the required grade, and smoothening out the ridges, are done with a *wood float* (*hand float*), *darby float,* or *bullfloat.* Floating helps fill in voids left by screeding and slightly embeds the coarse particles, leaving a thin layer of dense mortar on the surface. In addition, it prepares the surface for subsequent operations such as edging, jointing, and troweling.

When the surface can sustain foot pressure with no more than 1/4 in. (6.3 mm) indentation after floating, it can be troweled to produce a smooth surface. Troweling should not be done to a flat surface that has not been floated. A *steel float* or *trowel* is used for finishing the surface smooth. Additional troweling may be needed to improve the surface quality. There should be a lapse of time between successive trowelings to permit the concrete to become harder. Tooled joints and edges may have to be rerun after troweling to maintain true and uniform lines. To finish the edges of a slab, an *edger* is used; a *groover* is used to make and finish the joints in floor slabs, sidewalks, and pavements.

3.7.3 Finishing and Types of Finishes

A concrete surface—mostly flatwork—is finished to fulfill the requirements of aesthetics or service. Finishing can be done by hand or machine. Fresh concrete should be allowed to stand until the water disappears from the surface before the start of the finishing operation. The time needed to attain this stage depends on the brand of cement, amount of water, and atmospheric conditions (such as humidity, temperature, and wind velocity).

Power-operated machines for screeding, floating, and troweling are commonly used in large jobs. Excessive finishing creates a weakened layer at the top surface. Water should be added during finishing only in emergencies—such as under rapid drying conditions that may cause plastic shrinkage cracks.

A number of finishes can be obtained in freshly placed concrete. The finished appearance can vary from a smooth and polished appearance to a rough gravel look; in fact, variations in texture and finish are limited only by the imagination and skill of the craftsperson. A *smooth finish* is obtained by using a relatively hard finishing or steel trowel on the surface. A *swirl-float finish* is produced for visual interest and surer footing, by working the float—wood, aluminum, or canvas resin—flat on the surface in a swirling or circular motion, applying pressure. The result is a surface with a fine-textured mat-like finish, with ridges approximately 1/16 in. (1.6 mm) thick. *Broom finish* is an attractive nonslip texture obtained by pulling a damp broom across the freshly troweled surface. It is suitable for traffic areas such as driveways and sidewalks. For best results the broom should be rinsed in water and tapped to remove excess water after each pass. A *rock salt finish* is produced by sprinkling rock salt—of coarse gradation—on the surface of the troweled concrete surface, giving a decorative surface finish. After five days of curing under waterproof paper or plastic sheeting, the surface is washed with water and brushed to dislodge or dissolve the salt grains, leaving pits or holes on the surface.

A finish similar to the surface texture of some cut stones is created through mechanical spalling and chipping with a variety of hand and power tools. This technique, called *brushhammering,* involves removing a layer of freshly hardened concrete while fracturing the aggregate particles at the surface. The finish can vary from a light scaling to a deep and bold texture, achieved with the help of jackhammers fitted with single-point chisel. An *exposed-aggregate finish* is one of the most popular decorative finishes for concrete slabs, patios, tilt-up panels, and sidewalks.

3.7.4 Curing

Curing is any procedure that maintains proper moisture and temperature to ensure continuous hydration. It was indicated previously that hydration—reaction between compounds of cement and water—sets off when cement grains come into contact with water and proceeds until all the grains are hydrated, which takes a very long time—more than a few years. Properties of concrete improve with age as long as conditions are favorable for continued hydration. The rate of hydration is rapid in the beginning but continues rather slowly for an indefinite period of time—a growth mimicking that of living things like trees.

When all the ingredients of concrete are mixed together, the wet mix begins to set. Final set is the stage at which concrete loses its ability to be formed into shapes, and turns into a hard mass. As it hardens, concrete gains in strength. In most concrete that is allowed to hydrate normally, close to 90 percent of the cement will hydrate in about 90 days. If we value the 90-day strength of concrete as 100, the strength measured at 28 days will be about 80; at 14 days, about 70; at 7 days, about 45; and at 3 days, around 30; assuming uninterrupted hydration.

Concrete is usually made with enough mixing water to assure complete hydration. Although a water-to-cement ratio of about 0.35 is adequate to guarantee complete hydration, mixes with water-to-cement ratio of less than 0. 5—without suitable admixtures—tend to be harsh, stiff, and unworkable. Thus, most concrete mixes in practice are proportioned with high water-to-cement ratios, high enough to ensure uniform hydration. If such concrete is prevented from drying out—such as by keeping it in a humid environment—the hydration can prolong without interruption. On the other hand, if the water is allowed to evaporate—as is the case in most structures—the hydration ceases; in addition, the concrete shrinks. The rate of drying depends on the ambient temperature and humidity, which means that the continuance of hydration is feasible only through a proper control of the temperature and moisture of the concrete mass. The process that maintains this favorable condition of moisture and temperature is curing.

Concrete will not attain the desired properties—such as compressive strength, watertightness, and durability—without proper curing. Figure 3.15 shows the rate of strength development during the first 28 days after mixing in two types of samples: one properly cured (moist-cured) and the other left to dry out (air-cured). Since hydration is a process that can continue only in the presence of water, any loss of mixed water signals a slowdown and eventually the total stoppage of the reaction. Laboratory studies have verified that concrete must be maintained at humidity above 80 percent for hydration to proceed at an appreciable rate. As a corollary, below about 80 percent humidity hydration slows down or ceases completely.

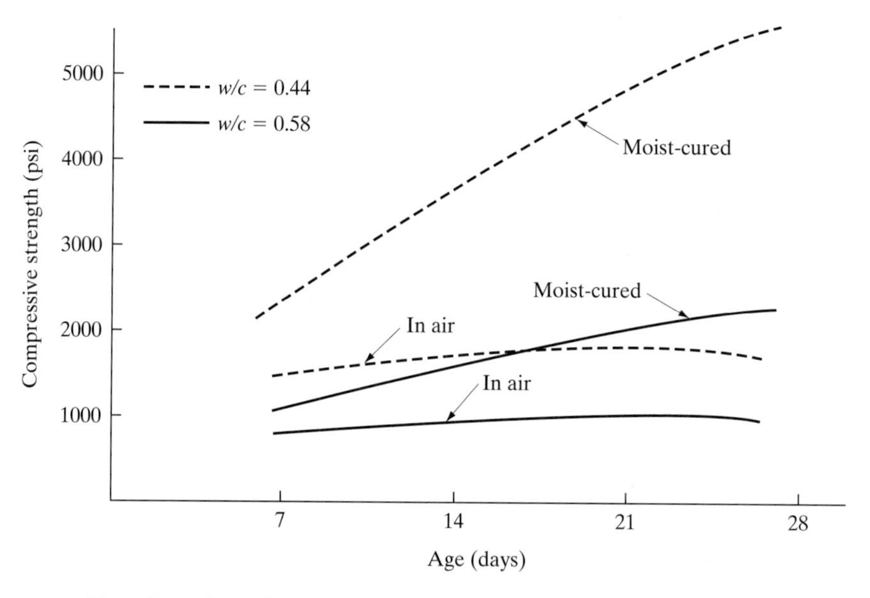

Figure 3.15 Rate of strength development in concrete (based on tests using 3 × 6-in. cylinders).

Concrete made outside the laboratory will be exposed to constant heat and wind, which can drive the moisture from inside the concrete pores. If loss of water—from evaporation or absorption by the surrounding ground—is not compensated, the hydration cannot go on. Curing can thus be looked upon as a process in which evaporation loss is minimized or compensated for. The ACI code (318) specifies that concrete must be maintained above 50 °F and in a moist condition for at least the first 7 days after placement, and high-early-strength concrete for at least the first 3 days.

Methods of Curing. The available methods of curing can be divided into three types:

- Methods that maintain the presence of water
- Methods that prevent the loss of mixing water
- Methods that accelerate hydration by supplying heat and moisture

In the first method, concrete after it is finished is kept under a layer of water or saturated continuously until the desired properties are achieved. The second method involves techniques and materials to prevent the evaporation of the mixing water. The third method involves the supply of both heat and water, accelerating the rate of hydration. The first method involves operations such as ponding, fog-spraying, or sprinkling, with or without the use of a cover such as burlap and plastic sheeting. This and other procedures will be explained in the following paragraphs.

Curing can be accomplished using four materials:

- Water
- Mats or blankets
- Waterproofing paper or plastic sheeting
- Liquid membrane-forming compounds

Water is the ideal curing medium, and water curing can be accomplished by ponding, spraying, or sprinkling. Ponding is accomplished by building earth or sand dikes around the perimeter of the concrete slab to retain a pond of water within the enclosed area. The method requires considerable labor and supervision and may not be practical (except for slabs). Fog-spraying or sprinkling are excellent methods of curing when the temperature is above freezing, but are costly and require an adequate supply of water. In addition, the process must be systematized to prevent the concrete from drying out between the applications of water.

Covering concrete with wet mats or blankets (such as burlap and cotton curing-mat) slows evaporation loss and supplies moisture to the concrete. Masking the concrete with a 2-in. layer of moist sand, earth, hay, or sawdust (wet once a day) can also be an effective curing technique for pavements and floors. Burlap must be free from substances that are harmful to concrete or can cause discoloration. All wet coverings should be kept continuously moist so that a film of water remains on the surface throughout the curing period.

Concrete cured using water or wet coverings is called *moist-cured* or *water-cured* concrete. Water curing may be impractical and expensive in countries where labor is costly, for the watering should be done at least two to three times a day. Consequently, other methods of curing have been developed and practiced, but are less efficient. These methods, which were developed to minimize the evaporation loss, include protecting with plastic sheeting or waterproof paper and, most commonly, forming a skin of a liquid membrane-forming compound.

Waterproof paper or plastic film (or impervious sheeting) should be applied as soon as the concrete has hardened and after the concrete has been thoroughly wet. The sheeting should lay directly on the concrete surface, and the overlapping edges should be a minimum of 12-in. (30 cm) wide. The primary advantage of this method is that periodic addition of water is not required; another advantage is that the concrete surface is not discolored. However, keeping the thin film in position and preventing the wind from blowing it away may be difficult. Black plastic film is satisfactory under normal conditions, but white film is good in hot weather, as it reflects the sunlight.

Liquid curing compounds or liquid membrane-forming compounds (also called *seal coats*) are used most often because of their versatility, ease of application, convenience, and economy. They are relatively inexpensive, and provide an effective means of preventing evaporation from flat slabs and pavements, assuming that they are applied as soon as the concrete is finished. These compounds are the pigmented or clear solvent solutions of resins or waxes, and are available in spray-on, roll-on, and brush-applied forms. The pigment helps to indicate the area covered by the compound. After application, the solvent evaporates, leaving behind a thin continuous film (membrane) of resin on the surface of the concrete, which seals in most of the moisture. The film remains intact for a month or so, after which it becomes brittle under the action of heat and thereafter peels off.

Curing compounds should not be applied to a dry surface, which may absorb the compound, causing stains. In some cases, the surface may have to be fog-sprayed before applying the compound. If proper care is taken in the selection of the most appropriate compound, it can be very effective.

As a rule, the curing compound should be nontoxic and not restricted to outdoor application. Certain curing chemicals can make the surface slippery, presenting a safety issue. Some may prevent satisfactory bonding between hardened concrete and any new concrete or floor finish that may subsequently be applied. Curing compounds are not recommended for use in late fall in locales where deicers are used to melt ice and snow, as they may prevent proper air-drying of concrete; before the application of deicers, air-drying is necessary to enhance the resistance of concrete to scaling.

Curing compounds can be applied by brushing, rolling, or spraying, generally in two coats in a continuous operation, with the second coat at a right angle to the first. The application should be done immediately after the finishing of concrete, when the concrete is still wet.

Effects of Temperature. The length of the curing period depends on such variables as atmospheric conditions (relative humidity and temperature), type of work, characteristics of concrete, and expected strength. The longer the curing period, the greater the final strength. Concrete kept moist, under normal conditions, will develop about 75 percent of its final strength in roughly 28 days. Concrete structures are usually cured for 28 days, and at a minimum for 7 days.

The rate of hydration of site concrete depends on the relative humidity and ambient temperature. In addition to keeping the concrete moist, the curing process should ensure that its temperature remains normal. The temperature range considered most desirable with conventional methods of curing is 70–90 °F (21.6–32.2 °C).

The temperature at which concrete is cured affects the development of strength with time. At temperatures below 68 °F (20 °C), the hydration rate as well as the strength gain decreases. Below 23 °F (−5 °C), some hydration may still take place, but very little water is available for reaction with the cement, due to freezing. After the formation of ice, fresh concrete can lose about half its potential strength. If new concrete freezes while its compressive strength is less than 500 psi (3.5 MPa), or remains in a saturated condition, it may be damaged permanently. Frozen concrete must be removed and replaced.

During cold weather, little or no external supply of moisture is required for curing. But when heavy frost or freezing is expected at the job site, concrete should be protected from freezing or be insulated for the first few days. Sidewalks and other flat work exposed to freezing at night and to melting snow during the day should be made with air-entrained concrete, and protected from freezing until it has attained a strength of some 3500 psi.

When concreting in cold weather, the goal should be to maintain a temperature sufficient to allow the concrete to gain strength and prevent it from freezing. If the mixing water freezes before the cement has set, no water will be available for the chemical action of setting and hardening. The concrete, as a consequence, will not set until the ice melts, and this factor should be taken into consideration before removing the formwork. When, however, the setting process has begun before the temperature drops below the freezing point, the water within the pores expands, and the resulting pressure destroys the binding property of the paste, forcing the concrete to crumble.

To maintain a high enough temperature for the hydration to proceed, the top exposed surface of slabs should be covered with straw or tarpaulin. Polystyrene insulation for the sides and blanket insulation for the top helps to keep the temperature of concrete fairly high. If the temperature dips to near freezing level, propane heaters can be employed to keep the temperature high during the curing period. At the end of the protection period, concrete should be cooled gradually to minimize cracking from differential strain between the interior and exterior. Forms should be left in place longer in cold weather.

In hot weather, exposed concrete must be kept cool, both during mixing and throughout the curing period. Problems in hot-weather concreting are mostly due to the faster rate of hydration and increased rate of evaporation. The latter depends on the concrete temperature, relative humidity, wind velocity, and air temperature. If all other conditions remain the same, for instance, an increase in temperature from 60 to 90 °F (16 to 32 °C) increases the evaporation rate by nearly 300 percent. When evaporation is not controlled, large sections of concrete are prone to thermal cracking.

Lowering the temperature of coarse aggregate will decrease the concrete temperature. Keeping the aggregate cool in mass-concrete work in hot climate permits reductions in water and cement without affecting the strength and workability.

Shrinkage is also reduced, due to the reduction in mixing water. Sprinkling of the coarse aggregate, to keep it cool through evaporation, must be sufficient to keep the particles continuously wet. When the water is warm, the sprinkling must be limited to the smallest quantity of aggregate needed for the batch. The fine aggregate cannot be cooled by sprinkling, for the sand grains accumulate too much water, thereby dewatering becomes very difficult. Use of fly ash or ground blast-furnace slag helps to slow down the rate of hydration and setting of concrete in hot weather.

The common methods for cooling concrete in hot weather application are:

- Cooling the mixing water with chipped ice
- Shading the aggregate storage and water tank
- Painting the truck mixers and water tanks white
- Avoidance of overmixing
- Keeping the humidity high and air temperature low with fog sprays around the area where concrete is placed

The rate of gain in strength increases with the increase in curing temperature (Fig. 3.16). *Steam curing* is a technique for increasing the hydration rate by

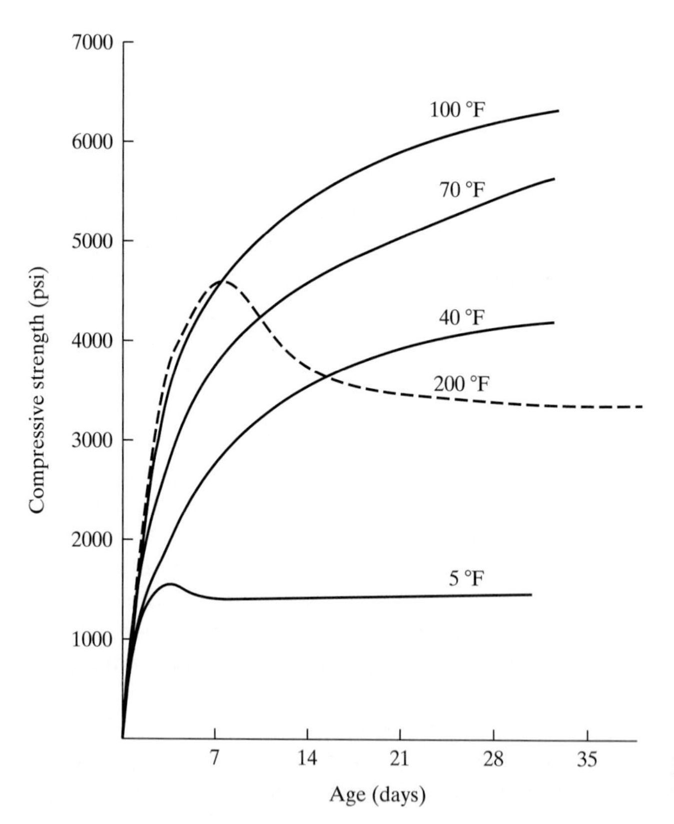

Figure 3.16 Effect of curing temperature on compressive strength.

supplying steam, generally under pressure. Precast concrete sections and concrete masonry blocks can reach 70 percent of their normal 28-day strength when cured by properly controlled wet steam at a temperature of about 150 °F (65.5 °C) for about 15 h. Full 28-day compressive strength can be achieved through steam curing for 24 h.

Two methods of steam curing are employed for early strength development. Curing in live steam at atmospheric pressure is used for enclosed cast-in-place structures, and curing using high-pressure steam autoclaves is utilized for small manufactured units. The former is done in a steam chamber or under other enclosures such as tarpaulins, to minimize the moisture and heat loss. The steam is applied for at least 2 h after the final placement of concrete. The maximum steam temperature inside the enclosure should not exceed 180 °F (83 °C). High-pressure steam curing in autoclaves derives its advantage from higher temperatures—in the range of 325–375 °F (160–190 °C); hydration is faster at elevated temperatures and pressures.

Proper curing of concrete improves its quality and performance. Apart from being stronger and more durable, concrete cured sufficiently is more resistant to traffic wear and is less permeable. In winter months, however, no curing is necessary, for the rate of evaporation is very low in a cold and humid atmosphere. Effects of curing temperature on the compressive strength are further explained in the following sections. Even though steam curing allows for faster gain in strength, it may decrease the 28-day strength by as much as 30 percent.

3.8 PROPERTIES OF HARDENED CONCRETE

The properties of fresh concrete just described are of interest for just a few hours after the ingredients are mixed together; beyond that, and for the remainder of the life of the structure, the properties of hardened concrete dominate the service scene. The major properties of hardened concrete are:

- Strength
- Modulus of elasticity
- Durability
- Creep
- Shrinkage
- Watertightness (impermeability)

Of these the strength in compression is generally considered to be the most significant—because of the most common requirement to endure compressive stresses in reinforced concrete construction, and the dependency of most other strength properties on the compressive strength. But there are many situations where other properties are considerably more important than the compressive strength. For example, durability should be the most important consideration in designing the concrete mix for ground-supported slabs such as walkways and pavements. Similarly, watertightness is the most important property of concrete designed for liquid-retaining structures. Although an increase in strength generally improves all other properties, there are exceptions to this rule. For instance, an increase in the quantity of cement may increase the compressive strength but at the same time may bring out more shrinkage cracks and make the concrete less resistant to freeze-thaw cycles. Thus, the attainment of maximum strength should not be the sole criterion in the design of concrete mixes.

Ordinarily the properties of concrete depend on:

- Mix proportions
- Curing conditions
- Environment

The role of curing has been discussed. Curing is a process of maintaining the progress of hydration, without which concrete will not attain any of the desired properties. Environmental factors influence the rate of strength development, shrinkage, and other properties. An offshore concrete structure may not be as durable as its counterpart built inland, because of the salt-laden, moist environment. Structures supported directly on soil, such as a slab-on-grade, may deteriorate from exposure to moisture and salts faster than those that remain above ground. However, the proportions and types of ingredients—cement, aggregates, water, and admixtures—have the greatest effect on the properties of concrete.

The following sections describe in detail various properties of concrete and the parameters that influence them. To reemphasize an important point: dry cement does not possess the cementing or binding property; the chemical reaction between cement and water—the hydration—yields products that provide this property after hardening.

3.8.1 Compressive Strength

Compressive strength is perhaps the most important property of hardened concrete, and is generally considered in the design of most concrete mixes. Although a compressive strength as high as 12,000 psi (82.7 MPa) is feasible, concrete for ordinary construction is made so as to possess a compressive strength in the range of 3000–6000 psi (20.7–41.4 MPa). Cast-in-place construction makes use of concrete that is close to the low range; precast construction, which routinely follows a rigorous quality control program, demands higher-strength concrete, strength nearer the high range.

It is customary to estimate the strength properties of concrete in a structure using tests performed on small samples, made from the fresh concrete as it is placed in the structure, which are cured in the laboratory in a standard manner. Cylindrical specimens of 3-, 4-, or 6-in. (7.5-, 10-, or 15-cm) diameter and of height equal to two times the diameter are used as test specimens. In addition, in many countries (other than the United States), cube samples of size 2, 4, and 6 in. (5, 10, and 15 cm) are used for compression tests. Since the compressive strength of concrete is affected by many variables, including environmental factors and curing conditions, the actual strength of concrete—the strength of concrete in the structure—will not be the same as the capacity measured from the test specimens. Most U.S. codes stipulate that concrete should be sampled for strength tests at each 100–150 yd^3 volume (77–115 m^3). A common requirement is for two 6×12 cylinders (6 in. diameter and 12 in. height) for each of the 7- and 28-day tests.

Effects of the Type and Amount of Cement. A number of factors—such as quantity of cement, amount of water, types of ingredients, mix proportions, curing, temperature, age, size and shape of specimen, and test conditions—affect the compressive strength of concrete. By visualizing concrete as a conglomerate of particles of different sizes held together by the products of reaction between cement and water, it follows that the strength of the concrete matrix is a measure of the strength of the particles and of the bonding agent.

As discussed in Chapter Two, most aggregates used in concrete have compressive strengths exceeding 10,000 psi (69 MPa) and probably close to 25,000 psi (173 MPa). When the desired strength of concrete is less than the strength of the aggregates—which is true in most construction—the strength of the concrete is mostly a function of the properties of the paste or the products of hydration. In other words, if we replace a 10,000-psi aggregate with a 20,000-psi aggregate in a 3000-psi concrete, the expected variation in strength is insignificant. If, instead, we alter the quantity and makeup of the paste by varying the amounts of cement and water, the compressive strength will definitely change.

At a given age of concrete, the volume of hydration products per unit volume of concrete depends on the amounts of cement and water. But water cannot be idealized as a material that possesses prominent compressive strength, and therefore it follows that the compressive strength of properly cured ordinary concrete—at a given age—is directly related to the amount of cement in the mix, or the quantity of cement per unit volume of concrete.

Figure 3.17 shows the gain in compressive strength with increase in the amount of cement in typical mixtures of equal consistency (measured with a slump test). With increase in the amount of cement a nearly linear growth in strength can be seen. The rate of hydration is not the same for all types of cement; consequently, the strength of concrete is also dependent on the type, composition, and fineness of cement. Cements containing a high percentage of tricalcium silicate—like Type III—gain strength much faster than do cements containing more dicalcium silicate. This means that concrete made with Type III cement has higher early strength than that made with Type I or Type II cement. However, at later ages, the differences in strength between concrete made with Type III cement and that made with other types is very small (Fig. 3.18). In fact, tests have shown that although the 1-day and 28-day strengths do appreciate with increased amounts of tricalcium silicate or tricalcium aluminate in the cement, the 3-month and later-age strengths decrease. After about 90 days, one can expect the strength of concrete made with Type III cement to fall below that made with Type I cement, and significantly less than that made with Type II cement.

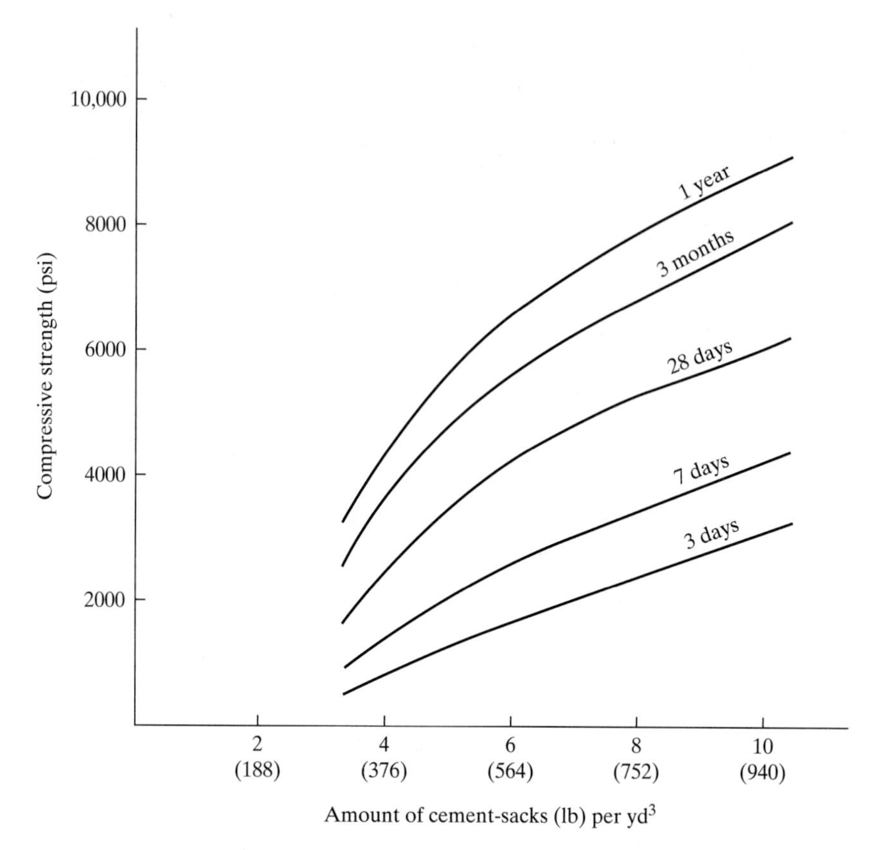

Figure 3.17 Effect of cement content on compressive strength of concrete.

Effects of Aggregates. The strength of an aggregate may not influence the strength of concrete made with it, if the desired strength of concrete is significantly lower than the aggregate strength. But several physical properties of aggregates, and the amount used, affect the properties of concrete both during mixing and when hardened.

The strength of concrete of a given consistency improves with an increase in the fineness modulus of the fine aggregate, which represents the average size of the sample. A higher number means a coarser gradation; a lower number, a finer gradation. The fineness modulus of graded sand for use in concrete construction lies between 2.2 and 3.2. (Natural sand of fineness modulus less than 1.5 is seldom found.) With an increase in fineness modulus, the surface area of particles goes up, requiring less mixing water at the same consistency. A decrease in the amount of water improves the compressive strength of concrete. Figure 3.19 shows a plot of the variation in compressive strength due to changes in the fineness modulus of mixed aggregates—mixtures of fine and coarse aggregate. The increase in strength that comes from coarser gradation is ascribable solely to the accompanying reduction in water requirements. However, coarser gradation lowers the surface or bond area for the same

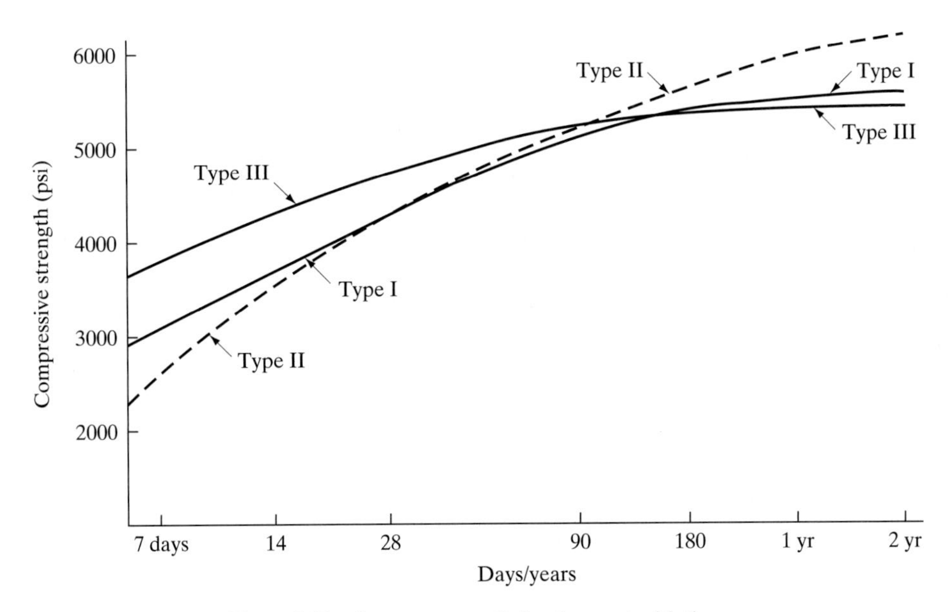

Figure 3.18 Concrete strength development with time.

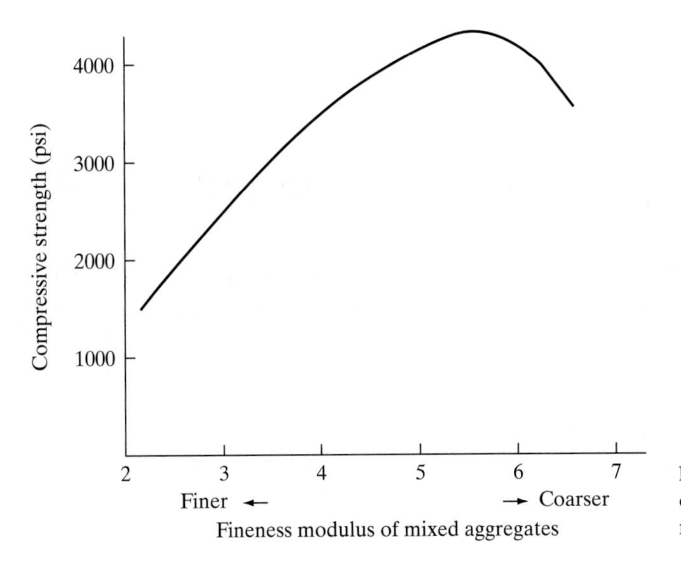

Figure 3.19 Effect of size of aggregates on compressive strength (consistency remaining constant).

volume, contributing to a decrease in compressive strength. Therefore, the increase in strength, expected from the higher fineness modulus of fine aggregate, could be achieved only when a corresponding reduction in water content is made. If, instead, the water content is kept unchanged, the increase in fineness modulus may in fact lower the strength of concrete (Fig. 3.20). Moreover, an increase in fineness modulus beyond a certain value may make the mix harsh and unworkable.

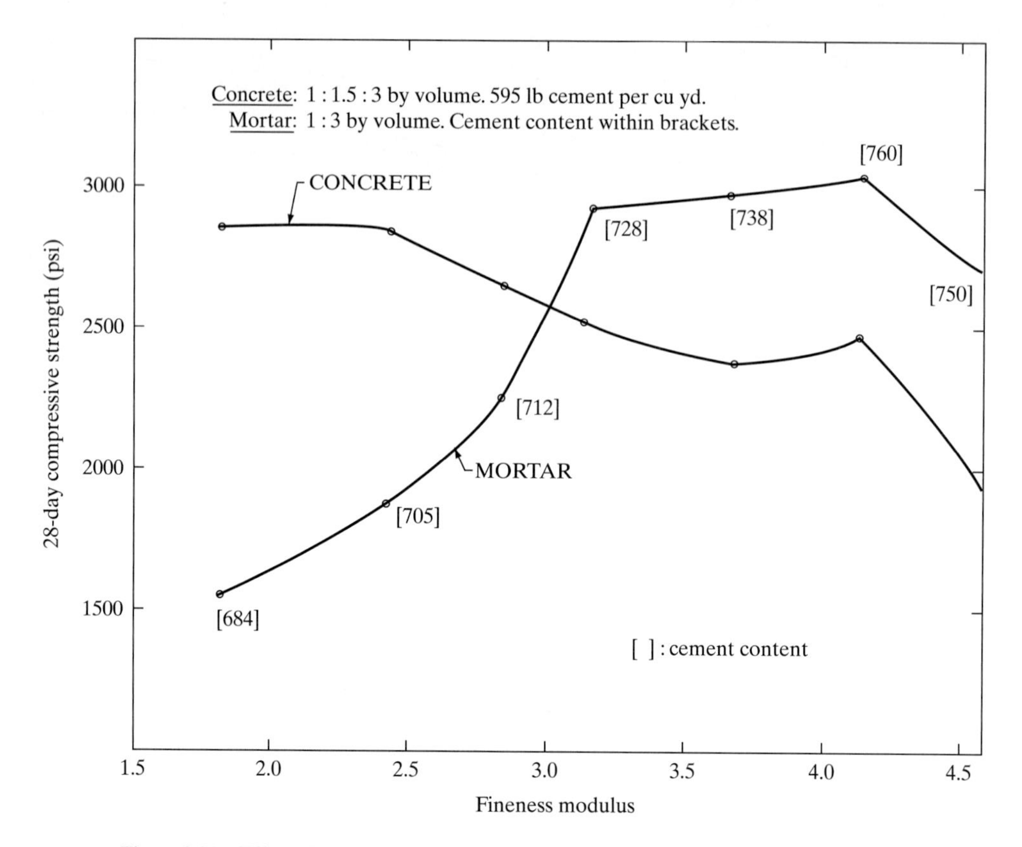

Figure 3.20 Effect of fineness of sand on compressive strength of mortar and concrete.

The strength of concrete may also be affected by the type and size of coarse aggregate. Because of the rough surface texture and angular shape of particles, limestone and granite aggregates may contribute to an increment in compressive strength of up to 20 percent, compared to concrete made with river gravel at the same water-to-cement ratio. The strength of concrete made with lightweight aggregates is lower than that of normal-weight concrete of equal consistency.

The larger the maximum size of coarse aggregate, the smaller the surface area and, therefore, the lower the water requirement at the same consistency. The reduction in the water-to-cement ratio improves the strength of the concrete. However, the use of larger aggregate per se—without a complementary decrease in the amount of water—decreases the compressive strength, due to a reduction in the surface area per unit mass.

Rich mixes—mixes with high cement content—are adversely affected by the use of larger coarse aggregates. In lean mixes, however, increase in the nominal maximum size of coarse aggregates lowers the water requirement and, possibly, increases the compressive strength. Typically, at any water-to-cement ratio, the coarse aggregate of the smallest nominal maximum size produces the highest strength. For example, at a water-to-cement ratio of 0.55, concrete made with a coarse aggregate of nominal maximum size 3/8 in. (10 mm) has higher strength than concrete made with coarse aggregate of nominal maximum size 2 in. (5 cm). High-strength concrete is made mostly with 3/8 or 1/2 in. (10 or 12.5 mm) coarse aggregate. Larger coarse aggregates make the mix harsh and unworkable, but can still be used with the help of admixtures. From the strength standpoint, little improvement can be expected from the use of aggregates larger than 3/4 in. (19 mm).

Effects of Water-to-cement Ratio. The water-to-cement ratio is the ratio between the weights of water and cement in a concrete mix. For proper hydration, this ratio (commonly called the w/c ratio), should be about 0.35, assuming no contribution to hydration from external water sources. But in practice, a much higher w/c ratio, typically in the range of 0.55 to 0.65, is used so that the concrete remains workable.

At a given cement content per unit volume of concrete, the maximum strength is expected when the mix contains water just sufficient for hydration. However, such a mix—with very little water—tends to be dry and harsh due to lack of plasticity (and homogeneity is hard to accomplish with a harsh mix). Yet when the amount of mixing water is increased, keeping the cement content constant, the void content increases and the concrete strength drops.

The compressive strength of concrete is commonly estimated from the water-to-cement ratio, and the relationship between the water-to-cement-ratio of the mix and the compressive strength of concrete is called *Abram's law,* after Duff A. Abrams who in 1918 developed a method of proportioning concrete based on the amount of water in the mix. (According to this rule, which is plotted in Fig. 3.21, the compressive strength of concrete decreases with increase in the w/c ratio.) The w/c

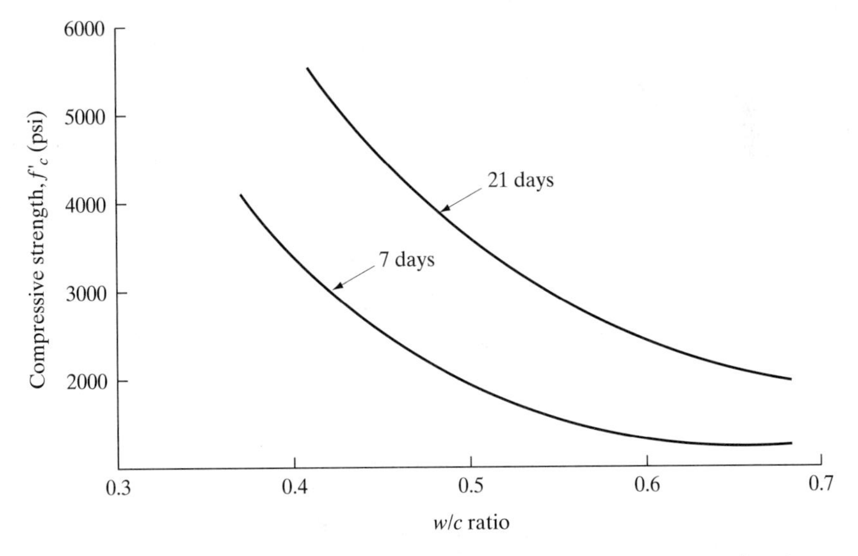

Figure 3.21 Typical relationship between f_c' and w/c ratio at 7 and 21 days (based on tests on moist-cured 3 × 6-in. samples).

ratio has been modified in recent years to be the *water–to–cementitious materials ratio,* to take into consideration the effects of other types of cement or cement-like materials, such as fly ash, on the compressive strength. However, the term w/c ratio in this text stands for water-to-cement ratio unless otherwise indicated. In short, Abram's law states that when the cement content is maintained constant and the amount of mixing water is increased, the compressive strength decreases (Fig. 3.22).

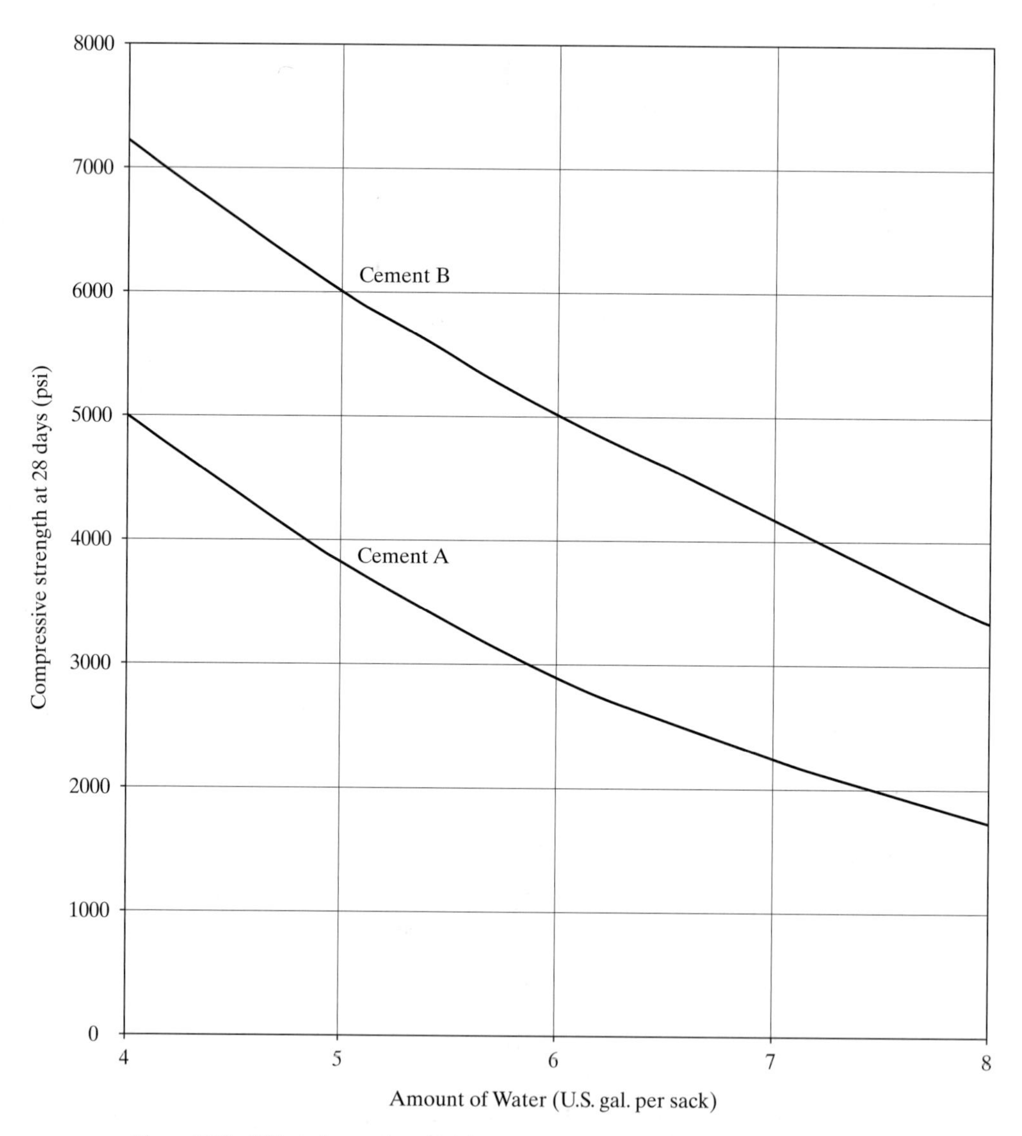

Figure 3.22 Effect of quantity of mixing water on compressive strength of concrete.

Although the w/c ratio is a major variable in the design of concrete mixtures, the relationship between this ratio and the compressive strength is not unique. The same strength cannot be maintained when the cement-aggregate proportions are varied, while keeping the water-to-cement ratio unchanged. For example, the strength of concrete of a mix using 1 part cement, 2 parts fine aggregate, and 3 parts coarse aggregate (1C : 2FA : 3CA), at a water-to-cement ratio of 0.55, will not be the same as that of concrete of mix 1C : 1FA : 2CA, at the same w/c ratio. Further, the strength varies if the maximum size of coarse aggregate, the fineness modulus of fine aggregate, or the type of coarse aggregate is changed. At the same w/c ratio, mortar, which does not have coarse aggregate, is generally stronger than concrete. Nonetheless, the w/c rule does provide a convenient method of controlling the quality of concrete: at any cement content, increase in the ratio decreases the compressive strength.

Influence of Voids. Increase in the water content increases the voids in concrete, lowering the durability, watertightness, and (clearly) the compressive strength. Early theories on the strength development of concrete established a clear-cut connection between the strength and the void content or void/cement ratio. When plotted, these relationships look similar to that between compressive strength and w/c ratio, which underscores the significance of adequate cement content and low void space in achieving concrete of superior quality and performance. Good, dense concrete requires a sufficient amount of cement to achieve strength, suitable gradation to minimize the void content, and proper consolidation to remove air bubbles trapped within the mass.

The amount of water should be just enough to guarantee the hydration of all cement grains. Any excess water in the mix—water that does not participate in the hydration—hikes the amount of voids (which will be filled with air or water, depending on the moisture content). Increase in the voids, again, diminishes the quality of the concrete. On the other hand, holding back on cement decreases the quantity of binder and severely hurts the strength capacity and quality of concrete. In summary, although the w/c ratio provides a convenient tool in the mix design process, and lowering it benefits the overall performance of concrete, good quality concrete calls for a sufficient amount of cement, well-graded aggregates, ample compaction and, of course, minimum mixing water.

Benefits of Curing. Human beings mature with age and care and, similarly, concrete ripens and grows stronger with age and curing. Curing—controlling the moisture and temperature—and its effects on hydration have been explained. Figure 3.16 (page 122) showed the rate of strength development with age and temperature.

Plastic or fresh concrete has negligible strength; a walk on freshly placed concrete (albeit unlikely) would leave behind footprints, yet the same concrete, once hardened, is able to bear the wheel loads of heavy traffic and machinery with no visible marks or scratches. The strength of properly cured concrete at 1 day after mixing is about 10–15 percent of its 28-day strength; at 7 days, it is about 50–60 percent. It can be seen from Figs. 3.16 and 3.23 that moist-cured specimens continue to gain

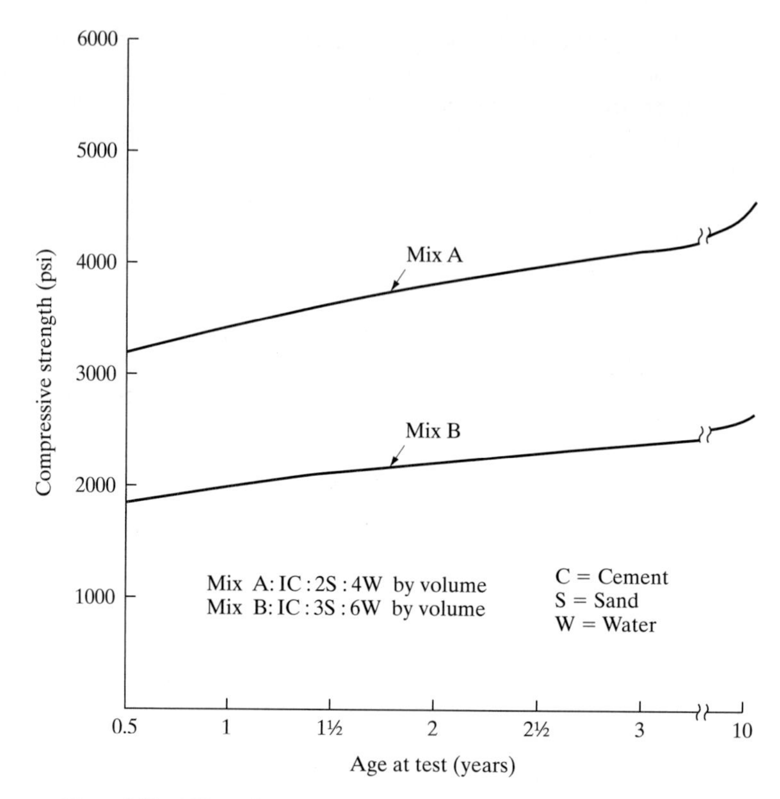

Figure 3.23 Effect of age on compressive strength of concrete. (Based on University of Wisconsin tests, *Journal of the American Concrete Institute,* Vol. 27, Feb. 1943.)

strength even after several years. On the other hand, when specimens dry out, hydration and strength development ceases. Improvement in strength beyond a year is small. Increase of water temperature, either at the mixing stage or during curing, augments the rate of gain in strength. The use of steam for early curing dramatically builds up the strength.

Role of Air-entrainment. Concrete structures exposed to freeze-thaw cycles, such as ground-supported slabs, are made with air-entraining admixtures, which introduce many tiny air bubbles into the concrete mass. The total amount of this air, called *entrained air* (as opposed to *entrapped air* that results from inadequate consolidation) lies usually between 3–8 percent. Air-entrained concrete has lower compressive strength than similar concrete made without the air-entraining admixture (Fig. 3.24). Air-entrained concrete, without adjustment in its w/c ratio, can be expected to lose about 5 percent of strength for each 1 percent (by volume) of added air. In practice, however, the w/c ratio and sand content of air-entrained concrete are normally reduced to maintain the strength level.

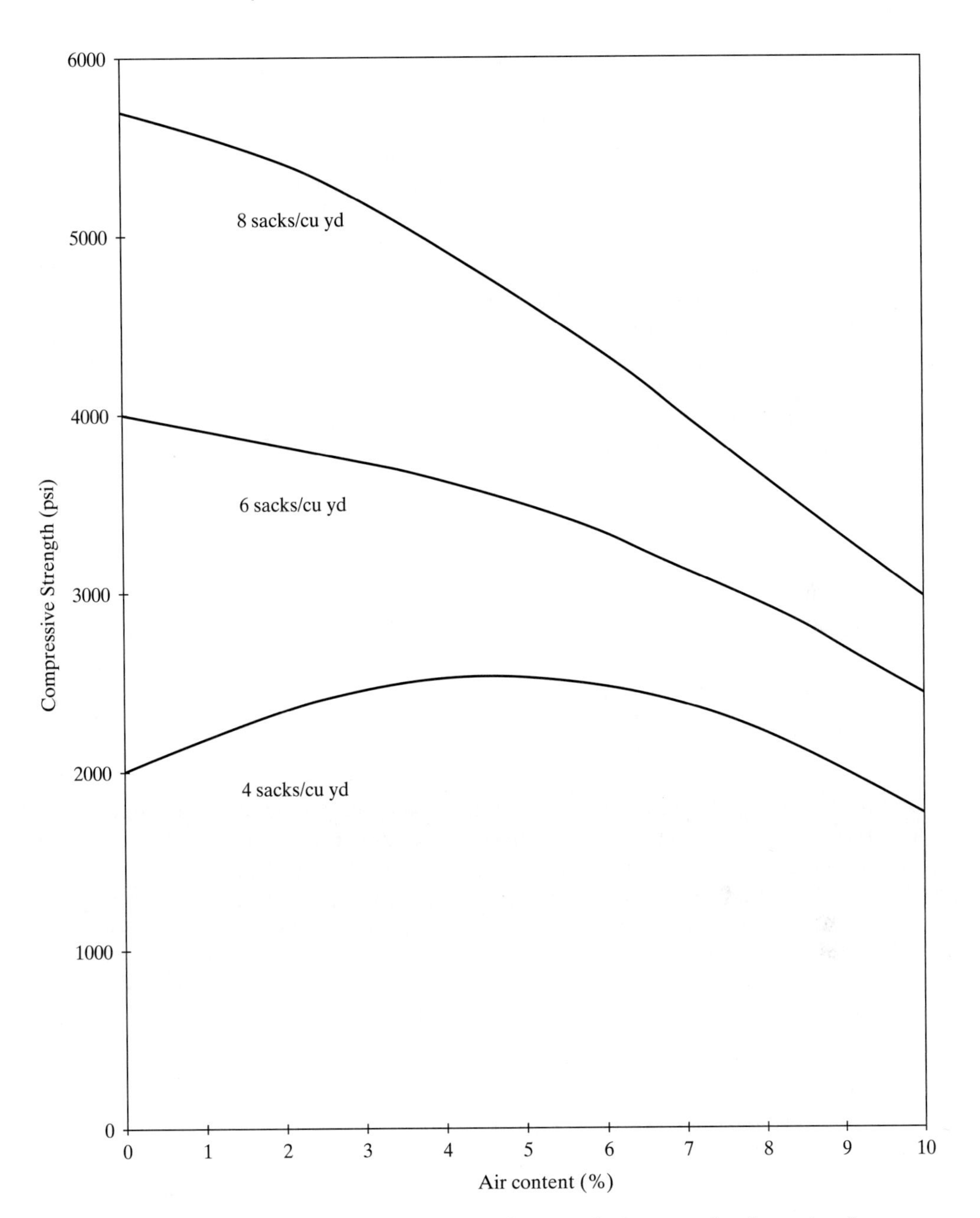

Figure 3.24 Effect of air content on compressive strength of concrete (maximum size of aggregate = 1 in., slump = 3 to 4 in.).

Compression Testing. In the United States, the compressive strength of concrete is established from compression tests on cylindrical specimens of 6-in. diameter and 12-in. height (ASTM C39, "Compression Strength of Cylindrical Concrete Specimens"). Other sizes, with a minimum of 2-in. (50-mm) diameter and 4-in. (100-mm) height, can also be used. The nominal maximum size of coarse aggregate should not be larger than one-third the diameter of the specimen.

Empty cardboard or metal cylinders are filled with fresh concrete using a standard procedure (ASTM C192, "Standard Practice for Making and Curing Concrete Test Specimens"). When sampling from truck-mixers, the concrete must be obtained before any additional water is added at the job site. After 24 hr, the specimens are taken out of the molds and moist-cured for 28 days. At the end of the curing period, the ends of the specimens are capped and they are tested in a moist condition. The failure load, (obtained from the testing machine) divided by the cross-sectional area is called the 28-day cylinder compressive strength, and is normally identified using the notation f'_c.

$$f'_c = \frac{P}{(\pi/4)d^2}$$

where P = failure load and d = diameter of the cylinder.

The compressive strength thus determined is affected by many variables, including the size and shape of the specimen, and moisture conditions. The greater the ratio of height to diameter of the cylinder, the lower the measured compressive strength. In specimens of height equal to twice the diameter, the compressive strength decreases with increase in the diameter. A 4-in. diameter cylinder, of 8-in. height, indicates approximately 5 percent higher strength than does a cylinder of 6-in. diameter and 12-in. height. Strength measured from cubes or cubic samples is higher than that measured using cylindrical samples of diameter equal to the size of the cube. The strength measured from a 6 × 12-in. cylinder is on an average about 80 percent of the strength measured using a 6-in. (150-mm) cube.

The moisture content of specimens also affects the compressive strength. Air-dry specimens—those that are air-dry at the time of testing—have been shown to fail at strengths higher than those of saturated specimens, on the order of 20–25 percent. In addition, the strength of concrete is influenced by the speed of testing; a slower rate will show a reduced strength. In the laboratory, the rate of loading should be such that the failure takes place within 2–3 minutes (or at a speed of 20–50 psi/s or 0.14–0.34 MPa/s).

Acceptance Criteria. The ACI code (318) recommends the following criteria for accepting (or rejecting) concrete placed within forms, based on 6 × 12-in. cylinders tests:

1. Frequency of samples, taken on a random basis, for each class of concrete placed each day, should not be less than:
   - once a day;
   - once for each 150 yd^3; and
   - once for each 5,000 ft^2 of surface area for slabs and walls.

2. A strength test is the average of the strengths of two cylinders, measured at 28 days, molded from the same concrete.
3. The average of any three consecutive strength tests should be equal to or greater than the specified compressive strength.
4. No individual strength test should fall below the specified compressive strength by more than 500 psi.
5. If either of the requirements in criteria 3 and 4 is not met, the concrete is said to have failed the strength requirement and to be unacceptable.

3.8.2 Tensile Strength

The tensile strength of concrete should be high enough to resist cracking from shrinkage and temperature changes, and is measured using one or more of the following procedures:

- Direct tension test
- Split-cylinder test
- Flexural test

Accurate direct tensile strength measurements, using axial tension specimens, are difficult, and are not usually done. It is common practice to assess tensile strength using either the *flexural* or the *split-cylinder* test.

In the split-cylinder test (ASTM C 496, "Standard Test Method for Splitting Tensile Strength of Cylindrical Concrete Specimens"), a cylindrical specimen of minimum 2-in. (50-mm) diameter, placed with its axis in a horizontal plane, is subjected to a uniform line load along the length of the specimen, as shown in Fig. 3.25. Knowing the load at which the specimen splits into two parts (P), the tensile strength (f_t), can be calculated as follows:

$$f_t = \frac{2P}{\pi l d}$$

where l is the length and d is the diameter of the cylinder.

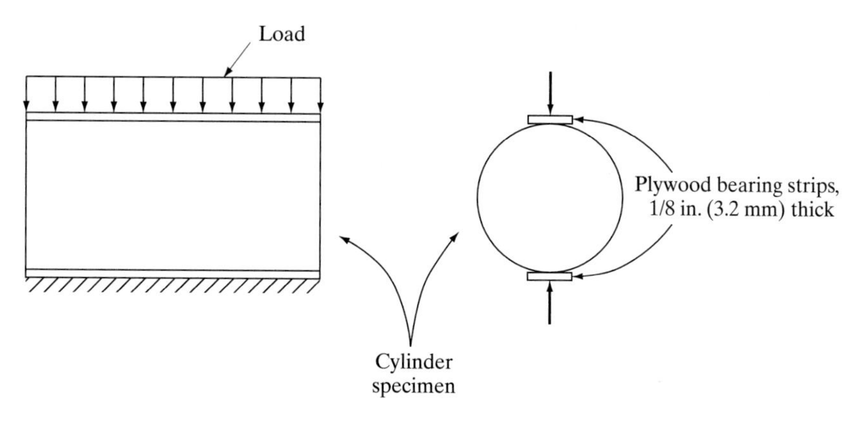

Figure 3.25 Split-cylinder test.

The tensile strength of concrete can also be estimated from its compressive strength using empirical equations. For normal-weight concrete, it can be computed (in psi) as:

$$f_t = 6.7 \times (f_c')^{0.5}$$

(with compressive strength in psi units). As can be confirmed from the equation, the tensile strength of concrete increases with its capacity in compression. As a general rule, the split-cylinder strength (splitting tensile strength) is about 10–15 percent of the compressive strength; the lower the compressive strength, the higher the percentage. Direct tensile strength is approximately 25 percent lower than the strength measured from the split-cylinder tests. The expected variation in tensile strength from changes in w/c ratio is similar to the relationship between the w/c ratio and the compressive strength. With increase in the fineness modulus of aggregate—or as the aggregate gradation becomes coarser—the tensile strength increases. The type and shape of coarse aggregate particles also affect the tensile strength, as well as cracking, principally from high tensile stresses. Concrete made from crushed stone can withstand larger drops or fluctuations in temperature during service without cracking than concrete made with gravel. The older the concrete, the greater its tensile strength at any given compressive strength.

3.8.3 Flexural Strength

A flexure test is the most common method for measuring the tensile strength of concrete. A concrete beam with span length equal to three times the beam depth (the length of the beam should be at least 2 in. (50 mm) larger than the span) is subjected to *third-point loading* (ASTM C78, "Standard Test Method for Flexural Strength of Concrete"). This produces tensile stresses at the bottom of the beam and compressive stresses at the top. Since concrete is weaker in tension than compression, the specimen fails—literally breaks into two—following the formation of a nearly vertical crack, called a *flexural crack,* near the section of maximum moment. From the failure load, the tensile strength, called the *modulus of rupture* (MOR), is then calculated with the help of the bending equation (Fig. 3.26):

$$\text{bending stress} = \frac{Mc}{I} = \text{strength in tension}$$

where M is the maximum moment, I is the moment of inertia, and c is the distance from the neutral axis to the extreme fiber in tension.

For a rectangular cross-section of width b and depth d subjected to third-point loading, the bending strength or MOR in the preceding equation turns out to be:

$$\text{MOR} = \frac{PL}{bd^2}$$

where L is the span length and P is the failure load.

The MOR is affected by a number of variables, including the w/c ratio, the age at test, and the curing, as shown in Fig. 3.27. In addition, the amount and properties

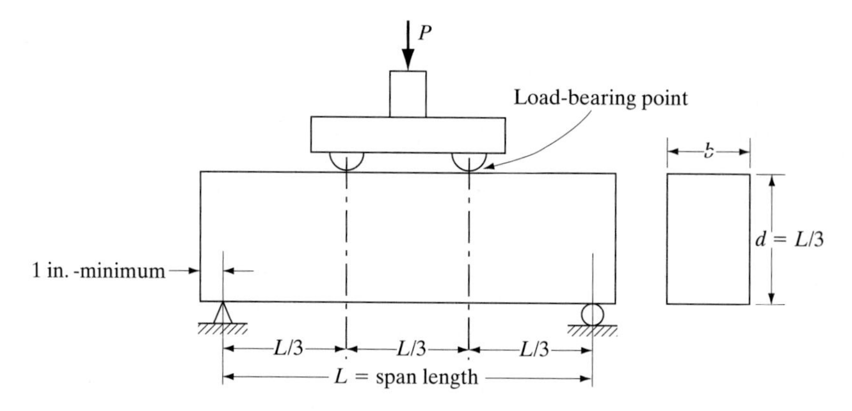

Figure 3.26 Flexure test (third-point loading).

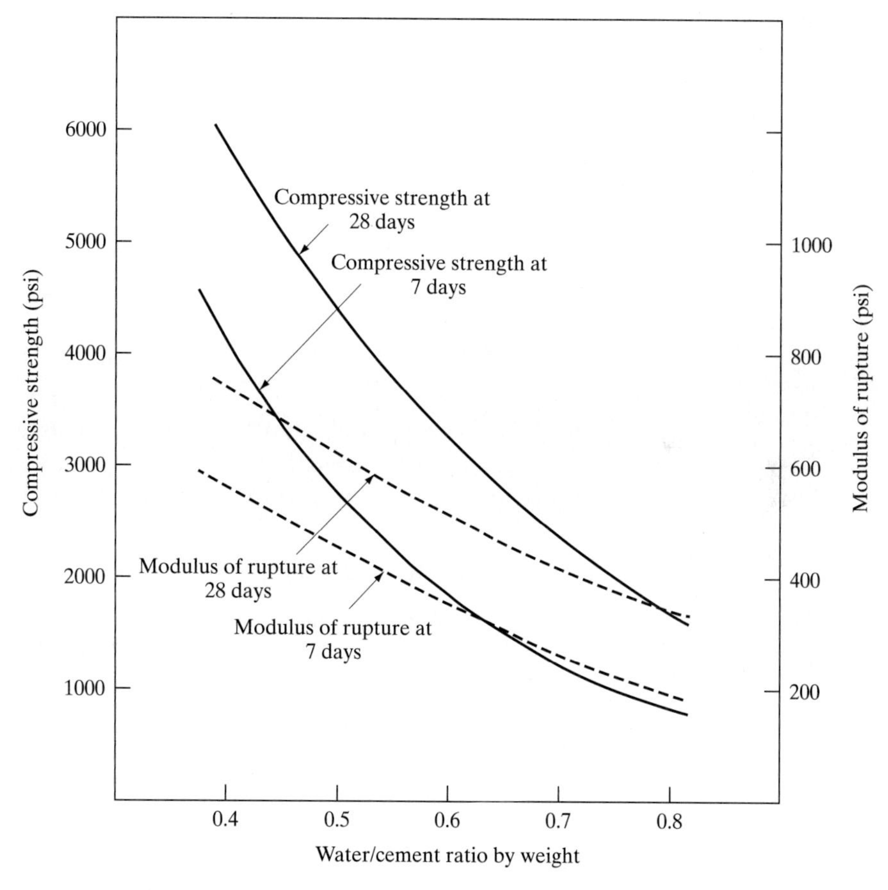

Figure 3.27 Typical variation of concrete strengths and modulus of rupture with w/c ratio.

of aggregates and the amount of cement also influence the bending strength. Moisture content of the sample during testing considerably affects the bending capacity; uniformity of moisture condition throughout the test specimen is essential to achieve reliable results. Laboratory tests have shown that the lowest MOR belongs to concrete made with aggregates of lowest flexural and bond strengths. The flexural strength of normal aggregates is significantly higher than that of the paste.

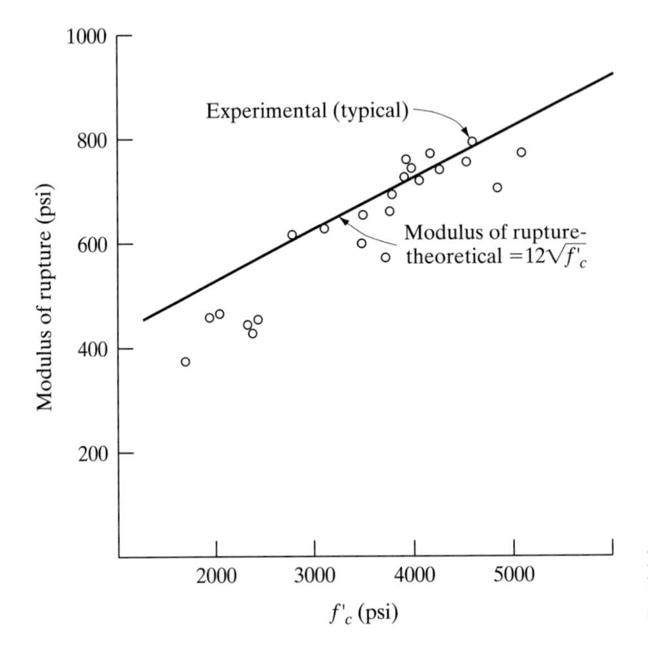

Figure 3.28 Relationship between compressive strength and modulus of rupture.

Typically the MOR of concrete is about 10–23 percent of its compressive strength, and the lower the compressive strength the higher the percentage. The following empirical relationship is commonly used to predict the flexural strength of concrete (Fig. 3.28).

$$MOR = 12 \times (f_c')^{0.5}$$

where MOR and f_c' are in psi units. The flexural test overestimates the tensile capacity of concrete because the propagation of a crack, which sets out at the extreme fiber in tension, is blocked by the less stressed material nearer the neutral axis.

3.8.4 Stress-strain Diagram and Modulus of Elasticity

When a structural element is subjected to a force, it deforms: a rubber band elongates when stretched out; a sponge contracts when squeezed. The change in length—or deformation—resulting from the applied force depends on the magnitude of the force and the type of material (in particular, the stress-strain characteristics of the material).

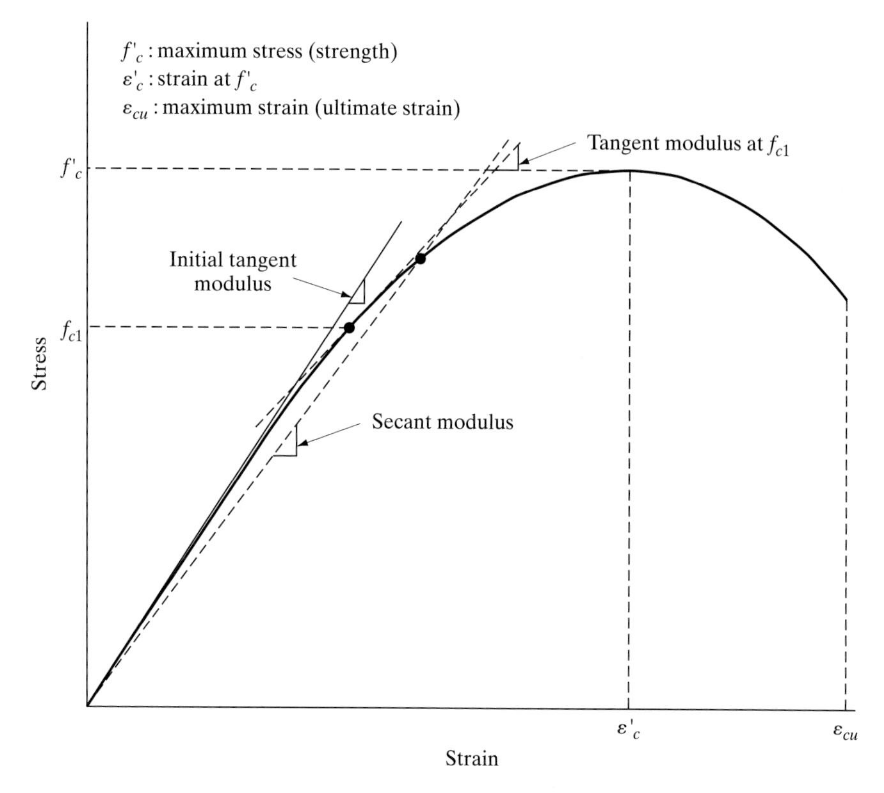

Figure 3.29 Typical stress-strain diagram for concrete.

Figure 3.29 shows a typical stress-strain diagram for concrete in compression. It constitutes an elastic range at the onset, followed by a nonlinear or inelastic response of stress increment for gain in strain, up to a maximum stress. Thereafter the stress decreases, inelastically, while the strain grows, until the failure takes place at a maximum strain, called *failure strain of concrete in compression*.

Various concretes exhibit differing stress-strain relationships (in terms of elastic range, maximum stress, and failure strain), but the shape of all the diagrams and the characteristics of the stress-strain relationship are about the same. Figure 3.30 shows typical stress-strain diagrams for normal-weight concrete of various compressive strengths. It can be seen that with increase in the compressive strength the length of the elastic portion of the curve gets bigger and that the failure strain diminishes, indicating that the concrete becomes more and more brittle. Although the failure strain, or the maximum strain, is dependent on the type (normal- or lightweight) and strength of concrete, it is assumed as constant, its value equal to 0.003, in the design of concrete structures.

The modulus of elasticity of concrete, which represents the slope of the elastic portion of the stress-strain diagram, depends primarily on the modulus of the

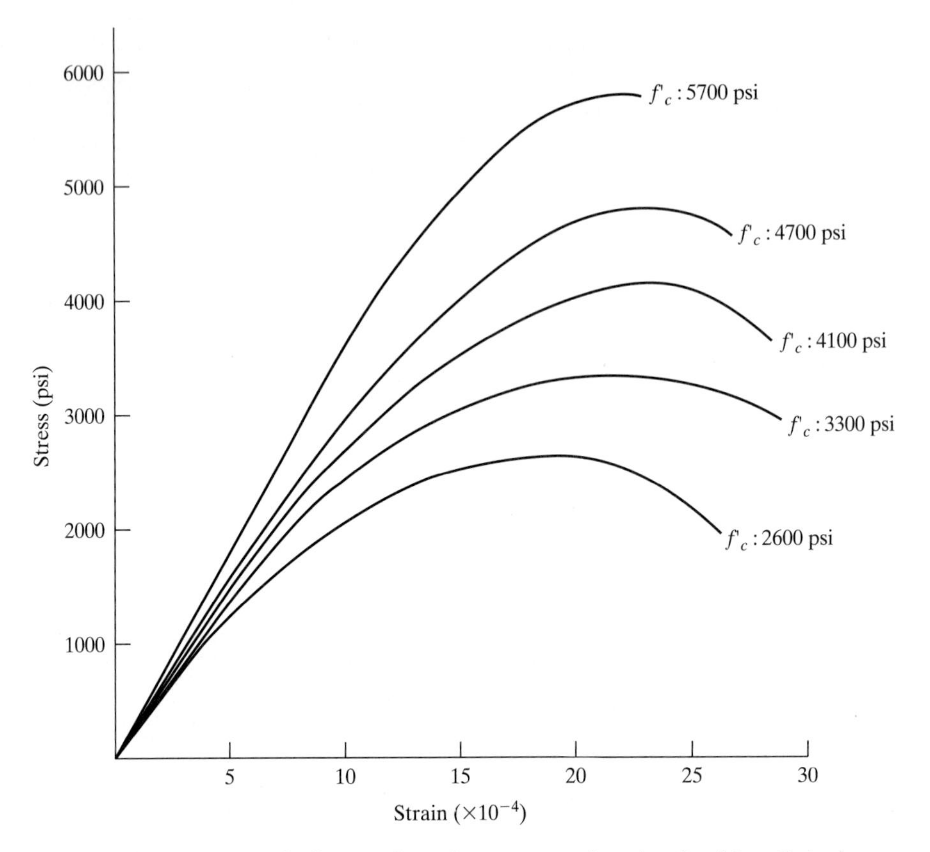

Figure 3.30 Stress-strain diagrams for various concretes (based on 3 × 6-in. cylinders).

aggregates, modulus of the paste, and the relative amounts of aggregates and paste in the mix. Both the paste and the aggregate exhibit linear stress-strain relationship up to failure, with the modulus of aggregate being larger than that of paste. The modulus of paste varies between 2.5–3.5 × 10^6 psi (17,250–24,150 MPa). The larger the amount of coarse aggregate—with elastic modulus higher than that of concrete—the greater the modulus of elasticity of the concrete, which ranges between 3–5 × 10^6 psi (20,700–34,500 MPa) for normal-weight concrete. Concrete made with lightweight aggregates (such as expanded shale) has much lower modulus than concrete made with normal-weight aggregates (like gravel). The modulus of lightweight concrete in general is about 40–80 percent of that of normal-weight concrete.

The shape and surface characteristics of the aggregate particles also influence the modulus and the shape of the stress-strain curve. Tests done on concrete made with porous limestone aggregates from Florida; dense, rough-textured, and angular limestone aggregates from Alabama; and smooth, round, river gravel have shown that both strength and stiffness are significantly affected by the type of aggregate.

Dense, angular limestone aggregates produced concrete of higher strength and stiffness than the other two types. In general, the modulus of elasticity of moist-cured concrete samples increases with density, age, and strength of concrete. At a given strength, the modulus is largely dependent on the type of aggregate and the age at test. Continuous air-drying lowers the modulus, probably from internal micro-cracks that result from shrinkage.

As indicated, the two components of concrete, cement paste and aggregate, when tested individually in compression, exhibit linear stress-strain relationships. The development of micro-cracks through the interface between aggregate particles and paste suggests a possible explanation for the nonlinear stress-strain relationship of concrete.

Methods of Measurement of Modulus. The modulus of elasticity (or Young's modulus) is the constant of proportionality between stress and strain. It is obtained by measuring the slope of the straight-line portion of the stress-strain diagram. When no straight-line portion exists (or beyond the elastic limit), the modulus of elasticity can be interpreted to be the slope of the tangent to the curve at any point on the diagram. When the modulus of elasticity is represented as the slope of either the straight-line portion or the tangent at a point, it is called the *tangent modulus.* When the tangent modulus is from the tangent drawn to the curve at the origin, it is referred to as the *initial tangent modulus*.

The modulus of elasticity can also be measured as the slope of a line connecting two points on the curve, known as the *secant modulus.* The secant modulus is the most commonly used in practice, as the initial tangent modulus may not be accurate owing to errors in reading very small strains at the start of the test. Because of the nonlinearity of the stress-strain diagram, the secant modulus decreases continuously with increase in stress beyond the proportional limit; as a consequence, the upper stress point should be indicated when specifying the secant modulus (the lower point is normally next to the origin). According to ASTM C469 ("Test Method for Static Modulus of Elasticity and Poisson's Ratio"), the modulus of elasticity of concrete, E, can be calculated as:

$$E = \frac{S_2 - S_1}{\varepsilon_2 - 0.00005}$$

where S_2 is the stress corresponding to 40 percent of the ultimate load, S_1 is the stress at a strain ε_1, 50 millionth inch per inch, and ε_2 is the strain produced at stress S_2.

According to ACI 318, the modulus of elasticity of concrete can be calculated using an empirical equation:

$$E = w^{1.5}(33) \times f_c'^{(0.5)}$$

where E is the modulus in psi, w is the unit weight of concrete in pcf, and f_c' is the compressive strength of concrete in psi. For normal-weight concrete, of density

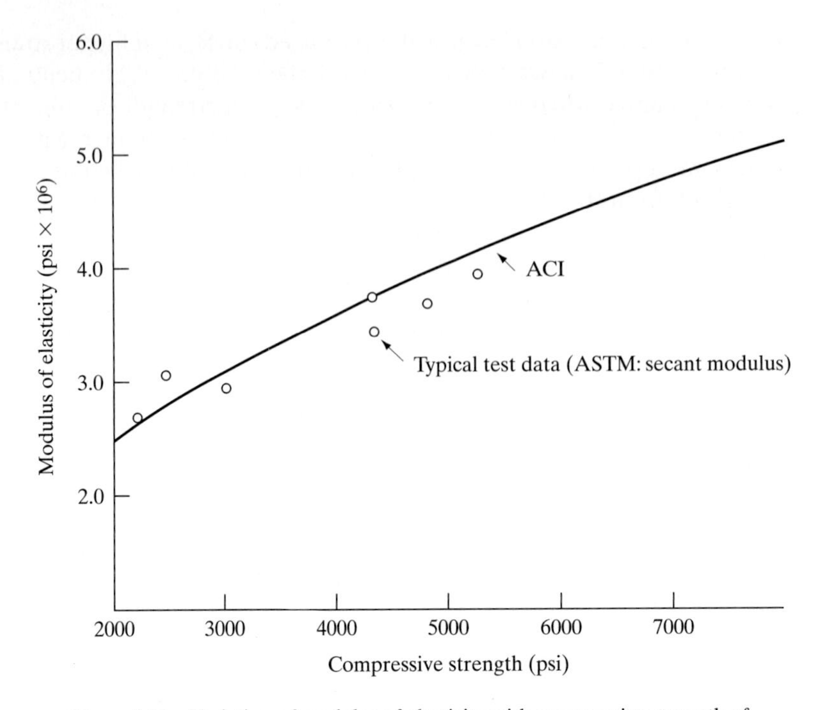

Figure 3.31 Variation of modulus of elasticity with compressive strength of normal-weight concrete.

around 144 pcf, this equation simplifies to:

$$E = 57{,}000 \times f_c'^{(0.5)}$$

Figure 3.31 shows typical relationship between the modulus of elasticity and compressive strength of normal-weight concrete. The modulus of elasticity in tension, obtained from direct tension tests, is about the same as the modulus in compression.

3.8.5 Examples

Example 3.1

Results from strength tests on concrete of specified compressive strength equal to 25 MPa are shown. Verify if the concrete satisfies the acceptance criteria.

Strength test no.	1	2	3	4	5
Strength of Cylinder 1 (MPa)	27	26	24	24.2	24.2
Strength of Cylinder 2 (MPa)	26.6	24.6	25.4	25.6	24.8

Solution The average of two cylinders is the strength test value. For the five strength tests, the strength test values are as shown.

Strength test no.	1	2	3	4	5
Value (MPa)	26.8	25.3	24.7	24.9	24.5

The two relevant conditions in the acceptance criteria are:

1. The average of any three consecutive strength tests should be equal to or greater than the specified strength.
2. No individual strength test should fall below the specified strength by more than 500 psi (3.45 MPa).

Per the second criterion, the strength value should be greater than or equal to $(25 - 3.45) = 21.55$ MPa. The minimum strength test value is 24.5 MPa, which is more than 21.55 MPa; it satisfies this condition.

Per the first condition, the three possible averages are:

$$\text{average of strength tests No. 1–3} = (26.8 + 25.3 + 24.7)/3 = 25.6$$
$$\text{average of strength tests No. 2–4} = (25.3 + 24.7 + 24.9)/3 = 25$$
$$\text{average of strength tests No. 3–5} = (24.7 + 24.9 + 24.5)/3 = 24.7$$

The last average is less than 25 MPa (the specified strength); thus, the concrete is not satisfactory.

Example 3.2

Estimate the tensile strength, the modulus of rupture, and the modulus of elasticity of normal-weight concrete with $f'_c = 3500$ psi.

Solution The average density of normal concrete is 145 pcf. The ratio of tensile strength to compressive strength varies with the w/c ratio, decreasing with decrease in the w/c ratio. The splitting tensile strength ranges from 0.10–0.15 times the compressive strength.

$$\text{estimated splitting tensile strength} = 0.15(3500) = 525 \text{ psi}$$

$$\text{modulus of rupture} = 12(f'_c)^{0.5} = 12(3500)^{0.5} = 710 \text{ psi}$$

$$\text{modulus of elasticity} = 57{,}000(f'_c)^{0.5} = 3.37 \times 10^6 \text{ psi}$$

3.8.6 Shrinkage

Concrete has the largest volume at the time of mixing or when it is placed freshly in forms. As it sets, the concrete slowly decreases in volume or, simply, it shrinks. Shrinkage or volume reduction, also called contraction, is the result of displacement of water from within to the surface, and the loss of water to the surroundings.

When concrete is still plastic, the aggregate particles (which are heavier than water) settle, displacing water and air to the top surface. Aided by conditions prevailing in the environment (such as high temperature and humidity), this surface water is lost through evaporation, drawing even more water to the surface. Both processes—settlement of solids and evaporation of water from the fresh mix—give rise to shrinkage of wet concrete. This reduction in volume of plastic concrete (the volume reduction while the concrete sets), typically during the first twelve hours after placement, is called *plastic* shrinkage. The shrinkage due to drying of hardened concrete is called *drying* shrinkage.

Plastic Shrinkage. When portland cement reacts with water, the system—cement plus water—undergoes a net reduction in volume, although the volume of solid matter (which consists of the products of reaction) increases. The volume occupied by saturated cement gel is about 2.2 times the volume of dry cement. When this cement paste is stored under water it expands by up to 0.1 percent in three months. As a hypothetical experiment, if neat cement paste is sealed in a glass tube, with a slightly excess of water on the top, it may eventually expand enough to crack the glass tube.

When it is continuously kept wet or remains saturated, concrete increases in size, though by a smaller amount than the expansion of the paste. But when it is stored under water or maintained in saturated condition only for short periods, or at different intervals, the initial shrinkage or reduction in volume from drying exceeds any subsequent expansion from the hydration. On the other hand, concrete that is allowed to dry continuously through evaporation decreases in volume. After a subsequent wetting period it expands, but the expected increase in volume may be less than the initial volume reduction.

The decrease in volume of the cement-water system while the concrete is still plastic is the plastic shrinkage or initial shrinkage. When the plastic concrete is forced to dry rather quickly, the plastic shrinkage is aggravated, leading to stresses and eventual surface cracks, which are short irregular cracks distributed throughout the surface. Typically, they extend barely a few inches into the mass, but can penetrate deeper with additional drying.

The magnitude of plastic shrinkage (which takes place mostly during the first few hours and entirely within 24 h after mixing) can be very large, as much as 5–10 times the amount of drying shrinkage, and is thus very significant in terms of cracking. Up to about 80 percent of the water loss from evaporation can occur within the first 24 h after mixing. Plastic shrinkage cracks, visible mostly in horizontal surfaces, happen usually around the time when the water sheen disappears from the surface.

Plastic shrinkage cracks are more common in slabs and pavements, as the large surface area contributes to high evaporation loss. When the evaporation loss is faster than the rise of water from bleeding, the water is forced out of interior much more rapidly and the concrete dries up swiftly. Tensile stresses that accompany this shrinkage differential between the interior and the surface can induce plastic shrinkage cracks. Hot weather conditions intensify plastic shrinkage, thanks to excessive evaporation loss. A larger than nominal cracking affects the durability performance of concrete.

The amount of plastic shrinkage and the resultant cracking depend largely on:

- Type of cement
- w/c ratio
- Quantity and size of coarse aggregate
- Consistency of the mix

Stiffer mixes have lower initial shrinkage than more fluid concrete. Generally, wet mixes settle over a longer period than dry mixes do, and thus shrink more. Dry spots in the subgrade pull more water out of fresh mixes in ground-supported slabs, leading to differential settlement of aggregates and consequent cracking.

Reduction in the temperature of the fresh mix lessens the amount of plastic shrinkage. A decrease in cement temperature of 8–10 °F, water temperature of 4 °F, or aggregate temperature of 1.8 °F lowers the concrete temperature by about 1 °F. Thus, in order to decrease the concrete temperature, it is easier to cool the aggregates or mixing water.

A number of techniques can be followed to minimize plastic shrinkage cracks. Revibration prior to floating aids in preventing their onset. Spraying the coarse aggregate pile with cold water or using cold water for mixing may offset their development. Concrete can be cooled by as much as 10 °F (6 °C) if the mixing water is chilled. When ice—crushed or flaked—is incorporated into the mixing water, the concrete can be cooled by as much as 20 °F (11 °C). Any procedure that minimizes or eliminates evaporation loss decreases plastic shrinkage cracks. Erection of sunshades and windbreaks also helps. Evaporation loss can also be minimized by applying curing compound or by covering the freshly placed concrete with wet burlap or plastic sheets. Water-reducing and air-entraining admixtures are also helpful in reducing plastic shrinkage cracks.

Crazing, which is a hexagonal pattern of surface cracks emerging at an early age in concrete, is primarily due to improper finishing or curing. Excessive floating and troweling is a major cause of crazing. Such finishing techniques bring excessive amounts of cement, water, and fines to the surface, thereby weakening the surface layer. Aided by immoderate evaporation loss, the top surface consequently cracks.

Drying Shrinkage. The reduction in volume from drying of hardened concrete is the drying shrinkage, and is attributed to the loss of water from the cement gel. The amount of drying shrinkage can be as large as 1500×10^{-6}, and for most concrete it is in the general range of $350–650 \times 10^{-6}$. The average value is

500×10^{-6}, or 0.05 percent of the length of the member. A 10-ft (3 m) member is expected to shrink by an average of 1/16 in. (1.5 mm). A 30 $\times$ 80-ft slab can be presumed to shrink by about 0.12–0.23 in. in the short direction and 0.34–0.62 in. in the long direction. When this shrinkage is restrained, internally or externally, the slab cracks. Provision of control joints minimizes the extent of these cracks. Free or unrestrained shrinkage does not cause cracks, but the restraint to shrinkage does.

Drying shrinkage is gradual, and the rate of shrinkage decreases with time. Between about 40–80 percent of the 20-year shrinkage takes place within about three months. At the end of one year, the rate of shrinkage drops to nearly one-half the initial value. When concrete that has been allowed to dry comes into contact with moisture, it swells. And concrete that is maintained in 100 percent relative humidity does not shrink, but expands.

Drying shrinkage depends on a number of factors, such as type and amount of cement, mix proportions, size and shape of structure, curing, environmental conditions (temperature and relative humidity), and reinforcement. When the w/c ratio goes up, drying shrinkage increases. Raising the ratio by 100 percent (from, say, 0.35 to 0.7) may increase the shrinkage three to four times.

Decrease in the maximum size or quantity of coarse aggregate increases the amount of drying shrinkage. Overabundance of sand is found to significantly increase the shrinkage, probably from requiring too much water in the mix. On the other hand, too much coarse aggregate restrains the shrinkage deformation, resulting in excessive tensile stresses. Hard aggregates offer increased resistance to compressive stresses and enhance the restraint to shrinkage, thereby lowering the amount of shrinkage. Generally, shrinkage of normal-weight concrete is less than that of comparable lightweight concrete. Concrete made with gravel or sandstone shrinks more than that manufactured with granite or limestone. Mixtures composed of highly absorptive aggregates shrink more, and those made with high-density, less absorptive aggregates shrink less.

As the cement content increases, the shrinkage (as well as the cracking tendency) increases. Decrease in humidity causes rapid moisture loss and increases the shrinkage. However, the most important factor that affects shrinkage is the amount of aggregates. Increase in the mass of aggregates or decrease in the quantity of paste diminishes the shrinkage (Fig. 3.32). Typically, mortar shrinks three times—and concrete seven times—less than cement paste.

Shrinkage Reduction. Shrinkage can be minimized by keeping the water per unit volume of concrete as low as possible, or by keeping the aggregate content as high as possible. Curing lowers the shrinkage. High-pressure steam-cured samples are found to shrink less than normally cured samples. The use of water-reducing admixtures decreases the amount of shrinkage; retarding admixtures may give rise to more shrinkage.

Shrinkage cracks run perpendicular to the length of pavements and walkways; retaining walls and basement walls may crack vertically. As noted, the restraint provided by reinforcement decreases the magnitude of shrinkage, but instead induces tensile stresses in concrete, causing it to crack. Thus, the extent of shrinkage in

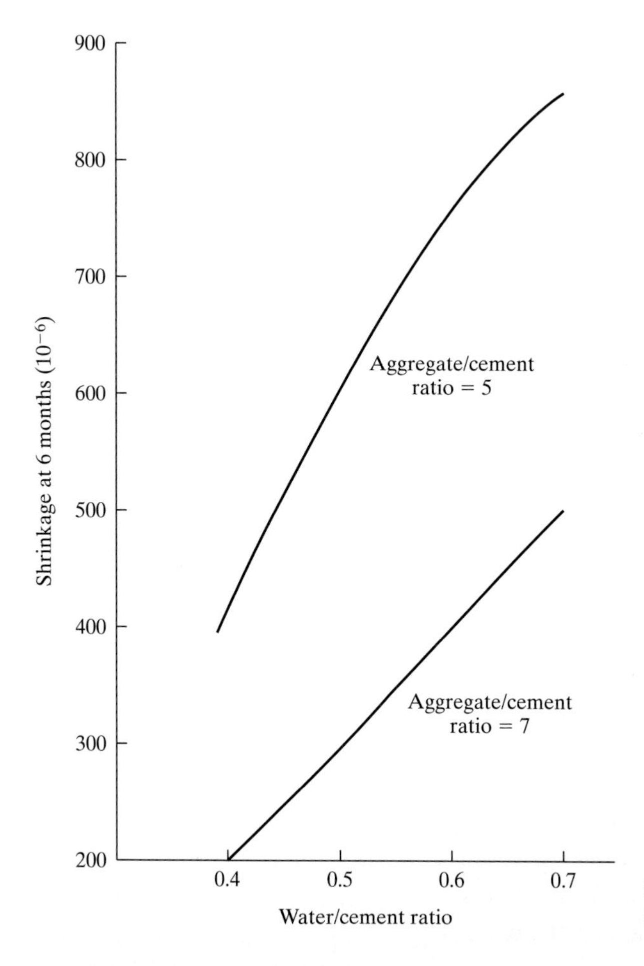

Figure 3.32 Typical variation of shrinkage in concrete specimens.

reinforced concrete is less than that in plain concrete. Reinforcement is provided in slabs-on-grade (or ground-supported slabs) to control the shrinkage. But it is important to recognize that this reinforcement does not prevent cracks from occurring; it merely controls the location and width of cracks resulting from both shrinkage and temperature effects. And it helps to hold the edges of the cracks together tightly. The use of synthetic fibers in concrete (fiber-reinforced concrete) helps to control shrinkage cracks.

Various parts of a structure my shrink differently. Foundations may not shrink much because they may not dry out. Walls may shrink more because of quick drying, and floors may shrink much more because of indoor heat. In slabs-on-grade the top surface dries out and shrinks faster than the bottom. The *differential shrinkage* between various elements in a building may lead to cracking, especially at the junction between the elements. Differential drying shrinkage between the top and bottom surfaces of a slab-on-grade may lead to *curling* or *warping*. The exposed top surface of the slab shrinks faster than the bottom, resulting in a shrinkage differential; a large differential forces the slab to curl up, eventually causing cracks. Placing the control joints at closer intervals, so that the total movement of each portion of the slab is small, may prevent curling. Besides, the finishing operation should be started only after the bleed water has disappeared. Casting the slab on a porous bed of sand or gravel, without the use of a vapor barrier, helps to equalize the loss of moisture between the top and the bottom and thus minimize curling. Furthermore, low w/c ratio, less cement content, and high-modulus aggregate may be used to prevent this. Oversanded aggregates, which require a high w/c ratio to obtain a satisfactory workability, give rise to more cracking.

A decrease in temperature also causes contraction in concrete. A drop in temperature of 100 °F (37 °C) may cause a reduction in length of about 2/3 in. per 100 ft (17 mm per 30 m). The net reduction in length of a concrete element is thus due to shrinkage and reduction in temperature.

Contraction and Construction Joints. When concrete is allowed to shrink without any restraint, no cracks will form. But most structures have restraint to shrinkage from reinforcement, soil, foundations, or adjacent members, causing cracks.

Shrinkage cracks are generally accommodated in large structures, such as pavements, by forming surface grooves every 10 ft or less, called *contraction joints* (Fig. 3.33). These joints, also called *dummy joints* or *control joints,* provided to accommodate movement from temperature changes, drying shrinkage, and creep, are meant to direct the cracks to the location of the groove. Without these joints the concrete member will crack in a random manner.

In general, joints in concrete elements are provided to accomplish the following:

- minimize undesirable cracking
- accommodate differential movement of adjacent elements
- provide a natural plane of weakness
- prevent bonding of adjacent elements

Control joints, provided in large structures such as sidewalks, floor slabs, driveways, and walls, are made by tooling, forming, scoring, or sawing partway through the concrete to form a weakened plane, so that when the member cracks it will do so along these predetermined lines (Fig. 3.34). In forming, the joint is accomplished by placing a thin wood strip or premolded joint material at the joint location, which acts to separate the elements on the two sides of the joint. The grooves are formed to a depth of about one-fourth the slab thickness, with a minimum of one-fifth the thickness. In reinforced concrete slab construction, half the horizontal bars—in alternation—should be cut at the joints. Control joints in walls may be spaced every 20 ft (6 m).

Construction joints (also called *isolation joints* or *expansion joints*) are provided to separate a slab from other parts of a structure, such as beams and columns, to prevent bonding and permit horizontal and vertical movement (Fig. 3.35). They are also introduced for the convenience or necessity of the construction process. The joints used at points of restraint, including junctions between similar or dissimilar elements of a concrete structure, are isolation or expansion joints. For example, a joint

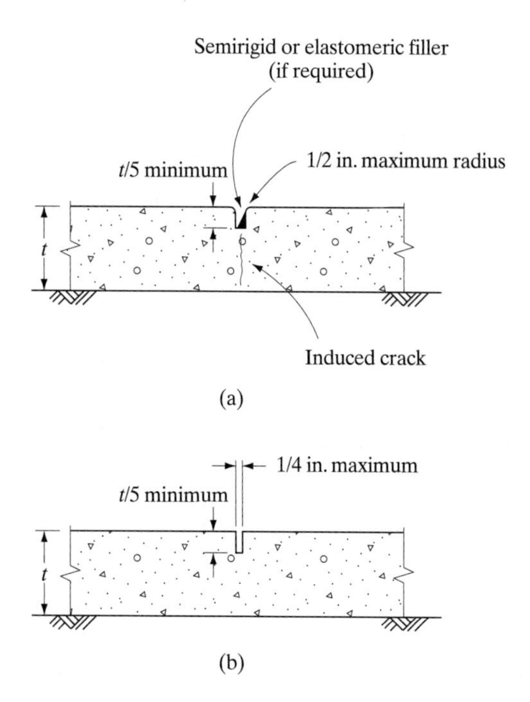

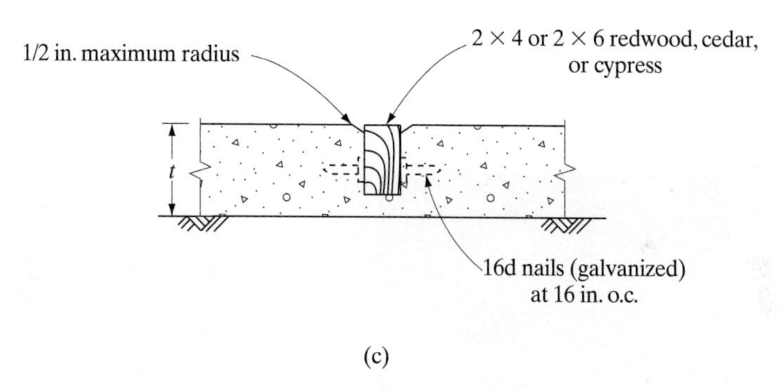

Figure 3.33 Control joints: (a) grooved joint; (b) sawed joint; (c) wood strips.

may be created to separate walls and columns—vertical elements—from the floor, as shown in Fig. 3.36, to allow free movement between various parts of a building. The joint placed at the end of a day's placement or at the beginning of the next day's concreting is the construction joint.

The joint or the filler material may be as small as 1/4 in. (6.3 mm) thick, and for thin slabs a butt-type construction joint is satisfactory. For thicker slabs, a tongue-and-groove joint or doweled butt joint is preferable. Sealing of the construction joints is necessary to keep pebbles, debris, water, snow and ice away from the joints

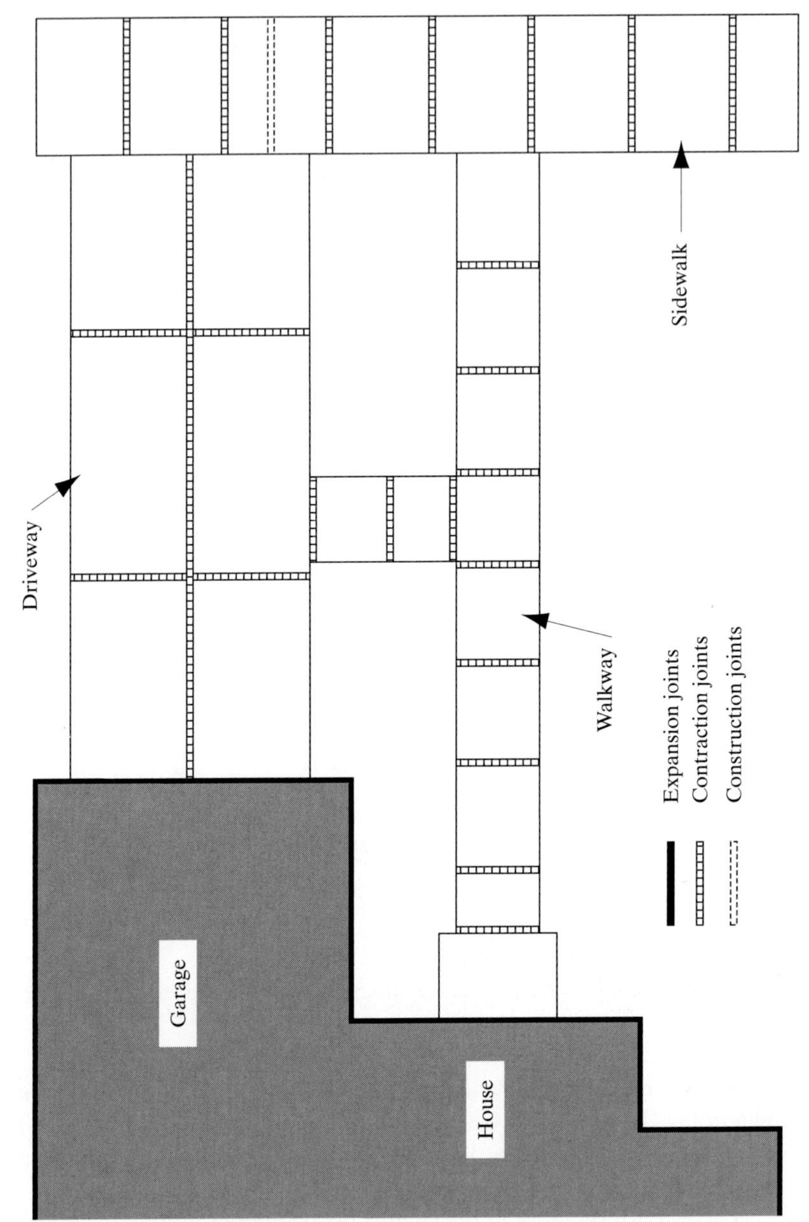

Figure 3.34 Typical locations of joints.

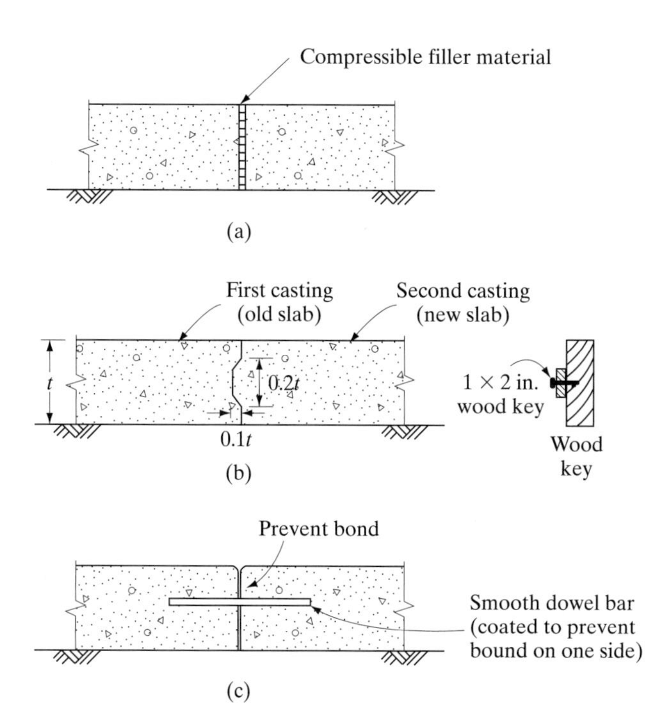

Figure 3.35 Construction joints: (a) compressible filler; (b) keyed; (c) dowel joint.

Figure 3.36 Isolation joint and control joint.

and to arrest the water from draining through the joint into the subgrade or soil. The entry of foreign materials may close a joint, leading to spalling and blowup when the slab expands. Similarly, the water draining through the joint can soften the ground, leading to loss of support. Elastomeric mastic compounds may be used for sealing.

Shrinkage-compensating Concrete. In situations where there is the potential for service problems from large shrinkage, *shrinkage-compensating concrete,* also called *expansive cement concrete,* can be used. Many bridge decks, water-retaining structures, slabs-on-grade, and pavements have been built using this special concrete. Water reservoirs, filtration plants, and sewage-treatment plants have been constructed using shrinkage-compensating concrete, with fewer control joints and a smaller number of waterstops. Using this concrete, buildings such as parking structures can be constructed with a more impermeable barrier against deicing chemicals than with conventional concrete. Slabs-on-grade can be built with control joints spaced at around 100 ft (30 m), compared to 10–15 ft (3–4.8 m) with ordinary concrete.

Shrinkage-compensating concrete is defined as expansive cement concrete that, when properly restrained by reinforcement, will expand by an amount equal to or slightly greater than the expected contraction from the drying shrinkage. In an ideal structure built with this concrete, a residual expansion will remain after the drying period, thereby eliminating cracking from shrinkage.

This concrete is manufactured using a special type of cement, called *Type K expansive cement,* which is an interground mixture of portland cement, anhydrous tetracalcium trialuminate sulfate (C_4A_3S), calcium sulfate, and lime. When this cement is mixed with water, along with other hydration products, a compound known as ettringite is formed, which causes expansion during the initial curing period. The reinforcement remaining in the concrete provides internal restraint to this expansion, causing tensile stresses in steel and compressive stresses in concrete. As the hydration proceeds, the shrinkage from drying of the concrete begins. The expansion produced from ettringite should nearly balance out the total shrinkage, so that slight residual expansion is left in the concrete (Fig. 3.37). In other words, the length of the concrete after shrinkage should be equal to or slightly greater than the original length.

To allow for the initial expansion, special attention should be given to the placement pattern of concrete. The sequence of placing should leave one edge of the freshly placed concrete member free to expand in each direction. Without sufficient space around it, the expansion potential of the member would not be realized. Dowel bars from adjacent slabs can prevent the expansion (as well as the shrinkage) parallel to the joint. The use of compressive material along the vertical face of the dowel (or a continuous horizontal strip of steel as the dowel) may solve this problem. Increase in the amount of reinforcement reduces the expansion as well as the shrinkage.

Type K cement is also used in cement-based grouts, to offset the volume changes resulting from shrinkage or to prevent settlement in the bedding of machinery, and in structural underpinning applications. This special cement is prone to aeration deterioration while in storage, reducing the expansiveness.

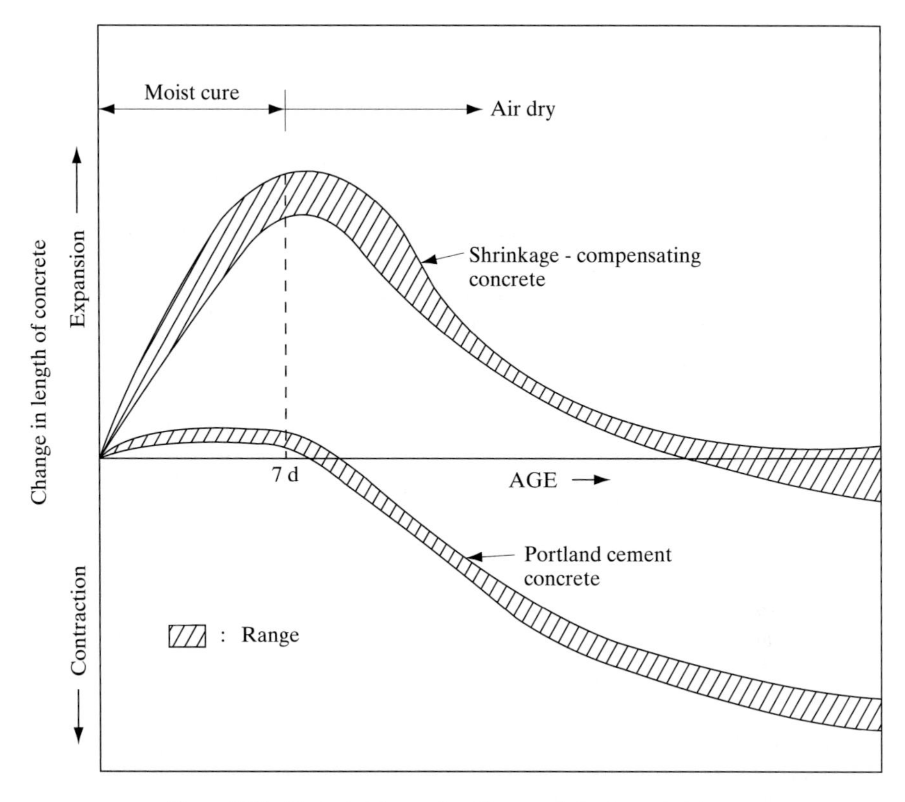

Figure 3.37 Typical deformation characteristics from shrinkage.

3.8.7 Creep

Another important property of hardened concrete is creep, which is the increase in strain (or deformation) with time. When subjected to an external load, a concrete member deforms elastically, and the resulting deformation is called the elastic or instantaneous deformation. But when the loading continues, or under sustained loads, this deformation increases with time, which is the creep component of the total deformation under the load. Thus, creep can be defined as the time-dependent increase in strain or deformation at a constant stress. This increase, which can be several times as large as the initial deformation, is on average about 0.1 percent per 1000 psi, or about three times the elastic strain.

A common meaning of "creep" is to sneak up on someone from behind, and in a way this meaning applies to the creep in concrete, for it is the component of deformation that "follows" behind the known elastic deformation.

Ordinarily, aggregates do not creep, and creep in concrete can be attributed to the characteristics of the cement paste. However, aggregate particles restrain the creep deformation, resulting in creep stresses. Concrete made with sandstone or gravel aggregates typically creeps more than that made with granite or limestone. The higher the modulus of elasticity of the aggregates, the greater the restraint offered to the creep. The higher the amount of aggregates, for a given w/c ratio, the lower the creep. The creep deformation of lightweight concrete is about the same as that of normal-weight concrete with similar mix proportions. For the most part, creep increases with increase in w/c ratio or volume of cement paste.

Creep occurs under all types of loading: compression, tension, and torsion. It appears rapidly following the application of loads, and roughly 50 percent of the one-year creep takes place within three to four weeks after loading. The rate of creep increases with ambient temperature. The earlier the age at which the loading is applied, the greater the creep. The creep is higher in a wet condition than in a dry condition.

Creep deformation is proportional to the stress level and is made up of two components:

- Reversible creep
- Irreversible creep

When a member, after being kept loaded for a length of time, is unloaded, a part of the total deformation (elastic plus creep) is immediately recovered, called *elastic recovery,* followed by a delayed recovery, termed *creep recovery.* Another part of the deformation remains permanent and cannot be recovered. The part of creep deformation that can be recovered (elastic recovery minus elastic strain) is the reversible creep, and that part which has become permanent is the irreversible creep.

Creep can also be thought of as a reduction in concrete stress under external loads. A concrete member in compression with its length held constant shows an increase in tensile stresses from shrinkage and a reduction in tensile stresses from creep. If a member carrying an external load is drying, creep and shrinkage deformations can be additive. The net increase in strain in such a member is due to shrinkage and creep strains. Creep and shrinkage, however, occur simultaneously, and separation of the two properties as independent components is only for computational simplicity.

3.8.8 Carbonation

Carbonation is the term for the reaction between the lime in concrete and the carbon dioxide from air, yielding calcium carbonate. This chemical reaction reduces concrete quality and its ability to protect reinforcement from corrosion (in an exposed environment), and results in additional shrinkage in the carbonated region. Resistance to carbonation can be improved by using high-quality dense concrete.

Carbonation is most active at around 50 percent relative humidity (RH), and it nearly disappears at RH below 25 percent or near saturation (for carbon dioxide cannot penetrate through pores filled with water). The reaction starts at the surface of concrete and slowly penetrates into the core; as a consequence, carbonation depends on the surface to mass ratio of the member. Poor quality concrete suffers carbonation earlier and deeper. Occasional wetting of a concrete surface prevents it from drying; for that reason, carbonation shrinkage is not a major source of deterioration in the ocean.

Carbonation causes no harmful effects but can contribute to significant weight gain and can cause a soft surface, dusting, and color change. But if the surface is exposed to salts and moisture, or in a salt-laden environment, carbonation may accelerate reinforcement corrosion. Carbonation is also a source of concrete shrinkage, called *carbonation shrinkage,* which is irreversible and can be as large as the drying shrinkage.

3.9 DURABILITY

A concrete structure will normally remain durable if

- the cement paste structure is dense and of low permeability;
- under exposed conditions, it has entrained air to resist freeze-thaw cycles;
- it is made with graded aggregates that are strong and inert; and
- the ingredients in the mix contain minimum impurities such as alkalis, chlorides, sulfates, and silt.

Durability is the capacity of concrete to resist deterioration from weathering (environment) and traffic. It normally refers to the duration or life span of trouble-free performance. An exposed concrete structure must be durable to resist freezing and thawing, heating and cooling, and action by chemicals such as deicers and fertilizers. Inadequate air content (entrained air) in slabs-on-grade, or outdoor flatwork exposed to moderate and severe climates, can lead to surface scaling, especially if deicers are used.

Permeability and ultimate compressive strength are two key factors affecting durability. Low strength and high permeability decrease durability. Well-graded mixtures perform better than poorly graded mixtures. Long-term strength (that is, strength beyond 28 days) is also important for superior durability. When fly ash is used with cement, the reaction between the silica in the fly ash and the lime in the hydrated cement paste produces compounds that fill the voids in concrete, contributing directly to lower permeability. Thus, fly ash and other mineral admixtures can be used to control deterioration from environmental factors and to improve durability. Low w/c ratio—less than 0.45—and continuous curing improve durability.

Good concrete can withstand general fluctuations in temperature and moisture without deterioration. However, under adverse exposure conditions, most concrete or reinforced concrete structures will eventually fail or crumble, largely from one or more of the following four causes:

- Alkali-aggregate reaction
- Sulfate attack
- Freeze-thaw cycles
- Reinforcement corrosion

The first three factors may occur individually or simultaneously, leading to expansion and development of cracks in concrete. Concrete is said to be durable if it can withstand the deteriorating effects of these and other factors.

3.9.1 Alkali-aggregate Reaction

In the presence of water, certain compounds in some natural aggregates react chemically with alkalis of portland cement released during hydration. This is followed by expansion or swelling of the aggregate particles, causing cracks and promoting disintegration of concrete, and this reaction is called alkali-aggregate reaction or alkali-silica reaction. The term *alkali* refers to the sodium and potassium hydroxides present in cement in relatively small proportions, expressed as sodium oxide equivalent (the sum of the percentages of Na_2O and 0.658 times the percentage of K_2O). The reaction involves the interaction of hydroxyl ions (associated with alkalis) in cement with certain siliceous constituents of the aggregates (reactive silica). The products of reaction, such as sodium silicate gel (also called *water glass*), form on the exterior surface of particles and can bring about excessive expansion, cracking, popouts, and general deterioration of concrete. (Popouts are conical craters left when a small portion of the concrete surface breaks away due to the internal pressure, typically generated by the permanent swelling within the concrete.) The aggregates that contain silica in an active state are hard-burned lime, hard-burned dolomite, siliceous limestone, chert, opal, and some other volcanic rocks.

The prerequisites for alkali-silica reaction are:

- Existence of alkali-reactive components in aggregate particles
- High alkali content in cement
- Presence of moisture

From this reaction, multidirectional cracking—called *map cracking*—takes place in horizontal surfaces, and horizontal cracking in walls and beams. Chemical analysis of reaction products—whitish amorphous deposits—shows them to be composed primarily of sodium and potassium silicates. Alkali-silica reaction does not occur if any one of the three factors listed above is not present.

The most direct solution to the problem of alkali-silica reaction is the avoidance of aggregates containing soluble silica. The use of low-alkali cement, whenever aggregates are suspected of containing reactive silica, can also prevent this reaction. Mineral admixtures such as fly ash or ground blast-furnace slag, which contains fair amounts of finely divided silica, can minimize the reaction.

3.9.2 Sulfate Attack

Sulfates in soil and seawater—primarily sodium sulfate, magnesium sulfate, and calcium sulfate—can react with free calcium hydroxide and aluminates in the cement gel to produce compounds that have volume greater than the initial volume. This process—called sulfate attack—causes expansion and cracking in concrete.

Ocean water contains, on the average, about 35 parts per thousand, or 3.5 percent, dissolved salts. The major cations are calcium, magnesium, sodium, and potassium, and the major anions are chlorides, CO_3, HCO_3, and SO_4. There are six principal elements present in solution that make up about 99 percent of the dissolved salts. These are chlorine (about 19,000 mg/l), sodium (about 10,600 mg/l), magnesium (about 1300 mg/l), sulfur (about 900 mg/l), calcium (about 400 mg/l), and potassium (about 380 mg/l). The effects of the chemical reaction of chlorides, sulfates, and alkalis of sodium and potassium should be considered when addressing the durability of concrete structures exposed to seawater. They may affect the long-term performance of cement and aggregates of concrete, and reinforcing steel.

Concrete that is totally submerged in water, even if the water contains dissolved salts, for the most part is in a protected state. Continuous immersion tends to reduce the potential for chemical reaction by removing changes in the degree of saturation, which is the mechanism for the flow of ions. But parts of concrete that are sometimes immersed in water (wetting) and at other times exposed to air (drying) undergo repeated volume changes. When these structures are situated where the temperatures dip below freezing, the partly exposed portions are prone to severe frost action.

The potentially aggressive constituents in seawater are sulfates, chlorides, carbonates, bicarbonates, alkali metal, and magnesium ions. Magnesium sulfate, rather than sodium or calcium sulfate, is the most common aggressive agent. The reaction between sulfates and calcium hydroxide produces gypsum, which reacts with calcium aluminate hydrate to from expansive compounds.

Sulfate attack can be minimized by using a cement that is low in tricalcium aluminate, like Type V and Type II cements. Fly ash and other mineral admixtures can react with lime—liberated during hydration—and reduce the amount of calcium aluminate hydrate; consequently, sulfate attack is reduced. Where high temperatures prevail, sulfate attack may not be a major problem. When sulfates are present along with chlorides, such as in seawater, deterioration from sulfate attack is insignificant, for the presence of chlorides inhibits expansion of concrete from sulfate solution.

3.9.3 Freeze-thaw Cycle

Freeze-thaw is the process by which water that is stored in voids within concrete expands from freezing temperatures. Consequently, the concrete cracks and deteriorates. If the top surface of concrete had water applied to it during floating or troweling, had bleed water worked into the surface during finishing operations, was floated too early or over-troweled, or is exposed to deicers, the top thin mortar layer may flake off from freeze-thaw effects, and the process of deterioration is called *scaling*. Natural mineral aggregates that are porous (such as cherts) are also susceptible to deterioration from freezing and thawing, in the form of popouts or cracking of the surfaces of concrete above the aggregate particles. Repeated freezing and thawing of aggregate particles may also produce cracks adjacent and parallel to the joints in ground-supported slabs (called D-cracking).

Entrained air and low w/c ratio provide good protection for concrete against expansion from freeze-thaw cycles. Entrained air guarantees empty spaces within which the compressed water from capillary pores can move and freeze. Resistance to freezing and thawing depends, in addition, on the w/c ratio, and permeability and degree of saturation of the paste. Concrete with a w/c ratio less than 0.4 has low permeability and probably does not require air entrainment. Partially saturated concrete does not suffer damage from freeze-thaw, for the necessary free space exists in it; in near-saturated concrete, however, air entrainment is necessary for protection.

3.9.4 Corrosion

Alkali-silica reaction, sulfate attack, and freeze-thaw deterioration apply to the durability of plain concrete. In concrete with reinforcing or prestressing steel, deterioration can also be due to corrosion of the reinforcement. Corrosion risk is higher for structures exposed to seawater or deicing salts. In the last decade, corrosion has emerged as the single most prevalent factor causing deterioration of reinforced and prestressed concrete structures all over the world.

Corrosion of steel is an electrochemical reaction involving four components of an electron cell: an anode, a cathode, an electrolyte, and a conductor. A portion of the bar becomes the anode, where ions go into solution and electrons are released, and the reinforcement functions as the conductor that permits the transfer of electrons to the cathode. Another portion of the bar turns into the cathode, where the electrons are consumed in the presence of moisture and oxygen, and the moisture present in the concrete is the electrolyte that permits the movement of ions. The electrochemical reaction involves two processes: oxidation at the anode, and reduction at the cathode, as shown in Fig. 3.38. Rust, which is made up of several iron compounds (such as $Fe(OH)_2$, Fe_2O_3, and FeO), gets formed and deposited around the anodic area. Due to different impurity levels in iron, different parts of the bar have dissimilar potential for the oxidation to occur, which can then set up a continuous galvanic cell. In the presence of moisture, oxygen, and chlorides, corrosion can proceed indefinitely, eventually causing the failure of the structure.

Corrosion affects the reinforcing steel by reducing its effective cross-sectional area, and also affects the concrete, for the corrosion products occupy larger volume than the original volume of steel, thereby generating enough pressure to cause cracking and spalling of concrete.

The alkaline environment of concrete provides sufficient protection against corrosion for reinforcement. The concrete cover should be thick enough for the alkalinity to be retained around the reinforcement. Also, the permeability of this cover should be fairly low to delay the onset of corrosion. Mix proportions, curing, reinforcement, and cracking affect the permeability. The type of cement also affects the corrosion resistance: cements with higher C_3A have been found to offer better resistance. Because of its dense impermeable microstructure, fly ash–blended

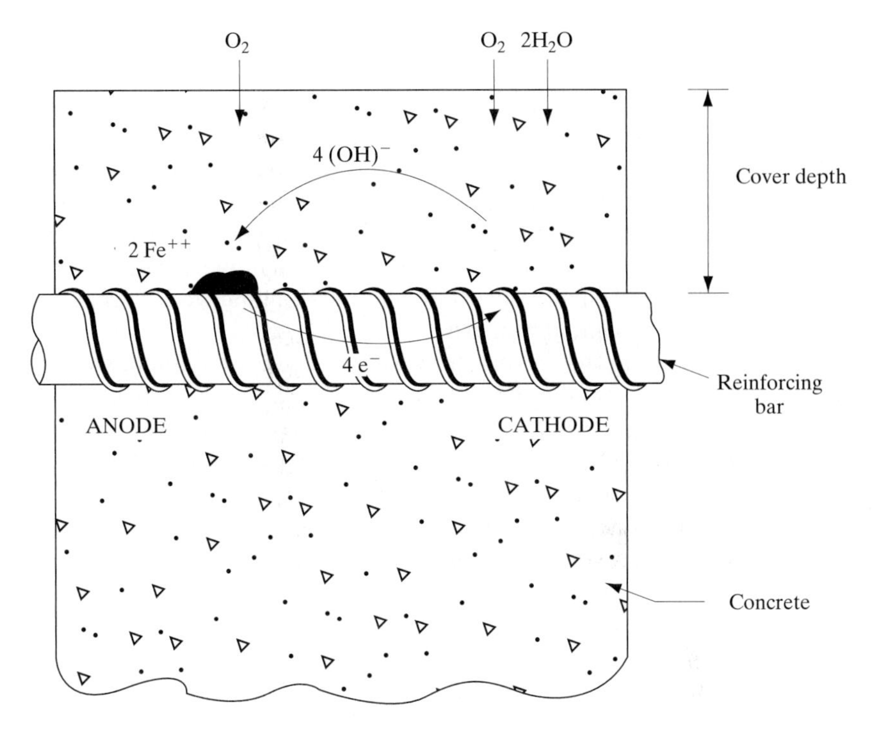

Figure 3.38 The corrosion process.

cement concrete is expected to perform better in exposed structures. Enhanced electrical resistivity of concrete also helps in reducing the potential for corrosion.

Corrosion comprises two distinct phases: corrosion initiation and corrosion propagation. The initiation phase defines the period from the placement of concrete to the point at which the steel is depassivated, marking the onset of corrosion. The propagation phase is the period during which the corrosion continues (along with cracking in the concrete). The initiation phase is much longer than the propagation phase, implying that once the corrosion sets in it goes on at a fast rate.

In normal noncarbonated concrete and when the chlorides are absent, steel reinforcement remains passive due to the high alkaline nature of the aqueous phase of concrete (pH greater than 12.5). The corrosion danger is nil in this state and until the passivity offered by the alkalinity is destroyed, either by carbonation or ingress of chlorides. With a minimum cover of 1–1.5 in. (25–38 mm) to the reinforcement, using quality concrete, carbonation is unlikely to endanger the passivity. Besides, CO_2 cannot penetrate pores that are filled with water or saturated. In concrete made with low w/c ratio, the depth of carbonation per year typically is less than 0.08 in. (2 mm) during the first year, and even less later on. With this low rate, carbonation will not be a serious contender for depassivation. On the other hand, chloride-related depassivation and subsequent corrosion is more widespread in concrete structures.

To protect reinforcement from chloride-induced corrosion, the depth of cover should be about 3 in. (75 mm), although tests have shown that even a 3/4-in. (19-mm) layer of dense, impervious concrete can provide excellent protection. Field observations on ferrocement structures have demonstrated that even a mere 1/8-in. (3-mm) cover of dense, rich mortar, with no voids around the reinforcement, can offer adequate protection. But adequate depth of cover of poor quality concrete will not be sufficient—even a 4-in. (10-cm) cover of poor quality, porous concrete will not stop corrosion. To obtain dense concrete, rich mixes with low w/c ratio should be used. Reinforcement congestion should be minimized, so that concrete can be placed and consolidated properly.

Parts of reinforced concrete structures that are continuously submerged under water or are outside the splash zone are generally free from corrosion. Prolonged immersion, even in ocean water, ensures a uniformity of environment with respect to both temperature and moisture, and reduces the potential for corrosive action.

3.10 MIX PROPORTIONING AND DESIGN

Mix proportioning refers to the method by which the most economical combination of materials for the desired quality—with respect to strength, durability, and workability or consistency—is established. It is normally the responsibility of the ready-mixed concrete producer.

Most concrete is mixed and delivered in revolving-drum truck mixers from ready-mixed or batch plants. The proportioning, batching, mixing, and delivery are all done by the concrete supplier. In order to obtain concrete of desired properties, one has to specify the required strength, exposure conditions, and the intended use. Before depositing the concrete, the ingredients should be mixed until the concrete is uniform in appearance and the materials are evenly distributed.

Concrete should be delivered and discharged within 90 min of mixing and before the drum has revolved 300 times. To ensure adequate time for consolidation and finishing, attention should be given to scheduling the concrete delivery and to providing easy access for the truck to the construction site. Residential wall footings and spread footings need to be reinforced only to support unusual loads or where unstable soil is encountered. Footings that cross over pipe trenches should be reinforced with at least two #5 bars that extend past the sides of the trench by about 1.5 times the trench width.

Slump tests are commonly carried out at the construction site to assess the consistency or workability of fresh concrete. The slump of concrete delivered through a pump frequently exceeds 5 in. (12.5 cm). A variable slump is likely to point to variation in the compressive strength. Concrete of insufficient slump may lead to poor consolidation and appearance. Unfortunately, slump is not subject to precise production or measurement. Therefore, it is more practical to specify a proper working limit on slump, with a reasonable tolerance of around 2 in. (5 cm). Temperature, humidity, and delay in placement can cause changes in slump.

3.10.1 Mix Design

The objective of concrete mix design is to determine the proportion of ingredients to produce concrete that is workable and durable, of required strength, and of minimum cost. The three parameters that form the basis of mix design are:

- Workability (consistency)
- Strength and durability
- Economy

The cost of concrete depends on the costs of materials, transportation, and labor for placing and finishing. In addition, the costs of inspection and future repair also contribute to the overall expense. Cement is the costliest ingredient in the mixture, and the lower the cement content, the lower the cost. Transportation of fine and coarse aggregates may also add to the price of concrete. The use of locally available stone, gravel, or sand can generally decrease the expenditure. When these materials do not satisfy the quality specifications, aggregates must be imported, adding to the overall expense.

Durability and strength depend on a number of factors; overall, the strength and watertightness improve with cement content, density, and curing time, and decrease with quantity of water. Air-entrained concrete offers improved resistance to deterioration from freeze-thaw cycles compared to normal concrete. Concrete structures that are exposed to freezing and thawing in a moist environment, or those need protection against corrosion of reinforcement, should be made with a minimum w/c ratio, preferably less than 0.4.

Generally, the selection of ingredients is based on the following principles:

- The mix should be workable
- As little cement as possible should be used
- As little water as possible should be used
- Coarse and fine aggregate should be proportioned to achieve a dense mix
- The nominal maximum size of aggregate should be as large as possible
- The water-to-cement ratio will determine the compressive strength

In most structures, the mix design calculations are determined by the required compressive strength of the concrete (Table 3.6). How ever, when concrete strength is not specified (as in residential foundations, pavements, walkways, and so on), the proportioning is based on an arbitrary selection of materials—primarily the amount of cement. The ingredients are chosen from experience and knowledge of general specifications. In most common forms of construction, only the cement quantity is varied between projects. The amount of water is controlled using workability or consistency requirements.

When the compressive strength is specified, the proportioning is generally done using a mix design procedure that is based on workability requirements, void content, and the relationship between strength and w/c ratio. The initial mix, called

TABLE 3.6 GUIDELINES FOR SELECTING CONCRETE COMPRESSIVE STRENGTH IN RESIDENTIAL CONSTRUCTION (MINIMUM AT 28 DAYS). ADAPTED FROM ACI COMMITTEE 332, *CONCRETE INTERNATIONAL,* SEPT. 1984, 13–20

	Compressive strength psi (MPa)			
		Exposure conditions		
Type of construction	Mild	Moderate	Severe	Slump in. (mm)
Basement walls and foundations not exposed to weather	2500 (17)	2500 (17)	2500 (17)	6 ± 1 (150 ± 25)
Basement slabs and interior slab-on-grade	2500 (17)	2500 (17)	2500 (17)	5 ± 1 (125 ± 25)
Basement walls, foundations, exterior walls, and others exposed to weather	2500 (17)	3000 (21)	3000 (21)	6 ± 1 (150 ± 25)
Driveways, curbs, walks, patios, porches, steps and stairs, and unheated garage floors exposed to weather	2500 (17)	3000 (21)	3500 (24)	5 ± 1 (125 ± 25)

Mild exposure: weathering index less than 100

Moderate exposure: weathering index between 100 and 500

Severe exposure: weathering index more than 500

(The weathering index is the product of the average annual number of freezing-cycle days and the average winter rainfall in inches)

a *trial mix* or *trial batch,* is then tested in a laboratory to determine its compressive strength. If the average strength determined from the test samples does not match the requirements relating to the expected or required strength, the mix proportions are adjusted accordingly. Different combinations of ingredients produce concretes that vary considerably in strength level attained at a given w/c ratio.

The mix proportions thus obtained using the trial batch, called the *dry mix proportions,* should be adjusted for the moisture content and absorption of aggregates. Wet sand may contain as much as 5–10 percent moisture, and moist sand about 3 percent. Most gravel and crushed stone absorb up to 2 percent moisture. The absorption capacity of sand on an average is 2–4 percent. Porous sandstone and lightweight aggregates may absorb a lot of moisture—about 7 percent or higher. The mix proportions adjusted to compensate for the moisture content and absorption of aggregates are called the *field mix.*

Required Strength. The ACI 318 recommends that the concrete mix proportions be based on field experience or laboratory trial batches. To meet the specified strength requirements, the proportion of ingredients should be such that the average compressive strength value, obtained from field tests, exceeds the specified compressive strength (or design compressive strength), f'_c, by an amount sufficiently high to make the probability of low tests small. In other words, the mix proportioning should aim at a strength that is much higher than the specified strength. For example, if the specified strength of concrete for a retaining wall is 3000 psi, the mix proportioning should be for a much higher strength, say 4200 psi, just to be on the safe side.

When field test results are used for mix proportioning, the *required average strength*, f'_{cr}, taken as the basis for the selection of the proportion of ingredients, is taken as the larger value calculated from the following equation:

$$f'_{cr} = f'_c + 1.34S$$

$$f'_{cr} = f'_c + 2.33S - 500$$

where S = sample standard deviation in psi.

When the mix proportioning is based on laboratory trial batches, the required average strength, f'_{cr}, is obtained from the code recommendations.

3.10.2 Mix Design Procedure

The trial mix design procedure is formulated to achieve a mix that will satisfy the strength, durability, and workability requirements of concrete as closely as possible. The procedure can be summarized as follows:

1. Workability (slump) is determined for the type of work
2. The maximum aggregate size is chosen based on the requirements of the job
3. Air content is determined from durability requirements
4. The water-to-cement ratio is selected to satisfy strength and durability
5. The amounts of water and coarse aggregate are chosen based on average workability.

Concrete used in all types of construction contains some entrapped air, usually between 0.5–2 percent. These air bubbles, which are relatively large and not distributed evenly within the concrete, do not improve workability and in fact lower the durability. On the other hand, an air-entraining admixture added to fresh concrete forms very tiny air bubbles that are distributed evenly within the volume of the concrete. In addition to improving the workability of fresh concrete, air-entrainment increases the durability of hardened concrete. Thus, any concrete that is exposed to freezing temperatures should be made with an air-entraining admixture, and the amount of air depends on the expected exposure conditions.

A stronger concrete is not always a better concrete. An excessively high strength achieved through unnecessary extra cement can increase cracking and internal stresses. High compressive strength is not usually required in most concrete structures, especially those that carry flexural stresses. The addition of some admixtures, such as pozzolans, may reduce the internal temperature of concrete because of comparatively little heat generation. Excessive amounts of fine aggregate increase the requirement for mixing water and also the drying shrinkage, and decrease the compressive strength.

The w/c ratio is normally selected based on the required compressive strength. However, French scientist R. Feret concluded in 1892 that the strength of concrete depends on the ratio of absolute cement volume to the space available to it. The higher the cement content, in a given volume, the greater the strength. An increase in the nominal maximum size of coarse aggregate lowers the surface area, reducing the water requirement. In other words, as the aggregate size increases, the amount of water required to achieving a certain level of workability decreases. But a smaller surface area means less bond area and, accordingly, a lower strength. The net effect on the strength from using a larger aggregate depends on the mix proportions, cement content, and water content, as shown in Fig. 3.39. For low-cement mixes, a larger aggregate may show an increase in strength, but for higher-cement mixes, the

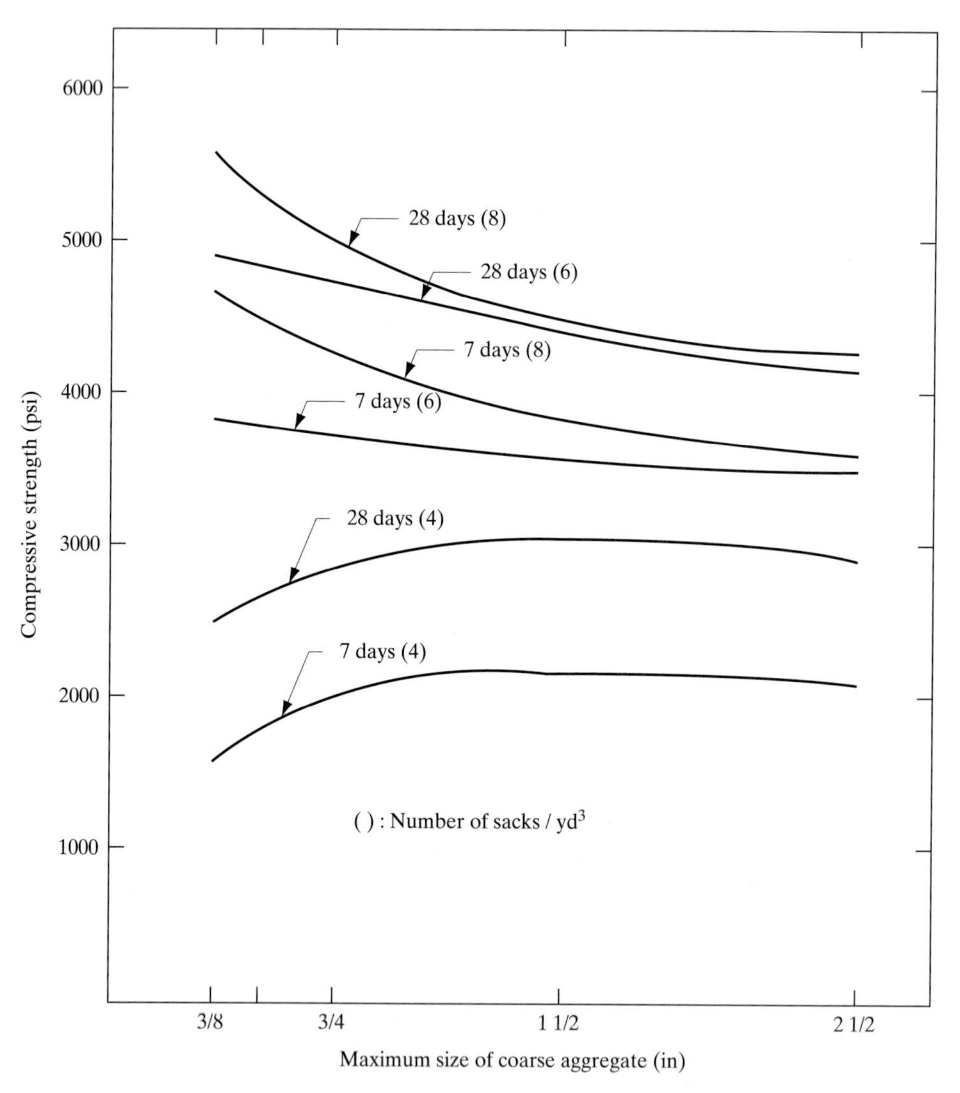

Figure 3.39 Effect of size of coarse aggregate on compressive strength of concrete.

strength reduces. An increase in the maximum size of coarse aggregate per se lowers the compressive strength. In general, the smallest size aggregate produces the highest strength for a given w/c ratio. Angular aggregates demand more water than do rounded aggregates. The workability of concrete is also affected by the ambient temperature; as the temperature increases, the water required to obtain a desired slump increases. About 2 lb of water is required (per cubic yard of concrete) for every 3 °F increase in temperature. Curing is essential for strength development.

Step-by-step Procedure. The mix design procedure is given in the following. Note that the process results in a trial mix, which may need appropriate modifications, depending on the test results.

Step 1: Select the slump.

If slump is not specified, a value appropriate for the work can be selected from Table 3.5 (page 103).

Step 2: Select the maximum aggregate size.

Choose the maximum possible size. The larger the maximum size of well-graded aggregate, the smaller the total void space. Such a mix requires less mortar per unit volume of concrete. Generally,

$$\text{maximum size} \leq \frac{\text{narrowest dimension}}{5}$$

Also,

$$\text{maximum size} \leq \frac{\text{depth of slab}}{3}$$

$$\leq 0.75 \text{ (clear spacing between individual bars or wires)}$$

Step 3: Estimate the mixing water and air content.

Table 3.7 provides an estimate of the mixing water for a given slump and type of concrete. Air-entrained concrete should be used when concrete is exposed and is required to possess adequate resistance to stresses resulting from temperature fluctuations. The amount of water required (for a given slump) decreases with increase in the maximum size of aggregate. The table also shows an approximate amount of air that can be expected in a well-compacted concrete.

Note that Table 3.8 provides the recommended air content for normal- and lightweight concrete in moderate and severe exposure conditions. A "moderate exposure" is one where, in a cold climate, the concrete will be exposed only occasionally to moisture prior to freezing, and where no deicing salts are used (for example, exterior walls, beams, girders, and slabs not in direct contact with soil). A "severe exposure" is one where, in a cold climate, the concrete may be in continuous contact with moisture prior to freezing, or where deicing salts are used (for example, bridge decks, pavements, sidewalks, parking garages, water tanks).

TABLE 3.7 APPROXIMATE MIXING WATER AND AIR CONTENT FOR TRIAL BATCHES

Range of slump[a] (in.)	Approximate mixing water (lb/yd^3 of concrete) for nominal maximum size of CA (in.)							
	3/8	1/2	3/4	1	$1\frac{1}{2}$	2	3	6
			Non–air-entrained concrete					
1–2	350	335	315	300	275	260	240	210
3–4	385	365	340	325	300	285	265	230
6–7	410	385	360	340	315	300	285	—
Approximate air content (%)	3	2.5	2	1.5	1	0.5	0.3	0.2
			Air-entrained concrete					
1–2	305	295	280	270	250	240	225	200
3–4	340	325	305	295	275	265	250	220
6–7	365	345	325	310	290	280	270	—
Recommended air content (%)[b]	4.5–7.5	4.0–7.0	3.5–6.0	3.0–6.0	2.5–5.5	2.0–5.0	1.5–4.5	1.0–4.0

Source: ACI 211.1-91, Table 6.3.3. Reprinted with permission.

[a]Refer to Table 3.5 [b]Refer to Table 3.8

TABLE 3.8 TOTAL AIR CONTENT FOR FROST-RESISTANT CONCRETE

Normal maximum coarse aggregate size (in.)	Air content (%)	
	Severe exposure	Moderate exposure
3/8	$7\frac{1}{2}$	6
1/2	7	$5\frac{1}{2}$
3/4	6	5
1	6	$4\frac{1}{2}$
$1\frac{1}{2}$	$5\frac{1}{2}$	$4\frac{1}{2}$
2	5	4
3	$4\frac{1}{2}$	$3\frac{1}{2}$

Source: ACI 318-89, Table 4.2.1, revised 1992. Reprinted with permission.

Example 3.3

$$\text{Slump} = 2 \text{ in.}$$

$$\text{Type} = \text{non–air-entrained}$$

$$\text{Maximum size} = 3/4 \text{ in.}$$

From Table 3.7:

$$\text{Mixing water} = 315 \text{ lb/yd}^3 \text{ of concrete}$$

Step 4: Select the water-to-cement ratio (w/c).

The w/c ratio (by weight) is selected by strength and durability requirements. When using Type I cement, approximate and conservative values can be estimated from Table 3.9. Note that this table shows the maximum permissible w/c ratios for *specified compressive strengths.* But the *required average strength* should be higher than the specified strength by the margin shown in Table 3.10 and Fig. 3.40.

TABLE 3.9 MAXIMUM PERMISSIBLE WATER/CEMENTITIOUS MATERIALS RATIOS BY WEIGHT FOR TRIAL BATCHES

Specified compressive strength f_c' (psi)	Absolute water/cementitious materials ratio by weight	
	Non–air-entrained concrete	Air-entrained concrete
2500	0.67	0.54
3000	0.58	0.46
3500	0.51	0.40
4000	0.44	0.35
4500	0.38	—
—	—	—

Source: ACI 318-89; Table 5.4, revised 1992. Reprinted with permission.

TABLE 3.10 REQUIRED AVERAGE COMPRESSIVE STRENGTH FOR TRIAL BATCHES

Specified compressive strength, f_c' (psi)	Required average compressive strength f_c' (psi)
<3000	$f_c' + 1000$
3000–5000	$f_c' + 1200$
>5000	$f_c' + 1400$

Source: ACI 318-89, Table 5.3.2.2, revised 1992. Reprinted with permission.

Example 3.4

The specified compressive strength of concrete or the design strength is 3000 psi. Find the w/c ratio.

Average strength of cylinder tested must be

$\geq 3000 + 1200 = 4200$ psi (Table 3.10)

Maximum w/c ratio (Table 3.9)

$= 0.58$ (normal concrete)

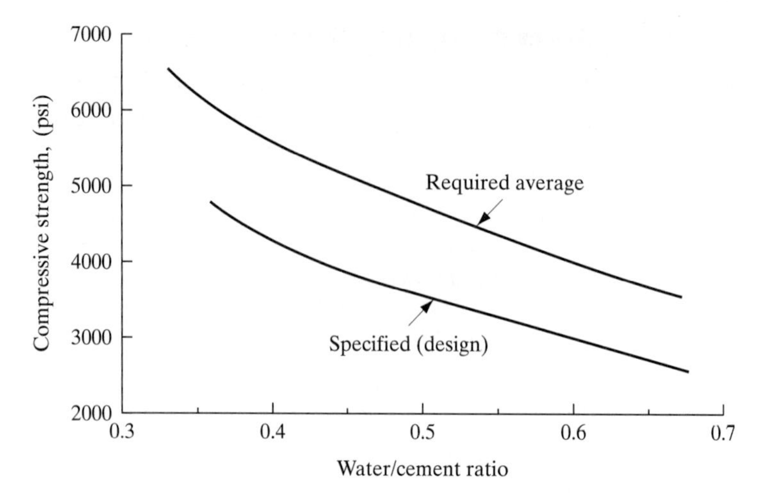

Figure 3.40 Specified and required average strengths.

This means that if the mix is made using a w/c of 0.58, the average strength expected is ≥4200 psi. When durability controls, the w/c ratio must be selected from Table 3.11.

TABLE 3.11 REQUIREMENTS FOR SPECIAL EXPOSURE CONDITIONS

Exposure condition	Maximum water/cementitious materials ratio, normal-weight aggregate concrete	Minimum f'_c, normal-weight and lightweight aggregate concrete (psi)
Concrete to have low permeability when exposed to water	0.50	4000
Concrete exposed to freezing and thawing in a moist condition	0.45	4500
For protection against corrosion of reinforcement, when concrete is exposed to deicing salts, brackish water, or seawater	0.40	5000

Source: ACI 318-89, Table 4.2.2, revised 1992. Reprinted with permission.

Example 3.5

For the concrete for the foundation of a bridge pier, the required average compressive strength is $f'_c = 4000$ psi. Find the w/c ratio.

The concrete is exposed to freezing and thawing in a moist condition. Maximum w/c ratio = 0.45 (Table 3.11). From Table 3.9, for air-entrained concrete, w/c = 0.35, which is less than 0.45. Use a w/c ratio of 0.35.

Step 5: Calculate the cement content.

$$\text{Required cement content, by weight} = \frac{\text{estimated mixing water}}{\text{w/c}}$$

Step 6: Estimate the coarse aggregate content.

Table 3.12 gives the volume of coarse aggregate (dry-rodded condition, per unit volume of concrete) to produce concrete of suitable workability. The volume is dependent on the fineness modulus of sand and the maximum size of coarse aggregate.

TABLE 3.12 VOLUME OF COARSE AGGREGATE PER UNIT VOLUME OF CONCRETE BASED ON WORKABILITY REQUIREMENTS

Nominal maximum coarse aggregate size (in.)	Volume of dry-rodded coarse aggregate per unit volume of concrete for fineness modulus of sand			
	2.4	2.6	2.8	3.0
3/8	0.5	0.48	0.46	0.44
1/2	0.59	0.57	0.55	0.53
3/4	0.66	0.64	0.62	0.60
1	0.71	0.69	0.67	0.65
$1\frac{1}{2}$	0.75	0.73	0.71	0.69
2	0.78	0.76	0.74	0.72
3	0.82	0.80	0.78	0.76
6	0.87	0.85	0.83	0.81

Source: ACI 211.1-91, Table 6.3.6. Reprinted with permission.

Example 3.6

Find the required volume of coarse aggregate to produce a workable mix when

$$\text{Fineness modulus} = 2.6$$
$$\text{Maximum size of aggregate} = 3/4 \text{ in.}$$

From Table 3.12, the required volume of coarse aggregate per unit volume of concrete = 0.64. For 1 yd^3 of concrete,

$$\text{Dry-rodded volume of coarse aggregate} = 0.64\,(27) = 17.28 \text{ ft}^3$$

$$\text{Weight of coarse aggregate} = (\text{dry-rodded wt}) \times (\text{dry-rodded density})$$

For a more workable mix, decrease the amount of coarse aggregate up to 10%.

Step 7: Estimate the fine aggregate content.

All ingredients have been estimated at the end of step 6, except the fine aggregate. The fine aggregate content is determined by calculating the difference between the total volume of concrete and the volume of the remaining ingredients.

Volume of fine aggregate = volume of concrete − (volumes of water, cement, coarse aggregate, and air)

$$V_{\text{fine aggregate}} = V_{\text{conc}} - (V_{\text{cem}} + V_{\text{CA}} + V_{\text{wat}} + V_{\text{air}})$$

$$V_{FA} = V_{\text{conc}} - (V_c + V_{CA} + V_w + V_a)$$

Volumes of ingredients are calculated by knowing their weights and specific gravities.

Example 3.7

The amounts of cement, coarse aggregate, and water are as given. The air content is 2 percent. The dry-rodded weight of coarse aggregate is 98 pcf. Find the amount of sand.

$$\text{Water} = 340 \text{ lb/yd}^3 \text{ of concrete}$$

$$\text{Cement} = 680 \text{ lb/yd}^3 \text{ of concrete}$$

$$\text{Volume of coarse aggregate} = 0.69/\text{unit volume of concrete}$$

Specific gravities of cement, sand, and coarse aggregate are 3.15, 2.65, and 2.68, respectively.

$$V_{FA} = 27 - \left[\frac{680}{3.15(62.4)} + \frac{0.69(27)98}{2.68(62.4)} + \frac{340}{62.4}\right] - \frac{2(27)}{100}$$

$$= 27 - 20.37 = 6.63 \text{ ft}^3$$

$$\text{Weight of fine aggregate} = 6.63(2.65)62.4$$

$$= 1096 \text{ lb}$$

Step 8: Find the field mix (based on moisture in fine aggregate and coarse aggregate).

Adjust the water content and the weights of fine and coarse aggregates depending on the moisture contents in the aggregates. Note that calculations in steps 1 to 7 are based on the assumption that aggregates do not release moisture to or absorb moisture from the mix. In practice, they are expected to hold some moisture, and the weights previously calculated must be increased to compensate for the moisture that is absorbed in and contained on the surface of the particles. In addition, the mixing water amount should be decreased by the amount of free moisture in the aggregates. Free moisture is the difference between the moisture content and the absorption.

Similarly, if the aggregates are dry, they will absorb moisture from the mixing water, and thus the mixing water amount should be increased to compensate for the water absorbed by the aggregates.

Example 3.8

Find the adjusted amounts for the data shown.

	Dry weight (lb/yd^3)	Moisture content (%)	Absorption (%)
Cement	580	—	—
Water	320	—	—
Coarse aggregate	1940	2	1
Fine aggregate	1110	6	1.5

$$\text{Required weight of coarse aggregate} = 1.02\,(1940) = 1979\text{ lb}$$

$$\text{Required weight of fine aggregate} = 1.06\,(1110) = 1177\text{ lb}$$

$$\text{Required weight of water} = 320 - \frac{(2-1)1940}{100} - \frac{(6-1.5)1110}{100}$$

$$= 320 - 19.4 - 50$$

$$= 250.6\text{ lb}$$

Step 9: Calculate the field mix proportions.

Find the proportions between cement, fine aggregate, coarse aggregate, and water by taking the weight of cement as one unit.

Cement : fine aggregate : coarse aggregate : water = C:FA:CA:W, by weight

Example 3.9

Use the data in Example 3.8 to calculate the field proportions.

$$\text{Field mix proportions} = 580\text{C}:1177\text{FA}:1979\text{CA}:250.6\text{W}$$

$$= 1\text{C}:2.03\text{FA}:3.41\text{CA}:0.43\text{W}$$

Step 10: Calculate the weights of individual ingredients required to make the desired amount of concrete.

The weight of each ingredient is calculated by equating the total volume of individual ingredients to the total volume of concrete.

Example 3.10

Find the weights using the following data.

Mix proportion = 1:2:3.3:0.5

Air content = 3%

Specific gravity of cement, fine aggregate, and coarse aggregate = 3.15, 2.65, and 2.7, respectively

Required volume of concrete = 2.2 ft^3

Let weight of cement required = W_c lb.

$$\text{Volume of concrete} = 2.2 = \frac{W_c}{62.4}\left[\frac{1}{3.15} + \frac{2}{2.65} + \frac{3.3}{2.7} + \frac{0.5}{1}\right] + \frac{3(2.2)}{100}$$

$$= \frac{W_c}{62.4}(2.794) + 0.066$$

$$W_c = 48 \text{ lb}$$

$$W_{FA} = 48(2) = 96 \text{ lb}$$

$$W_{CA} = 48(3.3) = 158 \text{ lb}$$

$$W_w = 48(0.5) = 24 \text{ lb}$$

3.10.3 Examples of Mix Design

Example 3.11

Determine the field mix proportions using the following data:

Concrete is used for foundation—mild exposure
Specified compressive strength of concrete = 3000 psi
Cement: Type I
Maximum size of coarse aggregate = $1\frac{1}{2}$ in.
Specific gravity of coarse aggregate = 2.69
Absorption capacity of coarse aggregate = 0.8%
Moisture content of coarse aggregate = 2%
Dry-rodded weight of coarse aggregate = 105 pcf
Specific gravity of fine aggregate = 2.65

Absorption capacity of fine aggregate = 6%
Moisture content of fine aggregate = 8%
Fineness modulus of fine aggregate = 2.6

Solution

Step 1: Select the slump. From Table 3.5, the range of slump for concrete in foundation or footing is 1–3 in. Use a slump of 3 in.

Step 2: Select the maximum aggregate size. Specified (in the data) maximum size is $1\frac{1}{2}$ in.

Step 3: Estimate the mixing water and air content. From Table 3.8, the range of air content is $4\frac{1}{2}$–$5\frac{1}{2}$ percent. Use air-entrained concrete. Assume 5 percent air. From Table 3.7, the water requirement is 275 lb/yd^3.

Step 4: Select the w/c ratio. From Table 3.9, for air-entrained concrete of specified compressive strength 3000 psi, w/c ratio = 0.46.

Step 5: Calculate the cement content. From step 4,

$$\frac{\text{weight of water}}{\text{weight of cement}} = 0.46$$

Thus, for 1 yd^3 of concrete,

$$\text{cement required} = \frac{275}{0.46} = 598 \text{ lb}$$

Step 6: Estimate coarse aggregate contents. From Table 3.12, for FM = 2.6,

$$\begin{aligned}\text{volume of CA} &= 0.73 \text{ yd/yd}^3 \text{ of concrete}\\ &= 0.73(27)105 \text{ lb/yd}^3\\ &= 2070 \text{ lb/yd}^3\end{aligned}$$

Step 7: Estimate the fine aggregate content.

$$\text{Volume of water} = \frac{275}{62.4} = 4.41 \text{ ft}^3$$

$$\text{Volume of cement} = \frac{598}{3.15(62.4)} = 3.04 \text{ ft}^3$$

$$\text{Volume of CA} = \frac{2070}{2.69(62.4)} = 12.33 \text{ ft}^3$$

$$\text{Volume of air} = \frac{5(27)}{100} = 1.35 \text{ ft}^3$$

$$\begin{aligned}\text{Total solid volume except FA} &= 4.41 + 3.04 + 12.33 + 1.35\\ &= 21.13 \text{ ft}^3\\ \text{Sand volume required} &= 27 - 21.13 = 5.87 \text{ ft}^3\\ \text{Weight of fine aggregate} &= 5.87(2.65)62.4\\ &= 971 \text{ lb/yd}^3 \text{ of concrete}\end{aligned}$$

Dry weight of materials for 1 yd^3 of concrete:

$$\text{Cement} = 598 \text{ lb}$$
$$\text{Fine aggregate} = 971 \text{ lb}$$
$$\text{Coarse aggregate} = 2070 \text{ lb}$$
$$\text{Water} = 275 \text{ lb}$$

Step 8: Find the field mix.

$$\text{Moisture content of CA} = 2\%$$
$$\text{Required weight of CA (moist condition)} = 1.02(2070) = 2111 \text{ lb}$$
$$\text{Moisture content of FA} = 8\%$$
$$\text{Required weight of FA (moist condition)} = 1.08(971) = 1049 \text{ lb}$$
$$\text{Absorption capacity of CA} = 0.8\%$$
$$\text{Free moisture in CA} = 2 - 0.8 = 1.2\%$$
$$\text{Absorption capacity of FA} = 6\%$$
$$\text{Free moisture in FA} = 8 - 6 = 2\%$$
$$\text{Net reduction in water} = \frac{1.2(2070)}{100} + \frac{2(971)}{100} = 44.3 \text{ lb}$$
$$\text{Net mixing water required} = 275 - 44.3 = 230.7 \text{ lb}$$

Step 9: Calculate the field mix proportions. Field mix quantities:

$$\text{Cement} = 598 \text{ lb}$$
$$\text{FA} = 1049 \text{ lb}$$
$$\text{CA} = 2111 \text{ lb}$$
$$\text{Water} = 230.7 \text{ lb}$$

Field mix proportions:

$$\text{cement: FA: CA: water} = 1\text{C}:1.75\text{FA}:3.53\text{CA}:0.39\text{W}$$

Example 3.12

Find the weight of all ingredients for 1.8 ft^3 of concrete using the field mix proportions from Example 3.11. Specific gravities of FA and CA are 2.6 and 2.7, respectively. The air content is 5 percent.

Solution From Example 3.11 field mix proportions:

$$\text{cement: FA: CA: water} = 1:1.75:3.53:0.39$$

Volume of concrete, V_{conc}, is 1.8 ft^3.

$$V_{\text{conc}} = \frac{W_c}{62.4}\left(\frac{1}{3.15} + \frac{1.75}{2.6} + \frac{3.53}{2.7} + \frac{0.39}{1}\right) + 0.05(1.8)$$

$$= 1.8 \text{ ft}^3$$

Therefore,

$$\text{Weight of cement, } W_c = 39.7 \text{ lb}$$

$$\text{Weight of FA} = 1.75\,(39.7) = 69.5 \text{ lb}$$

$$\text{Weight of CA} = 3.53\,(39.7) = 140.1 \text{ lb}$$

$$\text{Weight of water} = 0.39\,(39.7) = 15.5 \text{ lb}$$

3.11 ADMIXTURES

An admixture is a material other than water, aggregates, cement, and fiber, added to plastic concrete or mortar to change one or more properties at the fresh or hardened stages. Admixtures are introduced before, during, or (in some cases) after the mixing of the major ingredients. A number of different types of admixtures are available in the market to perform functions such as increasing the plasticity, accelerating the setting, improving the strength development, and reducing the heat of hydration.

Some admixtures are added to modify the workability characteristics and the setting (and hydration) rate of fresh concrete, whereas others change the properties of concrete both at the mixing and hardened stages. Normally, the use of one or more admixtures should be considered only when the desired modifications of properties could not be achieved through changes in the composition of the mix or proportioning of the primary ingredients.

Admixtures are generally divided into two groups:

- Chemical admixtures
- Mineral admixtures

Natural pozzolanic materials and some industrial byproducts (such as fly ash and blast-furnace slag) are mineral admixtures. On the other hand, any number of chemicals (chemical admixtures) are available in the market that can be added in small quantities to the mixture to develop special properties in fresh or hardened concrete. The effectiveness of an admixture in modifying the properties of concrete depends on many factors, such as mix proportions; type and size of aggregates; ambient temperature; type, brand, and amount of admixture; and type of cement.

3.11.1 Chemical Admixtures

A number of chemical admixtures, sold under different product names, have been employed successfully to adjust the properties of concrete. The ASTM identifies various types of chemical admixtures by their intended uses, as listed in Table 3.13. Brief descriptions of the most common chemical admixtures are given below.

TABLE 3.13 CHEMICAL ADMIXTURE

Type	Description	ASTM standards	Applications
A	Water-reducing admixtures	C494	To get dense concrete, to improve workability
B	Retarding admixtures	C494	To delay setting and hardening; hot-weather concreting; large structures
C	Accelerating admixtures	C494	To accelerate setting and early strength development; cold-weather concreting
D	Water-reducing and retarding admixtures	C494	Similar to those for types A and B
E	Water-reducing and accelerating admixtures	C494	Similar to those for types A and C
F	Water-reducing, high-range admixtures	C494	In high-strength concrete; to improve watertightness and workability
G	Water-reducing, high-range, and retarding admixtures	C494	Similar to those for types B and F
	Air-entraining admixtures	C260	To improve durability and workability
	Antifreeze admixtures		Cold-weather concreting; to minimize freezing of water in fresh concrete

Accelerating Admixtures. Hydration of concrete is affected by the ambient temperature. The setting rate and the early strength development are greatly minimized at temperatures below 40 °F (5 °C). Accelerating admixtures (ASTM Type C) are added to fresh concrete or mortar to accelerate the setting and early strength development, particularly in cold weather applications. They are also used to lessen the curing time and permit early removal of forms, especially in shotcreting. Accelerating admixtures speed up the early strength gain but do not prevent concrete from freezing. Calcium chloride ($CaCl_2$), the most widely used accelerator, can be used to expedite concreting operations and to permit the early removal of forms and opening of a project for service. A variety of soluble salts (such as chlorides, bromides, fluorides, carbonates, and silicates) are also used as accelerators. Calcium nitrite, which is a corrosion inhibitor, is also employed as an accelerator.

At the end of three days, the compressive strength of concrete made with an accelerating admixture is at least 25 percent higher than that of concrete made without an admixture. However, the use of an admixture to gain early strength is not common; a higher early strength can be achieved more effectively by other means—such as using Type III cement, employing a low w/c ratio, increasing the cement content, and steam-curing. Heating the mixing water or aggregate is one more procedure used to speed up the rate of hydration in cold weather applications.

Calcium chloride is added in the form of solution, as part of the mixing water. The amount of this admixture should be limited to 2 percent of cement, as a larger dose may cause severe corrosion and loss of strength at later ages. It is not recommended in hot weather applications and for prestressed concrete, and may be ineffective at temperatures below freezing, for concrete (water) may freeze even before it sets. In a moist environment, calcium chloride may promote the corrosion of embedded steel and produce greater alkali-silica reaction.

A number of problems are known to follow the use of accelerating admixtures. The heat of hydration rises, and the corrosion of reinforcing and prestressing steel is aggravated. In addition, shrinkage and creep are also found to increase following the addition of some accelerating admixtures, such as calcium chloride.

Retarding Admixtures. A retarding admixture (ASTM Type B) is added to concrete to delay its setting and hardening, especially in hot-weather applications or when the ambient temperature exceeds 90 °F (32 °C). High temperatures enhance the rate of hardening, making it difficult to place and finish the concrete. The addition of retarding admixture delays the setting and slows down the hydration. It is also used to keep the concrete workable long enough so that succeeding lifts can be placed without cold joints or discontinuities. In large structures and difficult situations, such as large piers and foundations, retarding admixtures are commonly recommended for delaying the initial set as well as for maintaining the workability through the entire placing period.

The chemical composition of retarding admixtures is similar to that of water-reducing admixtures, and many products available in the market exhibit both retarding and water-reducing characteristics. The use of these admixtures, in general, may lower the early strength of concrete.

Antifreeze Admixtures. More than 50 percent of the United States is hit by freezing temperatures at least one month each year. When fresh concrete is exposed to freezing temperatures, the strength development is delayed and the concrete may suffer damage from freezing.

An alternative to protecting fresh concrete against frost damage using insulation or by providing heated enclosures is to use an antifreeze admixture, which lowers the freezing point of pore water in concrete. Its use minimizes—but does not prevent—ice formation in concrete. These admixtures are especially beneficial to thin structural elements, as they can lose heat more rapidly. Thus, they allow the cement to hydrate at low temperatures. The long-term effects of this type of admixture on durability and other characteristics of hardened concrete have not been studied.

Water-reducing and High-range Water-reducing Admixtures. These types of admixtures (ASTM Types A and F) are used to reduce the quantity of mixing water, thereby increasing the compressive strength, and to produce concrete of desired consistency or high slump. A normal-range water-reducing admixture decreases the water requirement by 5–7 percent, and a midrange water-reducing admixture by up to 12 percent. A high-range water-reducing admixture lowers the water requirement by more than 12 percent, and up to 20–30 percent.

Sugar in any form is a water-reducing admixture as well as a retarder. Its effect is due to increased dispersion of cement particles, causing a reduction in the viscosity of concrete. But sugar is not used in present-day concrete construction, for it is difficult to control the water-reducing effect and the setting characteristics. Moreover, even a moderate amount of sugar may contain or completely stop the setting of cement. A number of compounds based on lignosulfonates, which are byproducts of the pulp and paper industry, are marketed as water-reducing (and retarding) admixtures.

A water-reducing admixture is added, typically, to improve the consistency or workability of concrete, to increase the compressive strength, or to attain savings in cement without compromising either strength or consistency. Consistency or workability, however, can also be improved by other means, such as the addition of more cement, the addition of more water, and proper mix proportioning.

High-range water-reducing admixtures, also called superplasticizers, are added in small dosages along with the mixing water—a typical dosage is 0.2 percent by weight of cement—to substantially increase the slump of concrete. Concrete with a slump greater than about 7 in. (178 mm)—which remains cohesive and does not bleed excessively—is called *flowing concrete*. It is used in areas that require maximum volume placement and also in areas congested with a lot of reinforcement. Ideal for pumping, flowing concrete reduces pumping pressure and increases both the pumping rate and the distance. The concrete mix usually must be reproportioned when flowing concrete is required. In order to assure sufficient fines, and prevent excessive bleeding and segregation, the fine aggregate/coarse aggregate ratio must be adjusted. Cement content may have to be increased, and the mix may require a pozzolan.

Superplasticizers can be added at the job site or at the batch plant. The dosage rate varies from one brand to another, and among different types of cement. Additional dosages may be added when the slump drops due to drying of the mix.

Modified lignosulfonates, polymelamine sulfonates, and polynaphthalene sulfonates are the most commonly used superplasticizers. They can be used either as high-range water reducers, for the production of concrete with normal consistency, or as plasticizers, for the production of flowing concrete. The defloculating and dispersing actions of these chemicals play an important role in the fluidification of concrete.

Superplasticizers help enable working with concrete of low w/c ratio, typically in the range of 0.3 to 0.45, which would be impossible to work with otherwise. A low w/c ratio yields concrete of very high compressive strength, called high-strength concrete, with strength in excess of 8000 psi (55 MPa). At such low ranges of w/c ratio, even a small deviation from the recommended quantity of water would significantly affect the strength. The amount of superplasticizer needed to attain a level of workability depends on the fineness and type of cement—particularly the amount of C_3A. Excessive dosages of these admixtures may render the concrete too fluid, causing severe segregation. Loss of entrapped air has also been observed with the use of superplasticizers.

In addition to the increase in strength, superplasticizers eliminate segregation of concrete and allow good dispersion of cement particles in water, accelerating the rate of hydration. The uniform distribution of cement particles is partly responsible for the higher early strength in concrete made with superplasticizer. The general nature of the effects of superplasticizers is explained in the following paragraphs.

Concrete mixtures that have a low w/c ratio are often difficult to place and work with. As they are stiff, such concretes demand a lot of labor to place and finish. Any additional water added to the mix will be accompanied by a loss of strength and increase of shrinkage cracks. In addition, the concrete will be more apt to have a powdery surface, prone to spalling and flaking. The addition of superplasticizer—without increasing the quantity of water—at a cost increase less than about 5 percent, results in very fluid concrete, with slump as high as 9 or 10 in. (23 or 25 cm), compared to the usual 2 to 3 in. (5 to 7.5 cm). The labor savings in placement and finishing due to the use of superplasticizer can be as high as 30 percent.

Plastic shrinkage cracks develop in concrete when water evaporates from the plastic concrete, thereby causing a reduction in volume. The more water in the concrete, the more water that will rise to the top surface and evaporate—and the greater will be the volume reduction of the poured concrete, and the ensuing shrinkage. Concrete made with a superplasticizer—and less water—has a smooth surface, is much less likely to chip and spall, and has fewer plastic shrinkage cracks, compared to concrete made without such an admixture.

In addition to high compressive strength, low permeability, and less shrinkage, the use of superplasticizer has a number of other advantages, such as decreased time to place and finish, accelerated curing period, and early removal of forms. In fact, the advent of superplasticizer is said to be the single most important development in concrete technology in more than three decades. Some tests, however, have shown a reduction in compressive strength from the addition of superplasticizer, for concrete made with Type V cement.

High-strength concrete, which is further discussed in Section 3.12.5, is being used more and more for the columns in high-rise buildings (for minimizing the size), bridges (to reduce their weights), and offshore structures (to make them less permeable). Concrete in the lower stories of Chicago's Water Tower Place had a compressive strength of 9000 psi (62 MPa), and the superstructures of the East Huntington Bridge across the Ohio River and the Annicis Bridge near Vancouver, British Columbia, were constructed with 8000 psi (55 MPa) concrete. The tallest building complex in Georgia, with a height of 842 ft (257 m), One Peachtree Center in Atlanta, was recently completed using concrete of strength 12,000 psi (83 MPa). The mix (with a water-to-cement ratio of 0.29) had a water-reducing

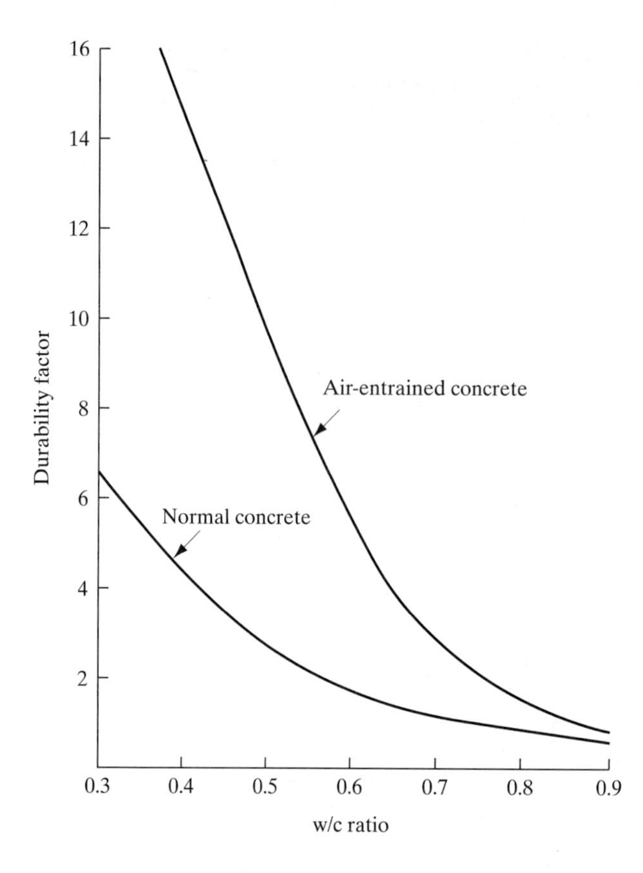

Figure 3.41 Relation between w/c ratio and durability factor (which is a relative measure of freeze-thaw resistance).

admixture, a retarding admixture, and fly ash and silica fume added with other normal ingredients.

Air-entraining Admixtures. The durability of concrete, measured by the number of freeze-thaw cycles it can withstand without disintegration, is dependent on the amount of entrained air (Fig. 3.41). Air-entrainment is a process by which a large number of microscopic air bubbles—diameter ranging from 0.0008–0.08 in. (20–2000 μm)—are dispersed throughout the concrete. These cause disruption of the continuity of capillary pores, and reduce internal stresses caused by expansion of pore water upon freezing.

A number of chemicals, called *foaming agents,* are marketed as air-entraining admixtures, which (when added in small amounts—typically around 0.05 percent by weight of cement) during mixing can entrain 3–10 percent air. The admixtures may consist of salts of wood resins, petroleum acids, animal and vegetable fats, and some synthetic detergents (ASTM C260). The amount of air needed to provide adequate freeze-thaw resistance depends on the maximum size of coarse aggregate and the level of exposure. A 1-in. (25-mm) nominal maximum size aggregate may require 4.5 percent air, compared to 6 percent for 3/8-in. (10-mm) aggregate.

Further advantages of air-entrainment are improvement in workability, increase in slump, reduction in segregation tendency, and control of bleeding. At equal slump, air-entrained concrete is considerably more workable than non–air-entrained concrete, except at higher cement content, when the mix becomes sticky and difficult to finish. High bleeding is responsible for the formation of minute cracks below the larger aggregate particles, and thus the addition of an air-entraining admixture may decrease the permeability of concrete. The reduction in compressive strength associated with these admixtures is generally proportional to the amount of entrained air.

Air-entrainment can also be accomplished by using air-entrained (Type A) portland cements. But adding an admixture separately during mixing makes it possible to adjust the quantity of the admixture, depending on the required amount of air.

3.11.2 Mineral Admixtures or Pozzolans

Mineral admixtures are natural pozzolanic materials or industrial byproducts that are commonly used in concrete to replace part of the cement or sand. When added as a substitute for a part of the cement, mineral admixtures are called *supplementary cementing materials*. The name "pozzolan" comes from the town of Pozzuoli, near Naples, Italy, which was a source of volcanic (pumice and tuff) ash. Natural pozzolanic materials were used in construction thousands of years ago, as in the construction of the Roman Aqueduct.

Mineral admixtures or pozzolans are generally added to concrete in relatively large quantities—in comparison to their chemical counterparts. Pozzolans are siliceous (or siliceous and aluminous) materials that possess little or no cementitious property, but in a finely divided form and in the presence of moisture, can react with the calcium hydroxide of concrete at ordinary temperatures to form cementitious compounds. They include raw and calcined natural materials (Class N) such as opaline cherts, opaline shale, tuff, and pumice, and manufactured products such as fly ash (Class F or C, ASTM C618) and blast-furnace slag (Table 3.14).

Pumice is the most commonly used natural pozzolan. Countries such as Greece, Italy, France, Germany, Turkey, Spain, China, Japan, India, and the United States have many deposits of natural pozzolans. Deposits of volcanic tuff, pumicites, diatomaceous earth, and opaline shales are found in Oklahoma, Nevada, Arizona, and California. The Los Angeles Aqueduct, built around 1910, was one of the first projects where large-scale use of portland-pozzolan cement—a mixture of equal parts of portland cement and rhyolite pumicite—was used. Siliceous shales of the Monterey Formation in Southern California have been mined since the 1930s for commercial production of natural pozzolan, used extensively in the past in

TABLE 3.14 MINERAL ADMIXTURES

Type/ class	Description	ASTM standard	Applications
N	Raw or calcined natural pozzolan	C618	To improve durability and impermeability; in low-heat applications; in hydraulic structures; to improve sulfate resistance and alkali–aggregate reaction
F	Fly ash, bituminous coal origin	C618	To have high later-age strength; low-heat cement, mass concrete; to improve sulfate resistance and alkali–aggregate reaction
C	Fly ash, lignite ash, or subbituminous coal origin	C618	
	Other pozzolanic materials, such as calcined clays and shales and volcanic tuffs		
	Silica fume		In high-strength concrete; to improve watertightness; in shotcrete

surrounding areas. Portland-pozzolan cement produced using calcined Monterey shale was utilized in the construction of the Golden Gate Bridge and the San Francisco–Oakland Bay Bridge.

The properties of natural pozzolan vary considerably depending on their origin. Most contain a large amount of silica or quartz—some 45–85 percent—and smaller amounts of alumina (10–20 percent), iron oxide (2–10 percent), and lime (3–10 percent). When mixed with lime, or with calcium hydroxide (a byproduct of the initial reaction between cement and water), pozzolans produce hydrates of calcium silicate, calcium aluminate, and calcium aluminosilicate. This reaction, between lime and pozzolan, called *pozzolanic reaction,* is much slower than the reaction between the compounds of cement and water. With time, there is a gradual decrease in free calcium hydroxide and increase in the formation of calcium silicate hydrate and calcium aluminosilicate. The end result is that the hardened cement paste will have less calcium hydroxide and more calcium silicate hydrate and other compounds—products of low porosity.

The physical properties (such as shape, fineness, particle size distribution, and density) and chemical composition of pozzolans influence the properties of the fresh and hardened concrete made with them. The molecular structure and the amount of silica determine the pozzolanic reactivity. Amorphous silica reacts with calcium hydroxide and alkalis generally more rapidly than does silica in crystalline form. Thus, the chemical composition alone of pozzolans does not truly determine the pozzolanic reactivity: clays and shales are not pozzolanic, for clay minerals do not readily react with lime unless their crystalline structure is destroyed by heat.

Many natural pozzolans can be used in the raw state; others, like clay and shale, should be calcined. When moist, particles in the raw state require drying and, usually, grinding to a fineness equal to that of cement, before use. They are used in the range of 15–35 percent, by mass, of the total cementitious materials in concrete.

The use of a pozzolan affects the water and cement requirements of concrete. Most pozzolans increase the water requirements due to their microporous character and large surface area. When the aggregates are deficient in finer particles, particularly those passing a No. 100 sieve (150 μm), the use of a finely divided mineral admixture can be advantageous, for it can reduce bleeding, decrease segregation, and increase strength. However, the amount of improvement depends on the particle shape, size, and gradation of pozzolan.

Because the pozolanic reaction is much slower than the initial cementitious reactions, the strength of concrete made with a pozzolan, at early ages, is lower than that of concrete made without the cement replacement by pozzolan. But the ultimate strength is nearly the same or, as in most cases, greater. However, the benefits of adding a pozzolan are in the ability to make concrete more durable and less permeable. In addition, the heat of hydration and the shrinkage are much lower than that of normal concrete. At a given amount of cement, the use of a pozzolan has little or no effect on the temperature of concrete. However, when the cement content is lowered, the use of a pozzolan may reduce the concrete temperature. Pozzolan is also beneficial in reducing or eliminating the expansion and map cracking resulting from reactive aggregates, or alkali-silica reaction. In mass concrete, the use of a pozzolan contributes to improved watertightness, due partly to decreased segregation, reduced bleeding, and reduction in water requirements. Portland-pozzolan cement, or blended cement, is ideally suited for use in hydraulic structures, where the wet condition improves the property continuously.

Fly Ash. A byproduct from the burning of powdered coal in electric generating power plants, fly ash (also called pulverized fuel ash or pfa) is collected in the dust-collection systems (electrostatic or mechanical precipitators) that remove particles from the exhaust gases. Some of the ash is withdrawn from the furnace as bottom ash and boiler slag, which can be used as aggregates in some applications but not for concrete. Fly ash ranks high in weight of the minerals produced yearly in the United States, behind stone, sand, gravel, coal, iron ore, and cement. It is a finer material than portland cement, and consists of small spheres of glass, with a complex composition of silica, ferric oxide, and alumina. It has been used in concrete construction and in marine structures since the late 1940s.

Fly ash is divided into two major classes:

- Class F fly ash
- Class C fly ash

Class F fly ash, also called low-calcium fly ash, which has lime (CaO) content less than 10 percent, is obtained from the burning of bituminous coal. Class C (or high-calcium) fly ash, in which the CaO content is more than 10 percent and can be as high as 30 percent, is produced from burning subbituminous coal or lignite. Class F fly ash possesses little no cementing property, but in a finely divided form and in the presence of water it can react with calcium hydroxide generated during the hydration of cement. Class C fly ash has some cementing property, in addition to pozzolanic properties.

Fly ash consists of a large proportion of SiO_2 (45–60 percent) and smaller amounts of Al_2O_3 (20–35 percent), Fe_2O_3 (3–13 percent), and CaO (2–8 percent). It is similar in size, color, and texture to portland cement, but the particles of fly ash are spherical in shape, whereas those of cement (produced by grinding) are generally irregular. This sphericity has a beneficial effect in the placement of concrete, especially in pumping. Fly ash is lighter than cement, with a specific gravity ranging between 2.0–2.5, and is finer than cement, and as a result—although it is not absorbent—the increased surface area demands more water (the water demand increasing with the fineness). Most fly ash is alkaline, a condition that is enhanced when added to portland cement.

Fly ash can be used in concrete either as an admixture or as an ingredient in blended cement. The addition of fly ash improves the placement of concrete—its workability and pumpability—because of the fineness of the material and its spherical particle shape. It helps to lower the heat of hydration and provides a higher long-term strength. Tests have shown a nearly linear relationship between the cement reduction with fly ash and the heat of hydration at 7 days. Low heat of hydration is an important requirement for mass concrete, such as in large foundations, piers, embankments, and bridges. Further, concrete made with fly ash shows less bleeding than control mixtures. The reaction between fly ash and calcium hydroxide, which produces a refined calcium silicate hydrate binder, lowers the permeability of concrete. Expansion resulting from alkali-aggregate reaction is greatly reduced when 30–40 percent of cement is replaced by fly ash. The silica component in fly ash combines with available alkalis in hydrated cement, reducing the potential for expansion from this reaction.

For more than 50 years, fly ash has been used in structures prone to sulfate attack. Generally, cements low in tricalcium aluminate are more resistant to sulfate attack. Type II cement, which contains less than 8 percent C_3A, and Type V, which has less than 5 percent, are used any time soil is suspected to contain sulfates. The addition of fly ash helps to decrease the total amount of C_3A in the mixture. The consumption of free calcium hydroxide in concrete by fly ash also decreases the potential for reaction between sulfates and calcium hydroxide. In addition to these benefits, the cost of fly ash is about one-third that of cement, thus the concrete made with it is also cost-effective. The use of fly ash—when proven effective either by tests or service record—is permitted in most transportation structures for protection against sulfate attack.

In summary, the general benefits of adding fly ash to concrete are:

- Increase in ultimate strength (or later-age strength)
- Reduced temperature rise
- Reduced alkali-aggregate reaction

- Improved resistance to sulfate attack
- Reduced permeability
- Improved workability and pumpability
- Economy

Typically, fly ash is used to substitute for up to about 20 percent of portland cement in building construction, albeit as much as 70 percent replacement has been used in the past. Up to about 20–25 percent replacement usually meets the normal strength requirements at 28 days. The vast majority of fly ash used in the United States is of Class F type. High-fineness fly ash that has a loss on ignition (LOI) no greater than 3 percent is permitted in high-strength concrete (maximum LOI for Class C or F fly ash is 6 percent).

A small amount of fly ash, usually less than 15 percent of the cement in the mix, is generally permissible in most concrete construction. For example, if a mix requires 550 lb/yd^3 (325 kg/m^3) of cement, between 55–82 lb/yd^3 (32–49 kg/m^3) of replacement with fly ash is acceptable. However, a simple replacement of cement by fly ash may not be the best method of mix proportioning. Since 1976, the Federal Highway Administration has required the state transportation agencies to permit the use of fly ash in concrete in transportation structures. Some states permit up to 30 percent maximum replacement (Class C or F); a few permit, a maximum of 15 percent. Replacement of up to 50 percent is allowed in mass concrete applications, where early strength is not required and ultimate strength is low (25 to 35 MPa). The replacement of sand (fine) by fly ash helps reduce the rate and magnitude of slump loss. Studies have shown fly ash to be a more effective retarder of the hydration of C_3A and C_4AF than an equivalent quantity of gypsum.

Problems may arise from the addition of fly ash because it can affect both the air content and the water demand of the mix. The water demand increases with the fineness; the finer the material, the more air-entraining admixture required for the desired air content. The early-strength development is slower, especially for high-volume fly ash concrete. As it is darker than normal portland cement, fly ash may affect the color of the concrete made with it.

Generally, all pozzolanic materials—natural or fly ash—may require sufficient moist curing to satisfy durability requirements. Fly ash concrete has been found to carbonate faster than concrete without fly ash, and may thus be more susceptible to corrosion. Carbonation, as explained, refers to the reaction between the concrete and the carbon dioxide in the environment. A decrease in the pH of the concrete, resulting from carbonation, may lead to early corrosion of reinforcing bars. However, the simple addition of fly ash to a concrete mixture—without a change in the cement content—is found to make the concrete less permeable, reducing the corrosion potential.

Silica Fume. Silica fume or condensed silica fume (also called fumed silica) is a byproduct from the electric-arc furnaces of the silicon metal and ferrosilicon alloy industries. The reduction of quartz to silicon at very high temperatures produces SiO vapors which, after oxidation, condense to tiny particles that are less than 0.1 μm in size, or about two orders of magnitude—that is 100 times—finer than cement. These primarily consists of noncrystalline silica—85–95 percent silicon dioxide, with small percentages of alumina, iron oxide, and lime. The specific gravity of silica fume is low, with an average of 2.2. It weighs about 9–25 pcf.

Due to its fineness and high glass content, silica fume is very highly pozzolanic; as a result, its contribution to strength development of concrete is at early ages, rather than later (as with fly ash). Typically, silica fume is added to produce very high-strength, dense, and durable concrete and shotcrete.

Silica fume is commonly available in three forms: water slurry, dry uncompacted powder, and dry densified (compacted) powder. The slurry form contains 50–55 percent silica fume (by weight), the balance being water. The handling of the uncompacted powder may pose a health hazard, as the particles are very minute—resembling tobacco smoke—and may end up in the human body.

The United States and Norway are the world's largest producers of silica fume, which was first tried as a concrete admixture around the 1950s. Since then it has been used successfully in structural concrete to improve strength and durability. An increase in compressive strength of 40–60 percent may be expected from the addition of about 8 lb (3.6 kg) silica fume for each sack of cement. Trump Place in New York City was the first large-scale project in the eastern United States where silica fume was used to obtain high-strength concrete. The reinforced concrete columns in this tower had a specified compressive strength of 12,500 psi (86 MPa). Concrete made with silica fume has also been found to have higher flexural strength, when compared with plain concrete. Silica fume is presently used to make high-strength concrete for bridge decks, columns, and walls. It is also used in shotcrete applications.

The addition of silica fume may increase the water requirements unless a water-reducing admixture is used. Concrete made with silica fume generally bleeds very little, for such mixes typically have low water content. Moreover, the ingredients in these mixes stick together, and the silica fume, which has a high surface area, renders the mix so impermeable that the limited water available cannot move up to the surface. Without bleeding, the concrete surface may begin to dry rapidly, leading to a weaker surface and more shrinkage cracks. Precautions must be taken to avoid rapid evaporation of water from the surface of concrete made with silica fume. The use of an evaporation retardant or fog spray may help, and curing should begin immediately following the finishing operations.

Blended Hydraulic Cement. The ASTM (C595) identifies a special class of cement, called blended hydraulic cement, which is made by blending portland cement with either cooled blast-furnace slag or a pozzolanic material such as fly ash. The ASTM specifications cover five classes of blended cement: portland–blast-furnace slag cement, portland-pozzolan cement, slag cement, pozzolan-modified portland cement, and slag-modified portland cement. The blending can be done by intergrinding portland cement clinkers with other materials or combining the materials in their final forms.

Blast-furnace slag is a nonmetallic product, consisting essentially of silicates and aluminates of calcium and other bases, which is developed in a molten condition simultaneously with iron in a blast furnace. Granulated blast furnace slag is the glassy granular material formed when the molten slag is rapidly chilled, as by immersion in water. The granulated slag is dried, partially ground, mixed with hydrated lime and then interground into a very fine powder to form slag cement, which is much more plastic than portland cement and better suited to foundations not exposed to severe weather. Between 35–40 percent of blast furnace slag is SiO_2, and about the same amount is CaO. When it is finely ground, this material exhibits the hydraulic property; however, the hydration rate lags behind that of compounds of portland cement. The addition of ground slag has been found to improve the workability of concrete.

Portland-pozzolan cement, produced by intergrinding portland cement clinkers and natural or artificial pozzolan, has better strength characteristics and watertightness than portland cement. In general, the properties of concrete made with blended cement are similar to those made with portland cement plus pozzolan. Tests have shown that the addition of silica fume, fly ash, or blast-furnace slag improves the strength, lowers the permeability, decreases the heat of hydration, and delays the onset of corrosion.

3.12 TYPES OF CONCRETE

Ordinary concrete possesses good compressive strength but exhibits weaker characteristics when subjected to tensile, flexural, or shear forces. In addition, all concrete structures suffer from volume changes resulting from shrinkage, creep, and thermal changes, producing cracks that are detrimental to strength and performance during service. For these reasons, and to impart ductility to structures, ordinary concrete is often used to form a composite along with materials that possess high tensile or flexural strength. Plain concrete is rarely used to form structural elements, except for ground-supported slabs. Concrete made ductile through coalition with high-modulus materials, such as reinforcing steel, is commonly called *structural concrete*. Some of the more common types of structural concrete are discussed in this section.

3.12.1 Reinforced Concrete

Reinforced concrete is a composite material made by combining plain concrete and reinforcing steel. It first appeared around 1850, when Jean-Louis Lambot constructed a reinforced concrete rowboat and flowerpots, which were exhibited at the Paris Exposition in 1855. Another Frenchman, Francois Coignet, is reported to have built a 12-in. thick roof, with 20-ft span of reinforced concrete in 1853, and Joseph Monier, a Parisian gardener, began making flowerpots, tubs, and tanks using a system of reinforcing steel and concrete around 1860. He made a framework of wires, around which the concrete was placed.

In principle, reinforced concrete is a system formed by combining two entirely different materials—concrete and steel—so that the system behaves as a unit. The reinforcing steel, primarily composed of nearly round bars—is located at all zones and sections where external loads cause tensile stresses. Various types of steels have been used in reinforced concrete ranging from soft steels, with yield strength of 30 ksi (207 MPa), to hard steels, with yield strength of 60 ksi (413 MPa).

Today, reinforcing steel is manufactured to meet the requirements of the following four ASTM standards:

- A615 ("Billet steel, Deformed and plain bars")
- A616 ("Rail steel, Deformed and plain bars")
- A617 ("Axle steel, Deformed and plain bars")
- A706 ("Low-alloy steel bars")

The size, yield strength, and tensile strength of these bars are shown in Table 3.15.

TABLE 3.15 MECHANICAL PROPERTIES OF ASTM REINFORCING BARS

Type of steel and ASTM No.	Identification mark	Size (Nos.) inclusive	Grade (min. yield)	Yield strength (min.; psi)	Tensile strength (min.; psi)
Billet steel, A615-84a	S, N	3–6	40	40,000	70,000
	S, N	3–11, 14, 18	60	60,000	90,000
		11–18	75	75,000	100,000
Rail steel, A616-84	I	3–11	50	50,000	80,000
	R	3–11	60	60,000	90,000
Axle steel, A617	A	3–11	40	40,000	70,000
		3–11	60	60,000	90,000
Low-alloy steel, A706-84a	W	3–11, 14, 18	60	60,000	80,000

Reinforcing steel is available in different strengths or grades. The grade refers to the specified yield strength. For example, grade 40 has a specified yield strength of 40,000 psi (276 MPa), grade 50 has a specified yield strength of 50,000 psi (345 MPa), and grade 60 has a specified yield strength of 60,000 psi (413 MPa). Bars of grade higher than 60 are also available but are not commonly used in reinforced concrete. The modulus of elasticity of all grades of steel is taken as 29,000 ksi (200×10^3 MPa). Plain bars in any grade have a smooth surface, whereas deformed bars are manufactured with surface deformations to achieve perfect bonding with concrete (Fig. 3.42).

Reinforcing bars are available in nominal diameters from 3/8–$1\frac{3}{8}$ in. in 1/8-in. increments, and also in two larger sizes: $1\frac{3}{4}$ in. and $2\frac{1}{4}$ in. diameter. They are identified

Figure 3.42 Reinforcing bar surface deformations.

TABLE 3.16 SIZE AND AREAS OF STANDARD REINFORCING BARS

Bar size (No.)	Nominal diameter (in.)	Nominal area (in.^{2})	Weight (pcf)
3	0.375	0.11	0.38
4	0.50	0.20	0.67
5	0.625	0.31	1.04
6	0.75	0.44	1.50
7	0.875	0.60	2.04
8	1.0	0.79	2.67
9	1.13	1.0	3.40
10	1.27	1.27	4.30
11	1.41	1.56	5.31
14	1.69	2.25	7.65
18	2.26	4.0	13.60

by number; the number corresponds to the number of eighth-inches in the nominal bar diameter. For example, a No. 8 bar has a nominal diameter of 1 in.; similarly, a No. 3 bar has a nominal diameter of 3/8 in. (Table 3.16).

All reinforcing bars have distinguishing marks that are rolled onto the surface of one side of the bar. These markings represent (1) the producer's mill designation or point of origin, (2) the bar number, (3) the type of steel (refer to symbols in Table 3.15), and (4) the number 60 or a single continuous longitudinal line offset from the center of the bar side in the case of grade 60 steel only (Fig. 3.43).

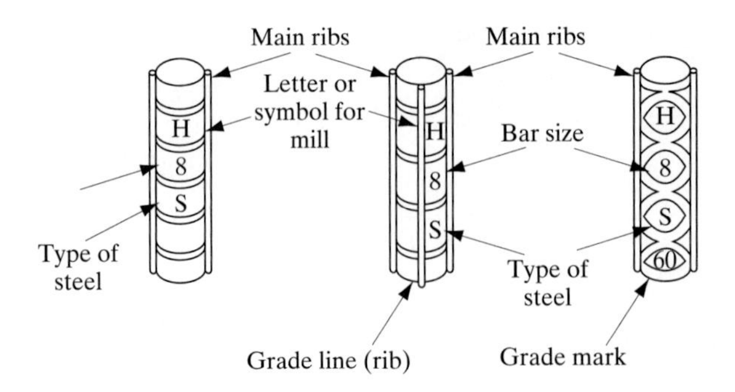

Figure 3.43 Identification marks on reinforcing bars.

Apart from groups of reinforcing bars, other forms of reinforcing, such as welded wire mesh, are also used in slabs and pavements. The mesh consists of longitudinal and transverse cold-drawn steel wires, placed at right angles and welded at all points of intersection. Additional details on reinforcing steel are given in Chapter 7.

Reinforced concrete construction is typically carried out by forming a cage of reinforcing steel and filling it with concrete. Typical cages are shown in Fig. 3.44 and consist of bars that are parallel to horizontal and vertical planes. For the development of composite action, it is essential that all individual bars be bonded to concrete, which means that every bar should have a minimum cover of concrete and the concrete mix should be homogeneous. To accomplish the objective of homogeneity, the consistency of the concrete mixture should be such as to prevent segregation and formation of voids. The proper amount of water, use of admixtures, and adequate vibration generally bring out acceptable results.

Reinforced concrete, the most common construction material in the world, is used in foundations, breakwaters, high-rise buildings, dams, and residential construction. With proper quality control, it can be built to withstand all types of weather conditions and loads. It is still the most durable construction material in adverse surroundings, such as soil and seawater. Compared to structural steel, reinforced concrete offers a number of advantages, such as reduction in floor vibrations, greater fire protection, and contribution to overall stiffness of the structure.

Moment Capacity. The moment capacity of a reinforced concrete flexural member depends on the cross-sectional dimensions, strength of concrete, grade of steel, and area of steel. Increase in any one of these variables increases the moment capacity. When all other variables remain unchanged, moment capacity increases with increase in steel area.

The failure moment, or *nominal moment capacity* of a reinforced concrete flexural member (beam or slab), M_n, can be predicted using the following equation:

$$M_n = A_s f_y (d - 0.5a)$$

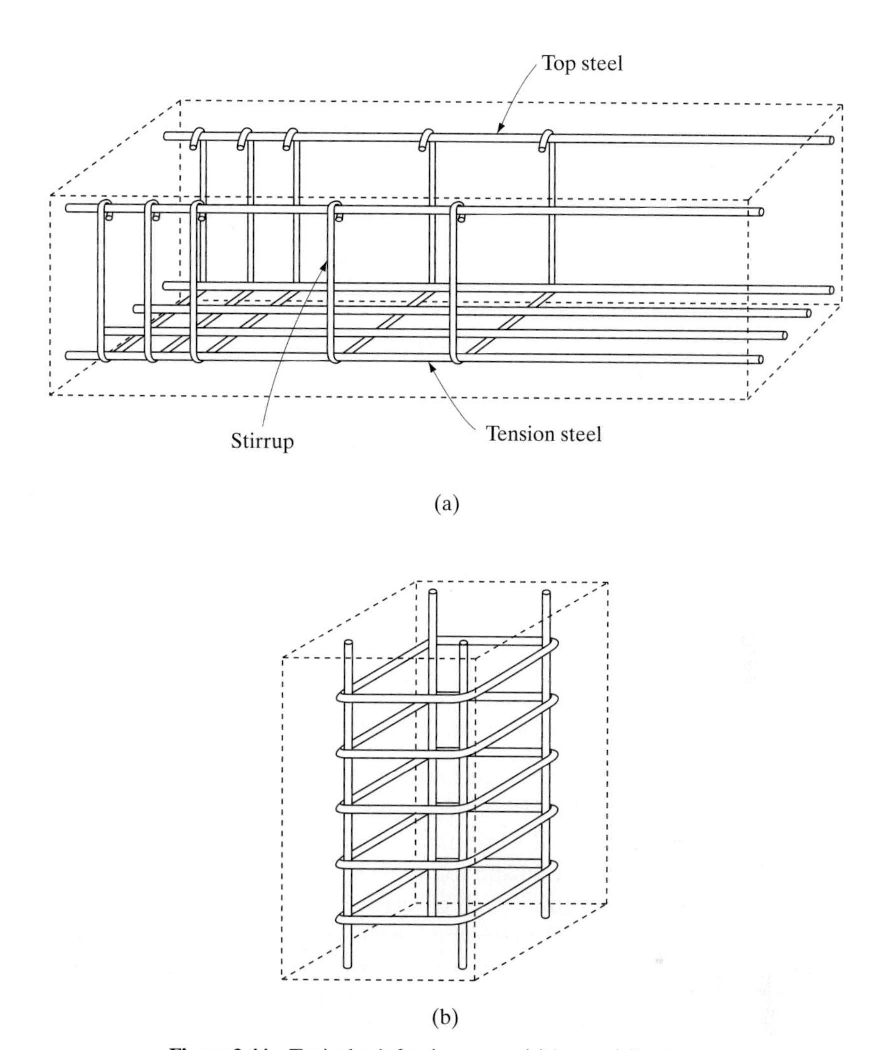

Figure 3.44 Typical reinforcing cages: (a) beam; (b) column.

where A_s is the area of cross-section of all reinforcing bars on the tension side of the member, f_y the yield strength of steel, d the depth of steel measured from the compression side of the member (called *effective depth*), and a the stress block depth, found as follows.

$$a = \frac{A_s f_y}{0.85 f'_c b}$$

Note that b is the width of the member. The theoretical relationship between moment capacity and steel ratio is plotted with some experimental values in Fig. 3.45.

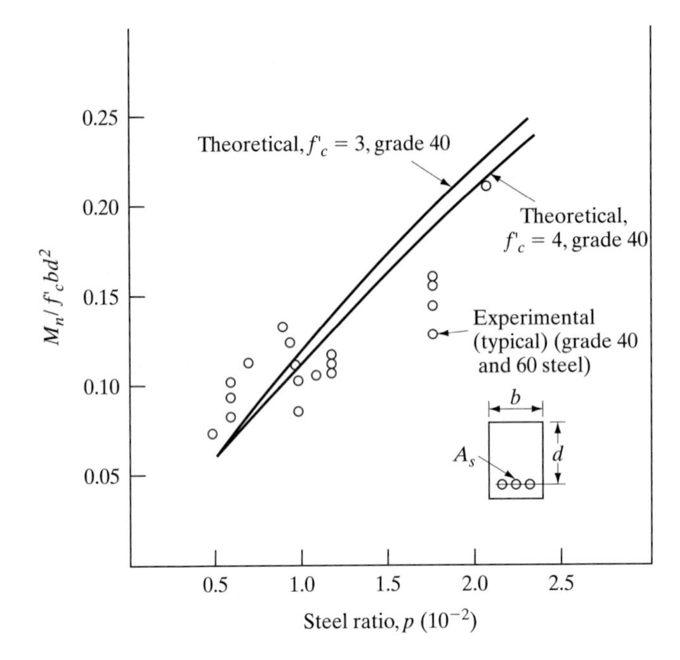

Figure 3.45 Moment capacity of reinforced concrete versus steel ratio.

3.12.2 Prestressed and Precast Concrete

The term *prestressing* refers to applying a preload on a structure or structural element before the application of design loads. *Prestressed concrete* is a structural composite material made with ordinary concrete and high-strength steel (prestressing steel) subjected to a pretensile force.

Concrete is essentially a high-compression material. Its strength in tension is substantially lower than the compressive strength. Provision of reinforcing bars to take up tensile stresses will result in a composite material (reinforced concrete) that has a high flexural capacity but in which the concrete cracks along the tension zone. These cracks, expected before the application of full service loads, do not ordinarily affect the structural integrity or capacity. However, they make the concrete more permeable and less durable. Application of a prestressing force (by pretensioning of steel or precompression of concrete) offers us a crack-free concrete and that is superior to normal reinforced concrete in terms of strength, deflection characteristics, and durability.

Renowned French engineer Eugene Freyssinet is credited with the development of prestressed concrete and the construction of a number of prestressed concrete bridges in France after the 1940s. At present, prestressed concrete is used in various types of construction, such as bridges, parking garages, foundation slabs, liquid storage tanks, and high-rise buildings.

All prestressed concrete members can be placed in one of two categories:

- Pretensioned concrete
- Posttensioned concrete

Pretensioned prestressed concrete (or pretensioned concrete, Fig. 3.46) members are produced by stretching the prestressing steel (called tendons) held between external anchorages before the concrete is placed. As the plastic concrete hardens it bonds with the steel. When the concrete attains the desired strength the jacking force is released, and this force is transmitted by bond from steel to concrete. In the case of *posttensioned prestressed concrete* (or posttensioned concrete), hollow conduits containing unstressed prestressing steel are placed in the forms, to the desired profile, before placing the concrete. When the concrete has gained sufficient strength, the steel is pulled using a jack at one end and by providing special fittings at the other end of the member. When adequate tension is in the steel, as measured by gauges, the steel is anchored against concrete at the jacking end with special fittings and the jack is removed (Fig. 3.47).

There are three common forms in which steel is used for prestressed concrete:

- Cold-drawn wires
- Stranded cables
- Alloy-steel bars

The round wires (ASTM A421), available in diameters of 0.192 in. (4.88 mm), 0.196 in. (4.98 mm), 0.25 in. (6.35 mm), and 0.276 in. (7.01 mm), are used for posttensioned construction. Stranded cables (or strands) are almost always used in pretensioned construction, and are also often used in posttensioned construction. They are fabricated with six wires wound tightly around a seventh of slightly larger diameter, and are often called uncoated seven-wire stress-relieved strand (ASTM A416). These strands may be obtained in a range of sizes from 0.25–0.60 in. (6.35–15.24 mm). Wires and strands can be obtained in tensile strengths ranging from 235,000–270,000 psi (1622–1860 N/mm^2).

Precast Concrete. Precast concrete can be defined as concrete that is cast in some location other than its final position in the finished construction. The casting location can be a precast concrete factory or job site. Precast concrete members are reinforced with mild steel reinforcing bars, prestressing steel, or a combination of the two.

Precast concrete is both a construction material and a construction method. High-strength concrete under good quality control is used in its manufacture. As a method of construction, precast concrete can greatly reduce construction time, since the elements are cast and stockpiled while other phases of the project are performed. Precast concrete is used in parking garages, buildings, and exterior facades of high-rise structures.

3.12.3 Fiber-reinforced Concrete

Fiber-reinforced concrete contains discontinuous discrete fibers. Ordinary concrete contains numerous micro-cracks, which are responsible for its low tensile strength. Fiber-reinforced concrete offers a solution to this problem of cracking by making

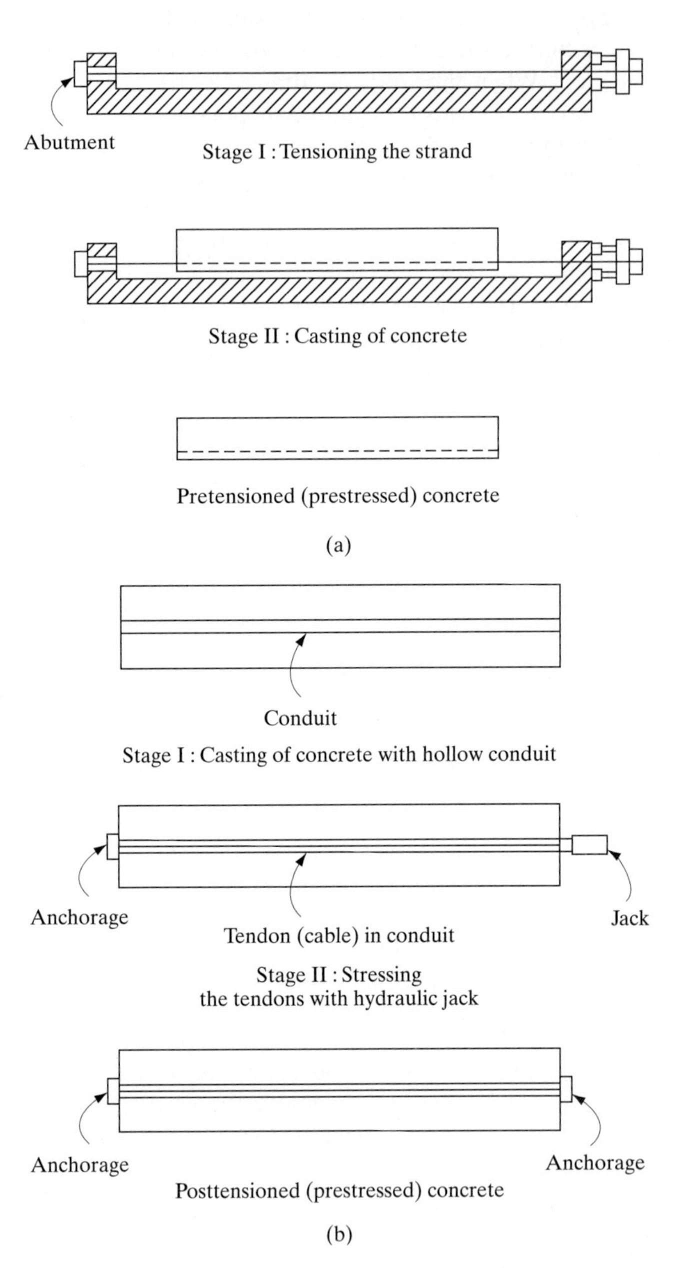

Figure 3.46 Methods of prestressing: (a) pretensioning; (b) posttensioning.

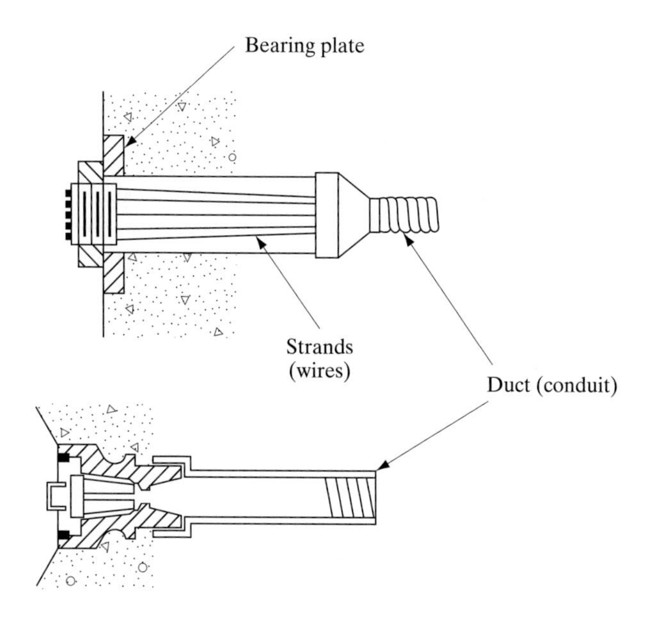

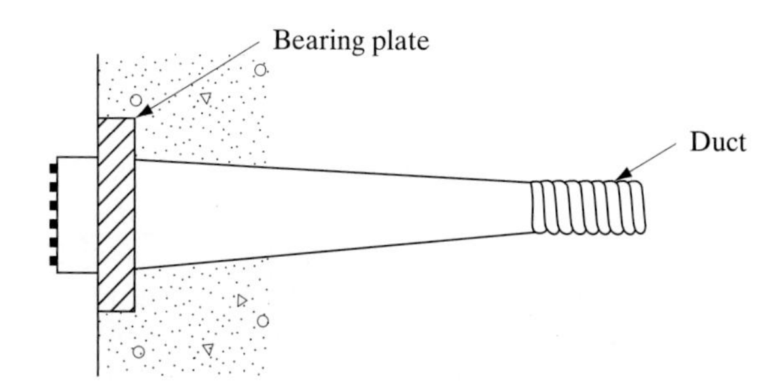

Figure 3.47 Posttensioning anchors.

the concrete tougher and more durable, by incorporating three-dimensional reinforcement within the concrete. *Toughness* is defined as the amount of energy needed to produce a specified damaged condition or complete failure of the materials. It measures the ability of the material to support loads even after the formation of cracks, and is influenced by tensile strength of fiber, fiber geometry, and fiber dosage. Fiber-reinforced concrete is much tougher and more resistant to impact than plain concrete.

Fiber-reinforced concrete—made with steel fibers—was developed in the United States in 1969 (J. P. Romauldi and Batelle Development Corp., Cleveland, Ohio). Various kinds of fibers can be used as reinforcement. The most common are steel, polypropylene, nylon, and glass. Strong steel fibers in long lengths or with surface deformations yield better toughness than short-length plastic fibers. Typical properties of these fibers are shown in Table 3.17.

TABLE 3.17 TYPES AND PROPERTIES OF FIBERS

Type of fiber	Diameter (in.)	Modulus of elasticity (ksi)	Tensile strength (ksi)
Polypropylene	0.004–0.008	725	65
Polyester	0.0004–0.003	1450–2500	80–170
Polyethylene	0.001–0.04	725–25,000	29–440
Carbon	0.0003	33,000–55,000	260–380
Acrylic	0.0002–0.0007	2600	30–140
Glass	0.0004–0.0005	10,400–11,500	360–500
Steel	0.004–0.04	29,000	50–250

Fibers come in various sizes and shapes. Round steel fibers, made from low-carbon or stainless steel, have diameters in the range 0.01–0.039 in. (0.25–1 mm). Flat steel fibers, produced by shearing sheets or flattening round wire, are available in thicknesses ranging from 0.006–0.016 in. (0.15–0.40 mm). *Crimped* and *deformed steel fibers* are available both in full length or crimped at the ends only. Some fibers are collated to facilitate mixing and placing. Glass fibers are commonly produced as chopped strand, each strand consisting of hundreds of filaments. *Aspect ratio,* defined as the ratio between the fiber length and diameter, is an important property that affects the performance of a fiber.

Steel fibers have been put to use since their development in the 1960s to increase the toughness, impact resistance, and flexural strength of concrete. Steel fiber–reinforced concrete is employed in airport pavements, bridge decks, industrial floors, hydraulic structures, and shotcrete applications. The amount of steel fibers typically ranges between 0.5–2 percent by volume. The addition of steel fibers to concrete may decrease its slump as much as 50 percent. The concrete may exhibit dryness caused by the fact that fiber-reinforced concrete does not exhibit as much bleeding as normal concrete. In general, fibers of all kinds tend to minimize segregation, which is beneficial in both pumping and placement.

The compressive strength is very slightly affected by the addition of steel fibers. (In fact, some studies suggest that the addition of fibers may decrease the compressive strength.) The increase in flexural strength compared to that of plain concrete is substantially greater than the improvement in tension or compression, and is about 50–70 percent. The addition of 1.5 percent (by volume) of fibers may increase the tensile strength by about 45 percent. Steel fibers have also been shown to increase shear capacity substantially. Corrosion of fibers may be a problem with steel fiber—reinforced concrete, although studies seem to indicate that corrosion does not propagate more than 0.10 in. (2.5 mm) below the surface.

Polypropylene and nylon (synthetic) fibers have resulted from research and development in the petrochemical and textile industries. They are characterized as "soft," because individual fibers have significantly less bending stiffness (elastic modulus) than either glass or steel fibers. When these fibers are applied at low volume percentages—less than 0.2 percent—no improvement in strength characteristics is seen. At higher percentages, especially above 2 percent by volume, all strength properties have been shown to increase. Studies have shown that synthetic fibers can control plastic shrinkage cracking and effectively reduce drying shrinkage cracking. Drying shrinkage cracking can be further decreased with higher fiber content. The addition of these fibers at more than 0.75 lb/yd^3 of concrete improves the toughness, but the pitfalls associated with the use of synthetic fibers include low bonding and slump loss.

Alkali-resistant glass fibers are used in the manufacture of architectural concrete panels and roofing shingles. These fibers have high tensile strength and modulus of elasticity, but the improvement in strength properties will be lost if alkalis in cement attack the glass fibers.

The primary benefit of adding fibers to concrete is the reduction in early shrinkage cracks. Fiber-reinforced concrete is used in a number of applications, such as tunnel lining (shotcrete), rock-slope stabilization, pavements, slab-on-grade, wall panels, and floor slabs. It is also used in the manufacture of architectural panels and roofing systems (tiles and shingles); the construction of bridge decks, storage tanks, parking floors, and driveways; as topping over deteriorated asphalt concrete pavements; and in precast elements. It is particularly useful when a large amount of energy has to be absorbed by the structural element, such as in structures designed to resist explosive loading. But present codes limit the use of fibers essentially to nonstructural and non–load bearing applications, except for use in slabs-on-grade and pavements.

3.12.4 Lightweight Concrete

Lightweight concrete is concrete with an air-dried unit weight not exceeding 115 pcf (1850 kg/m^3). Although there is more than one type of lightweight concrete, the term is generally used to identify structural concrete made with lightweight aggregates. When both fine and coarse aggregates are of lightweight type, the concrete is called *all-lightweight concrete,* and when regular sand is used with lightweight coarse aggregates, the concrete is called *sand-lightweight concrete.* According to ASTM C330, lightweight fine and coarse aggregates are those that have dry unit weights (dry-rodded weights) less than or equal to 70 and 55 pcf (1120 and 880 kg/m^3), respectively.

Various types of lightweight aggregates have been discussed in Chapter 2. The most common are pumice, scoria, expanded shale, and expanded clay. Pumice is a light-colored glassy rock of finely vesicular froth filled with elongated tubular bubbles. Scoria is a coarsely vesicular glassy rock containing more-or-less spherical bubbles. These two are commonly used in the manufacture of lightweight concrete masonry blocks, and the others are used to manufacture lightweight concrete in building construction. Structures from the dome of the Pantheon in Rome (using pumice aggregate) to the 75-story First Interstate World Trade Center building in Los Angeles (115-pcf concrete, using expanded shale coarse aggregate), have been built using lightweight concrete. The primary advantage of lightweight concrete is its low density, which reduces the loads on foundation and supporting structures to two-thirds or less.

Properties. Workability is a concern in the mix design of lightweight concrete. Lightweight aggregates, since they are very porous, absorb mixing water, and may continue to absorb for several weeks after construction. It is necessary to wet these aggregates before adding them to the mix.

As they are light, these aggregates tend to segregate and float to the surface. To improve workability and prevent segregation, it may often be necessary to limit the maximum slump (or limit the water content in the mix) and the use of air entrainment. Lightweight concrete is also more difficult to pump than concrete made with hard rock, because the porous aggregate particles absorb water and have a rougher exterior surface. Note that concrete pumping is the standard method of placement in high-rise concrete construction. Today, pumps have the capacity for a maximum vertical reach (in a single lift) of 1400 ft (420 m) or more at a volume of 170 yd^3 (130 m^3) per hour. Excessive mixing should be avoided in lightweight concrete manufacture because it tends to break up aggregate particles.

Lightweight concrete exhibits relatively high thermal insulating value (thermal conductivity can be half as much as that of normal-weight concrete). Concrete made with shale or clay is nearly four times more insulating than that made with normal aggregates. Concrete in exposed structures that suffer from alternating heating and cooling cycles should preferably contain porous coarse aggregate.

The compressive strength of lightweight concrete is low compared to normal-weight concrete of similar mix proportions. But by using high cement content and good-quality smaller-sized lightweight aggregates, it is possible to obtain strength in the range 3000–5000 psi (20.7 to 34.5 MPa). The modulus of elasticity of lightweight concrete is also somewhat lower than that of normal-weight concrete.

Compared to normal-weight concrete, lightweight concrete has higher drying shrinkage and higher creep. This is due to higher absorption and moisture movement in the concrete. However, the freeze–thaw resistance of air-entrained lightweight concrete is similar to that of air-entrained normal-weight concrete.

As pointed out, the use of lightweight concrete results in a substantial reduction in foundation load and usually contributes to a reduction in overall cost, even though the material will cost more than normal-weight concrete per cubic yard.

3.12.5 High-strength and High-performance Concrete

Concrete of strength less than 6000 psi (41.4 MPa) is normally referred to as ordinary concrete; when the compressive strength exceeds 6000 psi it is called high-strength concrete. When built in 1975, Chicago's Water Tower Place was the world's tallest reinforced concrete building, representing one of the largest uses of concrete of strength in excess of 6000 psi. But concrete with strength up to 11,600 psi (80 MPa) was used in the construction of the perimeter columns of Malaysia's 88-story Petronas Twin Tower, in Kuala Lumpur (which is currently the tallest concrete building in the world) and in the building of the Bioyette Tower in Bangkok. The term "high-performance concrete," which has nearly replaced the earlier "high-strength concrete," is not just a new name for concrete of high strength—in fact, it is much more. It can be defined as concrete that meets the requirements of strength, durability, density, ductility, and other parameters.

The use of high-strength concrete is not new, for concrete of strength 14,000 psi (96.5 MPa) had been used to fabricate telegraph poles in the United Kingdom in the 1930s. Chicago's Mercantile Building, built in 1982, had concrete also of 14,000 psi, and Seattle's Two Union Square Building, built in 1987, used 19,000 psi (131 MPa) concrete.

High strength is achieved primarily by carefully selecting, controlling, and proportioning all ingredients—which include (in addition to portland cement) aggregates and water, mineral admixtures (such as fly ash, blast-furnace slag, and silica fume), and chemical admixtures, especially high-range water reducing admixtures. High cement content, low w/c ratio, and smaller maximum size aggregate are some of the common factors in all high-strength mixtures.

Low w/c ratio requires the use of plasticizers (high-range water-reducing admixtures), and strong and uniform coarse aggregates are essential to produce workable—that is, with slumps of 3 in. (75 mm) or more—high-strength concrete mixtures. High-strength concrete of slump in excess of 7 in. (178 mm), which remains cohesive, is without excessive bleeding or segregation, and is of normal setting characteristics, is flowing concrete, discussed in Section 3.11.1. It is used in areas that require maximum volume placement such as in slabs, mats, and areas congested with lots of reinforcement. The low w/c ratio along with high amounts of water-reducing admixture in this type of high-slump mixtures may contribute to a greater than normal slump loss, which can be remedied by delaying the addition of plasticizer such as at the job site, right before pumping.

The choice of normal maximum aggregate size and the ratio between the fine and coarse aggregates require careful attention in the mix design of high-strength concrete. A large majority of experts recommend a maximum aggregate size of 0.4–0.5 in. (10–12 mm) or less, to minimize the cementitious materials content, and an optimum sand content somewhat lower than that which gives the minimum voids in the combined aggregates. The use of coarser sand, with a fineness modulus of 3.0, has also been recommended, as it tends to yield higher compressive strength than with sand made up of a high proportion of fines, for less water is necessary to wet the surface of all particles.

The ACI 211 suggests a maximum aggregate size not exceeding 3/4–1 in. (19–25 mm) for concretes of strength less than 9000 psi (62 MPa), and 3/8–1/2 in. (10–12 mm) aggregates for strength more than 9000 psi (62 MPa). In most high-strength concrete construction projects, however, coarse aggregates of maximum size less than 3/4 in. (19 mm) have been used.

High-strength concrete mixtures, which have a high cementitious materials content, are not dependent on the finer fraction of fine aggregate to supply the fine material required for lubrication and workability. Fine aggregate for use in ordinary concrete has 10–30 percent of particles passing a No. 50 sieve (0.30 mm) and 2–10 percent passing a No. 100 (0.15 mm). But in the proportioning of high-strength concrete, the percentage of particles passing a No. 50 sieve is no more than 20, and that passing a No. 100 is less than 5. Tests show that the optimum sand content in relation to workability and strength is between 40–50 percent by weight of total aggregates. The amount of coarse aggregate depends mainly on its nominal size; the larger the size, the higher the amount.

The quality and strength of concrete also affect the strength of high-cement mixtures. Angular aggregates such as crushed stone give better bond and higher strength than rounded particles of gravel.

To obtain high strength, concrete mixtures should have low w/c ratio, typically in the range of 0.32–0.4. The limit on water produces very stiff mixtures, with very low workability. The use of water-reducing admixtures is essential to make such concrete workable.

Type I cement is preferred over Type III, and mineral admixtures are commonly employed to favor the long-term strength gain. The fineness and composition of the cement, which vary significantly between brands, are two key factors for maintaining uniformity in concrete strength. Cements of high C_3S content and moderate fineness (Blaine fineness of about 350 m^2/kg) are preferred. The cement content in high-strength mixtures is high, in the range of 8.5–11.5 bags/yd^3 (475–640 kg/m^3). The strength increases, although at a decreasing rate, as the cement content increases. But at the same time, cement efficiency, which is the strength per pound (or kg) of the cementitious materials, decreases. When the cement content exceeds about 11.5 bags/yd^3, the strength may start to decrease.

When added at low proportions, fly ash has beneficial effects on both strength and workability. It also helps to lower the heat of hydration and control cracking. Silica fume (micro-silica) is added to increase the strength by increasing the density of concrete through pozzolanic action.

High-strength concrete—which is used in high-rise buildings, bridges, and precast applications—has a number of advantages. Both the favorable and unfavorable properties that accompany the higher strength are listed in the following:

- Reduction in cross-sectional area of structural elements, primarily columns in buildings and bridge piers
- Lower amount of reinforcing steel
- Increased usable floor space
- Low permeability and excellent durability
- Decrease in dead loads on foundations
- Low shear strength
- Large amounts of creep and shrinkage due to higher amounts of cement and lower amounts of aggregates (in comparison to ordinary concrete)
- Low ductility or small failure strain, and brittle failure
- Increased cost

3.12.6 EXAMPLES

Example 3.13

Find the failure loads in the following two simple beams when subjected to centerpoint loading. Use $f'_c = 3000$ psi and grade 50 steel. Span length 6 ft.

(a) Plain concrete beam of width 8 in., height 15 in.

(b) Reinforced concrete beam of width 8 in., height 15 in., reinforced with two No. 6 bars at a depth of 12 in. from the top.

Solution. (a) Plain concrete beam:

$$\text{Modulus of rupture} = 12(f'_c)^{0.5}$$

$$= 12(3000)^{0.5}$$

$$= 657 \text{ psi}$$

$$\text{Failure moment} = \frac{(\text{MOR})(I)}{c}$$

$$\text{Moment of inertia, } I = \frac{8(15)^3}{12}$$

$$= 2250 \text{ in}^4$$

$$c = 7.5 \text{ in.}$$

$$\text{Failure moment} = \frac{657(2250)}{7.5}$$

$$= 197{,}100 \text{ in.-lb}$$

In a simple beam subjected to centerpoint loading,

$$M = \frac{PL}{4}$$

where P is the load and L is the span length.

$$\frac{P(6 \times 12)}{4} = 197{,}100$$

$$\text{Failure load, } P = 10{,}950 \text{ lb}$$

(b) Reinforced concrete beam:

Depth of steel from compression face (top), $d = 12$ in.

$$\text{Area of steel, } A_s = 2(0.44)$$

$$= 0.88 \text{ in}^2$$

$$\text{Yield strength, } f_y = 50{,}000 \text{ psi}$$

$$\text{Width of beam, } b = 8 \text{ in.}$$

$$\text{Stress block depth, } a = \frac{A_s f_y}{0.85 f'_c b}$$

$$= \frac{0.88(50.000)}{0.85(3000)8}$$

$$= 2.16 \text{ in.}$$

$$\text{Failure moment} = 0.88(50{,}000)(12 - 0.5 \times 2.16)$$

$$= 480{,}480 \text{ in.-lb}$$

$$\text{Failure load, } P = \frac{480{,}480(4)}{6 \times 12}$$

$$= 26{,}693 \text{ lb}$$

3.13 OTHER CEMENTITIOUS MATERIALS

In addition to concrete, a number of other construction materials are manufactured using portland cement as a basic ingredient. Plaster (stucco), grout, mortar, and shotcrete are made using cement and various types of aggregates. The manufacture, properties, and uses of these materials are explained in this section.

3.13.1 Plaster and Stucco

Plaster is a fluid mixture of cement, lime, and sand, used as a finishing material. It should have the ability to adhere to a substrate such as concrete, masonry, or lath, and to itself (Fig. 3.48). *Stucco* is a common type of plaster, used on wood or masonry walls, which can be placed and formed in a variety of shapes, designs, and textures.

Plaster serves two primary functions:

- Appearance
- Protection

Moreover, a good plaster is incombustible, termite-proof, and possesses excellent insulating properties.

Stucco is applied in three coats: the *scratch coat,* the *brown coat,* and the *finish coat.* The scratch coat can be applied directly onto concrete or masonry walls, or onto a rough surface made from woven, welded, or expanded metal lath (metal reinforcement). *Woven wire lath* is fabricated from galvanized-steel wire, with or without stiffener wire backing. *Welded wire lath* is fabricated from copper-bearing, cold-drawn, galvanized-steel wire, by welding it into an intersecting grid pattern. *Expanded metal lath* is a fabric manufactured from coils of steel that are slit and then expanded, forming a diamond pattern. *Polyvinyl lath* is also available, but should not be used when extreme variations in temperatures are expected. Plaster strip should be installed wherever plaster or stucco terminates or abuts with dissimilar materials; control joints may be required over large surfaces.

Figure 3.48 Plaster ready for application.

For direct application over masonry walls, the surface should be rough, clean, and firm. Proper suction is essential for good bonding of stucco—on masonry or concrete work, and also between coats. Suction is obtained by uniformly dampening—but not wetting—the entire wall surface before applying the stucco.

The scratch coat is typically about 3/8 in. (10 mm) thick. Before it hardens, the scratch coat should be crosshatched, which will act as a key for the brown coat. The scratch coat should be kept continuously wet after it sets (for at least two days) and then allowed to dry before applying the brown coat.

Before spreading on the brown coat, which is also about 3/8 in. (10 mm) thick, the scratch coat should be dampened uniformly to provide suction. The surface of the brown coat is cross-scratched lightly to provide a key for the finish coat. The brown coat is kept wet for two days and then allowed to dry.

A seven-day interval should pass between the application of brown and finish coats. Before the finish coat, the brown coat surface should be dampened uniformly to provide suction. Finish color and texture are added to the mixture during the finish coat. The stucco should be cured; fogging is the preferred method of curing.

Fresh plaster should have good adhesion, cohesion, and workability. Adhesion is the capability to stick to a surface, which is influenced by aggregate, w/c ratio, and properties of the substrate. Cohesion is the ability of plaster to stick to itself, and is affected by the type of mix, aggregate properties (gradation and size), and mix proportions. Workability is the ease with which plaster can be placed, shaped, floated, and troweled, and depends primarily on cohesion and adhesion of the mix.

Well-graded sand is important in the application of plaster or stucco. For base coats, coarser aggregate particles within the allowable sizes are desirable. For the finish coats, finer sizes are suited. High-grade mineral pigments may be used in the finish coat for color effects. Normal portland cement with or without lime is used for the mix.

The proportioning standards for plaster are given in ASTM C926. Generally, 1 part cement, 3/4 part lime, and 4–8 parts sand (by volume) give a satisfactory mix. Water should be sufficient to produce a plastic consistency. Excessive water lowers the strength, increases the shrinkage, cracking, and crazing, and lowers the freeze–thaw resistance. The current trend is to use a special type of cement called plastic cement, which contains approximately 96 percent portland cement plus a plasticizing admixture.

3.13.2 Mortar

Mortar (or masonry mortar) is a mixture of cement, lime, fine aggregate, and water. There are two types of masonry mortar:

- Lime mortar
- Portland cement–lime mortar

Lime mortar is a mixture of sand, hydrated lime, and water. The consistency of the mixture should be such that it is cohesive and easy to trowel, and adheres to masonry units. As it is relatively weaker, lime mortar is generally used in temporary construction, from which masonry units can be salvaged and reused.

Portland cement–lime mortar is a mixture of portland cement, lime, sand, and water. The lime helps to improve the workability and increase the body (or fattiness) of the mortar. It also enhances the water retentivity of the fresh mix and the water-tightness of the hardened mortar. The flexure and shear strengths of masonry (which depend on the bond strength and the strength of the units) are increased with the addition of lime.

Alternatively, portland cement mortar can be manufactured using masonry cement in place of the regular portland cement and lime. The addition of proper amounts of sand and water to masonry cement is all that is necessary in ordinary construction. Masonry cement, which is specially manufactured for masonry construction, is portland cement containing supplementary materials selected to impart workability, plasticity, and water retention. Any one or combination of portland–blast-furnace slag cement, portland-pozzolan cement, natural cement, slag cement, or hydraulic lime may be used with normal portland cement. In addition, masonry cement usually contains one or more of the ingredients limestone, chalk, hydrated lime, and clay.

Mortar is used to bind structural masonry units together, and in so doing, acts as an adhesive and sealant. Whereas compressive strength is the most important property of concrete, the same is not true for masonry mortar. The primary function of masonry mortar is to develop a complete, strong, and durable bond with the masonary units. As a sealant, mortar should prevent the movement of water through the wall.

Specifications of sand for masonry mortar (ASTM C144) are as shown in Table 3.18. Further details on masonry mortar and its properties are given in Chapter 4.

TABLE 3.18 SPECIFICATIONS FOR MASONRY MORTAR

	Percent passing sieve no.						
Type	4	8	16	30	50	100	200
Natural sand	100	95–100	70–100	40–75	10–35	2–15	—
Manufactured sand	100	95–100	70–100	40–75	20–40	10–25	0–10

Source: ASTM C144. Copyright ASTM. Reprinted with permission.

3.13.3 Grout

Grout is a mixture of cement (or cement plus lime), fine aggregate, pea gravel (or finer coarse aggregate, generally 3/8 in. (10 mm) maximum), and water, with a consistency that allows for pouring without segregation. The term "grout" is derived from the Swedish word *groot,* which means "porridge." It is used to fill grout spaces in brick masonry 2 in. (5 cm) or more in horizontal dimension, and in cells of hollow block masonry.

The materials are generally mixed in a mixer and poured into the grout spaces. The mixture should be cohesive without segregation, and have a slump of about 9 in. The sand used is generally of finer gradation, and (as mentioned), coarse aggregate with a maximum size of 3/8 in. (10 mm) is normally used. When the coarse aggregate is left out, the grout is called fine grout, and mixtures containing coarse aggregate of up to one to two times the volume of cement are called coarse grout. The volume of fine aggregate is generally between two and three times that of cement. Lightweight aggregates can also be used in place of gravel. Lime, up to 10 percent by volume (of cement), is occasionally used to increase the fluidity and to provide water retentivity.

Grout serves two functions: (1) to provide bond, and (2) to increase masonry volume for adequate bearing and fire resistance. The bond refers to continuity between individual masonry units as well as to connection between the units and the reinforcement. Settlement and shrinkage, which may cause cracking in grout and masonry, can be minimized by the use of a stiffer mix.

3.13.4 Shotcrete

Shotcrete (also called gunite, pneumatically applied mortar or concrete) relates to mortar or concrete shot into place using compressed air. The procedure is carried out using one of two methods:

- Dry mix process
- Wet mix process

In the former method, the dry materials—cement and aggregates—are thoroughly mixed with just enough water to prevent dusting, and then forced through a hose by compressed air. The remaining water is introduced under pressure at the nozzle of the gun, and the intimately mixed mortar is jettisoned out the nozzle at a high velocity. In the wet process, all ingredients, including water, are premixed to produce mortar or concrete, and the mixture is then pumped or pneumatically conveyed down the hose by compressed air to the nozzle, where high-velocity compressed air is injected to increase the velocity. The amount of water can be regulated by a valve in the hand of the operator. The wet process has less rebound and dust, allows better control of the w/c ratio, and is most suited for continuous placement in large structures. On the other hand, the dry process (which is the more common of the two) gives higher strength, requires less equipment, produces a more consistent mix, and is readily suitable for smaller jobs. In the shotcreting operations, the dynamics of aggregate particles hitting a surface after traveling through air at speeds of 50–200 mph (80–320 km/h) presents problems due to the loss of these particles. The loss, resulting from rebound off the work surface (hence the name *rebound loss*) can be as high as 50 percent, and depends on the gun velocity. Proper control of the gun operation can minimize this loss.

Shotcrete is commonly made with mortar mix—without coarse aggregates—but in some cases, up to 25–30 percent coarse aggregate, of size not exceeding 3/8 in. (10 mm), is used. The addition of silica fume at a rate of 5–10 percent of the weight of cement improves the compressive strength and durability of shotcrete.

A number of procedures should be followed before the application of shotcrete. The surface or substrate on which the shotcrete is to be applied should be cleaned with water under high pressure to remove dust and sand. It should be saturated, at least overnight, prior to the application, to aid in the flow of cement into the parent material. Wet curing, which is very important to develop the strength properties, should be done for at least seven days after shotcreting.

Plain or unreinforced shotcrete is relatively brittle and may crack due to pronounced tensile stresses. Shotcrete mixes are generally oversanded and therefore may require a high amount of cement for proper workability, leading to increased shrinkage cracking. Shrinkage can be high and of the order of 900×10^{-6}, which is about twice that of conventional concrete. Steel fibers, when introduced into shotcrete, improve its crack resistance, ductility, and energy absorption. Density and other properties of the material are not much different from those of ordinary mortar or concrete.

Shotcrete permits the placement of concrete without any formwork. It is employed for tunnel and reservoir lining, and in swimming pools. It is increasingly used for repair and strengthening of buildings, dams, and bridge piers. Other uses for shotcrete are as ground support in mining shafts, underground bulkheads, and for ground stabilization. It is also used as a protective coating for structural steel, masonry, and rock. One of the newest uses of shotcrete is direct application, along with mechanical anchors and wire mesh, to slopes for arresting erosion.

3.13.5 Soil Cement

Cement is also used to stabilize foundation soil (in pavement, buildings, slopes, and so on). *Cement-stabilized soil* (also called *soil stabilization, cement-treated aggregates, rammed earth,* and *soil cement*) is produced by mixing and compacting water, soil, and portland cement. The mixture is then allowed to cure.

Soil cement can be described as a hardened material formed by compacting soil, cement, and water. The amount of cement used is between 5–10 percent by weight of soil. Even though almost any type of soil is suitable for soil cement or soil stabilization, a granular type of soil is preferred to clay soil. The primary application of soil cement is as a base material for flexible pavements or asphaltic pavements. The thickness of a soil cement base depends on a number of factors, including subgrade type, traffic, and thickness of wearing course. Most in-service soil-cement bases are 4–8 in. (10–20 cm) thick.

Soil cement is also used as an erosion-resistant facing for slopes (slope protection or slope stabilization), and as a low-permeability lining material in wastewater lagoons, settling ponds, and coal storage yards. In addition, it is used for foundation stabilization and embankments. Soil cement is also used in the manufacture of soil-cement bricks.

3.13.6 Pervious Concrete and Cement-bonded Particleboard

Cement is the main ingredient in the manufacture of *no-fines pervious concrete.* This type of concrete, which is made with little or no fine aggregate, is used for a type of pavement that allows rainwater penetration. Fairly large void content (about 15 percent) makes this concrete permeable. Some compaction may be required during the placement of no-fines concrete, and curing is needed. About 10–20 percent addition of sand (relative to the volume of coarse aggregate) may improve the compressive strength.

Cement-bonded particleboard is a panel material manufactured with cement, wood fibers, and very little water. The materials are mixed and pressed to produce panels comparable to particleboard or plywood. These boards are used as a covering material in wood frame construction and as a base for laying floor tiles.

3.14 TESTING

A number of laboratory tests are described in this section. Following is a list of concrete properties and the corresponding test numbers.

Property	Test no.
Consistency of concrete, slump test	CON-1
Consistency of concrete, ball penetration test	CON-8
Unit weight and yield of concrete	CON-2
Air content of concrete	CON-2
Mixing of concrete	CON-4
Compressive strength of concrete	CON-3
Flexural strength of concrete	CON-6
Splitting tensile strength of concrete	CON-7
Modulus of elasticity of concrete	CON-5
Capping procedure	CON-9
Compressive strength of cement using tests on mortar cubes	CON-10
Rebound number of concrete	CON-11

Test CON-1: Slump Test of Portland Cement Concrete

PURPOSE: To determine the slump of plastic concrete.

RELATED STANDARDS: ASTM C143, C172.

EQUIPMENT: Slump mold, tamping rod (5/8-in. diameter), pan, scale, shovel, hand scoop.

SAMPLE: Minimum 0.3 ft^3 of plastic concrete.

PROCEDURE:

1. Start the test within 5 min. after obtaining the final portion of the composite sample.
2. Dampen the mold and place it on a flat moist pan.
3. Hold the mold firmly in place during filling (by standing on the two foot pieces).
4. Fill the mold in three layers, each approximately one-third the volume of the mold.

5. Rod each layer with 25 strokes of the tamping rod. In filling and rodding the top layer, heap the concrete above the mold before rodding is started.
6. Strike off the surface by screeding and a rolling motion of the tamping rod.
7. Remove the mold immediately by raising it in a vertical direction. (The entire test, from the start of the filling through removal of the mold, should be completed within $2\frac{1}{2}$ min.)
8. Place the empty mold (upside down) adjacent to the concrete sample and measure the vertical difference between the top of the mold and the displaced original center of the top surface of the specimen. This is the slump.

REPORT: Record the slump in inches to the nearest 1/4 in.

Test CON-2: Unit Weight, Yield, and Air Content of Concrete

PURPOSE: To calculate unit weight, yield, and air content of fresh concrete (gravimetric basis).

RELATED STANDARDS: ASTM C138, C172.

DEFINITIONS:

- *Yield* is the volume of concrete produced per batch, cubic yard, or cubic meter.
- *Air content* is defined as the percentage of air voids in concrete.

EQUIPMENT: 0.2-ft^3 metal cylindrical measure (bucket), 5/8-in. tamping rod, balance, mallet (with a rubber or rawhide head) weighing approximately 1.25 lb, flat trowel. (*Note:* Bucket measure is for maximum size of coarse aggregate equal to or smaller than 1 in.)

SAMPLE: A minimum of 0.3 ft^3 of fresh concrete.

PROCEDURE:

1. Weigh the empty measure.
2. Fill the measure with concrete sample in three layers of approximately equal volume. Rod each layer with 25 strokes of the tamping rod. Add the final layer so as to avoid overfilling.
3. After each layer is rodded, tap the sides of the measure smartly 10–15 times with the mallet. (This procedure is required to release any trapped air bubbles.) After consolidation, the measure must not contain any excess

concrete protruding more than approximately 1/8 in. (3 mm) above the top of the mold.

4. Strike off the top surface with a sawing motion of the flat trowel (using little vertical pressure).
5. Clean all excess concrete from the exterior of the measure.
6. Weigh the measure with concrete.
7. Calculate the unit weight of concrete (W) as the ratio between net weight of concrete and volume of the measure.

$$W = \frac{\text{net weight of concrete}}{\text{volume of measure}} \text{ lb/ft}^3 \text{ (kg/m}^3\text{)}$$

8. Calculate the total weight of all materials batched (sum of the weights of the cement, the fine aggregate, the coarse aggregate, the mixing water, and any admixture) = W_1.

$$W_1 = W_c + W_s + W_g + W_w + W_a$$

where W_c, W_s, W_g, W_w, and W_a are the weights of cement, sand, gravel, water, and admixture, respectively.

9. Calculate yield as

$$\text{yield} = \frac{W_1}{27(W)} \text{ yd}^3\text{/batch}$$

10. Calculate the theoretical unit weight of concrete (on an air-free basis), T:

$$T = \frac{W_1}{V}$$

where V is the total absolute volume of the component ingredients in the batch, in ft^3 or m^3.

$$V = \left(\frac{W_c}{3.15} + \frac{W_s}{S_s} + \frac{W_g}{S_g} + \frac{W_w}{1} + \frac{W_a}{S_a}\right) \times \frac{1}{U_w}$$

where S_s, S_g, and S_a are the specific gravities of sand, gravel, and admixture, respectively, and U_w is the density of water (62.4 lb/ft^3).

11. Calculate the air content as

$$\text{air content} = \frac{T - W}{T} \times 100$$

REPORT: Report the values of unit weight, theoretical unit weight, yield, and air content.

Test CON-3: Compression Test of Concrete Cylinders

PURPOSE: To determine the compressive strength of concrete.

RELATED STANDARDS: ASTM C39, C192, C617.

EQUIPMENT: Concrete molds, tamping rod, mallet, trowel, scoop, compression test machine.

SAMPLE: Concrete mixed as in Test CON-4.

PROCEDURE:

1. Place the concrete in the molds using a scoop, trowel, or shovel. (Remixing in the pan or wheelbarrow to prevent segregation is allowed.)
2. Distribute the concrete inside the mold using a tamping rod prior to consolidation.
3. Fill the mold in three equal layers, rodding each layer 25 times using a 5/8-in. tamping rod.
4. Tap the sides of the mold gently with the mallet to close any voids.
5. Finish the top surface by striking it off with the tamping rod.
6. To prevent evaporation of water, cover the specimen immediately after finishing with a sheet of impervious plastic or wet burlap.
7. Remove the specimen from the mold not less than 20 h nor more than 48 h after casting.
8. Moist cure the specimens at 73.4 ± 3 °F (23 ± 1 °C) from the time of molding until the moment of test (by immersion in saturated-lime water or by storage in a moist room). Specimens are normally tested at 1, 3, 7, and 28 days, and 3, 6, and 12 months.
9. The ends of specimens that are not plane within 0.002 in. (0.05 mm) should be capped.
10. Compression tests are made as soon as practicable after removal from moist storage. The specimens are tested in moist condition.
11. Wipe clean the bearing faces of the upper and lower bearing blocks and of the test specimen.
12. Place the specimen on the lower bearing block.
13. Carefully align the axis of the specimen with the center of thrust of the spherically seated upper block.
14. Bring the upper block to bear on the specimen. Adjust the load to obtain uniform seating.

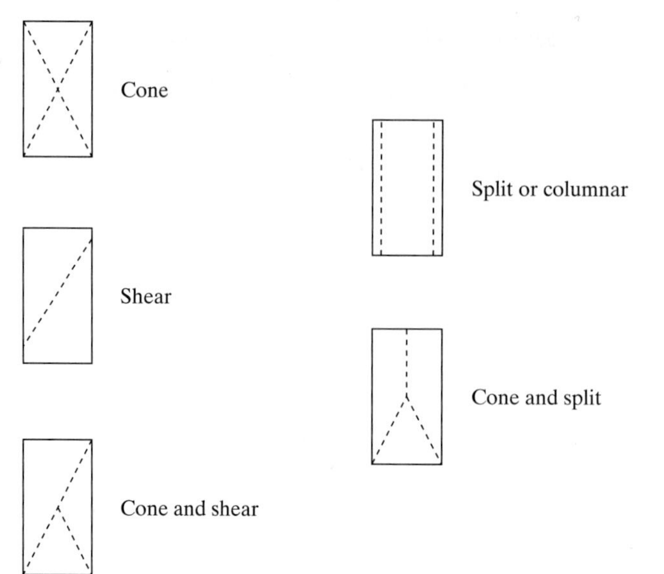

Figure 3.49 Types of concrete fracture.

15. Apply the load at a loading rate of 20–50 psi/s. (Time to failure for 3000-psi concrete is $1\text{–}2\frac{1}{2}$ m, and for 6000-psi concrete is 2–5 m.)
16. Apply the load (at the designated rate) until the specimen fails.

REPORT: Record the maximum load. Note the type of failure and the appearance of the concrete (Fig. 3.49). Calculate the compressive strength as the average of three specimens.

Test CON-4: Mixing of Concrete

PURPOSE: Procedure to mix concrete and obtain test samples.

RELATED STANDARDS: ASTM C192.

EQUIPMENT: Pan, tamping rod, mixer, shovel, scoop.

PROCEDURE TO MAKE CONCRETE SAMPLE BY HAND MIXING:

1. Hand mixing should be limited to batches of 0.25 ft^3 volume or less.
2. Mixing apparatus and accessories should be cleaned thoroughly.
3. Mix the batch in a damp, watertight metal pan.
4. Mix the cement, powdered admixture (if used), and fine aggregate.
5. Add the coarse aggregate and mix the entire batch without the addition of water until the coarse aggregate is uniformly distributed throughout the base.

6. Add water and admixture solution (if used), and mix until the concrete is homogeneous in appearance.

PROCEDURE TO MAKE CONCRETE SAMPLE BY MACHINE MIXING:

1. Add coarse aggregate, some of the mixing water, and admixture solution (if any) to the mixer.
2. Start the mixer.
3. Add fine aggregate, cement, and the balance of the water when the mixer is running. (If it is impractical to add these materials when the mixer is running, they may be added to the stopped mixer after permitting it to run a few revolutions following the addition of coarse aggregate and part of the mixing water.)
4. Mix for 3 min. followed by a 3-min. rest, followed by 2 min. of final mixing.
5. Tilt the mixer while it is running and pour the concrete into a clean and wet wheelbarrow.
6. Remove any concrete stuck in the mixer using scoop or trowel.
7. Remix concrete in the wheelbarrow using a shovel.

Test CON-5: Static Modulus of Elasticity and Stress-strain Curve of Concrete

PURPOSE: To determine Young's modulus of elasticity of a concrete cylinder by plotting its stress-strain curve.

RELATED STANDARDS: ASTM C469, C39, C192, C617.

DEFINITION:

- *Young's modulus of elasticity, E,* is the ratio between stress and strain within the working stress range (0–40 percent of ultimate concrete strength).

EQUIPMENT: Compression testing machine, compressometer.

SAMPLE: 6 × 12-in. concrete cylinders, capped and kept in moist condition.

PROCEDURE:

1. Determine the compression strength and unit weight of the sample using the compression specimens and procedure described in Test CON-3.
2. Attach the compressometer to the test cylinder.
3. Place the specimen (with the compressometer) on the lower platen of the testing machine.

4. Carefully align the axis of the specimen with the center of thrust of the upper block of the testing machine.
5. Lower the upper block slowly to bear on the specimen, and rotate the block gently by hand so that uniform seating is obtained.
6. Set up the compressometer for deformation reading.
7. Load the specimen at a rate of 35 ± 5 psi/s.
8. Upon reaching a load of about 20 percent of ultimate load, reduce the load to zero (at the same rate as the loading rate), and note the dial gage reading.
9. If the dial gage reading (deformation) is not zero, repeat steps 7 and 8 until the dial gage reading, upon unloading, is zero.
10. Start the final loading cycle and continue the loading until the maximum.
11. Record, without interruption in loading, applied load and longitudinal deformation at set intervals.
12. Calculate stress and strain as follows:

$$\text{Stress} = \frac{\text{load}}{\text{cross-sectional area of cylinder}}$$

$$\text{Deformation}, d = g \times I$$

where g is the dial gage reading and

$$I = \frac{e_1}{e_1 + e_2}$$

where e_1 is the eccentricity of the pivot rod from the axis of the specimen and e_2 is the eccentricity of the dial gage from the axis of the specimen

$$\text{Strain} = \frac{d}{\text{gage length}}$$

(*Note:* Gage length is the length between gage points, generally 6 in.)

13. Plot the stress-strain curve.
14. Calculate E to the nearest 50,000 psi (344.74 MPa) as follows:

$$E = \frac{S_2 - S_1}{\varepsilon_2 - 0.00005}$$

where S_2 is the stress corresponding to 40 percent of ultimate load, S_1 the stress corresponding to a strain of 0.00005, and ε_2 the strain at a stress of S_2.

REPORT: Report the compressive strength and unit weight of the concrete. Show the stress-strain diagram and measurement of E. Report the modulus of elasticity.

Test CON-6: Flexural Strength of Concrete

PURPOSE: To determine the flexural strength of concrete by the use of simple beam specimens with third-point loading.

RELATED STANDARDS: ASTM C78, C192, C39.

DEFINITION:

- *Modulus of rupture* is defined as the tensile strength of a material determined using a flexural specimen.

EQUIPMENT: Compression testing machine, loading apparatus (third-point).

SAMPLE: $6 \times 6 \times 20$-in. flexural specimen and concrete cylinders.

PROCEDURE:

1. Determine the compressive strength of the concrete and its unit weight using cylindrical specimens (Test CON-3).
2. Position the specimen under the bearing plate, and center the loading system in relation to the applied force.
3. Bring the load-applying block in contact with the surface of the specimen at the third points between supports (span length = 18 in.).
4. Apply the load continuously at a rate that constantly increases the extreme fiber stress between 125–175 psi/m (861–1207 kPa/m) until rupture occurs (about 2–6 m).
5. If the fracture occurs in the tension surface outside the middle third of the span length by more than 5 percent of the span length (about 1 in.), discard the result of the test.
6. Calculate the modulus of rupture (MOR), neglecting the beam weight, as follows:

   When the fracture initiates in the tension surface within the middle third of the span length,

$$\text{MOR} = \frac{Pl}{bd^2}$$

where P is the maximum load indicated by the testing machine, l the span length, d the beam depth, and b the beam width. For a beam with $b = 6$ in., $d = 6$ in., and $l = 18$ in.,

$$\text{MOR} = 0.0833P \text{ psi}$$

If the fracture occurs in the tension surface outside the middle third of the span length by not more than 5 percent of the span length,

$$\text{MOR} = \frac{3P(a)}{bd^2}$$

where a is the average distance between the line of fracture and the nearest support measured on the tension surface of the beam.

REPORT: Calculate the MOR value to the nearest 5 psi.

Test CON-7: Splitting Tensile Strength of Concrete

PURPOSE: To determine the tensile strength of concrete using cylindrical specimens.

RELATED STANDARDS: ASTM C496, C192.

EQUIPMENT: Compression testing machine, two 1/8-in.-thick plywood bearing strips (slightly longer than the specimens, and 1 in. wide).

SAMPLE: Concrete cylindrical specimens kept in a moist condition (Test CON-3).

PROCEDURE:

1. Determine the unit weight and the compressive strength of the concrete cylindrical specimens.
2. Draw diametral lines on each end of the specimen so that they are in the same axial plane.
3. Center one of the plywood strips along the center of the lower bearing block.
4. Place the specimen on the plywood strip and align so that the lines marked on the ends are vertical and centered over the plywood strips.
5. Place the second plywood strip lengthwise on the cylinder, centered on the lines marked on the ends.
6. Apply the load continuously at a constant rate of 100–200 psi/m of splitting tensile stress until failure (about 3–6 min./test).
7. Record the maximum load at failure, P.
8. Calculate splitting tensile strength, f_t, as

$$f_t = \frac{2P}{\pi l d}$$

REPORT: Report the type of concrete, compressive strength, unit weight, and splitting tensile strength.

Test CON-8: Ball Penetration of Concrete

PURPOSE: To find the depth of penetration of a metal weight into freshly mixed concrete (as a means of determining the workability).

RELATED STANDARDS: ASTM C360, C172, C143.

EQUIPMENT: Ball penetration apparatus, wood float.

SAMPLE: Concrete of minimum depth at least three times the maximum aggregate size and no less than 8 in. (20.3 mm) in a pan or wheelbarrow. The minimum horizontal distance from the centerline of the handle to the nearest edge of the level (concrete) surface on which the test is made is 9 in.

PROCEDURE:

1. Bring the surface of the concrete to a smooth and level condition (using a wood float).
2. Set the base of the apparatus on the leveled surface with the handle in a vertical position and free to slide through the frame.
3. Lower the weight to the concrete surface and release slowly.
4. After the weight has come to rest, read the penetration to the nearest 1/4 in. (6.4 mm).
5. Take a minimum of three readings. If the difference between the maximum and minimum readings is more than 1 in. (25 mm), make additional measurements until three consecutive readings have been obtained that agree within 1 in.

REPORT: An average of three or more readings (when they agree within 1 in.) is taken as the penetration value. Report the penetration value to the nearest 1/4 in.

Test CON-9: Capping Cylindrical Concrete Specimens

PURPOSE: To cap concrete cylinders (for compression testing) using sulfur mortar.

RELATED STANDARD: ASTM C617.

EQUIPMENT: Capping plate (mold), alignment device (guide bars), melting pot, exhaust fan.

SAMPLE: Moist-cured concrete cylinders (with no moisture on the surface).

PROCEDURE:

1. Sulfur mortar should have a minimum compressive strength, at 2 h, of 5000 psi (34.5 MPa).
2. Prepare sulfur mortar for use by heating to about 265 °F (130 °C). Note that the flash point of sulfur is approximately 440 °F (227 °C). Also, fresh sulfur mortar must be dry at the time it is placed in the pot, as dampness may cause foaming.
3. Oil the capping plate lightly.
4. Stir the molten sulfur mortar immediately prior to pouring each cap.
5. Dry the ends of the moist-cured specimens to preclude the formation of steam and foam pockets in the cap.
6. Pour the molten sulfur mortar in the plate, and lower the specimen, ensuring that the axis of the specimen is perpendicular to the plate.
7. The cap should have a thickness of about 1/8 in. (3 mm), but less than 5/16 in. (8 mm).
8. Remove the specimen from the plate using a slight twisting motion.
9. Cap both the ends of the specimen.
10. Maintain the specimen in moist condition between the completion of capping and the time of testing.

Test CON-10: Compressive Strength of Portland Cement Using Tests on Mortar Cubes

PURPOSE: To determine the compressive strength of portland cement by testing cement mortar cubes. (For Type I cement, minimum strengths of 1800 psi (12.4 MPa) at 3 days and 2800 psi (19.3 MPa) at 7 days are required).

RELATED STANDARDS: ASTM C109 and ASTM C150.

EQUIPMENT: Balance, compression testing machine, standard 2-in. (50-mm) cube molds, mixer, standard $1/2 \times 1$ in. (13×25 mm) wood tamper, trowel, dry cloth.

SAMPLE: Portland cement mortar samples of 1 part portland cement to 2.75 parts graded sand (ASTM C778) and 0.485 parts water (0.46 for air-entraining cement) by weight. (To make six samples of ordinary portland cement, take 500 g cement, 1375 g graded sand, and 242 g water).

PROCEDURE:

1. Coat the inside faces of the mold with light grease or release agent. Remove excess oil with a dry cloth.
2. Mix the ingredients and allow the mixture to stand for 90 s without covering.
3. Start making the specimens within a total elapsed time of not more than $2\frac{1}{2}$ min after completion of the original mixing.
4. Place a 1-in. (25-mm) layer of mortar in all of the cube compartments.
5. Tamp each compartment 32 times in 10 s to ensure uniform filling.
6. After tamping of the first layer in all of the cube compartments, fill with the remaining mortar.
7. Tamp the second layer as described for the first layer.
8. Cut off the mortar to a plane surface flush with the top of the mold, using a trowel.
9. Immediately after completion of molding, keep the specimens—in the molds on the base plate—in a moist room for 20–24 h, with the upper surface protected from dripping water.
10. After 24 h, place the specimens in saturated lime water in storage tanks.
11. Test the specimens immediately after their removal from the moist room or storage tanks. The rate of loading should be such that the maximum load is reached in not less than 20 s nor more than 80 s from start of load.

REPORT: Record the maximum load and calculate the compressive strength. The strength of the cement is the average strength of three cubes molded from a single batch of mortar and tested at the same age.

Test CON-11: Rebound Number of Concrete

PURPOSE: To determine the rebound number of hardened concrete (to assess the uniformity of concrete in situ, and to identify areas of poor quality concrete).

RELATED STANDARDS: ASTM C805.

EQUIPMENT: Rebound hammer: a spring-loaded steel hammer that, when released, strikes a steel plunger in contact with the concrete surface. The rebound distance of the hammer from the steel plunger is measured on a linear scale attached to the frame of the instrument.

SAMPLE: Rigidly supported concrete members of at least 4 in. (100 mm) thickness and of approximately the same age, temperature, and moisture content. Surfaces that are heavily textured, soft, or scarred with loose materials must be ground smooth with an abrasive stone (a silicon carbide stone should be supplied with the instrument). Troweled or smooth-formed surfaces are tested without grinding. Dry or carbonated surfaces must be kept moist for 24 h before testing.

PROCEDURE:

1. Hold the instrument firmly in a position that allows the plunger to strike perpendicularly to the surface tested.
2. Gradually increase the pressure on the plunger until the hammer impacts.
3. After impact, record the rebound number.
4. Repeat ten times for each test area. No two impact tests (points of impact) can be closer together than 1 in. (25 mm).
5. Disregard readings differing by more than 7 units from the average (of ten readings). If more than two readings differ from the average by 7 units or more, disregard the entire set of readings.

REPORT: Include details of structure, location of test area, description of test area, description of concrete, design strength, age, hammer type and serial number, and average rebound number for each test area.

PROBLEMS

1. What are the three main compounds in the raw materials for the manufacture of cement?
2. What is the purpose of adding gypsum to cement?
3. What are the four major compounds of portland cement?
4. Which is the most reactive compound in portland cement?
5. Name a nonhydraulic cement.
6. Which type of cement gives higher early strength?
7. Which is the major chemical compound in cement?
8. Name two hydraulic cements.
9. What are Type A cements?
10. Define consistency and workability.
11. What is slump and how is it measured?
12. What are the three considerations in mix design?
13. What is the maximum size of coarse aggregate in a 6-in. thick concrete wall.

14. How does the maximum size of aggregate affect the permeability of concrete?
15. If the size of aggregate is increased from 1/2 to 1 in., how does that affect the compressive strength?
16. Draw a typical relationship between compressive strength and water-to-cement ratio.
17. What is sand-lightweight concrete?
18. What is air-entrainment and when is it used?
19. What is freeze-thaw resistance, and how can it be improved?
20. Calculate the weight of cement required to make 2 yd^3 of concrete of mix proportions: 1C : 2.5FA : 3.5CA : 0.5W by weight with 2 percent air. (The specific gravity values of cement, FA, and CA are 3.15, 2.65, and 2.75 respectively.)
21. What is the number of sacks of cement per cubic yards in the preceding problem?
22. 14.3 lb of cement is used to make three 6 × 12-in. cylinders. How many sacks are required per cubic yards of concrete?
23. Mix design calculations show a dry mix of 1C : 2FA : 3.5CA : 0.55W by weight. The moisture content and absorption of coarse aggregate are 3 percent and 1 percent, respectively, and of fine aggregate, 8 percent and 3 percent, respectively. Find the field mix proportions.
24. Find the amounts of fine and coarse aggregates to make 2 ft^3 of concrete in the preceding problem. The specific gravity values of cement, FA, and CA are 3.15, 2.6, and 2.65 respectively.
25. The weight of water required for a trial batch is 270 lb/yd^3. If the w/c ratio is 0.55, what is the required amount of cement?
26. What are the two factors that affect the amount of water required per cubic yards of concrete?
27. In the mix design procedure, which property controls the w/c ratio?
28. In the mix design procedure, which property controls the amount of cement?
29. What is bleeding, and how is it controlled?
30. What is segregation, and how can it be prevented?
31. What is curing, and why is it done?
32. What are the various methods of curing?
33. What is the tensile strength of concrete of compressive strength 4000 psi?
34. What is the modulus of elasticity of concrete of compressive strength 6000 psi and density 125 pcf?
35. If the 28-day strength of concrete is 4000 psi, what is the expected strength at 7 days?
36. If the specified strength of concrete is 4000 psi, what is the expected average strength?
37. What is the primary reason for providing reinforcing steel in concrete?
38. Show the position of the reinforcing steel in a cantilever beam.
39. What are the two most important pieces of information inscribed on the surface of a reinforcing steel bar?
40. What is the value of the bulk density of cement?
41. Name two common chemical admixtures.
42. Name two common mineral admixtures.

43. What are the two classes of fly ash?
44. What are the advantages of adding fly ash to concrete?
45. What is the major compound in silica fume?
46. Give two reasons for adding a superplasticizer.
47. What is shotcrete, and what are the two methods of shotcreting?
48. Explain the purposes of adding fibers to concrete.
49. Explain the difference between a control joint and a construction joint.
50. Describe the effects of creep and shrinkage.

4

Masonry

The term "masonry" refers to a construction material formed by combining individual masonry units, such as stone and brick, with a binding material: mortar. Masonry is mainly used to build walls—vertical structural elements, thin in proportion to their length and height—that serve to enclose or divide a space and to support loads from other elements. It is one of the oldest materials of construction still in popular use. Examples of masonry can be found in historical landmarks all over the world, including the pyramids of Egypt, the Great Wall of China, the Taj Mahal of India, Roman and Greek ruins, the arches and vaults of Syria, and the great cathedrals of Europe. Today, masonry is used in various types of applications ranging from high-rise buildings to compound walls to water retaining structures (the world's largest masonry dam is the Roosevelt Dam in Arizona's Salt River Valley, built by U.S. Bureau of Reclamation engineers between 1903 and 1911).

Based on their location, masonry walls are divided into two types:

- Exterior type
- Interior type

Exterior walls are those exposed to the exterior environment on at least one side; interior masonry, as in a wall that divides two adjacent rooms, has both sides exposed to an interior environment.

Based on structural requirements, masonry walls are also divided into two types:

- Load bearing
- Non–load bearing

Load-bearing walls, also called structural walls, are those designed to carry loads from other members, whereas non–load-bearing walls carry their own weight only. For example, a partition wall that does not support floor or roof loads is a non–load-bearing interior wall.

Based on the method of construction, walls in general can be divided into two types:

- Solid wall or Hollow wall
- Framed wall

Solid and hollow walls (Fig. 4.1) are those built using masonry units (stone, brick, or block), whereas framed walls are constructed with timber or metal. Thus, a masonry wall is a hollow or solid wall built by combining individual masonry units (hollow or solid) with mortar, the binding material. Framed walls are constructed using thin sections of wood or metal joined together to provide strength and rigidity; they are finished by forming a skin of thin panel materials such as plywood, gypsum board, or stucco on both sides of the frame to fulfill functional requirements.

Each type of wall (solid and hollow, and framed) has different characteristics that satisfy its functional requisites. One type may have good resistance to fire but

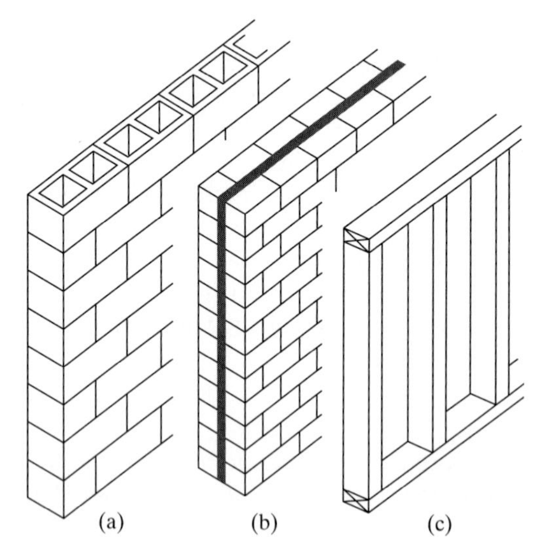

Figure 4.1 (a) Hollow and (b) solid masonry walls; (c) framed wall.

may possess poor insulating properties against heat (transfer of heat). Another type may have poor resistance to water penetration but good insulation against heat transfer.

Wall materials undergo dimensional changes (expansion and contraction) with changes in temperature and moisture. The stability of a wall may be affected by expansion if the movement is not accommodated in the building. Such walls should be provided with expansion joints along their length and also between the walls and the frames they are built into.

Masonry walls are generally built as single or double wythe (Fig. 4.2)—the *wythe* is the portion of a wall that is one masonry unit in thickness. *A bonded wall* is a wall in which two or more of its wythes of masonry are adequately bonded together to act as a structural unit. When the face and back wythes are completely separated except for metal ties that serve as cross ties or bonding elements, there exists a cavity between the wythes, and the wall is called a *cavity wall.* Neither of the two

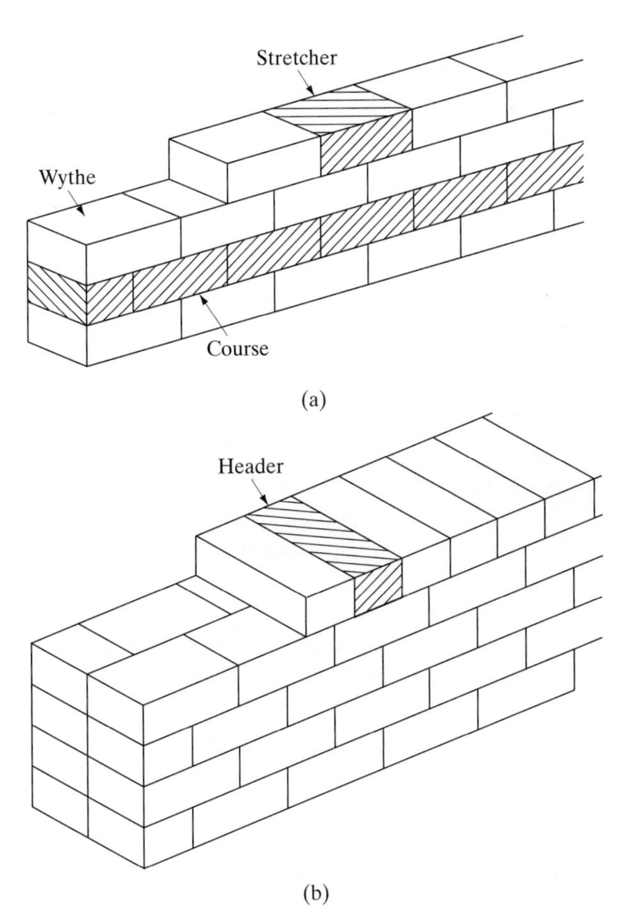

Figure 4.2 (a) Single-wythe wall; (b) double-wythe wall.

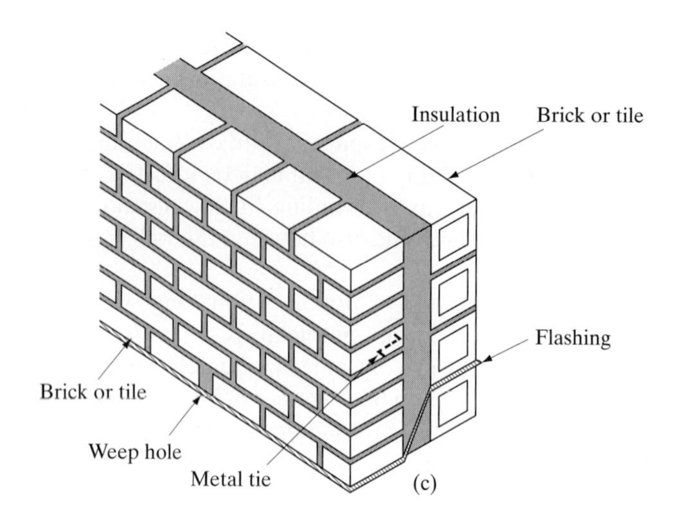

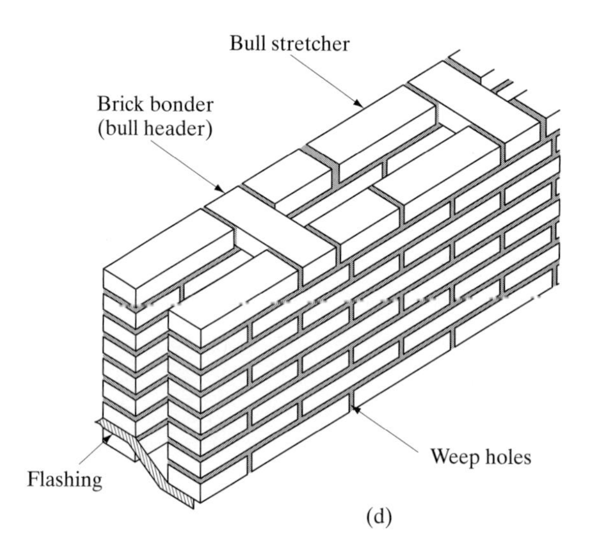

Figure 4.2 (*Continued*) (c) Cavity wall; (d) masonry-bonded hollow wall.

wythes can be less than 4 in. (10 cm) in thickness, except when both wythes are constructed with clay or shale brick, in which case the limit is 3 in. (7.5 cm). The cavity between the two wythes must be between 1 and 4 in. (2.5 and 10 cm) in width. A *veneered wall* is a nonstructural wall, exterior or interior, built using facing brick, concrete, stone, tile, metal, plastic, or similar material attached to a backing layer for the purpose of ornamentation, protection, or insulation. An *exterior veneer* is applied to weather-exposed surfaces, and an *interior veneer* is laid onto surfaces other than weather-exposed surfaces.

In the following sections, masonry construction and various masonry materials and their properties will be discussed.

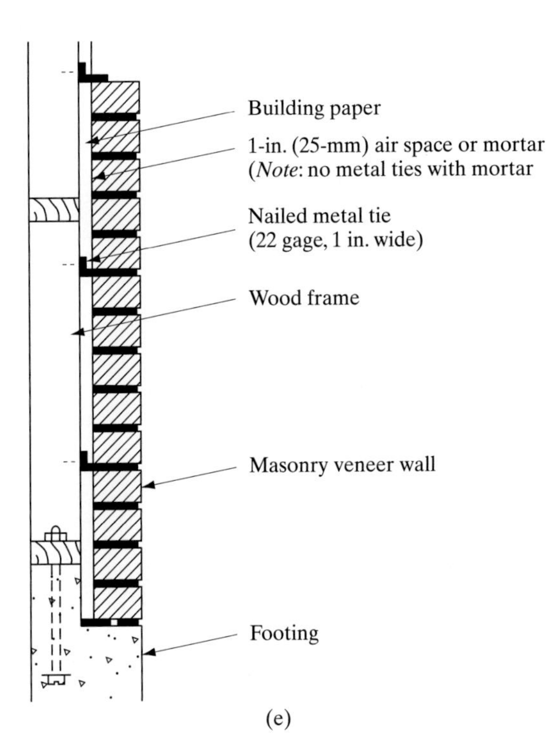

Figure 4.2 (*Concluded*) (e) Brick-veneered wall.

4.1 MASONRY UNITS

A masonry unit is a brick, tile, stone, glass block, or concrete block that conforms to certain ASTM product standards (Fig. 4.3). A *hollow masonry unit* is defined as having a net cross-sectional area in every plane, parallel to the bearing surface, less than 75 percent of the gross cross-sectional area in the same plane; a *solid masonry unit* has a net cross-sectional area in every plane, parallel to the bearing surface, 75 percent or

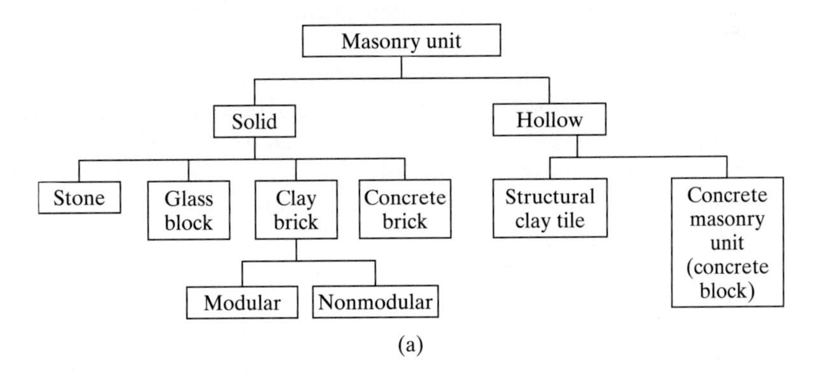

Figure 4.3 (a) Types of masonry units.

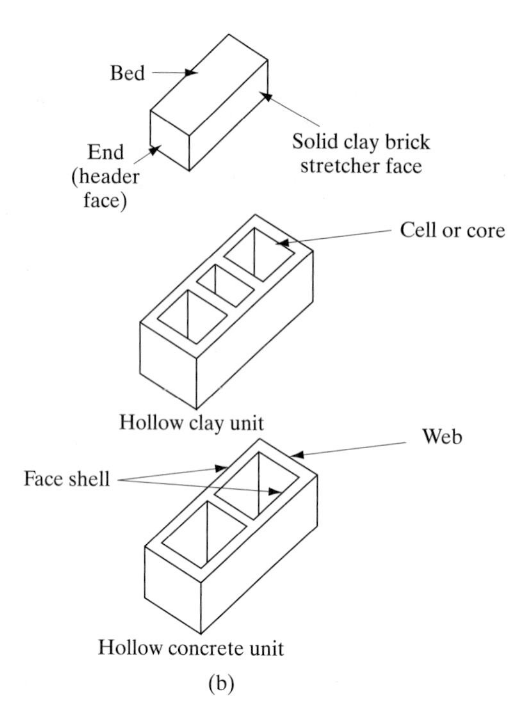

Figure 4.3 (*Continued*) (b) Solid and hollow masonry units.

more of the gross cross-sectional area in the same plane. Generally, clay units are solid and concrete units are hollow.

4.1.1 Clay Bricks and Structural Clay Tiles

A clay *brick* is a small solid block, usually rectangular, of burned clay (Fig. 4.4). Note that a solid block of concrete and sand-lime (calcium silicate) is also called a brick. A *structural clay tile* is a hollow clay unit, larger than a brick, developed for use where lightweight masonry is required—as in filler panels and partition walls. Structural tiles are also described as hollow burned clay masonry units with parallel cells or cores.

Bricks and tiles are manufactured from clay. The word "brick" comes from the French word *brique.* Brickmaking dates back to ancient times. Large brick kilns dating back to about 2000 B.C. have been found during excavations at Lothal, near Ahmedabad in the state of Gujarat in India. The famous Ishtar Gate in Babylon, built around the end of the seventh century B.C., was faced with kiln-baked bricks bedded in bitumen.

The art of brickmaking seems to have spread from Egypt to Greece and then to Rome. In Roman times, the size of a brick became more standardized. Many Roman bricks (well-burned broad and thin bricks—not more than $1\frac{1}{2}$ in. (38 mm) thick) from the middle of the first century B.C. onward bore a decorative motif, such as an animal head or bird. The use of bricks of nearly uniform size became common

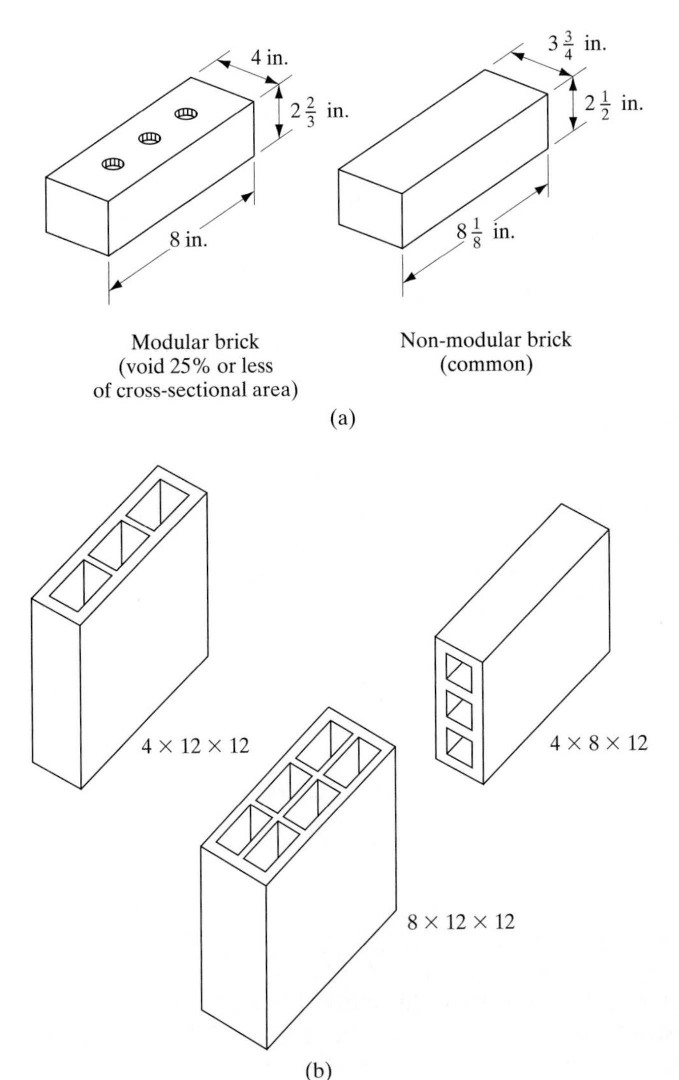

Figure 4.4 (a) Clay bricks (solid clay units). (b) Structural clay tiles (typical nominal dimensions are shown).

throughout Europe during the beginning of the thirteenth century. In England, there was a marked development in brick construction during the middle of the fifteenth century, in which period many castles were built; many manor homes and churches were erected during the latter part of that century.

The first extensive use of bricks in the United States was around the beginning of the seventeenth century, near Richmond, Virginia. A number of houses were built there, the first stories of which were of bricks made on-site by brickmakers brought from England. But it is only in the last century that bricks have found an important role in engineered construction in the United States. Until around 1900, possibly as a result of variations in the quality of bricks, clay bricks were employed primarily in the construction of sewers, bridge piers, and tunnel linings. An exception to this is the Monadnock Building in Chicago, built around 1891, which is 16 stories tall with stone and brick walls 6 ft thick at the base, and is generally cited as the last great building in the ancient tradition of masonry architecture.

Structural clay tiles are the most recently developed clay masonry units and were introduced in the United States in 1875. They were developed in response to the need for a lightweight backing material for the facing masonry used as cladding in high-rise buildings. Their webs can be laid in horizontal or vertical position, and the walls thus built can be employed in load-bearing and non–load-bearing applications. Structural clay tiles that can be plastered come in ribbed or wire-cut textures.

4.1.2 Manufacture of Bricks

Clays for brickmaking differ widely in composition from place to place. Even in the same site, clay deposits from one part or depth may differ significantly from those from another part or depth. Clays are composed mainly of silica (grains of sand) and alumina. Alumina is the soft, plastic part of the clay, which readily absorbs water, makes the clay plastic, and melts when burned. Present in all clays, in addition to these two compounds, are lime, iron, manganese, sulfur, and phosphates. The proportion between the compounds varies from place to place. Iron is useful in improving the hardness and strength of bricks. Lime present in clays will decompose during burning and promote shrinkage and disintegration when let to remain in bricks.

Types of Clay. There are three types of clays that are used in brick manufacture:

- Surface clays
- Shales
- Fire clays

Surface clays are found near the surface of the earth. They are unconsolidated and unstratified material, with a high oxide content (about 10 to 25 percent). Shale is also clay in its natural state, but as a result of compression due to heavy soil above is quite firm and has compressed flaking characteristics. Most shales are not soluble in water except when ground, becoming plastic with the addition of water. They are costlier to remove from the ground and contain large amounts of fluxes.

Fire clay is a material that occurs at greater depths than either surface clays or shale. It has more uniform physical characteristics and chemical composition, and is able to withstand high temperatures. Fire clays contain less oxide (2 to 10 percent), which raises their softening point much higher than that of surface clays or shales. This gives refractory qualities to bricks manufactured with fire clays, and allows them to withstand higher temperatures.

The best material for brickmaking is clay containing about 30 percent sand and silt, because the presence of sand reduces the shrinkage occurring during the burning of soft clays. To be molded and shaped, clays must have plasticity when mixed with water. Further, they should have sufficient tensile strength to retain their shape after forming.

Chemical analyses of clays show that they are composed principally of hydrated oxides of aluminum and iron, and hydrated silicates of aluminum. Gypsum present in clay in crystal form, as an impurity, leads to efflorescence and may also cause bricks to crack. In brick manufacture, clay is ground or crushed in mills, mixed with water (to make it plastic and moldable), shaped to the size of a brick, textured, dried, and finally fired. Mixing of clay with water is done in pug mills. After the units are burned, they are cooled and stored for shipment.

The majority of clays burn to a red color when fired at a temperature of 900–1000 °C (1650–1830 °F). Above this temperature, the color turns to dark red or purple, and then to brown or gray at about 1200 °C (2200 °F). Some clays may melt at this temperature. When the temperature is low, the color turns to pink or dull brown. A high iron content makes the brick salmon pink at around 900 °C (1650 °F), changing to darker red at higher temperatures. When lime or chalk is present in the clay, the brick may turn to white or cream.

Brick Manufacture. There are three methods of manufacturing bricks and tiles:

- Stiff-mud process
- Soft-mud process
- Dry press process

In the stiff-mud process (also called the wire-cut process), clay containing a minimum amount of water, generally 12–15 percent by weight, is forced through a die. The die molds the mass into desired shapes and sizes for bricks, tiles, and other products. The continuous band of clay that is forced out is later cut into bricks by a wire frame. If the cross-section of the band is the same as the end of the bricks, the bricks are called *end cut,* and if the cross-section is the same as the side of the bricks, they are called *side cut.* Most bricks and structural clay tiles are manufactured using the stiff mud process, which produces the hardest and most dense of the machine-made bricks.

In the soft-mud process, which is well suited to clays containing too much water in their natural state, ground clay is hydraulically pressed in steel molds. This process is used for brickmaking only and is the oldest method in brick production. All handmade bricks are made using this process. When the insides of the mold are sanded to prevent sticking of clay, the product is *sand-struck brick.* When the molds are wetted to prevent sticking, the product is called *water-struck brick.*

The dry press process, which is suited for clays possessing low plasticity, consists of dropping the moist clay (mixed with about 7–10 percent water) into dry press forming machines, where the bricks are molded under low operating pressures. The method uses the least water in tempering and makes bricks in the same way as concrete blocks. It produces the most accurately formed bricks but it is no longer widely used because it is more expensive, due to labor costs.

Bricks coming out of the forming machines contain about 7–30 percent moisture. After molding in any of the three processes, the units are sent to a dryer, where they are subjected to temperatures of 110–300 °F (43–150 °C) for 24–48 h. Modern brick plants use dryer kilns supplied with waste heat from the exhaust of the firing kilns. After they are dried, the units are baked in burning kilns at a very high temperature [temperatures as high as 1315 °C (2400 °F)] for about 40–150 h. During this process, parts of the clay melt, fusing the whole mass of the brick into a hard durable unit, due to alteration of the chemical structure of the clay. Burning is followed by a cooling cycle, which takes about 48–72 h.

Variation in the composition of clays and the burning temperatures results in a wide variety of bricks. Bricks can be obtained in various colors, ranging from white to black, and can be manufactured at different densities.

4.1.3 Grades and Types

Bricks may be classified into different types, according to their uses, such as building brick (also called common brick), facing brick, floor brick, and paving brick. Building brick is produced for construction, without great concern for appearance (texture and color), and used as a structural material where strength and durability are the most important requirements.

Facing brick is made especially for facing purposes. Manufactured from selected clays and available in all sizes and colors, it is smooth, fine, medium, or coarse textured. Facing bricks are used for exposed areas where appearance is an important design criterion. The selection of facing brick is based on color, dimensional tolerances, surface texture, uniformity, and limits on the amount of cracks and defects.

Floor brick is smooth, dense, highly resistant to abrasion, and is used on finished floor surfaces. Paving brick is a low-abrasion vitrified brick, generally furnished with spacing lugs and produced in smooth or wire-cut surface finishes, and used for roads, sidewalks, patios, driveways, and interior floors. The appearance of a finished floor depends on the color, size, texture, and bond pattern of bricks. Such bricks are usually uncored, and the color ranges from red to gray and brown. Used bricks (also called salvaged bricks) have a weathered appearance and come in a broad ranges of colors.

Building bricks are manufactured in three grades (durability grades):

- Grade SW
- Grade MW
- Grade NW

Grading is based on some physical requirements (minimum compressive strength, maximum water absorption, and maximum saturation coefficient), and is directly related to durability and resistance to weathering (Table 4.1). Grade SW has the highest minimum compressive strength requirement and lowest maximum water

TABLE 4.1 PHYSICAL REQUIREMENTS FOR BUILDING BRICK

Grade	Minimum compressive strength, gross area [psi (MPa)]		Maximum water absorption by 5-h boiling (%)		Maximum saturation coefficient	
	Average of five bricks	Individual	Average of five bricks	Individual	Average of five bricks	Individual
SW	3000 (20.7)	2500 (17.2)	17.0	20.0	0.78	0.80
MW	2500 (17.2)	2200 (15.2)	22.0	25.0	0.88	0.90
NW	1500 (10.3)	1250 (8.6)	No limit	No limit	No limit	No limit

Source: ASTM C62. Copyright ASTM. Reprinted with permission.

absorption; grade NW has the lowest minimum compressive strength requirement and no limit on the water absorption. (The letters SW, MW, and NW stand for severe weathering, moderate weathering, and negligible weathering, respectively.) The *Annual Book of ASTM Standards* (Vol. 4.05) provides a map of the general areas of the United States in which brick masonry is subject to severe, moderate, and negligible weathering.

The choice of a grade (of clay bricks) in wall construction should be based on the location of the construction in this weathering index map. (The weathering index is the number of freeze–thaw days in a year multiplied by the winter rainfall in inches.) At locations where brick masonry is subject to negligible weathering, as indicated in the map, grade MW is recommended; SW grade bricks should be used in locations that are subject to moderate and severe weathering.

In general, grade SW bricks are intended for use where high resistance to frost action is desired, such as at or below ground level. In relatively dry locations, even when temperature below freezing is expected, grade MW bricks are recommended. Bricks in the exterior faces of walls (above or below ground) that remain dry should be of MW grade. Note that bricks exposed to the exterior environment may take up moisture by capillary action. Grade NW is for use in interior construction, where no freezing occurs.

Facing bricks are manufactured in three types, based on factors that affect their appearance:

- Type FBS
- Type FBX
- Type FBA

Type FBS, which stands for *face brick standard,* is for general use in exposed masonry construction, and most facing bricks are manufactured to meet the standards for this type. Type FBX, which stands for *face brick extra,* is for general use in exterior or interior masonry construction where a higher degree of precision and a lower permissible variation in size is required than that permitted for type FBS. Bricks of this type are used where a high degree of mechanical perfection, narrow color ranges, and minimum permissible variation in size are required. Type FBA, which stands for *face brick architecture,* is manufactured to produce characteristic architectural effects resulting from nonuniformity in size and texture of the individual units. When no type is specified, type FBS should be used. Note that these three types of facing bricks are manufactured in two durability grades, SW and MW.

4.1.4 Sizes of Bricks

Like concrete blocks, bricks and tiles are designated by their *nominal dimensions.* Their *specified dimensions* (or modular dimensions) are such that nominal sizes, measured center to center of mortar joints, will be equal to one or more standard 4-in. modules: for example 4 in. thickness and 8 in. length, 4 in. thickness and 12 in. length, and so on (Fig. 4.5). The specified dimensions are also the measurements to which the masonry units are physically required to conform. The nominal dimension is greater than the specified dimension by the thickness of the mortar joint, a maximum of 1/2 in. (12.5 mm). The actual sizes of bricks depend on the nominal sizes and the shrinkage occurring during the burning process.

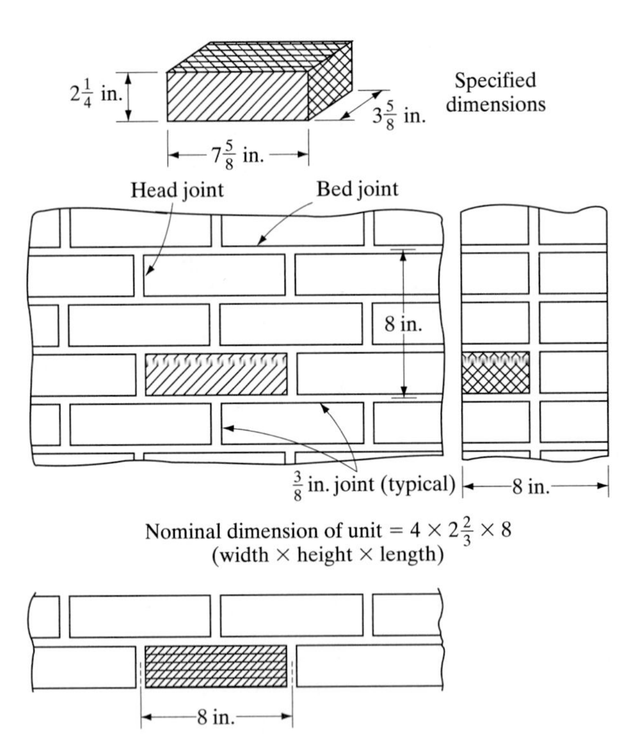

Figure 4.5 Nominal and specified dimensions.

Clay bricks are available in nominal widths (or thickness) ranging from 3–12 in., height or depth ranging from 2–8 in., and lengths up to 16 in. Brick dimensions are generally specified in a sequence of width times height times length (Fig. 4.5). For example, a nominal $4 \times 2\frac{2}{3} \times 8$ brick has a nominal width of 4 in., height of $2\frac{2}{3}$ in., and length of 8 in. The specified or modular dimension of this unit, with a 3/8-in. mortar joint, is width of $3\frac{5}{8}$ in., height of $2\frac{1}{4}$ in., and length of $7\frac{5}{8}$ in.

Most clays shrink some 4–15 percent during drying and burning. The amount of shrinkage depends on the fineness and composition of the clay, the amount of water in the mix, and the kiln temperature. Allowance for this reduction in cross-sectional dimensions is made during the molding process. Due to variations in the shrinkage between different clays, ASTM specifications permit deviations in dimensions of 3/32–3/8 in. (2.4–9.5 mm).

The thickness of mortar joints in brick masonry varies between 3/8–1/2 in. (10–12.5 mm). Facing bricks are generally laid in joints of thickness ranging between 3/8–1/2 in., and joints of building bricks are generally 1/2 in. thick.

Two size types are available: modular (conforming to a 4-in. grid system) and nonmodular [Fig. 4.4(a)]. Modular bricks come in a variety of core designs (depressions), depending on the manufacturer. Cores, originally developed to reduce the weight of a solid unit, facilitate ease of forming and handling and improve grip and mortar bond.

4.1.5 Properties of Bricks

The physical properties of clay bricks and structural clay tiles are:

- Color
- Texture
- Size
- Density

The engineering or mechanical properties are:

- Compressive strength
- Modulus of rupture
- Modulus of elasticity
- Tensile strength
- Absorption
- Thermal conductivity
- Fire resistance

A good brick should have plane faces, parallel sides, and sharp edges and angles. The texture should be uniform, and bricks should give a clear ringing sound when struck a sharp blow with a hard object. A good brick should absorb less than 20 percent of water by weight. It should have high compressive and bending strengths.

The color of bricks depends on the composition of the raw materials (presence of metallic oxides) and the degree of burning. The standard color of bricks is terra-cotta, or brownish-orange. Lime may be added deliberately to provide white color to bricks. Clays containing iron oxides turn red following burning. Various mixtures may be added prior to firing to produce variously colored bricks. The colors of bricks manufactured in the United States range from off-white to purple and black. Metallic oxides (iron, calcium, and magnesium) act as fluxes to promote fusion at lower temperatures, and give burned clay the necessary strength and hardness. Higher burning temperatures bring in darker color as well as an increase in compressive strength.

The texture of bricks—the surface appearance—can be flat and smooth, or irregular. A wire-cut texture appears on two faces of bricks manufactured using the stiff-mud process; a scored finish is produced by grooving the brick surface as it emerges from the die; a combed finish is produced by penetrating the face with parallel scratches.

The density or weight of a clay unit depends on the specific gravity of the green (unfired) clay, the method of manufacture, and the degree of burning. The specific gravity of clays and shales ranges from 2.6–2.8. The density of the burned material exceeds 100 pcf (1600 kg/m^3), averaging 125 pcf (2000 kg/m^3).

Clay bricks are very durable, and require very little maintenance. They have very good fire resistance and moderate insulating properties (against transfer of heat). Noncombustibility is an important property of bricks, and the primary reason why they are used in the construction of chimneys in masonry as well as wood buildings. Being porous, brick is a poor conductor. Well-built brick houses are cooler in summer and warmer in winter than are houses built of most other materials.

Absorption. Water absorption greatly affects the durability of brick (measured by its resistance to frost action). Very soft, underburned bricks may absorb water to as much as one-third of their weight; whereas good, hard bricks may absorb less than 10 percent water. The smaller the amount of absorption, the greater the durability. The saturation coefficient, defined in the following, is also a measure of freeze–thaw resistance. A large saturation coefficient indicates relatively fewer and smaller pores in the brick. The smaller voids accommodate expansion resulting from the freezing of water in larger voids. Consequently, bricks with a higher saturation coefficient are expected to have less resistance to damaging action from frost than units having a lower saturation coefficient.

The absorption of bricks (total water absorption) is defined as the increase in the weight of brick due to water, expressed as a percentage of the dry weight, and can be calculated as

$$\text{absorption} = \frac{(\text{weight of water absorbed after 24 h in cold water})}{(\text{dry weight of unit})} \times 100$$

The saturation coefficient, also called the C/B ratio, is defined as the ratio between absorption after 24 h in cold water and absorption after boiling for 5 h, and is calculated as

$$\text{saturation coefficient} = \frac{(\text{absorption after 24 h in cold water})}{(\text{total absorption after boiling for 5 h})} \times 100$$

$$= \frac{W_2 - W_1}{W_3 - W_1} \times 100$$

where W_1 is the dry weight of the unit, W_2 the saturated weight of the unit after 24 h submersion in water, and W_3 the saturated weight of the unit after 5 h submersion in boiling water.

Highly absorptive bricks can cause efflorescence and other problems in masonry. ASTM standards limit average absorption to a maximum 17 percent for grade SW and 22 percent for grade MW bricks. Most bricks have absorption of 4–10 percent. Small pores in bricks function as capillary pores, which try to draw water into the unit. This action is called the initial rate of absorption (also called suction), and affects the bond. High-suction bricks remove water from mortar, affecting its curing and bond strength.

ASTM C67 illustrates a laboratory procedure to determine the initial rate of absorption, which should be less than 30 g/min. per 30 in.^{2}. Suction values of clay units may range between 2–60 g/min. Use of bricks that have an excessive initial rate of absorption results in weak and permeable joints. Minimum water penetration and maximum bond strength are obtained with bricks having suction less than 20 g/min. High-suction bricks, those having a suction exceeding 30 g/min., should be thoroughly wetted prior to installation, preferably 3–24 h prior to their use (so as to allow time for the distribution of moisture throughout the unit). These bricks are undesirable in load-bearing construction because they cause rapid drying of mortar, and are responsible for poor bond strength and increased water penetration in the masonry.

$$\text{initial rate of absorption} = \frac{\begin{array}{c}(\text{weight of brick after 1 min. in 1/8 in. water} \\ -\ \text{dry weight of unit})\end{array}}{(\text{length of unit}) \times (\text{width of unit})} \times 30$$

$$= \frac{(W_2 - W_1) \times 30}{L \times B}$$

where W_1 is the dry weight of the unit, W_2 the weight of the unit after partial submersion for 1 min., L the length of the unit, and B the width of the unit.

Strength. The compressive strength of clay units depends on:

- Composition of the clay
- Method of manufacture
- Degree of burning

Compressive strength is determined by testing individual units flatwise (load applied in the direction of the height of the unit; ASTM C67) and is calculated

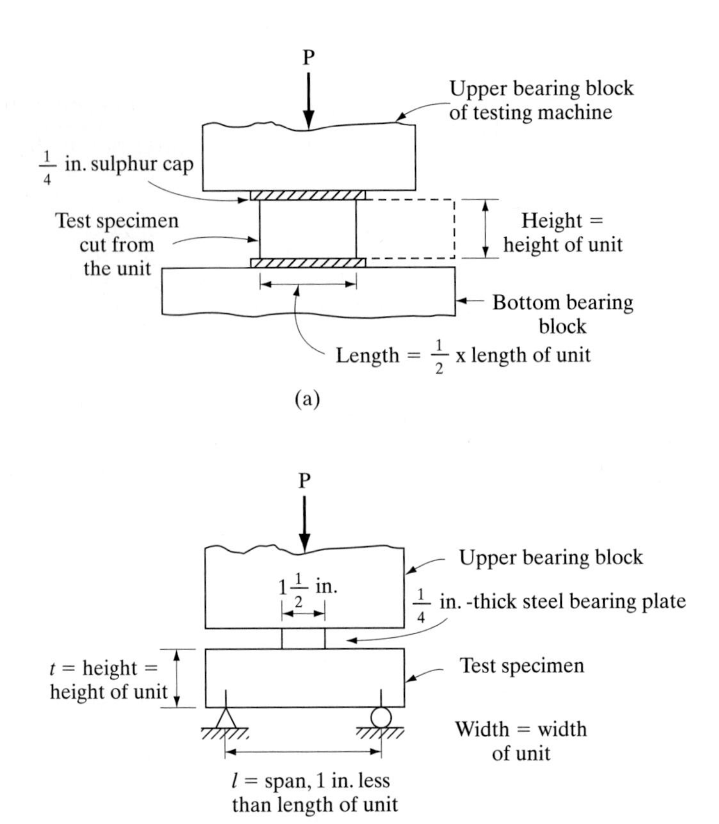

Figure 4.6 Test details: (a) compressive strength test; (b) flexure test.

[Fig. 4.6(a)] as

$$\text{compressive strength} = \frac{(\text{failure load})}{(\text{net cross-sectional area})}$$

If the net area is more than 75 percent of the gross area, the latter is used to calculate the compressive strength. The minimum average compressive strength of grade SW bricks is 3000 psi (20.7 MPa), those of grades MW and NW bricks are 2500 and 1500 psi (17.3 and 10.4 MPa), respectively.

The modulus of rupture (MOR) is determined by supporting individual units flatwise, and applying the load at the midspan, in the direction of the depth of the unit, on a span approximately 1 in. (25 mm) less than the basic unit length [Fig. 4.6(b)]:

$$\text{MOR} = \frac{1.5Pl}{Bt^2}$$

where P is the failure load, l the span length, B the unit width, and t the unit height.

Most clay bricks have flexural strength anywhere between 500–3800 psi (3.5–26.2 MPa). Good, hard bricks are expected to have a MOR in excess of 2500 psi (17.3 MPa). Soft bricks may have a MOR close to 500 psi (3.5 MPa). The tensile strength of bricks ranges from 30–40 percent of the modulus of rupture. Shear values are in the range 30–45 percent of the net compressive strength.

The modulus of elasticity of bricks lies anywhere between 1.5×10^6 and 5×10^6 psi (10.3×10^3 and 34.5 MPa). Increase in the burning temperature or the burning period increases the modulus of elasticity. The coefficient of thermal conductivity varies between 2.8×10^{-6} to 3.9×10^{-6} per °F. The thermal expansion of brick units is between 0.30–0.40 in. per 100 ft for 100 °F increase in temperature.

4.1.6 Concrete Masonry Units

Concrete masonry units (also called cinder blocks, hollow blocks, and concrete blocks) are solid or hollow and made from concrete. There are two types of concrete masonry units:

- Concrete building bricks
- Load-bearing concrete masonry units

A load-bearing concrete masonry unit or concrete block (Fig. 4.7) is a solid or hollow masonry unit made from cement, water, and mineral aggregates with or without the inclusion of other materials. The solid units have net cross-sectional area, in every plane parallel to the bearing surface, which is equal to or greater than 75 percent of the gross cross-sectional area measured in the same plane. A hollow unit is a unit whose net cross-sectional area in every plane parallel to the bearing surface is less than 75 percent of the gross cross-sectional area in the same plane. Its cross-sectional properties should conform to the minimum thickness requirements given in Table 4.2.

A concrete building brick is a solid masonry unit made from portland cement, water, and suitable lightweight or normal-weight aggregates, with or without the inclusion of other materials. Note that this is a completely solid unit, similar in shape and size to a clay building brick. The compressive strength of these units is considerably higher than that of the load-bearing masonry units previously discussed.

Both types of units—load-bearing concrete masonry units and concrete building bricks—are manufactured in three weight classifications:

- Normal-weight units
- Medium-weight units
- Lightweight units

Normal-weight units weigh more than 125 pcf (2000 kg/m^3) when dry, medium-weight units are in the weight range of 105–125 pcf (1680–2000 kg/m^3) when dry, and lightweight units have weights between 85–105 pcf (1360–1680 kg/m^3). Well-graded sand, gravel, and crushed stone are used for the production of normal-weight blocks.

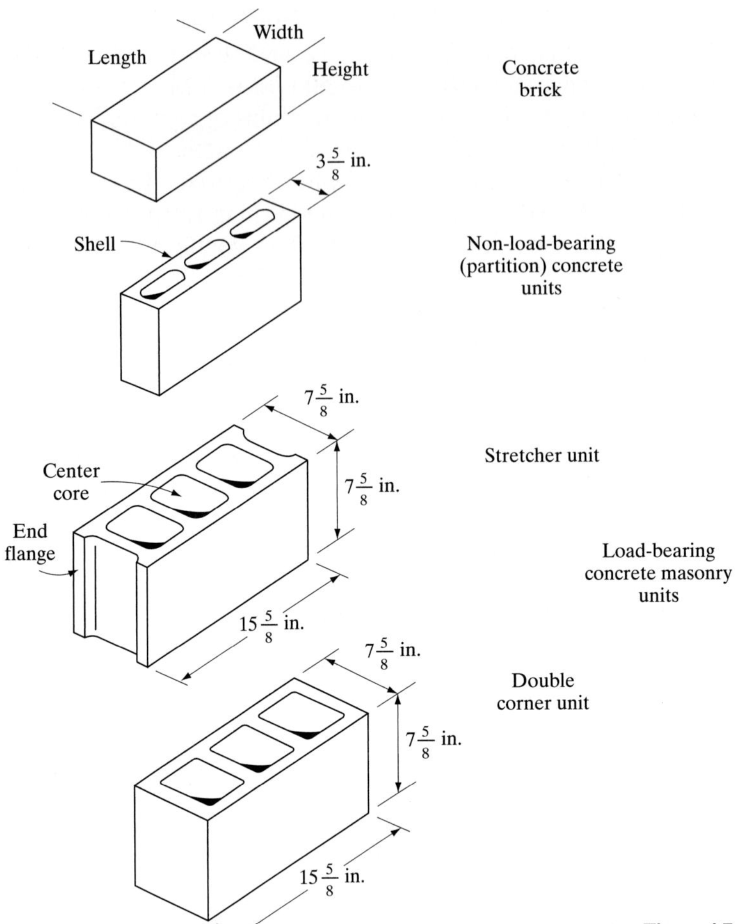

Figure 4.7 Concrete masonry units.

TABLE 4.2 MINIMUM THICKNESS OF FACE SHELL AND WEBS OF HOLLOW MASONRY UNITS

Nominal width of units (in.)	Minimum face-shell thickness (in.)	Minimum web thickness (at the thinnest point) (in.)
3	3/4	3/4
4	3/4	3/4
6	1	1
8	$1\frac{1}{4}$	1
10	$1\frac{3}{8}$	$1\frac{1}{8}$
12	$1\frac{1}{2}$	$1\frac{1}{8}$

Lightweight units (the most common in masonry construction) are made using lightweight aggregates such as pumice, scoria, cinders, expanded clay, and expanded shale. They should be dry during placement because wet units will shrink substantially due to drying, causing cracks in the masonry.

Concrete masonry units are made from a relatively dry (zero-slump) mixture of cement, aggregates, water, and admixtures. Aggregates account for as much as 90 percent of the composition. The units are molded under pressure, cured under controlled conditions, and stored to continue the hydration process. Curing is done using high-pressure or low-pressure steam; most units in the United States are produced using low-pressure steam curing.

Both the raw materials and the method of manufacture influence the strength, appearance, density, texture, and other properties of the units. In addition to strength, density, and shrinkage, the thermal, sound, and fire resistance properties of the units are affected by the use of lightweight aggregates. Lightweight blocks have higher thermal and fire resistance properties and lower sound resistance compared to normal-weight units.

Type I cement is preferred in the manufacture of concrete masonry units; Type III cement is also used, to obtain faster strength. The air-entraining cements, Type IA and IIIA, improve workability, compaction, and molding characteristics. Type A cement will also provide increased resistance to freeze–thaw cycles, but may decrease the compressive strength. High-strength units are made using higher amounts of cement and more water (but with zero slump).

4.1.7 Types and Grades of Concrete Masonry Units

Load-bearing concrete masonry units are manufactured in two types:

- *Type I:* moisture-controlled units
- *Type II:* non–moisture-controlled units

Type I units are required to comply with the moisture content provisions of ASTM C90 (Table 4.3). Type II units are manufactured without special consideration to controlling moisture content. The requirements shown in Table 4.3 recognize that

TABLE 4.3 MOISTURE CONTENT REQUIREMENTS FOR TYPE I UNITS

	Moisture content (avg. max.), percent of total absorption for humidity zone at job site (mean annual relative humidity)		
Linear shrinkage (%)	Humid (>75%)	Intermediate (50–75%)	Arid (<50%)
<0.03	45	40	35
0.03–0.045	40	35	30
0.04–0.065	35	30	25

Source: ASTM C90. Copyright ASTM. Reprinted with permission.

the inherent linear shrinkage properties of concrete in the masonry units will vary depending on material properties, manufacturing methods, and atmospheric conditions. The purpose of the moisture content requirement is to provide a method of limiting drying shrinkage of units in the masonry wall, regardless of the shrinkage properties of the units (up to a maximum of 0.065 percent). A lower moisture content in the units is required when they are to be used in less than 50 percent average relative humidity (arid environment), and a higher moisture content is permitted when the units are to be placed in an environment with a relative humidity above 75 percent (humid environment). ASTM C90 provides a map showing the mean annual relative humidity in the United States.

The primary goal of moisture-content grading is to limit the shrinkage of the units from moisture loss. In arid areas (such as inland areas of California and Nevada), shrinkage will rise if more moisture, when available, will leave the units; to limit this shrinkage, a lower moisture content (in the units) is required. In humid areas (such as in the coastal areas of the United States), units with higher moisture content are required.

Selection of a Type I unit is based on the moisture content in the units as delivered to the job site. This means that these units should be protected from rain or other moisture at the job site before they are placed in the wall. Units that are stored properly at the job site will achieve climactic balance and perform satisfactorily with minimum shrinkage. Units that are unprotected from rain or snow may not meet the requirements for Type I units.

Type II units, manufactured without special consideration given to controlling moisture content, are used more extensively in construction. These units should not be so moist as to cause excessive shrinkage cracks. They should be stored long enough to achieve climactic balance, which depends on the materials used, the moisture content, and the humidity conditions. Exposed Type II units may require closer spacing of control joints or increased horizontal reinforcement.

In addition to the moisture content requirements, both types of units should satisfy the strength and absorption requirements given in Table 4.4. Note that both Types I and II units require an average minimum compressive strength of 1900 psi (13.1 MPa) and an average maximum water absorption—which is a measure of total water-fillable void content in the unit—of 13–18 pcf (206–288 kg/m^3), or about 10–20

TABLE 4.4 STRENGTH AND ABSORPTION REQUIREMENTS OF CONCRETE UNITS

Minimum compressive strength based on net area [psi (Mpa)]		Water absorption, max. [lb/ft^3 (kg/m^3)] by weight classification; oven-dry weight [lb/ft^3 (kg/m^3)]		
Average of three units	Individual unit	Lightweight <105 (1682)	Medium-weight 105–125 (1682–2002)	Normal-weight >125 (2002)
1900 (13.1)	1700 (11.7)	18 (288)	15 (240)	13 (208)

Source: ASTM C90. Copyright ASTM. Reprinted with permission.

percent moisture content. Water absorption is determined as follows:

$$\text{water absorption} = \text{saturated weight} - \text{oven-dry weight}$$

It can be noted that porous aggregates, such as lightweight aggregates, have greater absorption than dense and nonporous aggregates; as a result, lightweight units are permitted higher absorption than medium-weight or normal-weight units.

The compressive strength of units, based on the net cross-sectional area, must be at least 1900 psi (13.1 MPa). In manufactured units, it varies between 1900–6000 psi (13.1–41.4 MPa), depending on the mix proportions. The average net area of a unit is calculated by establishing the net volume from water displacement tests described in ASTM C140. The gross area of a unit is the total area of a section perpendicular to the direction of load, and can be calculated as follows (Fig. 4.8):

$$\text{gross area} = \text{actual width} \times \text{actual length}$$

The net area can be calculated as

$$\text{net area} = \text{gross area} \times \text{percentage of solid}$$

$$\text{percent of solid} = (\text{net volume}/\text{gross volume}) \times 100$$

Concrete building bricks are manufactured in two grades and two types, based on strength and absorption requirements:

- Grade N (types I and II)
- Grade S (types I and II)

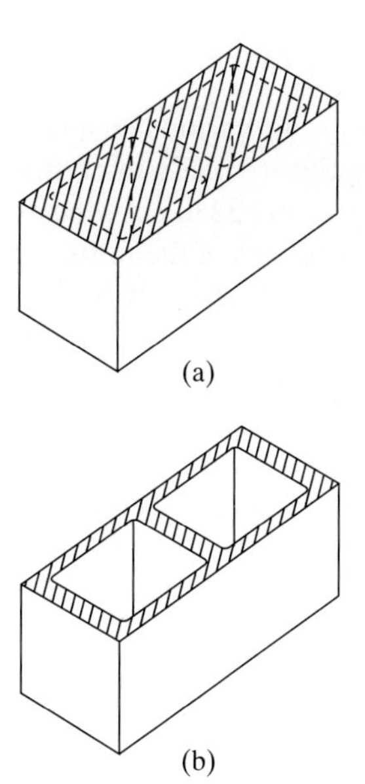

Figure 4.8 (a) Gross and (b) net areas.

TABLE 4.5 STRENGTH AND ABSORPTION REQUIREMENTS FOR CONCRETE BRICK

	Minimum compressive strength [psi (MPa)] (based on average gross area, brick tested flatwise)		Maximum water absorption [lb/ft³ (kg/m³)] by weight classification; oven-dry weight [lb/ft³ (kg/m³)]		
Grade	Average of three bricks	Individual brick	Lightweight <105 (1680)	Medium-weight 105–125 (1682–2002)	Normal-weight >125 (200)
N-I, N-II	3500 (24.1)	3000 (20.7)	15 (240)	13 (208)	10 (160)
S-I, S-II	2500 (17.3)	2000 (13.8)	18 (288)	15 (240)	13 (208)

Source: ASTM C55. Copyright ASTM. Reprinted with permission.

The strength and absorption requirements for these two grades are shown in Table 4.5.

Grade N units are suitable as architectural veneer or facing units in exterior walls, and for use where high strength and resistance to moisture penetration and severe frost action are required. The average maximum compressive strength of grade N bricks, when tested flatwise, should be equal to or higher than 3500 psi (24.1 MPa). Grade S units are for general use, where moderate strength and resistance to frost action and moisture penetration are required. Their average maximum compressive strength should be at least 2500 psi (17.3 MPa).

These two grades of concrete bricks come in two types: Type I, moisture-controlled units, and Type II, non–moisture-controlled units. The requirements for the units are the same as those for load-bearing masonry units, as shown in Tables 4.3 and 4.4.

4.1.8 Properties of Concrete Masonry Units

Concrete masonry units come in various shapes, sizes, colors, and textures. Load-bearing concrete masonry units are made in nominal 4-, 6-, 8-, 10-, and 12-in. widths. They come in lengths of 12, 16, and 24 in., but the common length is 16 in. The height of the blocks is 4 or 8 in. Actual sizes of the units are usually 3/8 in. (10 mm) less than the nominal dimensions. For example, an $8 \times 8 \times 16$ in. block has dimensions $7\frac{5}{8} \times 7\frac{5}{8} \times 15\frac{5}{8}$ in. Most load-bearing walls are 8 in. thick, and partition walls can be 4 in. Concrete bricks are produced in two nominal sizes: standard and modular. The actual size of a standard concrete brick is $3\frac{3}{4} \times 2\frac{1}{4} \times 8$ in., and that of a modular concrete brick is $3\frac{5}{8} \times 2\frac{1}{4} \times 7\frac{5}{8}$ in.

A *split block* is a hollow concrete unit that has the appearance of rough stone texture and the look like that of stone masonry. It is made by breaking a double-wide unit into two pieces, and comes in various colors. A *slump block,* which has the appearance of an adobe brick, is made with a concrete mixture that is relatively more fluid. During the production, the unit is jolted and, as a result, the wet mix sags and slumps when removed from the mold. The units vary in height and surface texture, but have a uniform earthtone color.

Water absorption of a unit is frequently taken as a measure of its durability. Total absorption is measured following a 24 h period of immersion. Linear shrinkage is the change in length of a unit from a wet to a dry condition, and is limited to 0.065 percent. In masonry walls, the amount of shrinkage depends on the moisture content in the units and the environmental humidity. The greater the moisture content, the higher will be the shrinkage if the units are allowed to dry out in service. The amount of shrinkage is also dependent on the size of the unit; the larger the size, the greater the shrinkage. The amount of potential shrinkage will be a factor in predicting the amount of cracking that may occur in the walls during service.

Since the amount of shrinkage increases with the moisture content in the units, the concrete units should not be wet prior to being placed in the wall. But in extremely dry weather and hot temperatures, the units should be wet, using a light spray, just before placement.

As explained, the compressive strength of concrete masonry units is in the range 1900–6000 psi (13.1–41.4 MPa). The tensile strength lies between 250–500 psi (1.7–3.5 MPa). The modulus of elasticity varies between 1.4×10^6 and 4.5×10^6 psi (9.7×10^3 and 31×10^3 MPa).

4.2 MORTAR, GROUT, AND PLASTER

Masonry walls are built using masonry units (clay or concrete) and mortar, with or without grout. Mortar is a plastic mixture of cementitious materials, sand, and water capable of binding together individual masonry units. The early Egyptians used a cementing material, obtained by burning gypsum, in masonry construction. Lime was used in the preparation of mortar at a very early period by the Greeks.

The primary function of the mortar is to bond individual units together so that the masonry will act as a single larger unit. In addition, it serves as a bedding material for the masonry units and allows the level placement of the units. As a binding material, the mortar is partly responsible for the strength characteristics of the masonry. Mortar joints also can be used to project some aesthetic features, such as color and appearance.

Grout is a high-slump concrete that is employed to fill the cores or voids in masonry. It is a mixture of cement, sand, fine gravel, and water. The primary function of grout is to bind reinforcing steel and individual wythes into a composite structural system.

Plaster is a portland cement–based cementitious mixture that is used as a finishing material for walls. It is used for masonry walls as well as framed (wood) walls.

Mortar. Mortar is a mixture of cement, lime, sand, and water. When made with lime, sand, and water it is called *lime mortar*. With the addition of cement to lime mortar, it becomes *cement-lime mortar* or simply *cement mortar*. In general, mortar is the term applied to any material used for bedding, jointing, and rendering brickwork, stonework, and concrete blockwork.

Cement, also called portland cement (Chapter 3) is manufactured by burning a mixture of lime, silica, alumina, and iron oxide. Generally, Type I or II cement is used in mortar.

The term "masonry cement" refers to a hydraulic cement for use in mortars for masonry construction, and containing one or more of the following materials: portland cement, portland–pozzolan cement, natural cement, slag cement, portland–blast-furnace slag cement, or hydraulic lime; and usually incorporating one or more additional materials such as hydrated lime, limestone, chalk, calcareous shell, talc, slag, or clay. It can be used with or without the use of other cements or hydrated lime. It is combined with sand and water at the job site to make mortar that meets specified standards. The quality and performance as well as the strength of mortar made with masonry cement may not be as superior as that made with portland cement.

The weakest part of masonry is the mortar; hence the less the mortar, the better the performance of the wall. The mortar serves two important purposes: to bond various units, and to form a cushion to distribute pressure uniformly over the surface. A thin layer of mortar is stronger in compression than a thick one. Bricks and blocks should not be merely laid, but should be rubbed and pressed down so as to force the mortar into the pores of the bricks to produce maximum adhesion.

Before it begins to set, portland cement–sand mortar has poor cohesion or bonding properties. It may fail to stick to the edge of the brick or block. Addition of a small percent of lime makes the mortar "fat" or "rich" and more pleasant to work with, as it becomes more plastic. The addition of lime in small amounts does not decrease the compressive strength of the mortar or the time to set.

4.2.1 Lime and Lime Mortar

The mineral calcite (calcium carbonate, $CaCO_3$) occurs naturally as marble, chalk, limestone, and cave deposits—such as those in Carlsbad Caverns, New Mexico. When it is burned at about 1650 °F (900 °C) in rotary kilns, releasing the carbon dioxide, quicklime (CaO) is produced—an off-white lumpy material having a specific gravity of about 3.3 in pure form. Slaking (satisfying the thirst) of quicklime with minimum water results in hydrated lime, $Ca(OH)_2$, as quicklime rapidly absorbs nearly a quarter of its weight in water. The absorption is accompanied by large rise in temperature and the evolution of heat. The bursting of quicklime into particles that are finally reduced to powder follows the reaction. Water at about $2\frac{1}{2}$ to 3 times the volume of lime is needed for complete slaking. The volume of this powder is about 250–350 percent of the original volume of lime and is called *slaked lime,* which can be used in mortar for masonry.

Hydrated lime (a nonhydraulic cement) can be described as a dry flocculent powder resulting from a minimum hydration of quicklime. The treatment is carried out in large tanks at lime manufacturing plants, with a limited amount of water, followed by screening and grinding. The ASTM defines hydrated lime as a dry powder obtained by treating quicklime with water sufficient to satisfy its chemical affinity for water under the conditions of its hydration. It consists essentially of calcium hydroxide, or a mixture of calcium hydroxide and magnesium oxide or magnesium hydroxide (or both). It has a specific gravity ranging between 2.1–2.4, an average bulk density of 35 pcf (560 kg/m^3), and is sold in bags of 50 or 40-lb. capacity. The chemical processes leading to hydrated lime and mortar are given below:

$$\underset{\text{(limestone)}}{CaCO_3} + \text{heat} = \underset{\text{(quicklime)}}{CaO} + \underset{\text{(carbon dioxide)}}{CO_2}$$

$$\underset{\text{(quicklime)}}{CaO} + \underset{\text{(water)}}{H_2O} = \underset{\text{(hydrated lime)}}{Ca(OH)_2}$$

Combining hydrated lime with sand and water produces lime mortar, in which the hydrated lime paste loses its water slowly through evaporation and simultaneously absorbs carbon dioxide from the atmosphere—in the presence of moisture—to form calcium carbonate, or limestone. The process is slow, taking six months or more, and depends on the levels of humidity and carbon dioxide in the atmosphere:

$$Ca(OH)_2 + CO_2 = CaCO_3 + H_2O$$

Quicklime (also called unslaked lime) can be purchased crushed, ground, and pulverized (based on the size of the particles), and must be slaked prior to use. Bagged quicklime may be stored for a long time in a dry place. If exposed to atmosphere of moderate to high relative humidity, it begins to harden slowly; this process is called *air-slaking*. The absorption of CO_2 from the atmosphere causes disintegration, producing a mixture of hydroxides, oxides, and carbonates.

Finer quicklime is easier to slake. Care is necessary when adding water, to prevent splattering, which has the potential for serious burns. When quicklime is slaked or dry hydrate is soaked in water, the volume expands due to the inherent ability of lime to absorb and retain a considerable amount of free water in addition to the chemically combined moisture. This form of lime hydrate in aqueous suspension, containing a large amount of free water, is called *lime putty*. Following the addition of water, the lime putty is allowed to age from a few hours to many days, to allow the slaking of all particles. The ability of lime putty to hold and retain moisture—with appreciable resistance to suction, even in the presence of sand—is one of the most important properties of lime. Increase in plasticity improves water retentivity. Soaking or aging lime putty—from 12–24 h—also increases the water retentivity.

Hydrated lime can be handled more conveniently than quicklime, and can be safely transported and stored. It is a material ready for mixing with water, sand, and cement for making mortar or plaster. *Finishing hydrated lime* is hydrated lime suitable for use in plaster, and is the major constituent of the finish coat in plaster. Two types of finishing hydrated lime are manufactured:

- Type N
- Type S

Type N is called normal hydrated lime, and Type S, special hydrated lime. Both are suitable for finishing purposes. Type N has no maximum limit on the amount of unhydrated oxides, and most of the MgO is unhydrated. Type S lime has an upper limit of 8 percent unhydrated oxides content.

Mason's hydrated lime is hydrated lime suitable for mortar and concrete, and can also be used in stucco and the base coat of plaster. It comes in four types:

- Type N
- Type S
- Type NA
- Type SA

Type N is called normal hydrated lime, and Type S, special or pressure-hydrated lime. Type NA is normal air-entraining and Type SA is special air-entraining hydrated lime. Types N and NA have no upper limits on combined unhydrated oxides of calcium and magnesium, whereas Types S and SA have a limit of 8 percent. The highly hydrated products, Types S and SA, have higher water retentivity and can develop high early plasticity. Types N and NA should be soaked in water (or as a putty) overnight prior to use, to improve plasticity and workability. The soaking is done by sifting the hydrated lime evenly in a pan half-filled with clean water. The lime is allowed to settle into the layer of water without stirring or mixing. The spreading of powdered lime is continued until a thick paste is formed, after which the mixture is left to soak until ready for use. The film of water protects the lime from coming into contact with the environment.

Hydraulic and Nonhydraulic Limes. Lime can be hydraulic or nonhydraulic. Nonhydraulic lime (described in the preceding section) is high in calcium and is produced from burning nearly pure limestone (high-calcium limestone, marble, and dolomitic limestone), shell, or white chalk. These raw materials produce a nearly white lime after slaking, and when used in mortar along with sand or sand and portland cement, nonhydraulic lime can gain strength very slowly by taking up carbon dioxide from the atmosphere to form calcium carbonate.

Hydraulic and semi-hydraulic lime are produced by burning siliceous and argillaceous limestone (containing alumina and silica in the form of hydrated aluminum silicate) and contain, after calcination, rather high percentages of calcium silicate. When mixed with water, this and other compounds in hydraulic lime form hydration products that are insoluble and of cementitious value. Mortar produced with hydraulic lime sets under water and gains considerable strength through hydration. Chemically, hydraulic lime might be classified as being intermediate between portland cement and hydrated lime. Unlike portland cement, it possesses a considerable amount of free lime (CaO and MgO), so that the product slakes in water. Higher silica content improves the hydraulic properties.

Lime Mortar. Lime mortar is a mixture of lime, fine aggregate, and water. The fine aggregate or sand should be graded, with most of the particles passing the

No. 4 sieve and about 10 percent passing the No. 200 sieve. A coarser sand lowers workability, and a finer sand decreases water retentivity. Combining all ingredients in a mixer produces a more uniform mixture. Workmanship, more than any other factor, affects the permeability of mortar and masonry. Lime with low plasticity produces mortar with low water retentivity, which in turn increases the permeability of masonry. Mortar joints that are not completely filled are more permeable than full mortar joints (created by shoving bricks into position); tooled joints are less permeable than cut joints. A decrease in the amount of water reduces the shrinkage and improves the watertighness. If the mortar spreads easily and smoothly, it is plastic. If, instead, it sticks or drags under or cracks and drops behind the trowd, it is nonplastic.

The compressive strength of lime mortar is generally in the range of 100–400 psi (0.7–2.8 MPa) at one year; the tensile strength varies from 40–150 psi (0.3–1.0 MPa). Mortars made with finer sand have higher strength than those made with coarser sand. Due to low strength, lime mortar is not generally recommended for permanent masonry construction. Addition of portland cement to lime mortar increases its strength. A higher percentage of cement would be used where strength is required at an early age or when ultimate strength must be fairly high. But for mortars that are used in places where freedom from cracking, good bond, or resistance to rain penetration is paramount, the proportion of cement to lime should be kept low.

4.2.2 Mortar for Unit Masonry

Masonry is constructed using solid or hollow masonry units and mortar, which performs the following functions:

- It bonds the individual units together
- It serves as a seating material for the units
- It allows for leveling and seating of units and seals irregularities
- It provides strength to the wall
- It can offer aesthetic qualities

There are four standard types (grades) of mortar, identified by every other letter of the word masonwork:

- Type M
- Type S
- Type N
- Type O

All four standard types of mortar are made by blending cement, hydrated lime, sand, and water. Type M has the lowest amount of hydrated lime (highest cement content), whereas type O has the highest amount, as shown in Table 4.6. Either cement plus lime or masonry cement can be used in mortar preparation.

TABLE 4.6 PROPORTION SPECIFICATION REQUIREMENTS OF MORTAR FOR UNIT MASONRY

		Proportions by volume (cementitious material)					
		Portland cement or blended cement	Masonry cement (Type)			Hydrated lime or lime putty	Aggregate content (measured damp, loose condition)
Mortar	Type		M	S	N		
Cement–lime	M	1	—	—	—	1/4	Not less than $2\frac{1}{4}$ and not more than 3 times the sum of the separate volumes of cementitious materials
	S	1	—	—	—	over 1/4 to 1/2	
	N	1	—	—	—	over 1/2 to $1\frac{1}{4}$	
	O	1	—	—	—	over $1\frac{1}{4}$ to $2\frac{1}{2}$	
Masonry cement	M	1	—	—	1	—	
	M	—	1	—	—	—	
	S	1/2	—	—	1	—	
	S	—	—	1	—	—	
	N	—	—	—	1	—	
	O	—	—	—	1	—	

Source: ASTM C270. Copyright ASTM. Reprinted with permission.

In design or construction, mortar can be prescribed using one of two alternative types of specifications:

- Proportion specifications
- Property specifications

In proportion specifications, any one of the four standard types of mortars, conforming to the mixing requirements listed in Table 4.6, is designated. The specifications limit the maximum amount of constituent materials by volume measurement. There is no limit on the water content, and the amount may be adjusted by the mason to provide the required workability. Aggregate content is $2\frac{1}{4}$ to 3 times the sum of the separate volumes of cementitious materials used. For example, in type S mortar, with a volume of lime equal to ½ of cement volume, the volume of sand is equal to $3\frac{3}{8}$ to 4.5 times the volume of cement. The mixture is not required to satisfy the compressive or any other strength requirements.

Mortar conforming to property specifications is made as a mixture of cement, lime, sand, and water to satisfy the strength requirements listed in Table 4.7. Samples of mortar should be tested in the laboratory to determine its compressive strength, water retention, and air content. The compressive strength is usually determined by making 2-in. cubes and, after curing for 28 days, testing them in compression. There is no requirement on mix proportions, although the volume of fine aggregate should be between $2\frac{1}{4}$ and $3\frac{1}{2}$ times the net volume of cementitious materials. Property specifications are used mainly for research purposes, to determine the physical characteristics of mortar. When a project specification

TABLE 4.7 PROPERTY SPECIFICATION REQUIREMENTS OF MORTAR FOR UNIT MASONRY

Mortar	Type	Average minimum compressive strength at 28 days [psi (MPa)]	water retention (min %)	Air content (max. %)	Aggregate content (measured damp, loose)
Cement–lime	M	2500 (17.2)	75	12	
	S	1800 (12.4)	75	12	
	N	750 (5.2)	75	12–14	Not less than
	O	350 (2.4)	75	12–14	$2\frac{1}{4}$ and not
Masonry cement	M	2500 (17.2)	75	18–no limit	more than $3\frac{1}{2}$ times the
	S	1800 (12.4)	75	18–no limit	sum of the separate volumes
	N	750 (5.2)	75	18–no limit	of cementitious materials
	O	350 (2.4)	75	18–no limit	

Source: ASTM C270. Copyright ASTM. Reprinted with permission.

does not indicate the type of specification to be used, the proportion specification governs.

4.2.3 Mixing and Properties of Mortar

Mortar can be mixed by hand or (preferably) in a paddle-type mixer. Hand mixing is permitted for small batches. The procedure is to mix all the dry materials with a hoe, working from one end of the wheelbarrow and then from the other. (For ease in mixing, sand and cement can be spread in alternate layers.) Next, about two-thirds of the water is added and mixed until the mixture is uniformly wet. Additional water is added until the desired workability is obtained.

A *ready-mixed mortar* consists of cementitious materials, sand, and water (with or without an admixture for set control), which are measured and mixed at a central location and delivered to the construction site for use within a period of $2\frac{1}{2}$ hours.

Procedure for mixing mortar in a paddle-type mixer consists of the following steps:

1. Place all the mixing water.
2. Add cement to the mixer.
3. Mix at low speed for 30 seconds.
4. Add lime and sand while the mixer is running.
5. Mix at medium speed for 30 seconds.
6. Rest for $1\frac{1}{2}$ minutes.
7. Run the mixer for 1 minute at medium speed and dump the mortar out.

Effects of Mix Proportions. Performance of mortar depends largely on its mix proportions. In addition, it is influenced by various mortar properties, such as workability, water retentivity, bond strength, compressive strength, and durability. Selection of a mortar type should be based on these properties, type of masonry units, and applicable building code requirements.

Portland cement contributes to the early hardening, compressive strength, and durability of mortar. Hydrated lime contributes to plasticity, workability, water retentivity, and reduced shrinkage characteristics. Water retentivity refers to a property of mortar that allows it to hold water longer, resisting the suction of dry porous masonry units. Additionally, lime provides for improved bond strength and decreased permeability, and allows the mortar to be spread smoothly. Retempering of mortar, with the addition of lime, is easier because lime is slow setting. Sand acts as a filler material and reduces shrinkage deformations. Water is required for plasticity.

Sand used for mortar can be either natural or manufactured. The latter may consist of crushed stone, crushed gravel, or air-cooled blast-furnace slag. Manufactured sand is characterized by particles of angular shape and rough texture that affect the workability of mortar. Both natural and manufactured sand should be graded as specified in Table 4.8. Concrete sand should not be used in mortar preparation because the maximum particle size is higher than that allowed in mortar and fewer smaller particles are available in the mix. Finer particles contribute to plasticity, water retentivity, and workability of the mix. There is no laboratory testing procedure to measure the workability of the mix; the best method of measuring it is the mason's own judgment. Typically, mortar should be as wet as possible but should stay stiff enough to support the units.

Properties. As discussed, the major function of the mortar is to bind the masonry units together; in so doing it acts as an adhesive and sealant. For this reason, the most important property of mortar is its ability to form a complete, strong,

TABLE 4.8 STANDARD SPECIFICATION FOR SAND FOR MASONRY MORTAR

Sieve size	Percentage passing	
[No. (mm)]	Natural sand	Manufactured sand
4 (4.75)	100	100
8 (2.36)	95–100	95–100
16 (1.18)	70–100	70–100
30 (600 μm)	40–75	40–75
50 (300 μm)	10–35	20–40
100 (150 μm)	2–15	10–25
200 (75 μm)	—	0–10

Source: ASTM C144. Copyright ASTM. Reprinted with permission.

and durable bond with the masonry units and with the reinforcement. The capacity to bond individual units is measured by the *tensile bond strength* of mortar, which is related to the force required to separate the units.

It is important to note that although concrete and mortar are made using the same ingredients (with and without the coarse aggregate for concrete and mortar, respectively), the two are not alike in terms of their functions and requirements. What is good for concrete may not be good for mortar. Increase in compressive strength, which is the most important property of concrete, is not accompanied by a proportionate improvement in tensile bond strength, which is the most important strength characteristic of mortar. Other important properties of mortar are its workability, water retentivity, compressive strength, tensile strength, and freeze–thaw resistance.

The compressive strength of mortar depends on the amount of cement in the mix. The shear and flexural strength of masonry are related to the mortar bond strength, and not directly to its compressive strength. As explained, hydrated lime improves the bond strength of masonry and thus influences its shear and flexural strengths. An increase in lime content by 100 percent can increase the shear and flexural strength by as much as 100 percent. Note that there are no code requirements for minimum bond strength of mortar, although there are limits on the compressive strength of mortars covered by property specifications.

Type M and S mortars have higher compressive strength than N or O mortars. Masonry subjected to high compressive loads, and excessive lateral loads from wind, seismic forces, or soil should be built using Type M mortar. Structures constructed below or against ground exposed to severe frost action should also use Type M mortar.

In general, Types M and S mortar can be used for exterior walls at or below grade, such as foundation walls, retaining walls, manholes, sewers, pavements, walkways, and patios. For exterior walls above grade, such as load-bearing or non–load-bearing walls and parapet walls, Types N, S, or M mortar can be used. In interior construction, such as load-bearing or non–load-bearing partition walls, any type of mortar can be used. Types N and O mortars are not recommended in heavy seismic zones.

A mortar is considered workable if it can be placed and spread with little effort, and it possesses the capability to adhere to masonry units immediately after being placed. Workability is affected by water retentivity, mix, flow, and aggregate characteristics.

The freeze–thaw resistance of mortar is improved by air entrainment. An increase in compressive strength is also found to increase freeze–thaw resistance and improve durability. Mortar that has less absorption (amount of water a hardened mortar will absorb) possesses better freeze–thaw resistance. Tests have shown that mortar made with masonry cement absorbs less moisture and thus can better resist freeze–thaw damage.

Flow. Mortar flow is a property measured using a flow test and flow mold. It is the percent increase in the diameter of a conical frustrum of mortar 4 in. in

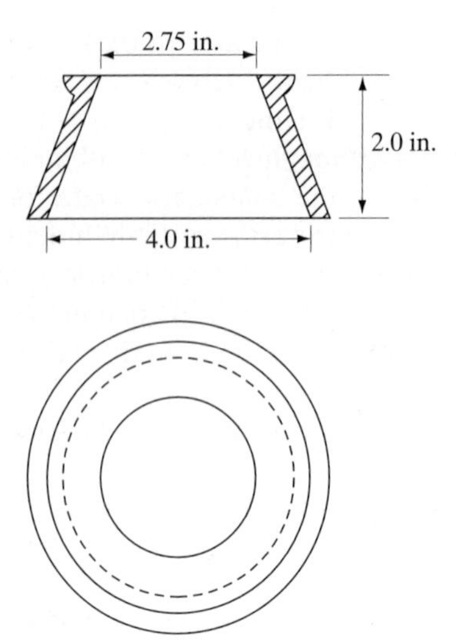

Figure 4.9 Flow cone (mold).

diameter at its base, after the flow table has been dropped through a height of 1/2 in. 25 times in 15 seconds. For example, if the final diameter of the flow cone after the test is 8 in., the flow is 100. The flow test is described in ASTM C109 and C230 (Fig. 4.9). Note that flow measurement is not always accurate, as it is sensitive to the techniques of the operator and the condition of the setup.

Tests have shown that the tensile bond strength of mortar—which estimates the ability of mortar to form a strong, complete bond with the units—is related to the flow, as shown in Fig. 4.10. An increase in flow is generally accompanied by an increase in tensile bond strength and a decrease in compressive strength. However, in practice, there is no ASTM requirement for flow or bond strength measurement of mortar.

Loss of moisture from freshly mixed mortar decreases its flow and thus reduces the bond strength. Retempering the mix with the addition of water does not seriously affect the bond strength. In ordinary mortar the bond strength is greater when brick suction is less than 20 g/min than when it is in the range 20–40 g/min.

Bond Strength. The tensile bond strength can be measured using the procedure described in ASTM C952. It ranges from 20–80 psi (0.14–0.55 MPa) for most mortars, and depends on brick suction, mix proportion, and flow. Mortars made with masonry cements are shown to exhibit excellent tensile bond strength, in excess of 100 psi (0.7 MPa). Addition of lime increases the tensile bond strength, but decreases the compressive strength as well as the modulus of rupture. There seems to exist an optimum lime content, somewhere between 1 and 1/4 parts of cement by volume, that will result in higher bond strength in a given mix. Entrained air in mortar improves the workability, water retentivity, and resistance to freeze–thaw cycles, but this improvement is generally at the expense of its tensile bond strength.

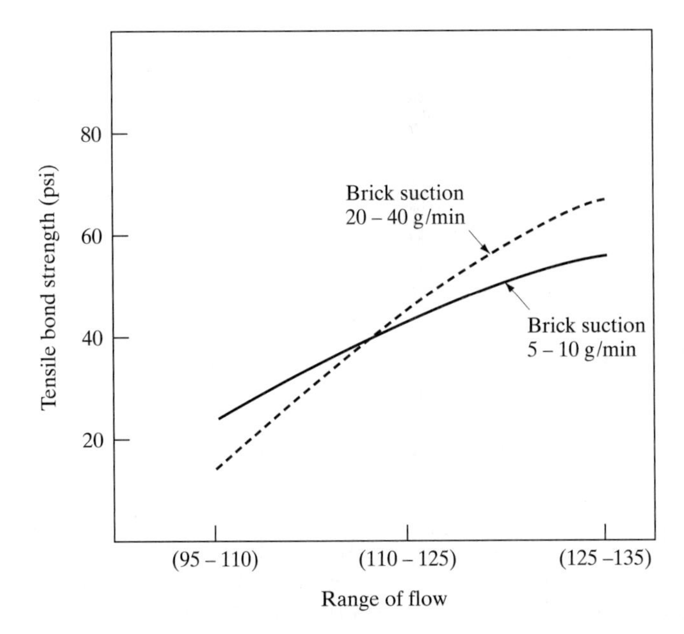

Figure 4.10 Typical variation of tensile bond strength with changes in flow.

Curing conditions may also affect the bond strength. When there is sufficient moisture in the mortar as bricks are laid, additional moisture from sprinkling (curing) will saturate the bricks, decreasing the adhesion between the mortar and the units. Damp curing of brick masonry is thus neither necessary nor desirable.

Lack of bond between mortar and masonry units allows them to separate. To check the bonding properties of mortar, lay a unit on the mortar bed and then pull it out immediately. If the mortar clings to the surface of the unit, it is an indication that the bond is adequate. Clay bricks should be wetted before placement, stone (stone masonry construction) should be dampened, and concrete masonry units should be dry to develop a good bond.

Workability and Water Retentivity. Workability of mortar is difficult to define and is even more difficult to measure. It represents the ability of a mortar to spread easily, to cling to vertical surfaces, and to provide resistance to flow during the placing of bricks.

Water retentivity, which is the ability of the mortar to retain its mix water when subjected to an absorptive force, is a very important property of mortar. Mortar possessing low water retentivity loses its water rapidly—from the mortar bed—making the laying of bricks difficult, whereas that with high water retentivity retains the water, making the task of bricklaying much simpler. Water retentivity can be improved by adding finely ground plasticizer or lime. It is affected by aggregate gradation, properties of cement and lime, and air content and can be measured using the procedure shown in ASTM C91.

4.2.4 Grout and Its Uses

As defined, grout is a fluid mixture of portland cement, lime, sand, fine gravel, and water. The fluid consistency should be such that it fills all voids in the grout space and completely encases the reinforcement. The open space between wythes or the cells in hollow units marked for grouting is called the grout space. The purposes of grout are:

- To bond individual wythes together to form a composite masonry
- To bind the reinforcement to the masonry so that the two can act as a composite material
- To increase the volume, which results in higher density, improving the overturning resistance and enhancing bearing area and fire resistance

The grout mixture is poured from the top of a wall, and on its way down will fill all voids, making a solid mass of the wall (Fig. 4.11). When too much water is added to the grout mixture, it becomes thin and results in porous or weak concrete. If the grout is dry or thick, it will fill only the upper portion of the voids.

Aggregates for masonry grout must conform to the specifications of ASTM C404. Generally, the maximum aggregate size is 3/8 in. (10 mm). High water content is required for ease of placement. One part cement, one-tenth part lime (hydrated lime or lime putty) by volume, and some sand and fine gravel are mixed with water (ASTM C476). Sand is typically $2\frac{1}{4}$ to 3 times the sum of the volumes of the cementitious materials. For fine grout, no gravel is used. For coarser grout, coarse gravel one to two times the sum of the volumes of cementitious materials is used. Hydrated lime should be of Type S. Air-entraining admixtures are not recommended for the grout in reinforced masonry construction.

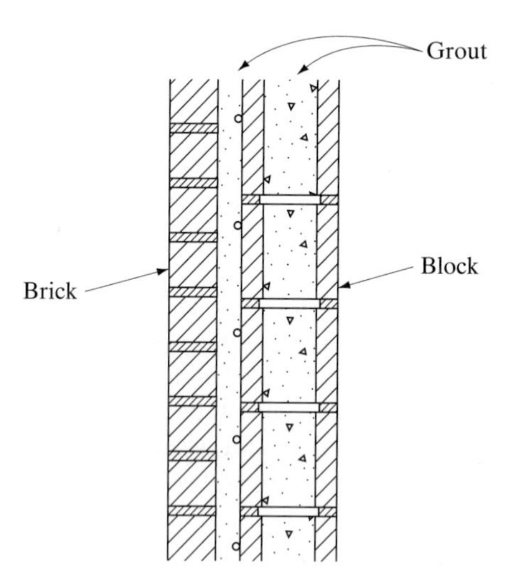

Figure 4.11 Cross-section of a composite wall.

4.2.5 Plaster

Plaster is a fluid mixture of portland cement, lime, and sand which is used as a finishing material for exterior or interior walls. Stucco is plaster used as an exterior wall covering.

Stucco is applied in three coats: scratch, brown, and finish coat (Fig. 4.12). The scratch and brown coats are prepared with coarser fine aggregate, and the finish coat has somewhat finer particles. Excessive fineness can cause crazing and cracking. High-grade mineral pigments such as black iron oxide, yellow iron oxide, and red iron oxide are added for color.

Each coat of stucco is made with 1 part of cementitious materials and 3 to 4 parts of damp loose sand. Average compressive strength of plaster is around 2000 psi (13.8 MPa) at 28 days. Uniform suction is necessary to achieve a good bond on masonry. This is achieved by dampening (not soaking) the walls evenly by fog spraying before applying the stucco. In addition, wetting of the surface before plastering releases the air trapped inside the surface and consequently helps to reduce

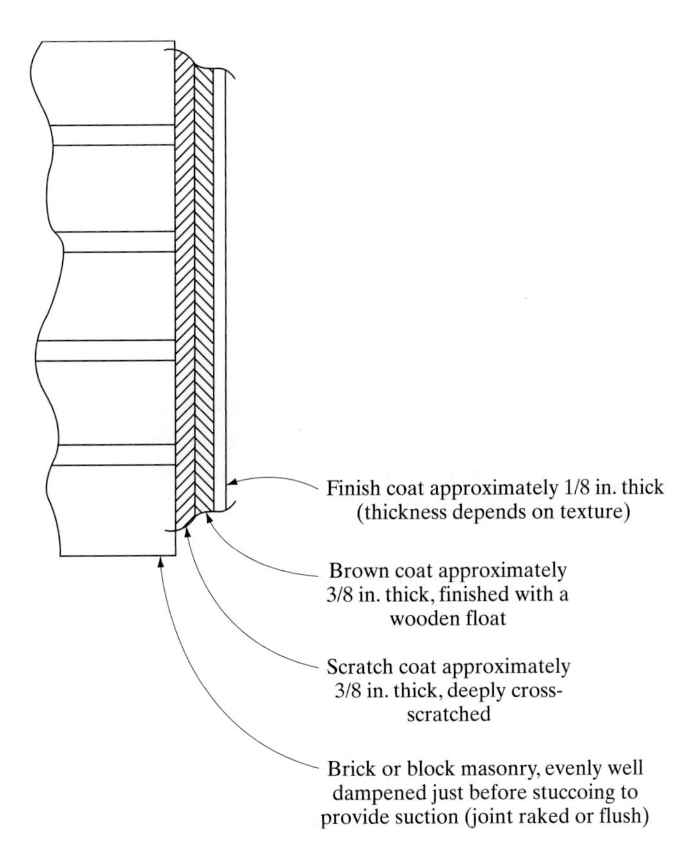

Figure 4.12 Plastering on masonry.

the suction; excessive suction can slow the hydration process. Brown and finish coats should be kept damp continuously for at least 2 days, starting with the hardening of the surface. Each layer should be allowed to dry thoroughly before the next coat is applied.

A mixture rich in cement should be avoided, as it tends to crack from excessive shrinkage. Efflorescence in bricks can cause severe surface damage to plaster. Shrinkage of concrete block walls may cause cracking and crazing in plaster. Crazing is the term used to identify small shrinkage (surface) cracks, forming a pattern on the surface. The crazed surface has an "alligator" appearance when it is wet. Too rapid drying and overworking can also cause crazing.

4.3 MASONRY CONSTRUCTION

Masonry is normally constructed by laying bricks or blocks in a regular pattern, so that each unit sits partly on two or more units. The particular procedure used for laying the units in a regular arrangement (for strength and decoration) is called the *bond* or *bonding*.

A masonry unit is laid on its larger face, called the *bed*. A *bed joint* is the horizontal joint between two courses of brickwork. A *head joint* is a vertical mortar joint placed between masonry units within the wythe at the time the units are laid (Fig. 4.5). The two sides normal to the bed are called the *stretcher face* and *header face* [Fig. 4.3(b)]. There are two stretcher faces and two header faces per unit. The header face of a clay brick is generally darker in color than the stretcher face. A *stretcher* is a unit with its long sides visible on the wall face, and a *header* is a unit with its ends exposed.

Procedure. The individual units are laid with mortar in a regular pattern. The mortar must be sufficiently plastic. (Note that no admixtures are generally allowed with mortar or grout.) The units must be placed with sufficient pressure to extrude mortar from the joint, to result in a tight joint. Mortar for block wall construction is stiffer than that for brick walls. Since blocks are heavier than bricks, a wet mortar may be squeezed out of the joints when the blocks are pressed down. This may result in very little mortar left under the units and joints thinner than 3/8 in. (10 mm). (The joints are typically 3/8 to 1/2 in. (10–12.5 mm) thick). For a wall of a given length, the mason makes adjustments in the thickness of head joints so that a round number of full and half units will make up the wall length. Concrete blocks laid directly on a footing get *full mortar bedding,* and those laid on top of other blocks get *face shell bedding* only (Fig. 4.13). In full mortar bedding, mortar is spread onto the full thickness of the wall, and every portion of the block in the bottom row rests on mortar. In face shell bedding, the mortar is spread along the two outer edges only (face shell), and no mortar is spread along the webs.

The thickness of the initial bed joint must not be less than 1/4 in. (7 mm) and not more than 1 in. (25 mm). Subsequent bed joints must be between 1/4 and 5/8 in.

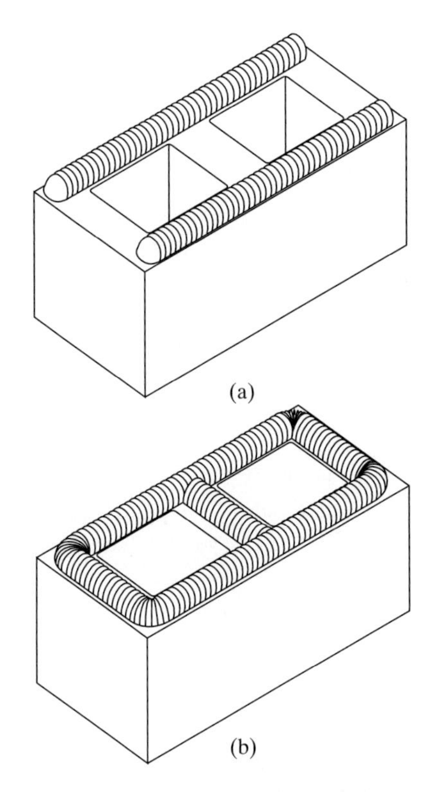

Figure 4.13 Mortar bedding: (a) face shell bedding; (b) full mortar bedding.

(7 and 16 mm) thick. Solid masonry units must have full bed and head joints, and hollow masonry units must have face shell bed and head joints. Improper placement of mortar at the ends of the bricks (buttering) results in weaker head joints.

Mortar and grout mixed at the job site should be mixed for a period of not less than 3 and not more than 10 minutes in a mechanical mixer following the addition of water sufficient to provide the desired workability. Mortar and grout must be used within $2\frac{1}{2}$ hours and $1\frac{1}{2}$ hours, respectively, after the initial mixing. Mixtures that are more than $2\frac{1}{2}$ hours old should be discarded. Within this period, mortar that is stiff can be retempered (in this process the mortar is mixed with additional water to compensate for evaporation loss).

Concrete blocks and clay bricks should be laid on concrete footing, which is generally about twice as wide as the wall thickness (Fig. 4.14). For example, for a brick or block wall that is nominal 8 in. (20 mm) wide, the width of the footing is 16 or 18 in. (40 or 45 mm). For retaining walls, the footing should be larger. The depth of the footing should be at least equal to the wall width. Every footing should extend below the frostline. For clay brick walls, the footing can also be made of clay masonry instead of concrete.

The footing should be dampened before spreading the mortar on its top. First, corner blocks are laid on the footing at each corner, followed by one or more corner

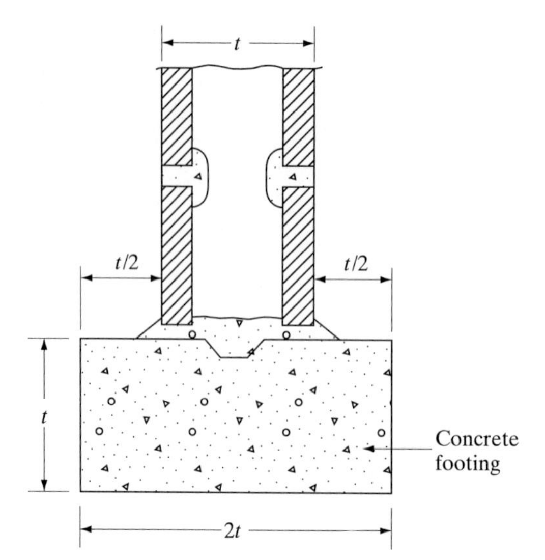

Figure 4.14 Footing dimensions (t = wall thickness.)

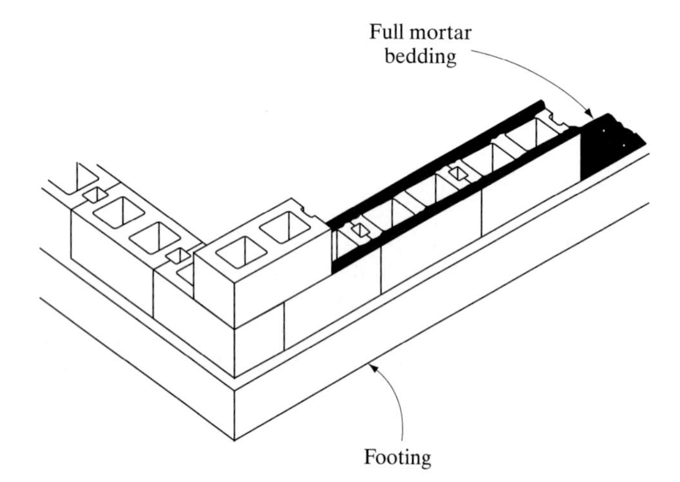

Figure 4.15 Masonry construction.

blocks on top of them (Fig. 4.15). Each successive course of corner blocks is laid so that the blocks overlap to form a running bond. Intermediate blocks are laid following the placement and leveling of two or more courses of corner blocks. The intermediate blocks must be buttered at their ends before being laid. All adjustments in positioning the blocks should be carried out while the mortar is still plastic; later adjustments will break the mortar bond.

To make watertight joints, concrete masonry units should have full mortar bedding on the face shells. The units should be pushed into place when laying so that excess mortar squeezes out at the head joint and along the sides of the wall. The head joint must be completely filled with mortar. After the unit is bedded, the excess mortar is cut off and used for the next end joint. Surplus mortar is thrown back into the mortar board. Before the mortar sets, the wall should be checked vertically, horizontally, and along an angle, to see that all the units are aligned, leveled, and plumb. If they are not, adjustment can be made to the offending units. After the mortar sets, units should not be moved; they have to be relaid. When building double-wythe walls, both wythes should be laid at the same time. The head joints of units should be staggered between the two wythes, and all courses should be level with each other (Fig. 4.5). Tool finishing of joints should be delayed until the mortar is sufficiently stiff to hold its shape.

At the site, masonry units should be stored so that at the time of use they remain clean and structurally suitable for the use intended. The rate of absorption of clay units, when held in 1/8-in.-deep water for 1 minute, should not exceed 0.025 ounce per square inch. Uniform suction of units is necessary to achieve a good bond with mortar. To control the suction, concrete units should be kept dry until they are built into the wall and clay units must be dampened. Clay units have greater affinity for water, and it is best to dampen them before laying, for if the mortar is stiff and the brick is dry, the latter will dry out the mortar before it can set or harden.

Partially completed walls should be covered at all times when work is not in progress. Prior to grouting, the grout space must be clean without any mortar projections greater than 1/2 in. (12.5 mm), mortar droppings, or foreign material. Grouting of any section of wall must be completed within 1 day, with no interruptions longer than 1 hour.

A brick-veneered wall [Fig. 4.2(d)], which is one-wythe thick, provides weathertightness and requires low maintenance. It needs a wider foundation than that just for a framed or solid brick wall (about $4\frac{1}{2}$ in. (11.5 cm) additional width). This type of wall is laid without headers but it should be tied to the frame. A common method of tying is to nail 22-gauge corrugated metal ties to the frame and bend them down into the mortar joints as and when the bricks are laid. One tie for every 2 ft^2 (0.19 m^2) of wall area is recommended, and the maximum spacing between the ties, both horizontally and vertically, is 24 in. (61 cm). Before laying the bricks, a layer of building paper should be stapled to the frame or the sheathing.

Grouting. Based on the extent of grouting, a block wall is classified as a partially grouted or solid grouted wall. In a partially grouted wall, grout is confined only to those cells that have horizontal or vertical steel (Fig. 4.16). The partial grouting technique reduces the wall dead load and the quantity of grout. Expanded wire mesh is placed on the bed joints below (and sometimes above) the horizontal steel to confine the grout to cells with steel reinforcement. In addition, to prevent the grout from flowing between joints in neighboring empty cells, cross webs of units should be bedded.

In a solid grouted wall, all cells are filled with grout, and the wall is a solid masonry wall. Due to the increased cross-sectional area, a solid grouted wall possesses greater load-carrying capacity than a partially grouted wall. It also has an increased fire rating and provides greater stability against overturning, but it is more expensive than a partially grouted wall and is heavier, requiring a wider foundation.

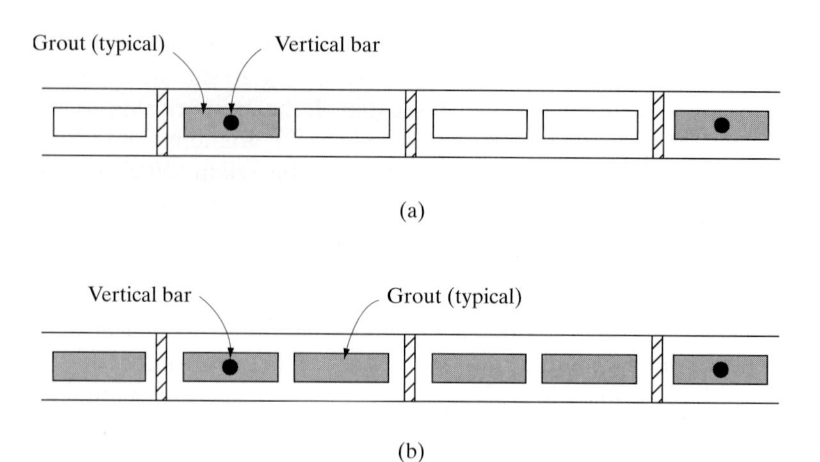

Figure 4.16 (a) Partially grouted wall; (b) solid grouted wall.

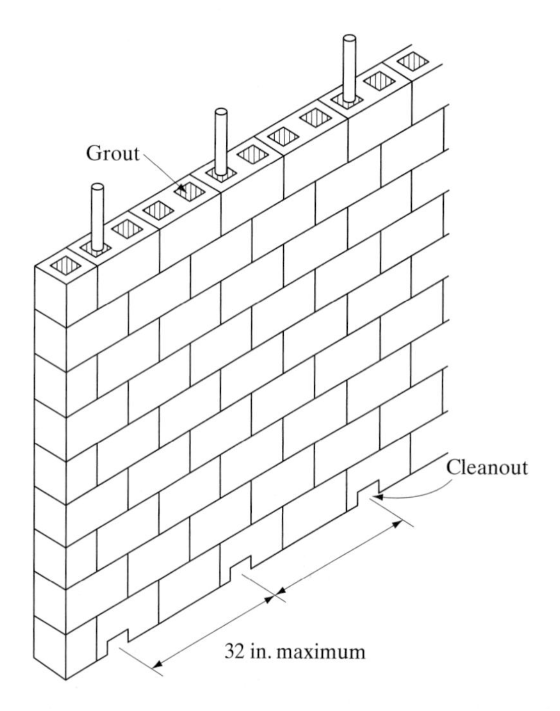

Figure 4.17 Cleanouts in a solid grouted wall.

Low-lift grouting is a method in which the height of grout pour is 5 ft (1.5 m) or less. Cleanouts are not required in this method of construction. In *high-lift grouting,* the height of masonry prior to grouting exceeds 5 ft (1.5 m), and cleanouts should be provided. Cleanouts are openings placed at the bottom of hollow masonry walls for the removal of mortar droppings and other debris (Fig. 4.17). They are formed by a number of means, such as leaving out a unit, removing face shells, or cutting holes in the face shells.

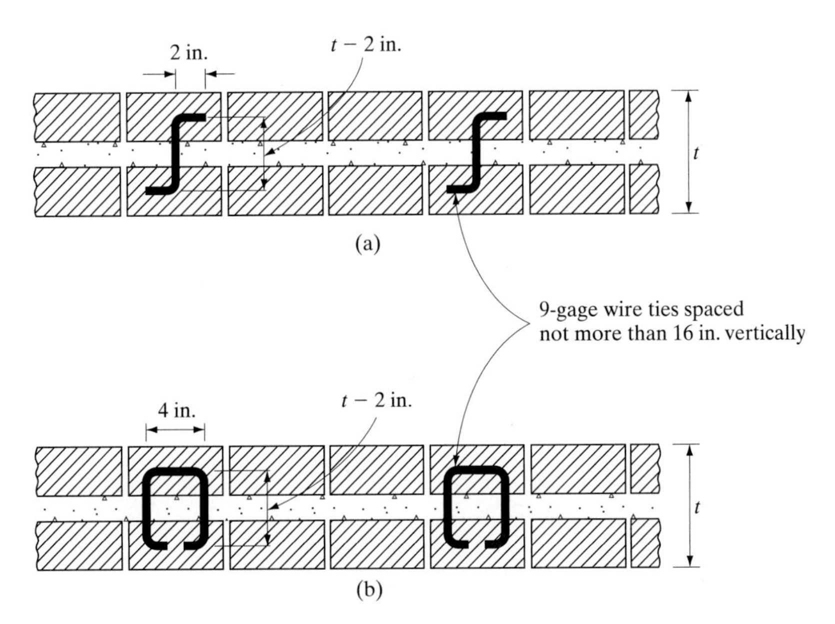

Figure 4.18 Bonding of double-wythe walls: (a) Z tie; (b) rectangular tie.

In two-wythe masonry construction, when the grout pour height exceeds 12 in. (30 cm), the wythes should be tied together with metal ties to prevent them from bulging (Fig. 4.18). This procedure is required in both low-lift and high-lift grouting techniques. The spacing of ties in running bond construction is 24 in. (61 cm) in the horizontal direction and 16 in. (40 cm) in the vertical direction. A hollow unit masonry does not require ties.

4.3.1 Types of Bond

As explained, the bond (also called *pattern bond*) refers to the type of procedure used in laying the units. When the bricks are laid on the bed with every brick showing a stretcher or long face on each side of the wall, it is called *stretcher bond* [Fig. 4.2(a)]. Note that the stretcher units bind the wall together lengthwise and the header units crosswise. When the bricks are arranged so that every brick shows a header or short face on each side of the wall, it is called *header bond.*

Running bond is the simplest system of laying bricks and consists of all stretchers or the stretcher bond (Fig. 4.19). In running bond, the head joints in successive courses are horizontally offset at least one-quarter the unit length. It is used commonly in cavity wall, veneer wall, and facing tile wall construction. Since there are no headers, metal ties should be used for tying.

A *block* or *stack bond* consists of no overlapping units, and all vertical joints are aligned. Due to inadequate lateral strength, this type of wall is usually bonded to the backing with rigid steel ties.

When a combination of header bond and stretcher bond is used, the procedure comes under *English bond, Flemish bond*, or *common bond* (Fig. 4.19). In English

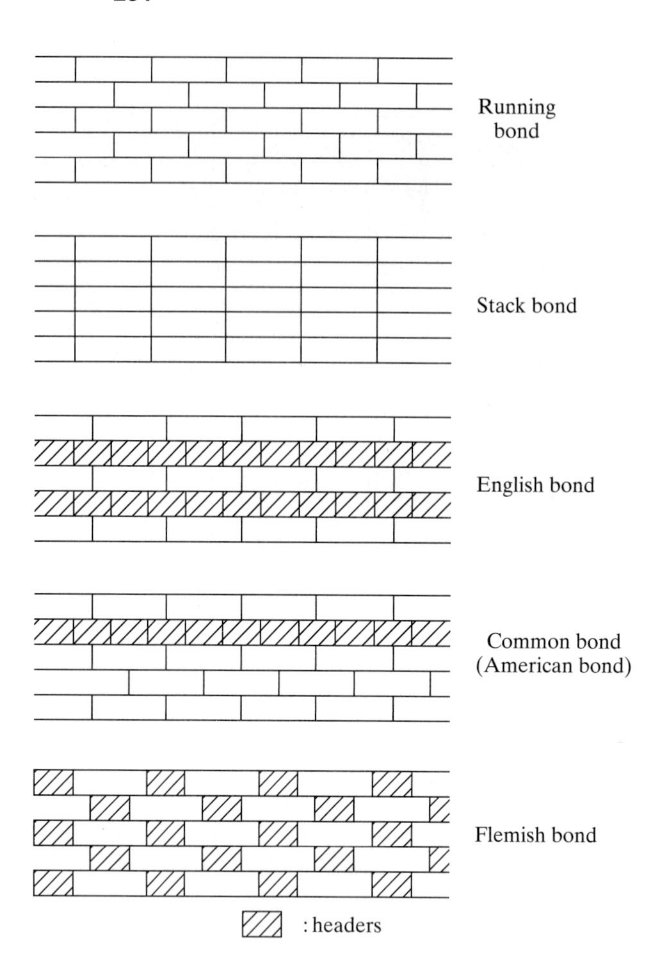

Figure 4.19 Types of masonry bonds.

bond, the bricks in one layer are laid on header bond and the courses above and below this are laid on stretcher bond. There are twice as many vertical or side joints in a course of headers as there are in a course of stretchers. This bond is very strong because of the absence of aligned joints within the vertical section of the wall, but the pattern is more difficult to lay and is more expensive than other bonds.

In Flemish bond, bricks in every course show, alternately, header and stretcher bonds. The outer end of each header lies in the middle of a stretcher in the course below. The number of vertical joints in each course is the same. Flemish bond is more decorative and is not as strong as English bond.

Common bond or *American bond* consists of a course of headers for every few courses of stretchers. The number of courses of headers and stretchers varies from wall to wall, but it is more common to lay one course of headers for every four to seven courses of stretchers. The header course provides lateral strength to the wall.

Varied color bricks can be used in Flemish bond to give color-combination effects. Common bond is considered to be one of the strongest structural bonds because the header course serves to bond the various wythes together. English bond provides exceptional strength across the width of the wall. The term *structural bond* refers to the system by which individual wythes are tied together or a single-wythe brick wall is tied to a backup. It involves laying header courses or the provision of metal ties in a cavity wall construction [Figs. 4.2(c) and 4.18]. For a veneered wall erected over a framed wall, structural bond is through the provision of metal straps.

Rat-trap bond (also called *Chinese bond*) is a variation of Flemish bond. It has alternate headers and stretchers in each course, but the bricks are laid on edge instead of on a bed. This arrangement results in a cavity between the two wythes. This bond is cheaper than other types of bond but is of low strength and is less weather resistant.

4.3.2 Types of Joints

In masonry construction, individual masonry units are bonded together using head joints and bed joints (Fig. 4.5). Tight mortar joints are essential for good performance of walls. All joints—horizontal and vertical—must be filled over the depth of the face shell. A number of types of joints can be formed after the mortar is stiffened sufficiently, as shown in Fig. 4.20.

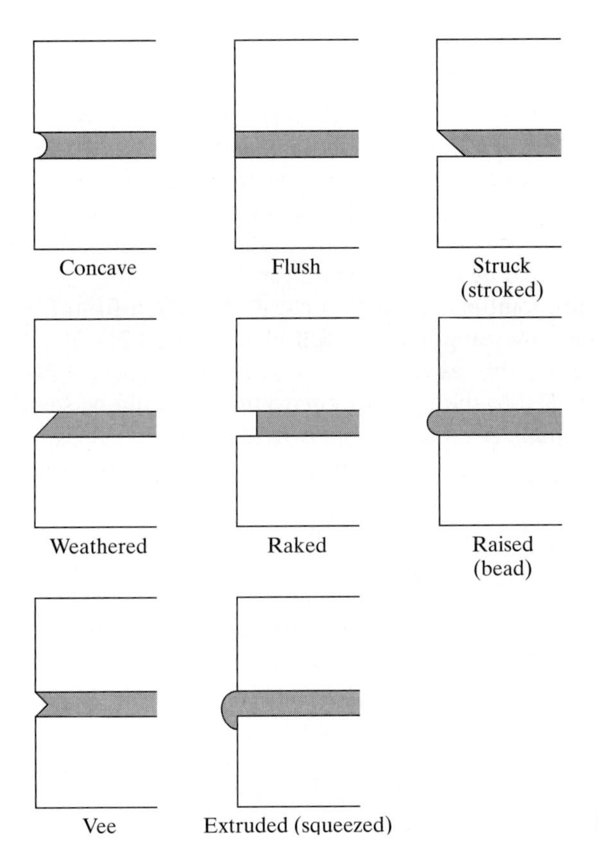

Figure 4.20 Types of joints.

The best weathertight joint is the *concave joint,* which is made with a special tool after the excess mortar has been removed with the trowel. *Flush joint* is made by keeping the trowel nearly parallel to the face of the wall while forcing the point of the trowel along the joint. It can also be done by rubbing the surface with a burlap sack. It is used when the wall is to be plastered or painted. *Raked joint* may promote leakage, and *concave* and *vee joints* offer the best protection against leaks. Both concave and vee joints produce stronger and more watertight bond.

Raked joints are made by removing the mortar with a tool while it is still wet, and *stripped joints* are made by placing wood strips in joints when the bricks are laid. Thin mortar joints produce stronger, watertight walls. When the thickness of joints exceeds 1/2 in. (12.5 mm), watertightness may be affected.

4.3.3 Control Joints and Expansion Joints

Concrete and brick masonry walls will undergo dimensional changes with changes in temperature and humidity. To provide for expansion and contraction, and to force cracks at predetermined locations, control joints or expansion joints should be provided in masonry walls every 20 ft (6 m) or more in length. Control joints are used to curb movement in concrete masonry walls due to shrinkage resulting from loss of moisture and temperature changes.

Expansion joints control expansion of clay masonry due to absorption of moisture and temperature. Dimensions of a clay brick are the smallest after firing. With gain in moisture and with rise in temperature, the brick expands. This increase in length is about 0.02 percent for each 1 percent increase in moisture, and 0.03 percent per 100 °F increase in temperature. If this expansion is prevented, high compressive stresses will develop, causing crushing or displacement of bricks. In both types of joints, forces resulting from the restraint to movement should be relieved to prevent cracking and this is accomplished by vertical separation.

A control or expansion joint is a straight vertical separation from the top to the bottom of a wall, formed by using half and full blocks [Fig. 4.21(a)]. All joints should be laid up in mortar joints as any other vertical joint [Figs. 4.21(b) and 4.21(c)]. Joints that are exposed to the exterior environment should be sealed with a caulking compound (Fig. 4.22). A recess can be provided for the caulking material by raking out the mortar to a depth of about 3/4 in. (19 mm) after the mortar has hardened.

Concrete units, as they are made with cement, are known to shrink with time (drying shrinkage) and to creep when subjected to compressive loads. Clay units, on the other hand, absorb moisture from the environment and expand with time. The average coefficient of lineal thermal expansion for clay units varies between 2.5–4.0 $\times 10^{-6}$ in./°F. For concrete units this value depends on the aggregate type, and varies between 3.0–5.2 $\times 10^{-6}$ in./°F. Cinder aggregate has the lowest expansion coefficient; normal-weight aggregates have the highest.

Because concrete units are susceptible to shrinkage, it was indicated that the ASTM provides limits to the moisture content of Type I units (Table 4.3). Note that

Type II units do not have to satisfy these limits. For (unreinforced) concrete walls using Type I units, the maximum spacing between control joints is twice the height of the wall or 40 ft (12 m), whichever is smaller. In walls using Type II units, this should be reduced in half.

Expansion joints in brick walls must be such as to permit relative movement due to thermal and moisture expansion. Spacing of these joints may vary between 32–150 ft (9.6–45 m). The width of the joint should be selected in such a way that it

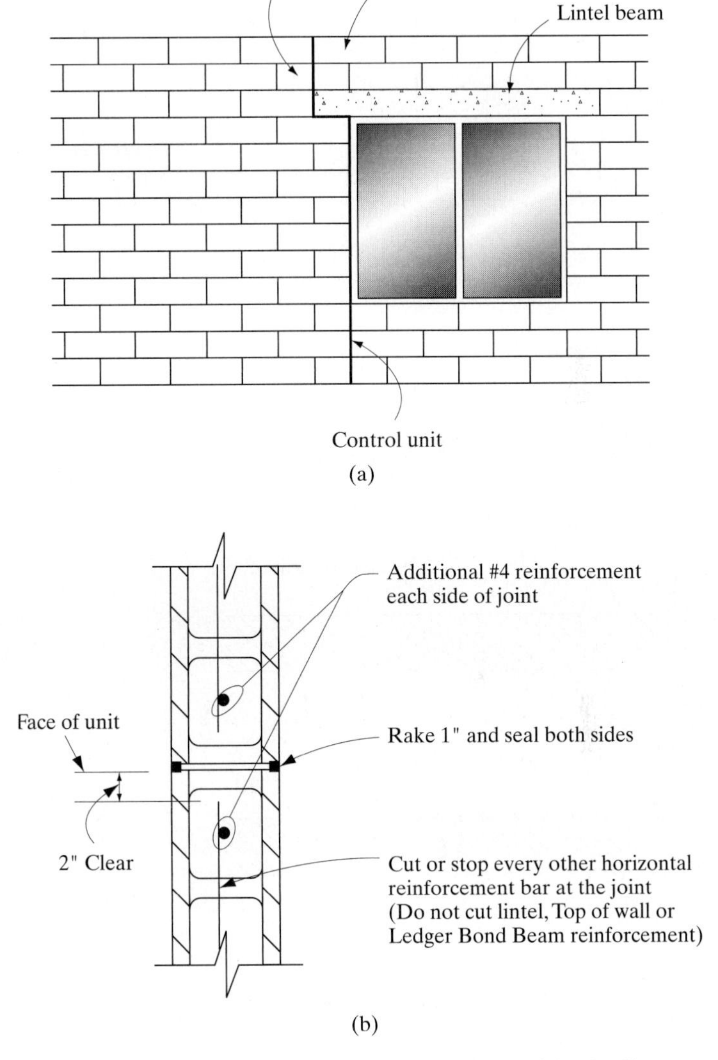

Figure 4.21 (a) Vertical control joint. (b) Control joint detail.

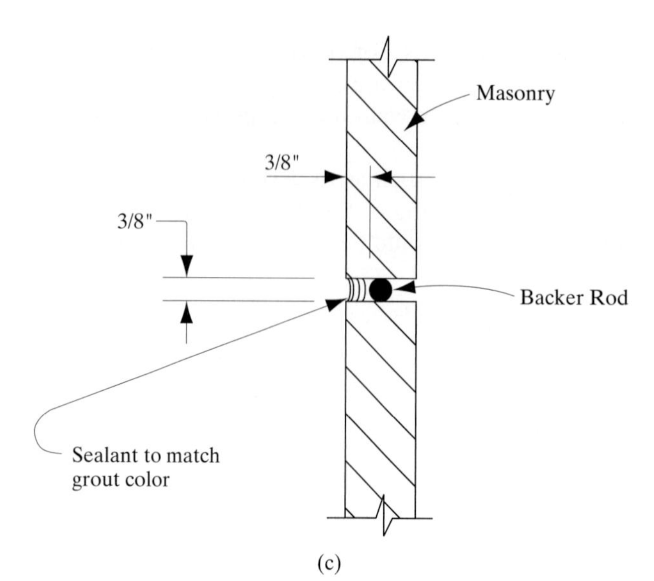

Figure 4.21 (*Concluded*) (c) Joint seal detail.

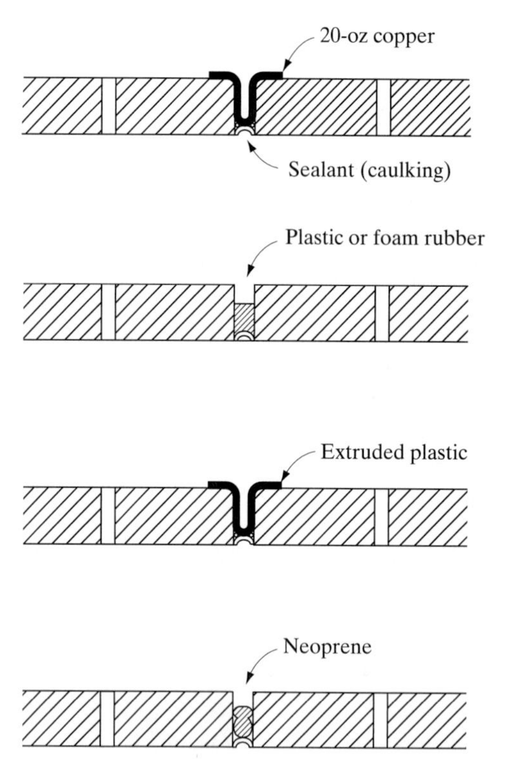

Figure 4.22 Expansion joints in brick-veneered wall.

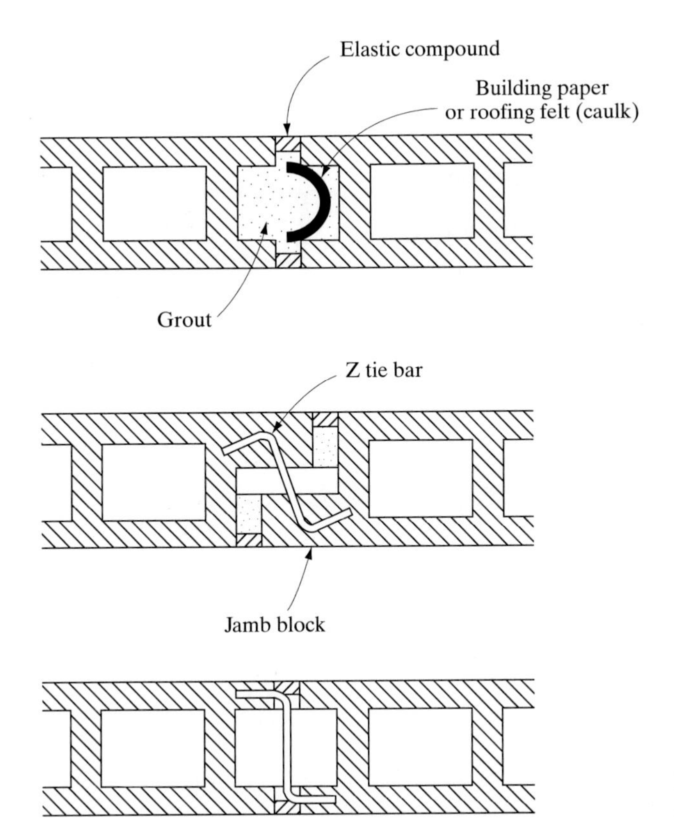

Figure 4.23 Lateral support across control joints in block wall.

will at least close partially under any expansion and is limited to 1 in. (which permits 0.5-in. expansion of the wall). The width and spacing of expansion joints can be calculated using equations given at the end of this section.

Lateral support for walls across the control joint can be provided using a number of techniques (Fig. 4.23). Placement of control joint blocks, which look like tongue-and-groove-shaped blocks, at the joints provides lateral support. Installing a noncorroding metal Z tie bar 2 in. (5 cm) narrower than the width of the wall in every other horizontal joint ensures lateral support for wall sections on each side of the joint. Control joints can also be made with building paper or roofing felt inserted in the end core of the block, extending to the full height of the wall. The end core is then filled with grout to provide lateral support. The function of the building paper is to prevent mortar from bonding to one section of the masonry. At any type of vertical joint, mortar is raked out to a depth of 3/4 in. (19 mm) before it hardens and is filled later with an elastic compound.

Control joints should also be provided when the wall thickness changes and at wall intersections (except for corner walls). Intersecting concrete walls should not be tied together in a masonry bond except at the corners. Instead, one wall should terminate at the face of the other wall with a control joint at that joint. To provide

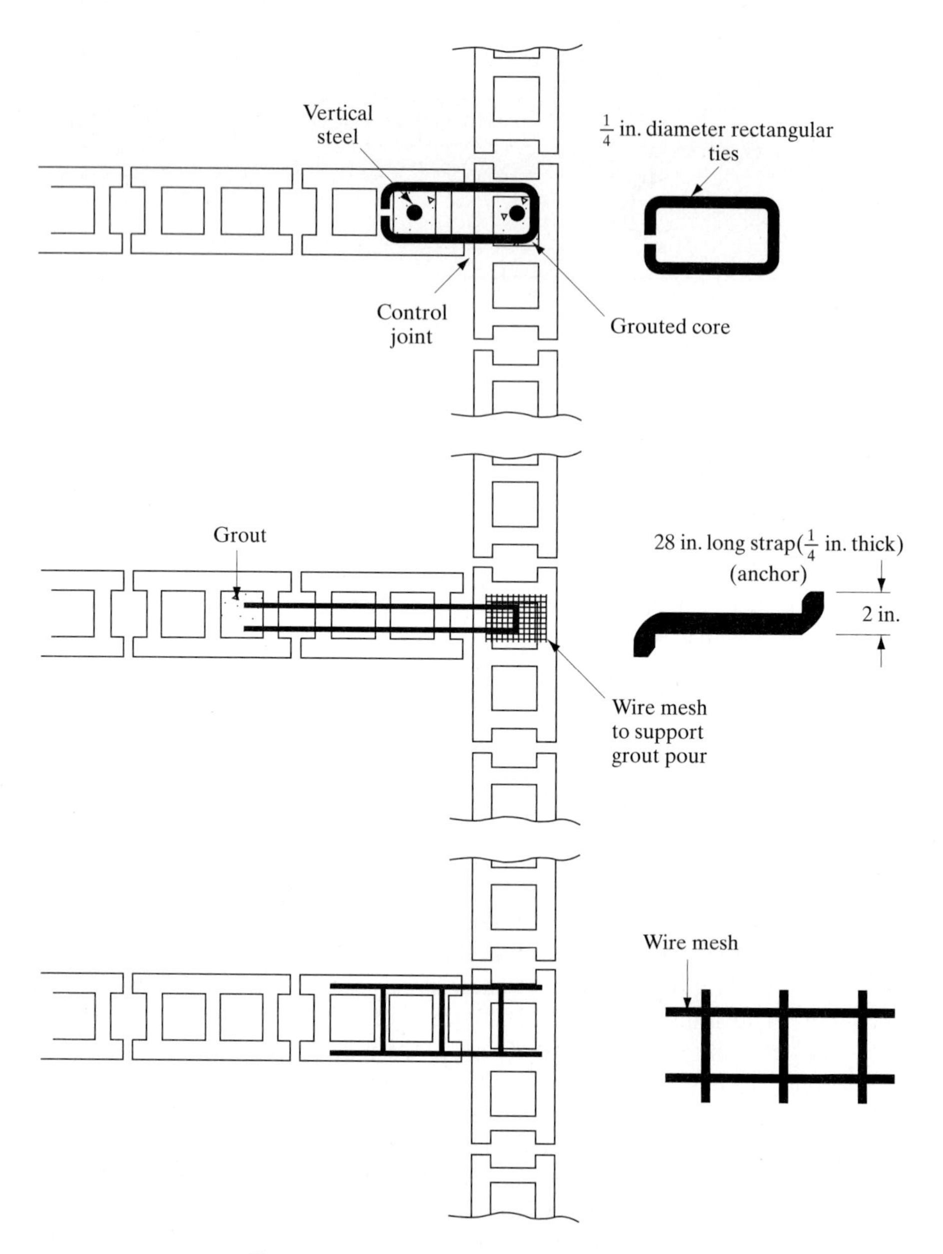

Figure 4.24 Control joints in intersecting walls.

lateral support for the intersecting walls, tie bars with two right-angle ends are embedded in joints and later grouted (Fig. 4.24). The tie bars are spaced at a maximum spacing of 4 ft (1.2 m) on centers vertically. Lateral supports can also be accorded using joint reinforcement (wire mesh), typically of No. 9 longitudinal wires.

In (unreinforced) concrete masonry, the horizontal spacing of control joints (using Type I units) is two times the height of the wall or 40 ft (12 m), whichever is smaller. For example, a wall of height 10 ft will have a maximum spacing of control joints as the smaller of (a) $2 \times 10 = 20$ ft and (b) 40 ft. Thus the maximum spacing is 20 ft. By incorporating reinforcement within the wall (reinforced masonry wall), this spacing can be increased.

For brick masonry, the width and spacing of expansion joints can be estimated as

$$\text{net wall expansion} = [C_m + C_t(T_{\text{max}} - T_{\text{min}})]L$$

where C_m is the coefficient of moisture expansion, C_t the coefficient of thermal expansion; T_{max} the maximum mean wall temperature in °F, T_{min} the minimum mean wall temperature in °F, and L the length of wall in inches. Average values of C_m and C_t are 0.0002 and 0.0000040, respectively.

$$\text{maximum spacing of expansion joints (feet)} = \frac{24{,}000}{T_{\text{max}} - T_{\text{min}}}(p)$$

where p is the ratio of opaque wall area to gross wall area. The spacing is generally between 150–200 ft (45–60 m) for a straight length of wall. Expansion and control joints may also be located at sections of plan irregularity. Typically, they are located at pilasters, at changes in wall height and thickness, over openings, and at wall intersections.

4.4 PROPERTIES OF MASONRY

Important properties of masonry are:

- Density
- Compressive strength
- Flexural strength
- Modulus of elasticity
- Durability
- Thermal conductivity

Density or unit weight of a wall depends on weight of units, wall thickness, and grouting. For example, the weight of a 4-in.-thick (nominal) clay masonry wall is 37 psf (lb/ft^2) (181 kg/m^2), and that of an 8-in.-thick concrete masonry wall (unit height 8 in.) is 58 psf using normal-weight units or 36 psf using lightweight units.

Compressive Strength. Compressive strength of masonry depends primarily on the compressive strength of units, the mortar mix proportions (amount of cement), grouting, the workmanship, and the type of bond. An increase in cement content (of mortar) increases the mortar strength. Studies have shown that

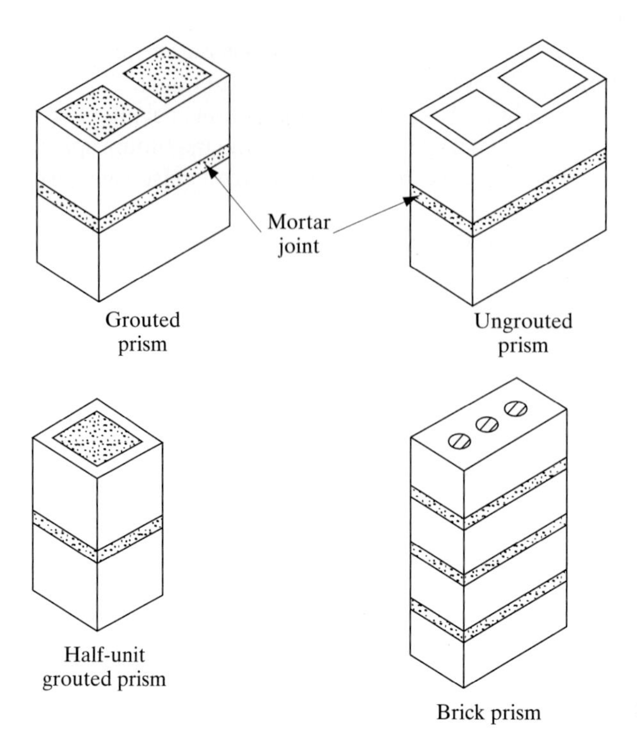

Figure 4.25 Masonry prisms.

proportions of cement or lime can be altered without seriously affecting the strength of the masonry, although considerable variations in the mortar compressive strength can be expected. This means that the strength of the mortar has a limited effect on the compressive strength of the masonry. Consequently, strong mortars are not recommended in masonry construction. Moreover, high-strength mortars shrink tremendously, which weakens the bond. In addition, they may tend to exhibit a high rate of efflorescence.

The compressive strength of walls can be predicted most accurately from measuring the strengths of masonry prisms constructed with similar material. Prisms contain two or more units bonded with mortar with or without grout, and are used primarily to predict the strength of the actual masonry wall (Fig. 4.25). The strength of the prism varies with its height, decreasing in strength with increase in height.

Testing of these masonry specimens in compression will provide the ultimate compressive strength of the masonry, f'_m. The allowable axial, flexural, shear, and bearing stresses in the wall are based on this compressive strength measurement.

When prism tests are not made, as is the case in most masonry construction, the compressive strength of masonry is assumed based on the listed empirical data provided in the building codes (Fig. 4.26). They are tabulated for various compressive strengths of the units and for different mortar mixes. Their values range from a low 875 psi (6 MPa) to a high 2400 psi (16.6 MPa) for concrete units, and from 530–4600 psi (3.7–31.7 MPa) for clay units. Note that the compressive strength of concrete units may not exceed 6000 psi (464 MPa) and that of brick units can be as high as 14,000 psi (96.6 MPa). An increase in the compressive strength of units does not necessarily correspond to a proportionate increase in the masonry strength due to weak mortar joints. For example, the expected compressive strength of brick masonry built with 12,000 psi units and Type S mortar is 4700 psi, whereas that with 6000-psi units and the same mortar is 2700 psi.

The allowable stresses in the masonry are calculated from the masonry compressive strength by applying a safety factor. The building codes allow higher stresses when special inspection by a building official or engineer is mandated during construction. This inspection is required during preparation of masonry wall prisms, sampling and placing of all masonry units, and placement of reinforcement. In addition, inspection is required of grout spaces, immediately prior to closing of

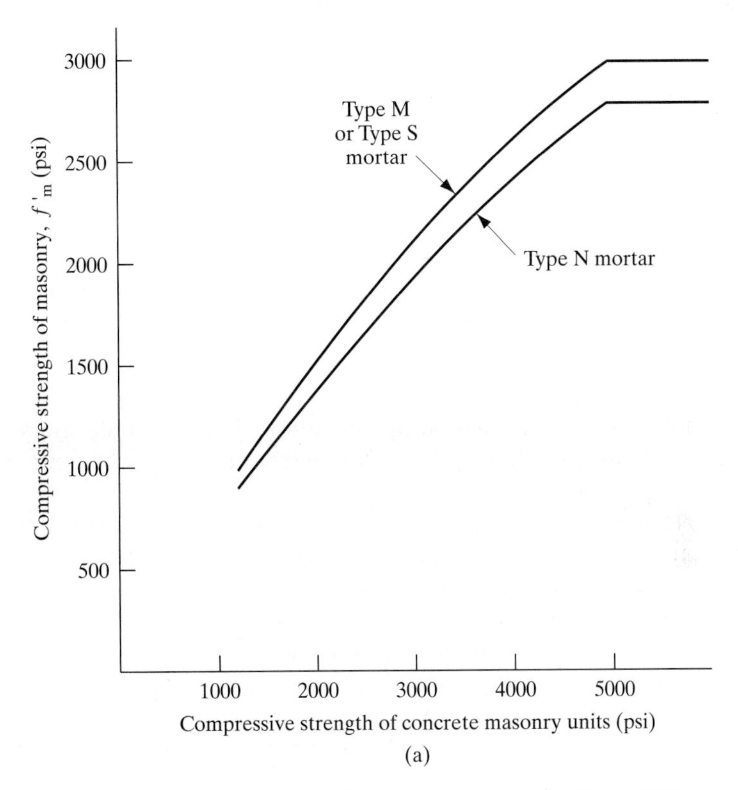

Figure 4.26 Compressive strengths of concrete masonry walls. (From UBC Table 24C. Reproduced from the 1991 edition of the *Uniform Building Code*, copyright © 1991, with the permission of the publisher, the International Conference of Building Officials.)

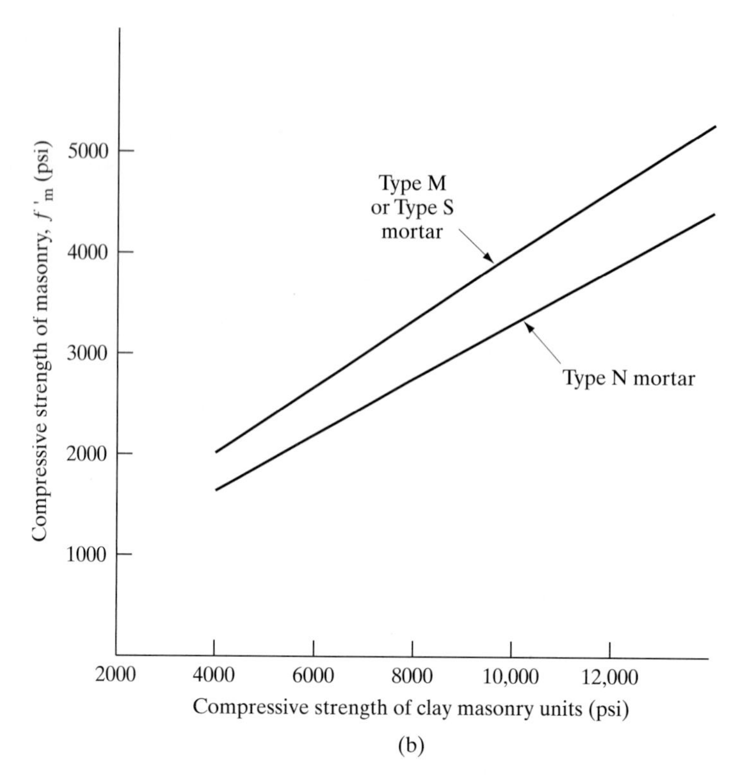

Figure 4.26 (*Concluded*) Compressive strengths of clay brick walls. (From UBC Table 24C. Reproduced from the 1991 edition of the *Uniform Building Code,* copyright © 1991, with the permission of the publisher, the International Conference of Building Officials.)

cleanouts, and during all grouting operations. The allowable stresses in masonry constructed with special inspection is twice that in masonry without inspection.

Modulus of Elasticity. There is no consistent relationship between the ultimate compressive strength of a wall and its modulus of elasticity. However, most of the measured E values lie between 1200 and 700 times the compressive strength of the wall, as shown in Fig. 4.27. The modulus of elasticity values are affected by the wall design, the strength of the unit, the mortar strength, and the workmanship. When the workmanship remains the same, the modulus of elasticity improves with the brick strength and the mortar strength, and decreases with the increase in total thickness of mortar joints per unit length, measured parallel to the direction of the compressive force.

Thermal Conductivity. Thermal conductivity is defined as the time rate of heat flow through a unit area of a material, from one surface to the other per unit

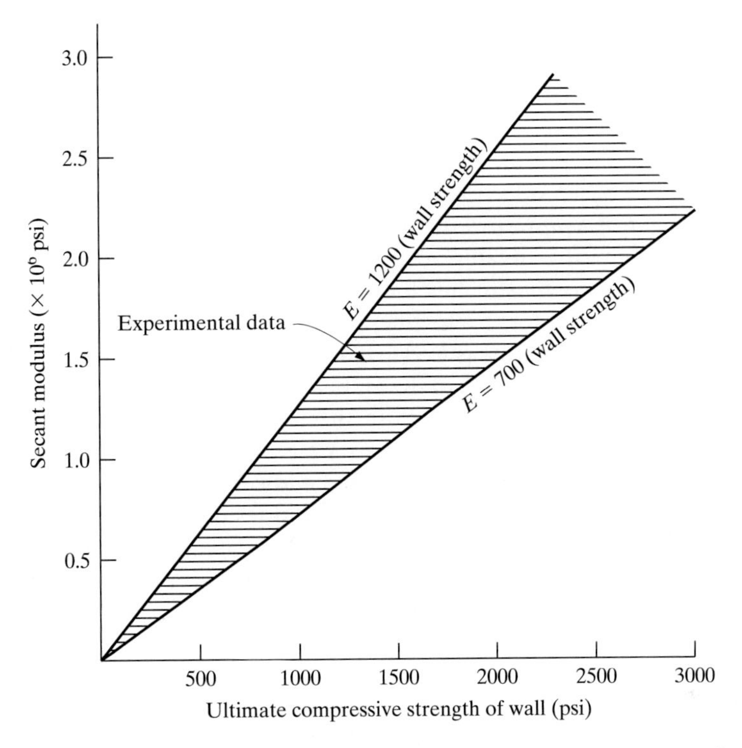

Figure 4.27 Relationship between modulus of elasticity and ultimate compressive strength of clay brick walls in cement–lime mortars.

temperature difference between the two surfaces. It is expressed in Btu (British thermal units) per hour per square foot per degree Fahrenheit temperature gradient per inch of thickness. The thermal conductivity of masonry units depend on the density, the amount of moisture present, and the type of unit.

In general, thermal conductivity increases with density and moisture content. Since air is a poor heat conductor, materials that are porous or contain a high proportion of air in their structure have low conductivity values. Conductivity of dense concrete is higher than that of common bricks, which is much higher than that of fiberglass insulation. In double-wythe masonry walls, the conductivity of the outer wythe may be higher than that of the inner wythe, due to a higher moisture content in the outer wythe.

Heat Transfer. The heat transfer through a wall depends on the difference in temperature between the two sides. Through a wall containing voids, the heat transfer takes place by convection, radiation, and conduction. Convection refers to the process whereby heat is transferred by the movement and mixing of liquids or gases (for example, convection ovens). Radiation is a process of transferring heat through invisible wave radiation, similar to the transmission of light (for example, solar radiation). Conduction is a process in which heat is passed between adjacent stationary particles of matter. From a microscopic standpoint, it refers to energy being handed down from one atom or molecule to the next. Heat transfer through a solid wall takes place only by conduction.

The overall transmission of heat from the air on one side of a structure to the air on the other side is referred to as the *heat transmission coefficient* (or *overall coefficient of heat transmission,* or *thermal transmittance*). Denoted the *U-value* of the structure, it is related to the conductivity of the material. Note that the heat transmission coefficient has the same unit as the conductivity, with the exception that the former refers to the thickness of the material.

The *U*-value is measured in Btu per hour per square foot per degree Fahrenheit difference in temperature of the air between the warm and cool sides of the wall. One Btu is the heat required to raise the temperature of 1 pound of water 1 degree Fahrenheit at sea level. Brick cavity walls have *U*-values in the range 0.25–0.35, and solid walls have *U*-values of 0.3–0.5; the lower the *U*-value, the thicker the wall. The plastering decreases the *U*-value and so does insulation. A lightweight hollow concrete wall has a *U*-value of about 0.33. The *U*-value of a wall can be calculated by knowing the thermal conductivities of all materials through the thickness of the wall.

The *U*-values just indicated assume a constant differential temperature of the air between the inside and the outside, and do not take into account the daily cycles of solar radiation, air temperature, and air velocity. These causes result in temperature fluctuations that produce a dynamic thermal response, which can differ substantially from the heat flow calculations based on standard *U*-values.

The massive masonry walls in many older buildings are more thermally stable than modern lightweight walls (Fig. 4.28). A solid masonry wall is heavy enough to store heat and retard its migration substantially. This characteristic, called *thermal storage capacity,* is a measure of the quantity of heat required to raise the temperature of a mass of the material. Two materials may have the same *U*-values or insulating values, but a heavier wall requires longer to heat up than does a thinner wall. Similarly, the latter will cool down faster than the former. Partition walls of high thermal storage capacity are capable of storing more heat during the night than are lightweight partition walls.

4.4.1 Efflorescence

Clay bricks contain soluble salts that migrate to the surface along with water. As water evaporates to outside air, these salts will collect as patches of white crystals on the face of brickwork (Fig. 4.29). These irregular and unsightly patches are called efflorescence ("powdery substance").

The most common salts are calcium and magnesium sulfates, although various salts of potassium and sodium are also found. The percent of calcium sulfate in bricks is generally very high compared to the other salts. These salts originate generally from the raw clay, and in some cases are formed in the burning process. Aggregates washed by sea-water may contain sodium chloride and other salts that

Figure 4.28 Thick walls of older buildings provide good thermal storage capacity.

contribute to efflorescence. Some curing compounds and air-entraining admixtures may also contain sodium chloride or calcium chloride solution, which accelerates the efflorescence. In coastal areas, efflorescence may be caused by sea salts.

The amount of efflorescence depends on the type and amount of salts. Some salts, such as calcium sulfate, are less soluble and take a longer time to leach out of bricks. The solubility of magnesium sulfate is greater and may cause flaking of the brick faces; this is the most destructive soluble salt in clay units.

Efflorescence is usually white in color. It occurs more often in winter months or following heavy rains. Without the presence of water or moisture in the wall, there will be no efflorescence. Cyclic changes in moisture—alternate wetting and drying cycles—contribute greatly to efflorescence.

Efflorescence is more pronounced in exposed walls such as parapets, chimneys, and basement walls. These white salts do not cause damage or deterioration but result in an unsightly surface. To prevent efflorescence, the wall must be protected from water migration. Any method to prevent water migration into the wall is helpful in controlling efflorescence. Bricks that have been saturated before they are laid are more prone to efflorescence than are those laid surface-wet. The units must be free of soluble salts, and the construction should allow for the drainage of rainwater away from the walls. The exterior walls can be capped and flashings may be needed to keep rainwater away from the outside face of the wall.

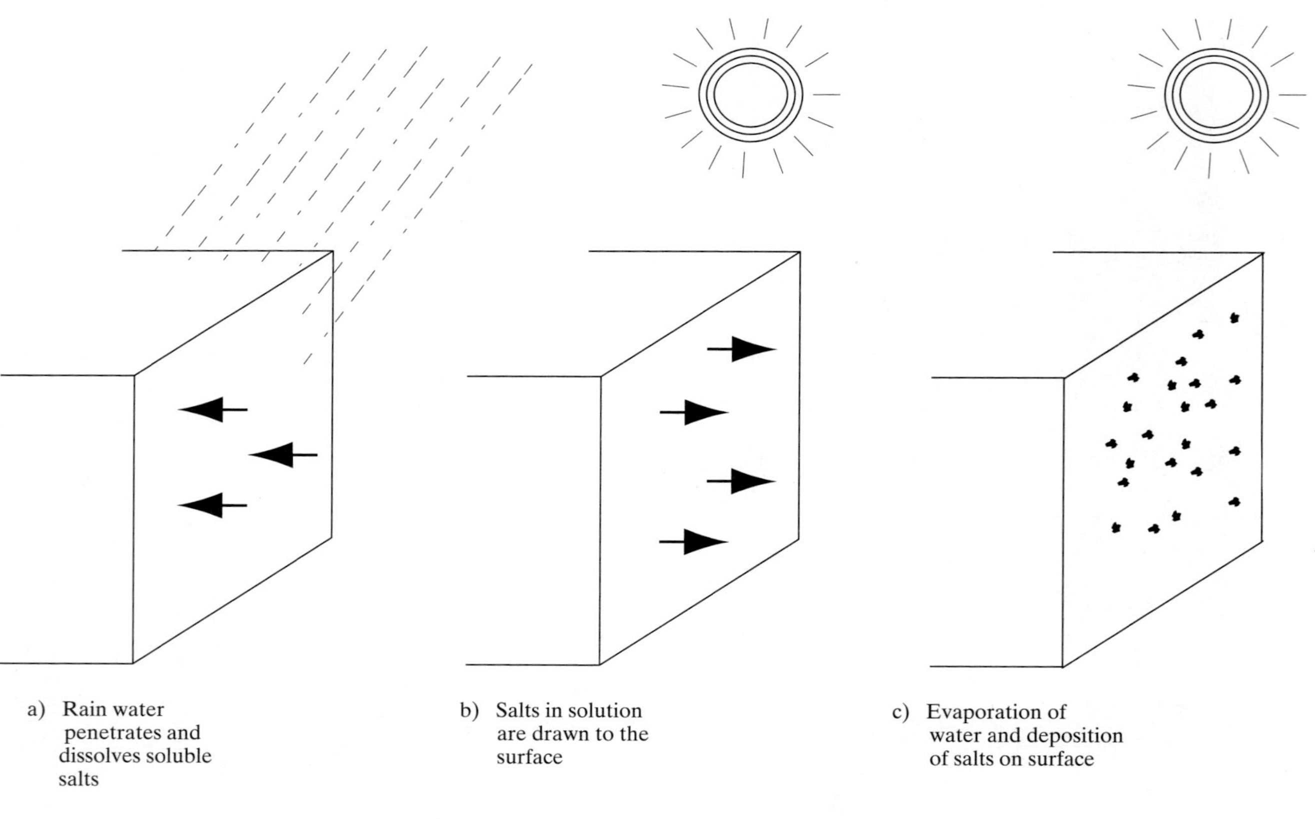

Figure 4.29 Stages leading to efflorescence.

4.5 REINFORCED MASONRY

In reinforced masonry, reinforcement acting in conjunction with the masonry is used to develop flexural strength and resist lateral loads (Fig 4.30). Reinforced grouted masonry is made with solid clay units (bricks) or solid concrete building bricks in which interior spaces of masonry are filled by pouring grout around reinforcement therein (Fig. 4.31). The minimum grout space (space between wythes) is $2\frac{1}{2}$ in. (6.3 cm) for low-lift grouted construction, and generally does not exceed 5 in. (12.5 cm). The thickness of the grout between bricks and reinforcement (cover) should be equal to or greater than one bar diameter. In high-lift grouted construction, the two wythes must be bonded together with wall ties (minimum No. 9 wire) spaced at a maximum of 24 in. (61 cm) horizontally and 16 in. (40 cm) vertically

Figure 4.30 A reinforced masonry wall.

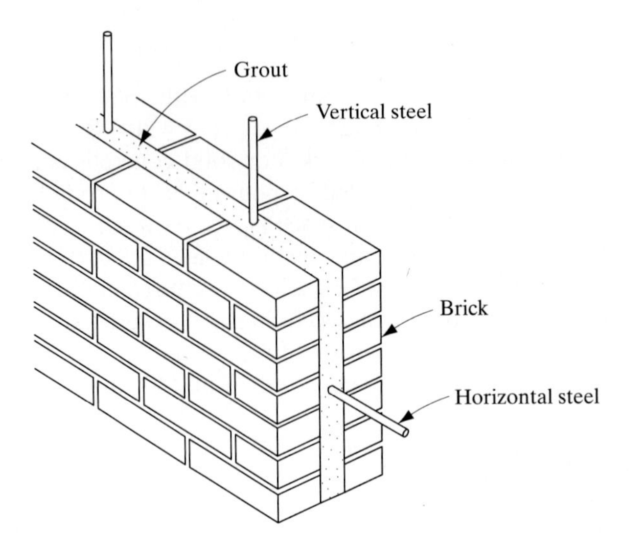

Figure 4.31 Reinforced brick wall (reinforced grouted masonry).

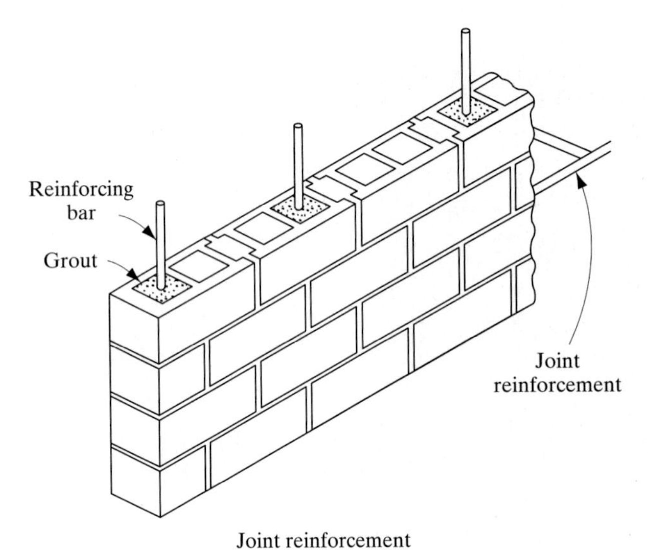

Figure 4.32 Reinforced hollow unit masonry.

(for running bond construction). Cleanouts are required at vertical reinforcement bar locations and at a maximum horizontal spacing of 32 in. (82 cm), as shown in Figure 4.17.

Reinforced hollow unit masonry is made with hollow masonry units in which cells are continuously filled with grout and in which reinforcement is embedded (Fig. 4.32). Generally, all cells are filled with grout. The method of construction can be low-lift or high-lift; in the former, the maximum height of construction laid before grouting is 2 ft (0.6 m), whereas in the latter, the full height of construction between horizontal cold joints is grouted in one operation. All reinforcement, horizontal and vertical, must be embedded in the grout. The horizontal reinforcement is placed on bond beam units with a minimum grout cover above the reinforcing steel of 1 in. (2.5 cm) for each grout pour. In the high-lift method of construction, cleanouts are required.

The design of these walls should conform to the relevant specifications of local building codes. Masonry is assumed to carry no tensile stresses, and reinforcement is assumed to be completely bonded to the masonry. Reinforced masonry construction is used for load-bearing walls, retaining walls, columns, and beams (Fig. 4.33).

Figure 4.33 Reinforced block masonry retaining wall.

4.6 MIX PROPORTIONING AND EXAMPLES

The amount of mortar required in a masonry wall can be calculated by knowing the size of the units and the thickness of the joints. Using modular bricks of nominal size $2\frac{2}{3} \times 4 \times 8$ in. and 3/8-in. joints, the approximate quantity of mortar required per 100 ft^2 of wall area is about 6 ft^3. Increase in the thickness of joints to 1/2 in. will increase this quantity by about 30 percent. The number of bricks required for the wall can be calculated by knowing the dimensions of brick.

Example 4.1

Find the number of clay bricks required for a wall of height 10 ft and length 25 ft. Use modular bricks of size $4 \times 2\frac{2}{3} \times 8$ in.

Solution

$$\text{Surface area of bricks} = \text{height} \times \text{length}$$
$$= 2\tfrac{2}{3} \times 8 \quad = 21.3 \text{ in.}^2$$
$$= 0.148 \text{ ft}^2$$
$$\text{Number of bricks per 100 ft}^2 \text{ of wall area} = \frac{100}{0.148} \quad = 676$$

Provide 5% allowance for wastage.

$$\text{Total number of bricks} = 710 \text{ per 100 ft}^2$$
$$\text{Area of wall} = 25 \times 10 \quad = 250 \text{ ft}^2$$
$$\text{Number of bricks required for wall} = \frac{710}{100} \times 250$$
$$= 1775$$

The amount of ingredients required for mortar can be calculated by knowing the mix proportions. Mortar can be mixed by weight or volume. Proportioning by weight is more accurate (due primarily to bulking of sand when moist, which results in incorrect volume measurement), but mortar is generally proportioned by volume. It is estimated that 1 ft^3 of damp loose sand will yield about 1 ft^3 of mortar. Note that 1 loose ft^3 of cement and lime weigh 94 lb and 40 lb, respectively. The specific gravities of cement and lime are 3.15 and 2.25, respectively.

Example 4.2

Using Type S mortar, determine the quantities of ingredients to mix 100 ft^3 of mortar.

Solution From Table 4.6, for Type S mortar, determine the mix proportions. Let

$$\text{Volume of cement} = V_c \text{ ft}^3$$
$$\text{Volume of lime, } V_l = 1/4 \text{ to } 1/2 \text{ of } V_c$$
$$\text{Volume of sand, } V_s = 2\tfrac{1}{4} \text{ to 3 times } (V_c + V_l)$$
$$= 2.81 \text{ to 4.5 times } V_c$$

Use the following mix proportions:

$$\text{Volume of cement} = V_c$$
$$\text{Volume of lime} = 0.5V_c$$
$$\text{Volume of sand} = 4.5V_c$$

Assuming a yield of 1 ft^3 of mortar for 1 ft^3 of damp loose sand, for 1 ft^3 of mortar,

$$V_s = 1 \text{ ft}^3$$
$$V_c = \frac{1}{4.5} = 0.222 \text{ ft}^3$$
$$V_l = \frac{0.5}{4.5} = 0.111 \text{ ft}^3$$

For 100 ft^3 of mortar,

$$V_s = 100 \text{ ft}^3$$
$$V_c = 22.2 \text{ ft}^3$$
$$V_l = 11.1 \text{ ft}^3$$

The weights of ingredients can also be determined from mix design based on weight. The unit weight of sand in a moist loose condition is taken as 80 pcf. For 1 ft^3 of mortar,

$$\text{Weight of sand} = 80 \text{ lb}$$
$$\text{Weight of cement} = (94 \text{ lb/ft}^3 \text{ of loose volume})$$
$$= 0.222(94) = 21 \text{ lb}$$
$$\text{Weight of lime (1 ft}^3 \text{ of lime weighs 40 lb)} = 0.111(40) = 4.44 \text{ lb}$$

Example 4.3

Find the absolute volume of all materials in Example 4.2. Assume a w/c ratio of 0.7 by weight (about $1\frac{1}{3}$ gal/ft^3 of mortar). Specific gravity of sand = 2.65; moisture content of sand = 7 percent; air content = 6 percent.

Solution

$$\text{Absolute volume of cement} = \frac{21}{3.15(62.4)}$$
$$= 0.11 \text{ ft}^3$$
$$\text{Absolute volume of lime} = \frac{4.44}{2.25(62.4)}$$
$$= 0.032 \text{ ft}^3$$
$$\text{Dry weight of sand} = \left(1 - \frac{7}{100}\right) \times 80$$
$$= 74.4 \text{ lb}$$
$$\text{Volume of sand} = \frac{74.4}{2.65(62.4)}$$
$$= 0.45 \text{ ft}^3$$

$$\text{Water in sand} = \left(\frac{7}{100}\right)80$$
$$= 5.6 \text{ lb}$$
$$\text{Mixing water} = 0.7(21)$$
$$= 14.7 \text{ lb}$$
$$\text{Total water} = 20.3 \text{ lb}$$
$$\text{Volume of water} = \frac{20.3}{62.4}$$
$$= 0.33 \text{ ft}^3$$
$$\text{Total volume of ingredients} = 0.11 + 0.032 + 0.45 + 0.33$$
$$= 0.922 \text{ ft}^3$$
$$\text{With 7\% air, net volume of mortar} = \frac{0.922}{0.93}$$
$$= 0.99 \text{ ft}^3$$
$$= 1 \text{ ft}^3$$

Note that 1 ft^3 of sand yields about 1 ft^3 of mortar.

4.7 TESTING

A number of laboratory tests are described in this section. Following is a list of masonry properties and the corresponding test numbers.

Property	Test no.
Absorption of bricks	MAS-3, MAS-4
Saturation coefficient of bricks	MAS-4
Suction of bricks	MAS-5
Compressive strength of bricks	MAS-2
Modulus of rupture of bricks	MAS-1
Mortar flow	MAS-6
Compressive strength of mortar cubes	MAS-7
Compressive strength of concrete masonry prisms	MAS-8

Test MAS-1: Modulus of Rupture of Brick

PURPOSE: To calculate the bending strength of clay brick specimens.

RELATED STANDARDS: ASTM C67.

DEFINITION:

- *Modulus of rupture* is the tensile strength of brick determined from a flexural test.

EQUIPMENT: Balance, universal testing machine, steel bearing plate ($1\frac{1}{2}$ in. wide and 1/4 in. thick), drying oven.

SAMPLE: Full-sized dry unit. The unit is dried at 230 °F (110 °C) for 24 h, and cooled for 4 h.

PROCEDURE:

1. Measure the average cross-sectional dimensions.
2. Support the specimen flatwise (load is applied in the direction of the depth of the unit).
3. The span length is 1 in. less than the basic unit length.
4. The specimen is loaded at the midspan.
5. Apply the load to the upper surface through the steel bearing plate.
6. The end supports should be able to rotate.
7. Apply the load at a rate of 0.05 in. (1.27 mm) per minute (crosshead movement).
8. Using the maximum load, calculate the value of MOR:

$$\text{MOR} = \frac{3WL}{2bd^2} \text{ psi}$$

where W is the maximum load in pounds, L the span length in inches, b the average overall width in inches, and d the average overall depth in inches.

REPORT: Report the average MOR value.

Test MAS-2: Compressive Strength of Brick

PURPOSE: To determine the compressive strength of brick samples.

RELATED STANDARD: ASTM C67.

EQUIPMENT: Universal testing machine, drying oven, capping pot and mold.

SAMPLE: Half-size dry unit (dried and cooled as in Test No. MAS-1). The length of the specimen is one-half the full length of the unit.

PROCEDURE:

1. Measure the dimensions at the top and bottom of the specimen.
2. Cap the specimen with sulfur (thickness of cap is about 1/4 in.).
3. Allow the cap to cool for a minimum of 2 h.
4. Test the specimen flatwise (load is applied in the direction of the depth of the brick).

5. Apply the load continuously so that the failure takes place within 2 to 3 min.
6. Calculate the compressive strength as:

$$\text{compressive strength} = \frac{W}{A}$$

where W is the maximum load and A is the average of the gross areas of the upper and lower bearing surfaces of the specimen.

REPORT: Report the average compressive strength.

Test MAS-3: 24-Hour Absorption of Brick

PURPOSE: To determine the amount of absorption of brick by immersion for 24 h.

RELATED STANDARD: ASTM C67.

DEFINITION:

- *Absorption* is the weight of water absorbed after 24 h of submersion in cold water expressed as a percentage of the dry weight of the brick.

EQUIPMENT: Drying oven, balance, pan or container.

SAMPLE: Half-brick specimen dried and cooled as in Test MAS-1.

PROCEDURE:

1. Obtain the dry weight of the specimen, W_d.
2. Submerge the specimen in clear water for 24 h.
3. Remove the specimen at the end of 24 h.
4. Wipe off the surface water with a damp cloth and weigh, W_s.
5. Calculate absorption as

$$\text{absorption} = \frac{W_s - W_d}{W_d} \times 100$$

REPORT: Report the absorption of all specimens (in percent).

Test MAS-4: Absorption by 5-Hour Boiling, and Saturation Coefficient

PURPOSE: To determine the absorption of brick samples by the 5-h boiling method and calculate the saturation coefficient.

RELATED STANDARD: ASTM C67.

DEFINITION:

- *Saturation coefficient* is the ratio between absorption after 24 h in cold water and absorption after boiling for 5 h.

EQUIPMENT: Balance, container.

SAMPLE: Use the saturated half-brick specimen from Test MAS-3.

PROCEDURE:

1. Submerge the specimen in clear water at 60–86 °F (15.5–30 °C). Note that water has to circulate on all sides of the specimen.
2. Heat the water to boiling and continue boiling for 5 h.
3. Allow it to cool by natural loss of heat.
4. Remove the specimen, wipe off the surface water with a damp cloth, and weigh, W_b.
5. Calculate the absorption as

$$\text{absorption} = \frac{W_b - W_d}{W_d} \times 100$$

$$\text{saturation coefficient} = \frac{W_s - W_d}{W_b - W_d}$$

Note that W_s and W_d are obtained in Test MAS-3.

REPORT: Report the percentage of 5-h absorption and saturation coefficient values.

Test MAS-5: Initial Rate of Absorption (Suction) of Bricks

PURPOSE: To calculate the initial rate of absorption of brick units.

RELATED STANDARD: ASTM C67.

DEFINITION:

- *The initial rate of absorption* is the amount of water absorbed by a standard brick (of surface area 30 in^2) after 1 min in 1/8 in. of water.

EQUIPMENT: Balance, container (tray), 1/4-in. square, 6-in.-long noncorrodable metal bars—two required (aluminum).

SAMPLE: Whole-brick specimen, dried and cooled as in Test MAS-1.

PROCEDURE:

1. Measure the length L (in.), and width B (in.), of the specimen.
2. Weigh the specimen, W_d (g).
3. Set the two metal bar supports in the tray.
4. Add water in the tray so that the depth of the water is 1/8 in. above the bottom of the test brick.
5. Set the test brick in place flatwise on metal supports without splashing.
6. Remove the brick at the end of 1 min.
7. Wipe off the surface water with a damp cloth and weigh, W_s (g).
8. Calculate the initial rate of absorption as

$$\text{initial rate of absorption} = \frac{30(W_s - W_d)}{L(B)}$$

REPORT: Report the average initial rate of absorption in g/min.

Test MAS-6: Determination of Mortar Flow

PURPOSE: To calculate the flow of mortar samples.

RELATED STANDARDS: ASTM C109, C230.

DEFINITION:

- *Flow* is the increase in the average diameter of the mortar mass in the flow table expressed as a percentage of the original base diameter.

EQUIPMENT: Flow table, flow mold, standard tamper ($1/2 \times 1 \times 6$ in., rubber or oak), trowel.

SAMPLE: Mortar sample of desired mix proportions.

PROCEDURE:

1. Wipe the flow table top clean and dry.
2. Place the flow mold at the center of the table.
3. Place a layer of mortar in the mold about 1 in. (25 mm) in thickness, and tamp 20 times with the tamper.
4. Fill the mold to the top and tamp 20 times.
5. Using a trowel, cut off the mortar to a plane surface, flush with the top of the mold.
6. Wipe the table top clean and dry.
7. Lift the mold away from the mortar 1 min. after completing the mixing operation.

8. Immediately drop the table through a height of 1/2 in. (12.7 mm) 25 times in 15 s.
9. Measure the diameter at four places.
10. Using the average base diameter, calculate the flow as

$$\text{flow} = \frac{\text{increase in average diameter}}{4} \times 100$$

11. Continue varying the water content in the mix until the desired flow is obtained.

REPORT: Report the amount of water (w/c ratio) required for the desired flow.

Test MAS-7: Compressive Strength of Mortar Cubes

PURPOSE: To determine the compressive strength of mortar of known mix proportions using 2-in. (50-mm) cube specimens.

RELATED STANDARD: ASTM C109.

EQUIPMENT: Cube molds, universal testing machine, tamper (Test MAS-6), mixer or mixing bowl, trowel.

SAMPLE: Use mortar from Test MAS-6. (If no mix proportions are given, use 500 g of cement, 1375 g of sand, and an initial water of 220 g. Increase the amount of water until the desired flow is obtained. The quantity of ingredients will make six cubes.)

PROCEDURE:

1. Apply a thin layer of mineral oil to the inside surface of the mold.
2. Remove the excess oil with a paper towel.
3. After mixing the mortar, fill the mold with mortar to a depth of about 1 in. (25 mm).
4. Tamp the mortar in all the cube compartments 32 times in about 10 s in four rounds, each round to be at right angles to the others.
5. Fill the compartments with the remaining mortar and tamp as specified for the first layer.
6. Smooth off the cubes by drawing the flat side of the trowel across them (with the leading edge slightly raised).
7. Cut off the mortar to a plane surface flush with the top of the mold by drawing the straight edge of the trowel across.
8. Store the specimens (covered with a plastic sheet) at room temperature for 24 h.

9. Transfer the specimens in the mold to the moist room.
10. Remove the specimens in the mold from the moist room. Remove the specimens from the mold.
11. Transfer the specimens to the moist room and keep them up to the testing age (7 or 28 days).
12. Test them immediately after removal from the moist room. Wipe the specimens to a surface-dry condition.
13. Apply the load to the specimen faces that were in contact with the true plane surfaces of the mold.
14. Loading rate should be such that the maximum load will be reached at about 30 s to $1\frac{1}{2}$ min.
15. Determine the compressive strength as

$$\text{compressive strength} = \frac{\text{maximum load}}{4} \text{ psi}$$

REPORT: Report the compressive strength of all specimens and the average value.

Test MAS-8: Compressive Strength of Concrete Masonry Prisms (Hollow)

PURPOSE: To calculate the compressive strength of masonry using concrete hollow blocks and known mortar.

RELATED STANDARD: ASTM E447.

DEFINITION:

- A *masonry prism* is an assemblage of masonry units, mortar, and sometimes grout used as a test specimen for determining properties of the masonry.

EQUIPMENT: Universal testing machine, capping mold and pot, trowel.

SAMPLE: Use half-blocks (length = width = 8 in.) to build a prism of nominal height = 16 in. Fully bed the face shells with mortar. Strike mortar joints flush with the face of the masonry without tooling.

PROCEDURE:

1. Cure the specimens for 28 days in the moist room.
2. Take the specimens out of the moist room and measure the average cross-sectional dimensions.

3. Cap the specimens as in Test MAS-2.
4. Apply the load at a uniform rate so that the maximum load is reached in 1–3 min.
5. Note the loading at which the first cracking sounds are heard.
6. Note (if observed) the load corresponding to the first crack.
7. Note the maximum load.

REPORT: Describe the mode of failure. Determine the compressive strength based on net area.

PROBLEMS

1. Define the following terms: header, stretcher, and wythe.
2. Name the two types of masonry units.
3. Discuss the difference between hollow and solid units.
4. What are the raw materials in the manufacture of clay brick?
5. Name the three grades of brick.
6. Name the two grades of concrete block.
7. What are the differences between portland cement and masonry cement?
8. What are the advantages of adding lime to mortar?
9. What are the differences between hydraulic and nonhydraulic cements?
10. Name the grades of lime.
11. What are the four types of standard mortar?
12. What is flow?
13. Describe the difference between face shell bedding and full mortar bedding.
14. Name three types of bond.
15. Name three types of mortar joint.
16. Describe control and expansion joints.
17. What are the two properties that determine the compressive strength of masonry?
18. What is the expected strength of concrete block masonry built with 2000-psi units and Type S mortar?
19. What is efflorescence?
20. What is a prism test and when it is used?

5

Wood and Wood Products

Wood is one of the oldest known materials of construction—and the only one that is naturally renewable. Historically, wood has been employed in many applications, from shipbuilding to bridges, railroad ties to flooring, and cabinets to nails. A number of factors—simplicity in fabrication; lightness; reusability; insulation from heat, sound, and electricity; esthetically pleasing appearance; resistance to oxidation, acid attack, and salt water; and environmental compatibility—have made this material one of the most popular in light construction. Many products made with wood as the principal raw material have been introduced into the construction market in the last twenty years or so, and these are presently utilized extensively, yet the predominant consumption of wood is still in the form of lumber—timber cut from tree trunks. Mature trees, primarily those bearing evergreen needlelike leaves, are the source of structural lumber; the trunk when cut is a *log,* from which lumber is sawed.

In this chapter, we review the structure of wood, lumber manufacture, grading of lumber, types of wood products, properties of wood and wood products, and applications of wood in construction. Many generalizations are made in the following pages in terms of these topics, but in fact the physical, mechanical, and other properties of wood vary considerably between species, and even within a single tree.

5.1 STRUCTURE OF WOOD

A cross-section of a log or trunk is shown in Fig. 5.1. A relatively thin, rough, and dense covering called *bark* surrounds the trunk. A thin (microscopic) layer of wood cells called *cambium* exists inside the bark, in which the growth of wood takes place continuously. This growth eventually results in a ring-like structure referred to as an *annual ring* (or *annular ring, growth ring,* or *tree ring*). Rapidly growing trees having

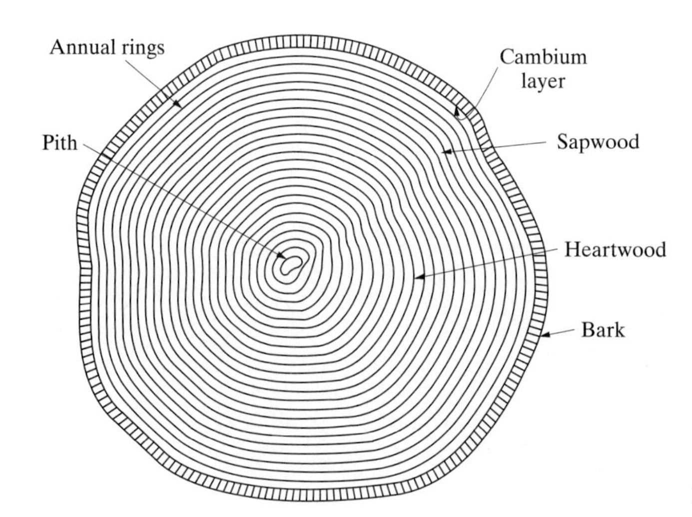

Figure 5.1 Section through a log.

wide annual rings produce coarse-grained wood, whereas slow-growing trees having narrower rings yield fine-grained wood. The width of a ring depends on the rate of growth of the tree. Old-growth softwood trees (slowly growing) have thinner rings, generally 7–12 rings per inch. Plantation trees (rapidly growing), which are harvested in short-rotation cycles, have wider rings, generally 2–4 rings per inch. Although growth rings are found in the trunks of most hardwood and softwood trees, some trees such as palm and coconut do not possess a ring-like structure. The center of the log, called the *pith,* is surrounded by the annual rings, the number of which approximately represents the age of the tree: a tree with 250 rings is roughly 250 years old. A number of redwood trees in northern California are more than 2500 years old, and some bristlecone pine trees in the high desert of California are older than 4000 years.

In most species, each annular ring appears to have an inner and an outer layer. The inner layer of each ring represents the more rapid spring growth, and is referred to as *springwood* or *early wood.* The outer layer represents a heavier, harder, and stronger material called *summerwood.* The proportion of summerwood to springwood affects the wood density, which in turn affects the strength.

The inner part of the trunk is made up of dead tissue (or cells), and the primary function of this part is to provide mechanical support to the tree. This part is called *heartwood,* which is darker, drier, and harder than the outer part, known as *sapwood,* which contains living cells. The differences in the density and the strength between the heartwood and the sapwood are rather small, but the sapwood is less durable and more permeable—which means that impregnation with preservatives is easier—than the heartwood.

5.1.1 Chemical Composition of Wood

Cells, which are the smallest organizational units in all plants and animals capable of carrying on the life process, make up the structural element of wood (Fig. 5.2).

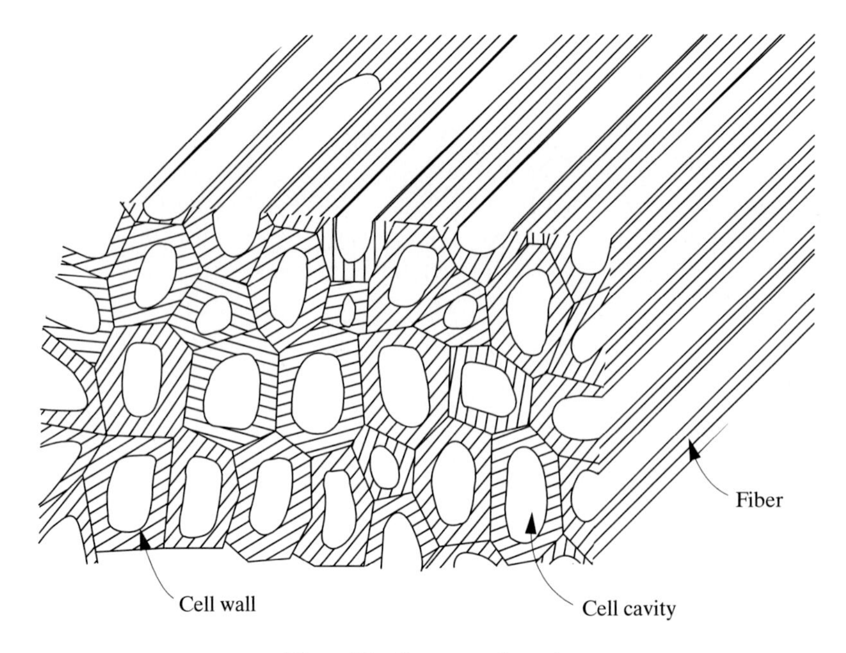

Figure 5.2 Structure of wood.

They come in a variety of sizes and shapes in wood, and are cemented together quite rigidly. In most types of wood, a large proportion of cells are found with their longitudinal axes nearly parallel to the axis of the tree. Some cells that lie with their axes perpendicular to the tree axis (in the radial direction) are called *radial cells* or *medullary rays.* Cells formed in the growing season (spring) differ from those produced in the later season (summer) in size, shape, and wall thickness. Cells developed in spring have no more than about 10 percent of gross area in cell walls (the balance is cell cavity), whereas the wall area of cells formed in summer is some 90 percent of the gross area.

Dry wood cells may be empty or partly filled with deposits such as gums and resins. The majority of wood cells are considerably elongated, and they are commonly called *fibers.* The length of a fiber varies; hardwood fibers average around 0.04 in. (1 mm) in length and softwood fibers, called *tracheids,* around 0.25 in. (6 mm). *Wood grain* refers to the arrangement of wood fibers. Each cell has a hollow center called the *lumen* or *cell cavity.* The density of wood is related to the ratio of cell wall to lumen volume.

Dry wood is made up of:

- Cellulose
- Hemicellulose
- Lignin
- Extractives

Cellulose, a carbohydrate of high molecular weight, is a major constituent of wood substance—approximately 50 percent by weight. Raw cotton is 91 percent cellulose, which can be described as a long-chain carbohydrate (glucose) having the general formula $C_6H_{10}O_5$ and found in all plant matter. A crystalline array of cellulose chains is called a crystallite. Lignified cellulose fibers of various kinds form the bulk of all woods. As the tree grows, the cellulose molecules are arranged into ordered strands, called *fibrils,* which in turn form the larger structural elements that make up the cell walls. Delignified wood fibers are mostly cellulose, and are used in paper production, textiles, films, and explosives.

Hemicellulose—various polymers built up of units of one or more species of sugar, such as glucose, galactose, xylose, and mannose—designates materials other than (but similar to) cellulose and found in wood. The nature and proportion of the hemicelluloses found in different woods vary. They exhibit some degrees of orientation and crystallinity, particularly when they are in close association with cellulose, but are largely amorphous. About 20–25 percent of wood is hemicellulose, and hardwoods contain generally more hemicellulose and less lignin than softwoods.

The natural cementing and rigidifying material forming the matrix in the cell walls and between cells, and holding them together to form various anatomical structures, is called *lignin*. It is an amorphous material composed of a very complex, cross-linked, three-dimensional polymer formed from phenolic units. The actual composition of lignin varies between species, and it is polymerized in plant cell walls from three primary precursor monomer alcohols. Comprising around 23–33 percent of softwood and 16–25 percent of hardwood, lignin is the third most abundant constituent of wood, after cellulose and hemicellulose. The biological process of depositing lignin within the cell walls is called lignification. The intercellular layer between wood cells, composed principally of lignin and lacking cellulose, is called the *middle lamella.*

In addition to the three major cell wall components just discussed, wood often contains a large number of substances in varying amounts called *extractives,* which include a wide variety of chemicals. Soluble substances such as fats and resins, and insoluble substances such as pectic and proteinaceous compounds form the bulk of extractives, their total amount not exceeding 10 percent of wood. Some inorganic materials, such as SiO_2 and MgO, are also present in some species. A few types of trees contain some organic crystals that have toxic or repellent effects on organisms that attack wood.

5.2 TYPES OF WOOD

Wood is broadly classified into two categories:

- Hardwood (from deciduous trees)
- Softwood (from conifers)

Trees with broad leaves that are shed in winter are hardwoods, and softwoods are any species that have needlelike leaves and are generally evergreen. Oak, Maple, Aspen, and Ash are examples of hardwoods, and Douglas Fir, Cedar, Larch, Pine,

TABLE 5.1 COMMON SOFTWOODS AND HARDWOODS

Softwoods	Hardwoods
Douglas Fir	Ash
White Fir	Basswood
Western Hemlock	Cottonwood
Western Larch	Aspen
Ponderosa Pine	Birch
Western White Pine	Elm
Redwood	Red Oak
Sitka Spruce	Black Walnut
Balsam Fir	Maple
Eastern Hemlock	White Oak
Eastern White Pine	
Eastern Red Cedar	
Southern Pine	

and Redwood are examples of softwoods. Table 5.1 lists some common softwood and hardwood species.

The terms "hardwood" and "softwood" refer to the botanical origin of the particular wood, and do not necessarily indicate the relative hardness or density of a specific species, although it is true that most species of hardwood are harder to work with and are darker than typical species of softwood. In general, conifers are easier and faster to grow, making them cheaper than many hardwoods. As a consequence, most of the structural lumber used in North America is derived from the softwood category.

Physical and mechanical properties of wood differ from species to species and also within each species. The conditions of growth—such as climate, density of the surrounding forest, character of the soil, the area in the log from which the lumber is derived, moisture content, and defects—influence these and other properties.

In an attempt to simplify the structural design process and limit the large number of variables that affect the properties of wood, the available species are divided into several groups of species that have very similar characteristics and mechanical properties (Table 5.2). For example, the species group "Eastern Softwoods" consists of a number of softwoods such as Balsam Fir, Black Spruce, and Eastern Hemlock. Similarly, a number of pines such as Loblolly Pine and Shortleaf Pine are grouped together under the name "Southern Pine." The most common groups of species used in general construction in the south and western parts of the U.S. are Southern Pine, Douglas Fir–Larch, and Hem-Fir. Pines are grown in the southern states from Virginia to Texas, and Douglas Fir is common along the Pacific Coast.

The mechanical and physical properties of groups of species differ. The species that belong to the Douglas Fir–Larch group are relatively strong but brittle. California Redwood species, which thrive along the Pacific Coast, are lightweight, soft, straight-grained, and very durable. The heartwood of species belonging to the

TABLE 5.2 SOFTWOODS LUMBER SPECIES AND SPECIES GROUP

Standard lumber name or species group	Official Forest Service species designation
Douglas Fir–Larch	Douglas Fir
	Larch
Eastern Softwoods	Balsam Fir
	Black Spruce
	Eastern Hemlock
	Jack Pine
	Red Pine
	Red Spruce
Hem-Fir	California Red Fir
	Grand Fir
	Noble Fir
	Pacific Silver Fir
	Western Hemlock
	White Fir
Northern Pine	Jack Pine
	Norway (Red) Pine
	Pitch Pine
Southern Pine	Loblolly Pine
	Longleaf Pine
	Shortleaf Pine
	Slash Pine
Virginia Pine–Pond Pine	Virginia Pine
	Pond Pine
Western Cedar	Alaska Cedar
	Incense Cedar
	Port Oxford Cedar
	Western Red Cedar

Southern Pine group is fairly durable even in contact with the ground. Some important physical and mechanical properties of common species of lumber are discussed in the following sections.

5.3 PHYSICAL PROPERTIES OF WOOD

Two of the most important physical properties that affect the strength and durability of clear wood are:

- Moisture content
- Specific gravity

In old-growth timber, these two properties are adequate to assess the mechanical and physical properties. But, to satisfy the increased demand for forest products, much of the current timber supply comes from managed plantations of rapidly grown trees. For example, timber is harvested from Douglas Fir trees when they are less than 15–20 years old, with an average diameter of 10 in., compared to old-growth trees more than 60 years old. Timber cut from these juvenile logs has the potential to be markedly low in strength and stiffness compared to lumber from mature trees, the magnitude of reduction varying between species.

5.3.1 Moisture Content

Wood is a hygroscopic substance, meaning that it has affinity for water in both liquid and vapor forms. The ability to absorb or lose moisture depends on the environmental conditions such as temperature and humidity. The moisture content (MC) is the weight of water in wood, expressed as a percentage of its oven-dry weight.

$$\text{Moisture content (\%)} = \frac{\text{weight of water}}{\text{oven-dry weight}} \times 100$$

The moisture content in a living tree varies with the species. South American Balsa wood, for example, which is very lightweight and porous, has a moisture content in excess of 400 percent and has very low specific gravity (about 0.2 or less). A few species, such as Black Ironwood (of Florida) and White Ash, have very low moisture content with very high specific gravity (about 1.15). Some old-growth Redwood and Western Cedar have high moisture content (>200 percent), whereas Western Larch and Eastern Hemlock have low moisture content (about 100 percent). Within any species, there is considerable variation in the moisture content, depending on the age and size of the tree and its location. Typically, trees contain water about two times the weight of their solid material.

The moisture content in a log depends on the part of the log: sapwood versus heartwood. The sapwood of softwood species, in general, has much higher moisture content than the heartwood. For example, sapwood of Western Red cedar has a moisture content in excess of 250 percent, whereas the moisture content in its heartwood is around 60 percent. The difference in moisture content between sapwood and heartwood is small for hardwoods. For example, sapwood of White Oak has an average moisture content of 78 percent compared to 64 percent for its heartwood. Heartwood of common softwoods and hardwoods has moisture content in the range 35–100 percent.

Water exists in wood in two ways: in the cell cavities as *free water*—similar to water in a measuring cylinder—and in the cell walls as *bound* or *adsorbed water*. (Note that *adsorption* refers to the attraction of water molecules to hydrogen-bonding sites in the cell material.) As long as there remains some water in the cavity, the cell walls remain saturated. The amount of water in cell cavities varies with the amount of drying; this water is the first to be lost when wood is dried. Further drying will remove water from the cell walls. A completely dry wood will absorb free water (in the cell walls) when exposed to liquid water.

When a log is processed into lumber or particles, the wood begins to lose some of its moisture to the surrounding atmosphere immediately. Lumber has moisture generally in excess of 50 percent at the time of its manufacture. It can take in or give off moisture depending on the relative humidity of its surroundings. As wood dries, water is driven off the cell cavities. A point is reached when the cavities contain only air and the cell walls are saturated with water, which is called the *fiber saturation point.* The moisture content at the fiber saturation point can vary from 25–30 percent depending on the species.

During the drying period, prior to reaching the fiber saturation point, the volume of wood remains constant as well as most of its mechanical properties, although its density decreases. Below the fiber saturation point, however, additional drying begins the removal of water from the cell walls. This stage of drying is marked by a reduction in the cross-sectional dimensions. Almost all properties of wood—physical and mechanical properties, resistance to deterioration, and dimensional stability—are affected by the amount of water present in the cell walls. Consequently, the removal of water from the cell walls affects nearly all properties of wood.

When lumber is dried, during service, below the fiber saturation point, it eventually assumes a condition of equilibrium, with the final moisture content dependent on the relative humidity, the ambient temperature, and the drying conditions to which it has been subjected. The moisture content at this condition is called the *equilibrium moisture content* (EMC). The EMC is also defined as the moisture content at which the wood neither gains nor loses moisture relative to the surrounding air. Its value ranges between 5–17 percent at 70 °F and relative humidity between 2–90 percent. (Relative humidity is defined as the amount of water vapor in the air compared to the amount of vapor it could hold at the same temperature if the air were fully saturated with moisture.) A 50-percent relative humidity indicates that only half the amount of water vapor is in the air compared to what it can hold at that temperature. Since the relative humidity of the atmosphere does not remain constant, the moisture content of lumber during service, or the EMC, fluctuates. Because moisture moves slowly through wood, daily changes in the EMC are barely noticeable. Seasonal variations in temperature contribute to fluctuations in the actual amount of moisture in the air—warm air holds more moisture, and is the cause of the cyclic seasonal swelling and shrinking of lumber.

5.3.2 Density and Specific Gravity

The weight of lumber depends on the species, growth of the tree, moisture content, and the part of the log. Green wood (or freshly cut lumber) is heavier than dry wood; green sapwood is heavier than green heartwood, whereas dry heartwood is heavier than dry sapwood. The *density* (or unit weight) of wood is defined as the mass or weight per unit volume. It is directly related to porosity, or proportion of voids.

Both density and volume of wood vary with the moisture content, and thus the two closely related physical properties of wood, density and specific gravity, can be measured in different ways. In many parts of the world, wood density is calculated using the total weight of wood, which includes the weight of wood substance and the weight of water. However, in the United States, the wood density is found using the oven-dry weight (without water) and the volume at the time of the test (dry weight and green volume).

$$\text{Wood density} = \frac{\text{oven-dry weight}}{\text{volume of green wood}}$$

It was explained that the volume of wood does not remain constant but changes with the moisture content. Accordingly, the density varies with the moisture content, although the oven-dry weight—the numerator in the relationship above—remains unchanged.

Specific gravity is the ratio between the density of wood and the density of water. For example, wood with a specific gravity of 0.5 has a density of 31.2 pcf. The specific gravity is calculated as the ratio between the weight of a given volume of oven-dry wood and the weight of an equal volume of water.

$$\text{Specific gravity} = \frac{W_s}{w_w(V)}$$

where W_s is the oven-dry weight, w_w the density of water, and V the volume of (green) wood.

Wood is composed of solid matter (or wood substance), water, and air. The volume V in the equation above is made up of solid volume, V_s, water volume, V_w, and air volume, V_a. When wood is dried in an oven, water is driven out, and the weight of wood is due to the weight of wood substance, and the volume is equal to V_s plus V_a. Thus, irrespective of the moisture content, the numerator in the preceding equation remains the same in a sample of wood. But the volume V in the denominator varies with the moisture content, and thus the definition of specific gravity must be qualified by its MC.

The specific gravity is frequently determined in three conditions:

- Green (moisture content in excess of 19 percent)
- Air-dry (moisture content about 12 percent)
- Oven-dry

The specific gravity of all wood substance (oven-dry weight, and solid volume V_s only) is approximately 1.55. But the specific gravity of a piece of wood or lumber varies from 0.07–1.30. Very heavy woods such as Live Oak and Black Ironwood have specific gravity nearly equal to or higher than 1.0 (which means that they may be heavier than water), but the specific gravity of very light woods such as Black Cottonwood, Cedar, and Balsam Fir is less than 0.35. Table 5.3 lists some common species of wood and their specific gravity values.

TABLE 5.3 AVERAGE SPECIFIC GRAVITY VALUES OF COMMON SPECIES

		Specific gravity	
Category	Species/species group	Green	Oven dry
Hardwood	Ash, White	0.55	0.62
Hardwood	Aspen	0.36	0.40
Hardwood	Birch, Sweet and Yellow	0.60	0.66
Hardwood	Cottonwood, Black	0.30	0.33
Hardwood	Maple, Black and Sugar	0.55	0.66
Hardwood	Oak, Red	0.53	0.67
Hardwood	Oak, Live	0.80	0.90
Softwood	Cedar, Northern White	0.29	0.31
Softwood	Douglas Fir	0.45	0.51
Softwood	Hem-Fir	0.35	0.42
Softwood	Balsam Fir	0.33	0.38
Softwood	Hemlock, Eastern	0.38	0.45
Softwood	Hemlock, Western	0.42	0.48
Softwood	Pine, Longleaf	0.54	0.59
Softwood	Pine, Ponderosa	0.38	0.49
Softwood	Southern Pine	0.50	0.55
Softwood	Redwood	0.35	0.42
Softwood	Spruce, Englemann	0.33	0.36
Softwood	Spruce, Sitka	0.40	0.43

As described, the moisture in wood has a very large effect on the specific gravity as well as the density. Since the volume of wood decreases with the loss of moisture content below the fiber saturation point (due to shrinkage), the specific gravity in green condition is less than that measured in air-dry or oven-dry conditions. Specific gravity in the oven-dry condition is the highest. The following equation is recommended in the National Specifications for Wood Construction to calculate the density of wood at any moisture content:

$$\text{density or weight (lb/ft}^3) = \frac{62.4G_o}{1 + 0.009(G_o)\text{MC}}\left[\frac{100 + \text{MC}}{100}\right]$$

where G_o is the specific gravity in oven-dry condition (from Table 5.3) and MC is the moisture content in percent.

As indicated, specific gravity is one of the two most important physical properties of wood, and nearly all mechanical properties are related to the specific gravity or the density of wood, which is related to the rate of growth, percent of summerwood, moisture content, and other factors. Strength and stiffness of wood increase with the specific gravity. The effects of changes in the density of wood on various strength characteristics are discussed in subsequent sections.

5.3.3 Examples

Example 5.1

A sample of wood has a moisture content of 24 percent. Its dimensions are $1.45 \times 3.45 \times 6$ in. The weights of the sample in green and oven-dry conditions are 0.61 lb and 0.55 lb, respectively. Find the specific gravity.

Solution

$$\text{Specific gravity (green)} = \frac{\text{oven-dry weight}}{62.4(\text{green volume})}$$

$$\text{Green volume} = 1.45(3.45)6$$
$$= 30.01 \text{ in.}^3$$
$$= 0.0174 \text{ ft}^3$$

$$\text{Specific gravity (green)} = \frac{0.55}{62.4(0.0174)}$$
$$= 0.51$$

Example 5.2

Find the density of Red Oak at a moisture content of 30 percent.

Solution. From Table 5.3,

$$\text{Specific gravity in oven-dry condition} = 0.67$$
$$\text{Moisture content} = 30\%$$

$$\text{Density} = \frac{62.4(0.67)}{1 + 0.009(0.67)30}\left[\frac{100 + 30}{100}\right]$$
$$= 46 \text{ pcf}$$

5.3.4 Defects

In addition to the two most important properties of wood just discussed, strength and durability of lumber are also affected by the presence of defects (imperfections). Defects are natural irregularities in the structure of wood. The common defects are cracks, knots, and slope of grain, and occur principally during the growing period and the drying process. They lower the strength and durability of lumber, and should be taken into consideration in design as well as construction. Some of these defects are shown in Fig. 5.3 and explained in the following.

Cracks occur in various parts of the tree and are given names such as checks, shakes, and splits. Most of these are the result of shrinkage that follows the drying of wood. A *check* is a lengthwise separation of wood occurring across or through the annual rings, usually as a result of seasoning, and may occur anywhere on a piece of lumber.

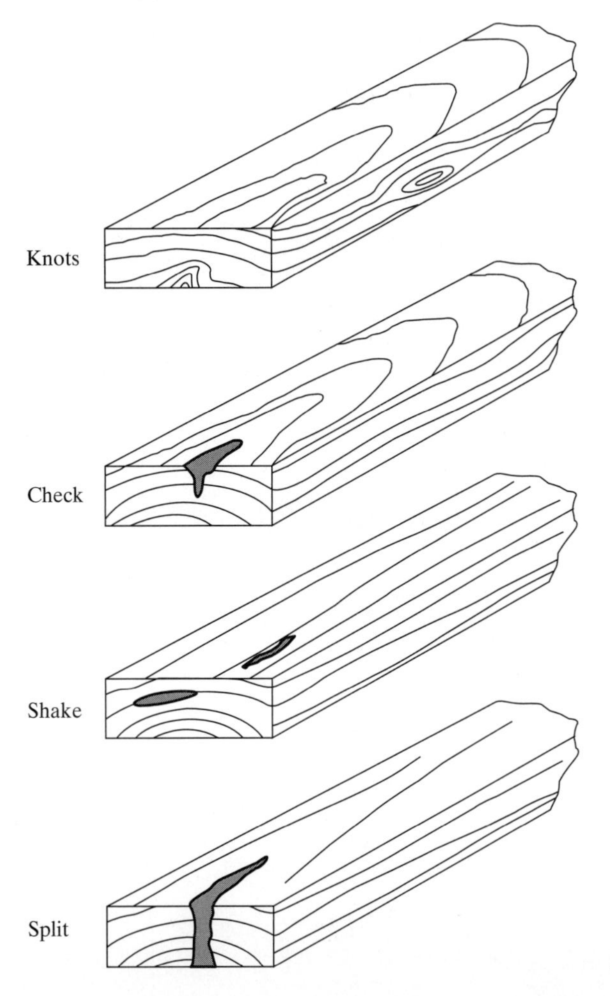

Figure 5.3 Lumber defects.

A *shake* is a lengthwise separation of wood occurring between and parallel to annual rings. A *split* is a complete separation of wood fibers, usually at the ends, throughout the thickness of lumber and parallel to the fiber direction. Both checks and shakes adversely affect the durability of timber because they readily admit moisture, which leads to decay.

A *knot* is a cross-section or longitudinal section of a branch that was cut with the lumber. Knots displace the clear wood and force the grains to deviate, thus affecting the mechanical properties of wood. They also allow stress concentration to occur. Moreover, knots are harder, denser, and possess different shrinkage characteristics than those of wood tissue.

The term "wood grain" refers to the arrangement of cells or fibers of the wood. Rapidly growing trees having wide annual rings produce coarse-grained wood, whereas those of slower growth produce wood with narrower rings or fine-grained wood. When wood elements do not run parallel to the axis of the piece, it is said to be *cross grained.*

The term "slope of grain" refers to the angle between the direction of fibers and the edge of the piece. For example, a slope of grain of 1 in 20 means that the grain deviates 1 in. from the edge in 20 in. of length along which deviation occurs (Fig. 5.4). The slope of grain is measured and limited at a zone along the length of the piece of lumber that shows the greatest slope.

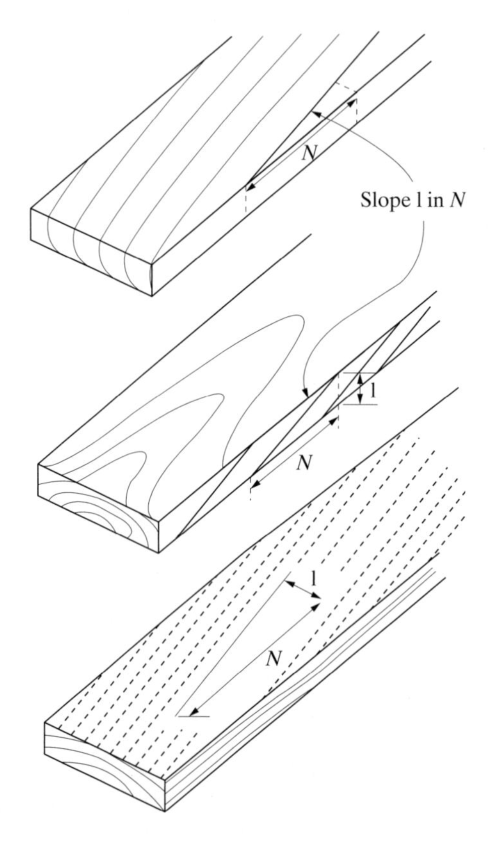

Figure 5.4 Grain pattern and slope of grain.

5.4 SHRINKAGE AND SEASONING

It was mentioned that wood is hygroscopic, which means that it can absorb or give out moisture. When a log is split into various sizes of lumber, the moisture content is normally in excess of 30 percent, and such lumber is called *green lumber.* The moisture is retained in both the cell cavities and, as bound water, in the cell walls. The effects of changes in moisture content on the cross-sectional dimensions, and the process of controlled drying, are discussed in the following sections.

5.4.1 Shrinkage

When water in the cell cavities of wood evaporates, no reduction in the cross-sectional dimensions of lumber can be observed. But when the wood dries further, water is driven off the cell walls, leading to decrease in the dimensions and shrinkage of the wood. The shrinkage, or the reduction in dimensions from drying, varies with the species, the thickness of cell walls, the arrangement of cells, and the grain pattern. Thick-walled cells shrink more than cells with thin walls, and this inequality in shrinkage in a particular species creates stresses in lumber.

Moisture evaporates more rapidly from the ends of a wood element than from its sides, developing bending stresses. When these stresses exceed the tensile strength of wood, longitudinal cracks extending across the annual rings (checks) are produced.

The drying of lumber below the fiber saturation point is accompanied by shrinkage; similarly, a dry piece of lumber absorbs moisture when it comes into contact with water, accompanied by an increase in volume. The absorption continues until the cell walls are saturated or the moisture content reaches the fiber saturation point (which varies with the species but generally lies between 25–30 percent). Repeated wetting and drying, plus the accompanying expansion and contraction, weakens the wood and, more importantly, promotes decay.

Typically, sapwood of any species shrinks more than does heartwood of the same species. Hardwoods shrink more than softwoods: for example, Oak shrinks by a considerable amount and is likely to check during the drying process. The shrinkage also depends on the rate of growth. Juvenile wood has excessively high shrinkage, more than ten times that of mature wood.

Shrinkage of a log can be characterized as occurring in three mutually perpendicular directions:

- Along its axis
- Along a radius (across annual rings)
- Along a tangent to the radius (along annual rings)

These three shrinkage directions are identified on a log in Fig. 5.5. The tangential shrinkage is the highest and is about twice as much as the radial shrinkage. The longitudinal shrinkage is negligible. The volumetric shrinkage, which is the sum of the shrinkages in all three directions, is about 1.6 times the tangential shrinkage.

It should be noted that shrinkage of each piece of wood is unique since the grain pattern varies from piece to piece. Although it is nearly impossible to predict exactly the nature of shrinkage of any piece of wood, it is fairly reasonable to assume that shrinkage continues in a nearly linear pattern until the lumber is completely dry (Fig. 5.6). This means that the rate of shrinkage is constant.

In most cases the volume of wood decreases, on an average, by about 12 percent when it is dried from the fiber saturation point to the oven-dry condition. This means that an average value of tangential shrinkage can be assumed as 1 percent for every 4 percent decrease in the moisture content. As an example: a piece of lumber

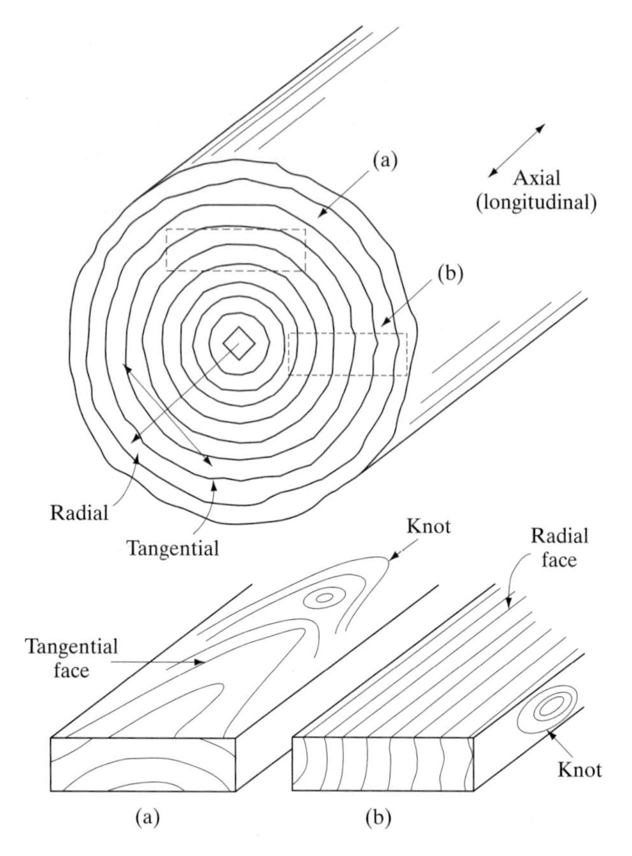

Figure 5.5 Manner of cutting a log: (a) flat-sawn; (b) edge-sawn (quarter-sawn).

with a cross-sectional width of 2 in. (5 cm) and a depth of 14 in. (35 cm) may decrease typically by about 0.1 in. (0.25 cm) along the width and 0.7 in. (1.75 cm) along the depth when the moisture content decreases from 30 to 10 percent during service. To accommodate this large reduction in the depth of the member, the floor supported by such a joist would have to settle by the same amount, which is not feasible. As a result—and from the restraint to shrinkage provided by other members—a number of shrinkage-related problems would arise in a finished structure. When this shrinkage movement—the differential shrinkage between different parts of the building—is not taken into consideration in design and construction, the potential problems can affect plumbing fixtures, electrical systems, mechanical systems, and finished surfaces. Cracks in walls, broken plumbing lines, and cracks in drywall ceilings are some of the defects that can be attributed to the shrinkage in wood.

In addition to shrinkage, changes in the moisture content also affect the mechanical properties of wood, and this topic will be addressed in subsequent sections. As a wood member dries, the modulus of elasticity increases, but the moment of inertia decreases due to reduction in the cross-sectional dimensions. The stiffness, which is the product of modulus of elasticity and moment of inertia, may increase or

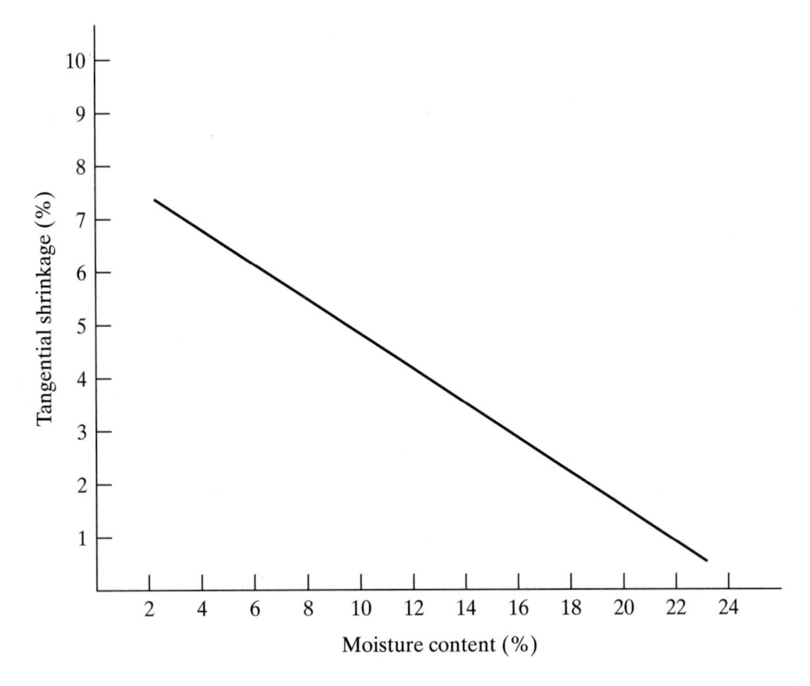

Figure 5.6 Average tangential shrinkage for Douglas Fir.

decrease depending on the relative changes in the two properties. However, tests have shown that the stiffness remains generally unchanged with changes in the moisture content.

5.4.2 Seasoning

The process of controlled drying of lumber to improve its structural properties is known as *seasoning*. In addition to an increase in strength characteristics, drying has other advantages, such as lowering of shrinkage in service, improvement in decay resistance, reduction in weight, and better workability. Seasoning is also required to prepare lumber for preservative treatment.

Two methods of drying are practiced.

- Air drying
- Kiln drying

In the air drying procedure, which is cheaper than kiln drying, lumber is stacked in open-sided sheds so as to promote drying without artificial assistance. Species that dry slowly are usually stacked in winter, so that the lumber will not be affected by summer heat until its moisture content is relatively low. Kiln drying employs a heated, ventilated, and humidified oven; the humidity is essential to control the rate of evaporation of moisture from within the lumber.

Whatever the method used to dry or season lumber, there are only three factors that affect the drying process: temperature, relative humidity, and air circulation. In the United States, seasoning is generally carried out for smaller-sized lumber, and the cost of seasoned lumber is relatively high. Bigger sizes of lumber are marketed as unseasoned or partially seasoned. Softwoods generally take a shorter time to dry than do hardwoods. The sapwood part of the log, which contains most of the moisture, dries out more rapidly than the heartwood part.

5.5 TREATMENT AND DURABILITY

Wood is a very durable construction material and lasts for centuries if it is well seasoned and kept in a dry place, immersed in water, or buried in the ground. However, when subjected to fluctuations in moisture or moderate heat, unprotected wood decays. The rapidity with which lumber decays depends on a number of factors, such as species, seasoning, and the environment. In exposed structures such as bridge posts, fence posts, and piles, decay always starts at the sills and the member bottom, where water and snow collects.

Although all species of wood possess some natural resistance to decay, some species are much more durable than others. Durability refers to long-term performance of the material. Most oaks, cedars, and redwood are highly durable, whereas aspen and fir are nondurable. The heartwood of all species has very good resistance, whereas the sapwood is susceptible to decay if not used properly in construction.

For most structural applications, ordinary sawn lumber can be used in the natural form, although coatings such as stains and paints may prolong service life in an exposed environment. Seasoning followed by the use of an effective waterproofing technique will render most wood species resistant to attack by wood-destroying elements. Paints, stains, and waterproofing chemicals are used to make lumber waterproof; stains and waterproofing chemicals penetrate farther into the wood than do paints, but none of these materials can be effective when wood is in contact with the ground.

Lumber that is in contact with the ground or is susceptible to decay should be protected through chemical treatment. Wood needs to be chemically treated mainly for two reasons:

- To prevent destruction from fungi and insects
- To inhibit combustion

Such treatment is carried out by injecting chemicals, under pressure, into the wood fibers, and is called *pressure treatment* or *preservative treatment.* It is generally carried out by incising the surface of lumber by toothed rolls to a depth of 1/2–3/4 in. (12.5–19 mm) and applying pressure until impregnation is complete.

Treatment to prevent destruction from fungi and insects is carried out with one of the three classes of chemicals:

- Pentachlorophenol (penta)
- Creosote
- Inorganic arsenicals (waterborne preservatives)

These chemicals penetrate deep inside the wood and remain for a long time. Incising is commonly done to provide deeper, more uniform penetration, although some species, such as Ponderosa Pine, can be treated without incising. Proper seasoning is required prior to the treatment.

Penta is a chemical that can be carried in either oil or light hydrocarbon solvent; generally 5 percent of the solution is made up of penta. Wood treated with penta is used for poles, posts, glue-laminated beams, bridges, and marine decking. Creosote solutions are derived from coal tar and carried in oil base. Creosote-treated wood is used for piles, poles, and railroad ties. Waterborne preservatives can be used for both interior and exterior applications. They leave the treated surface relatively clean, odor-free, and paintable. The most common waterborne preservatives are chromated copper arsenate (CCA), chromated zinc chloride (CZC), and fluorochrome arsenate phenol (FCAP).

Wood treated with penta or creosote should not be used where it will be in prolonged contact with skin (as for furniture and decking). In fact, penta- or creosote-treated wood should not be used in the interior of residential, commercial, or farm buildings. Wood members treated with either of these two chemicals are most suitable when they are in contact with ground. Lumber treated with waterborne preservatives may be used inside residential buildings and for patios or decks but should not be used for storage of food items and countertops. Treated wood should not be burned in fireplaces or stoves because of the production of toxic chemicals as part of the smoke or ashes; it is used, however, as a fuel in some industrial boilers and in the generation of electricity. Disposal in landfill is permitted.

The ease with which preservatives can be injected into wood depends on a number of factors, such as density and, structure of the wood, chemical composition of cell walls, cross-sectional dimensions of lumber, and moisture content. In most species, the sapwood is more easily impregnated and absorbs more preservatives than the heartwood. The preservative-absorption capacity of springwood may be different from that of summerwood. The seasoning facilitates the penetration of preservatives. Smaller-sized lumber can be impregnated more easily than can larger lumber; pines can be injected more readily than Douglas Fir.

5.5.1 Decay and Destruction

Wood structures that remain dry during service are generally durable. Wooden elements that remain close to the ground or are exposed to alternate wetting and drying cycles should be treated with a preservative. Untreated lumber that is not properly protected, under favorable conditions, may decay or be destroyed. Decay, or decomposition of the wood substance, is caused by attack from fungi; destruction, is by insects. Fungi are low forms of plant growth producing thin branching tubes that spread through the wood and use cell walls and lignin as food. The early stage of decay, called incipient decay, is characterized by a slight discoloration or bleaching of the wood, but the hardness is unaffected. Wood suffering advanced fungal attack becomes brittle or weak, and the destruction can be readily recognized. It is further characterized by a lack of resonance when stuck with a hammer, the capacity to absorb a large quantity of water, and unnatural odor and color.

Conditions necessary for fungal growth are:

- Proper temperature
- Moisture content over 19 percent
- Oxygen
- Food (wood fiber)

Elimination of any one or more of these will effectively control the growth of fungal organisms, but not all fungi thrive equally well under the same surroundings or conditions. For example, a type of fungus known as house fungus (also called dry-rot fungus) can live in seasoned lumber in somewhat dry conditions, and this type of decay is called *dry rot.* But for most decay fungi, wood must have a moisture content in excess of the fiber saturation point. At 20 percent or less moisture, fungi stop growing and become dormant, and it may remain dormant for years if the wood is kept dry. Similarly, if wood is too wet, or saturated, fungus growth stops because of lack of oxygen. Wood will not decay if it remains dry, however dry-rot fungi have the ability to transport moisture from a distant source of supply to the point of attack. In *brown rot,* the attack concentrates on the cellulose and associated carbohydrates rather than on the lignin; in *white rot* the attack is on both the cellulose and the lignin, producing a generally white residue.

The early stages of fungal decay are often characterized by dramatic decrease in some mechanical properties, with very little loss in component volume and limited changes in appearance. With some fungal attack, a linear relationship exists between the strength and weight losses, but strength loss occurs faster than the latter. Dry rot is the most serious of all fungi attack, because once established, it can spread rapidly, and the spread may be extensive before visible signs appear. *Wet rot* is fungal deterioration that occurs in excessively wet lumber. To minimize both dry and wet rot, all sources of moisture in a structure should be eliminated, followed by the drying out of the lumber (as well as the building), commonly with the aid of dehumidifiers.

Another principal cause of wood deterioration is attack by insects or marine borers. Some insects (such as horntail wasps, round-headed borers, and flat-headed borers) live in weakened or dying trees and, during the larval stage, use wood as their food source. Their activity can be eliminated through application of heat or steam. A second group of wood-boring insects can live and multiply in dry softwood lumber; the best known of these are termites. In addition, carpenter ants and white ants live in highly organized colonies and feed on the cellulose in wood. Commonly used oil-borne wood preservatives (preservative oils) such as penta, copper naphthenate, and creosote, and waterborne salts such as CCA and CZC are toxic to these insects and can be used as eradicants. Termite attack can be prevented by following precautionary measures during the construction stage, such as building concrete foundations, treating the wood that is in contact with the ground, and providing sheet metal shields for all the sill and foundation lumber.

5.6 LUMBER SIZES

Many commercially important softwood species are used in construction, and lumber from several of these species shares similar performance characteristics. These species are also similar enough in appearance that they can be grouped together into "marketing categories." For example, Douglas Fir and Larch are often alike and have similar properties, and as a result they are marketed under a single category or group of species: Douglas Fir–Larch. But shipments containing the Douglas Fir–Larch mark need not contain both species. The most common categories, and the species included in those species groups, were listed in Table 5.2.

Lumber can be broadly classified as:

- Framing lumber
- Appearance lumber
- Industrial lumber

Framing lumber (Fig. 5.7) includes the grades intended for structural applications in both conventional and preengineered framing systems. For example, lumber for joists, posts, and trusses is framing lumber. This type of lumber is graded based on its strength properties—it is structurally graded. *Appearance lumber* includes a variety of nonstructural grades intended for applications where strength properties are not

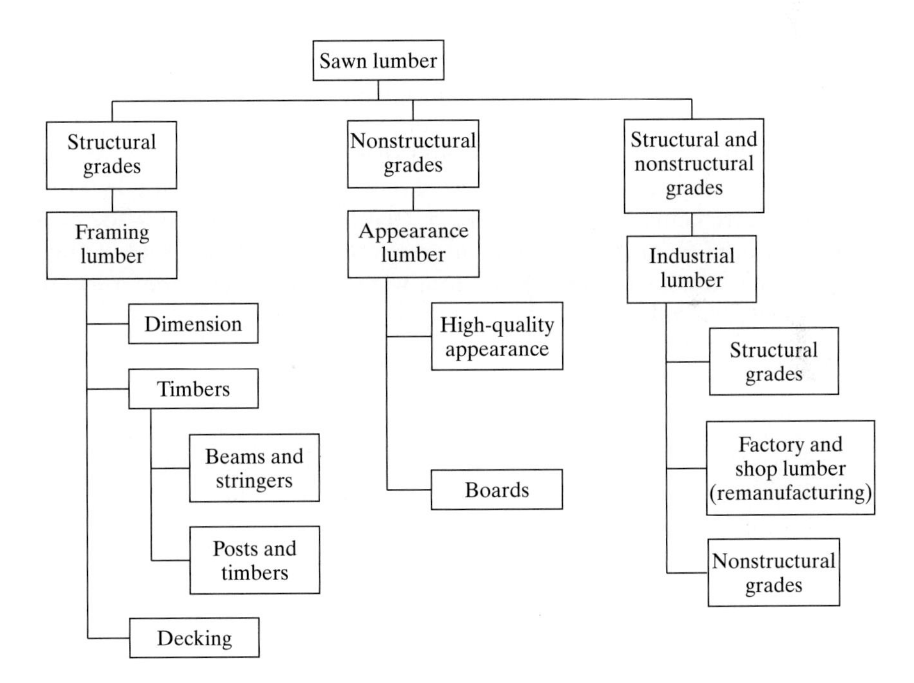

Figure 5.7 Sawn lumber categories.

the primary consideration. Many of the products used in applications for paneling and siding belong to appearance lumber. Its grade reflects a judgment of appearance and suitability to end use. *Industrial lumber* represents a broad category of structural and nonstructural grades intended for a variety of specific applications, such as mining, scaffolding, and foundations. Unless stated otherwise, any reference to lumber in the following sections should be assumed to refer to framing lumber. It should be noted that only lumber that is structurally graded should be used in engineered applications.

Softwood lumber is a manufactured wood, in that it is derived from a softwood log in a mill, and in a rough stage is sawed, edged, and trimmed at least to the extent that it shows saw marks on four longitudinal surfaces. *Rough lumber* has surface imperfections caused by the primary sawing operations, whereas *dressed lumber* is planed or sanded on combinations of sides and edges. The term S4S is used to identify lumber that is surfaced on all four sides. Such a piece of lumber is identified by its *nominal size,* which is a commercial size designation of the thickness and the width (or depth) of standard sawn lumber. The *dressed size* or *net size* is the size assumed to be obtained by dressing the lumber—using a planing machine on a combination of sides and edges—and is the size used in all structural calculations. This size is smaller than the nominal size by 7/32 to 3/4 in. (6 mm to 19 mm), depending on the size category and moisture content, as shown in Table 5.4. For example, lumber of nominal size 4×16 has dressed size (dry condition) $3\frac{1}{2} \times 15\frac{1}{4}$ in. The cross-sectional properties, such as area, moment of inertia, and section modulus are based on the dressed size (dry condition). Unseasoned lumber is manufactured oversized so that when it reaches the dry state its size is approximately the same as the dry-dressed size.

Generally, lumber is manufactured in standard sizes starting with nominal size 2×2 (thickness of 2 in. nominal and width of 2 in. nominal) and in increments of 2 in. (thickness as well as width). Thus common thicknesses are 2, 4, 6, and 8 in., and common widths are 2, 4, 6, 8, 10, 12, 14, and 16 in. In a typical wood construction, if a particular size is not feasible, one must choose the next-higher size in 2-in. thickness or width increments. For example, if 2×10 is found inadequate, a 2×12 may be tried.

Board foot is the unit of measurement for most lumber items. It is defined as a piece 1 inch thick (nominal) by 1 foot wide (nominal) by 1 foot long (actual), and is calculated as follows:

$$\text{board feet} = \frac{t \times w \times L}{12}$$

where t is the nominal thickness in inches, w the nominal width in inches, and L the actual length in feet. For example, 12 pieces of nominal 2×4 lumber 12 ft long measure

$$= \frac{2 \times 4 \times 12}{12} \times 12 = 96 \text{ board feet}$$

Untreated lumber may begin to dry in outdoor storage if stored in warm weather for an extended time, causing splits, checks, and warp. To prevent

TABLE 5.4 NOMINAL AND DRESSED SIZES OF SAWN LUMBER

Use category	Thickness (in.)			Width (in.)		
		Minimum dressed			Minimum dressed	
	Nominal	Dry	Green	Nominal	Dry	Green
Boards	1–$1\frac{1}{2}$	3/4	25/32	2	$1\frac{1}{2}$	$1\frac{9}{16}$
		$1\frac{1}{4}$	$1\frac{9}{32}$	3	$2\frac{1}{2}$	$2\frac{9}{16}$
				4	$3\frac{1}{2}$	$3\frac{9}{16}$
				5	$4\frac{1}{2}$	$4\frac{5}{8}$
				6	$5\frac{1}{2}$	$5\frac{5}{8}$
				7	$6\frac{1}{2}$	$6\frac{5}{8}$
				8	$7\frac{1}{4}$	$7\frac{1}{2}$
				9	$8\frac{1}{4}$	$8\frac{1}{2}$
				10–16	3/4 less than nominal	1/2 less than nominal
Dimension	2–4	1/2 less than nominal	7/16 less than nominal	2–4	1/2 less than nominal	7/16 less than nominal
				5, 6	1/2 less than nominal	3/8 less than nominal
				8–16	3/4 less than nominal	1/2 less than nominal
Timbers	5 and larger	—	1/2 less than nominal	5 and greater	—	1/2 less than nominal
Decking[a]	2–4	1/2 less than nominal	—	4–6	1 less than nominal	—
				8–12	$1\frac{1}{4}$ less than nominal	

[a]Dressed width shown is the face width.

deterioration, lumber in storage yards should be stacked on stringers off the ground. This procedure allows for air circulation and aids in handling. In a job site, lumber should be unloaded in a dry place. Untreated lumber should be placed on stringers or short pieces of lumber. To prevent excessive moisture gain, lumber should be covered with a plastic sheet or similar weather-protective materials.

5.7 USE CLASSIFICATION AND LUMBER GRADING

Softwood lumber is classified for market use by species, form of manufacture, and grade. The grade designates the quality of lumber or wood product. Generally, the grade of a piece of lumber is based on defects, imperfections, and other features that may lower its strength, durability, and utility value.

Hardwood lumber grading in the United States is carried out based on the Rules for the Measurement and Inspection of Hardwood and Cypress Lumber, drafted by the National Hardwood Lumber Association (Memphis, Tennessee). Some hardwoods, suitable for construction, are graded using the rules for softwood.

Softwood lumber grading is based on general guidelines set in the American Softwood Lumber Standards issued in 1970 (PS20-70). But a number of softwood lumber trade associations, such as West Coast Lumber Inspection Bureau (WCLIB), Western Wood Products Association (WWPA), and Southern Pine Inspection Bureau (SPIB), prepare their own grading rules that apply to their species. Hardwood and softwood grading rules differ in principle: hardwood is used predominantly for cabinetry, whereas softwood is used by carpenters in construction.

Softwood grade depends on the effect of defects and imperfections on the strength and deformation—mechanical properties—of lumber. It can be considered in the context of two major categories of use:

- Construction
- Remanufacture

Construction (Fig. 5.8) refers to lumber that will reach the consumer as graded and sized after primary processing (sawing and planing), whereas the term "remanufacture" refers to lumber that will undergo a number of manufacturing steps before it reaches the consumer. A piece of construction lumber does not undergo further grading once it leaves the sawmill.

The construction grade is divided into three general categories:

- Stress-graded lumber
- Non–stress-graded lumber
- Appearance lumber

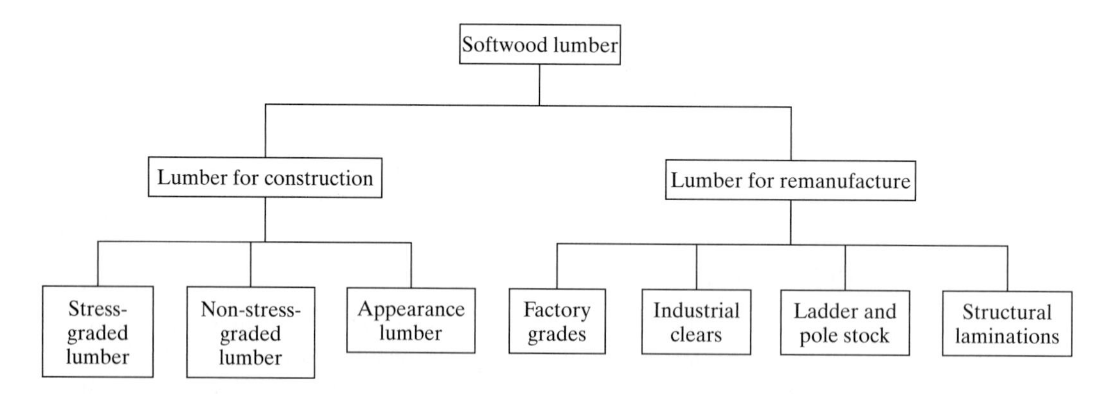

Figure 5.8 Softwood lumber grades.

Stress-graded lumber (also called framing lumber) and non–stress-graded lumber are used in applications where structural integrity is the primary requirement, and appearance lumber is for use in applications where the appearance (and not the structural integrity) is of primary importance. Stress-graded lumber or framing lumber is lumber of any thickness and width that is graded for its mechanical properties. Lumber retailers also carry yard lumber or common lumber, which is a general term for non–stress-graded lumber, which has an appearance quality lower than of select grade but is suitable for general construction and utility purposes (for example, "redwood con-common" and "con-heart"). Note that select lumber is a general term for lumber of good appearance and finishing qualities. The following discussions apply to stress-graded lumber only.

When a log is sawed into lumber, the various pieces differ in quality, strength, serviceability, and value. Individual pieces will have different characteristics with respect to freedom from knots, cross grain, shakes, and so on (Fig. 5.3). To standardize the mechanical properties of lumber, a stress-grading scheme has been established, which is referred to as stress grading or structural grading.

Under the stress-grading system of classification, the grade description of a piece of lumber depends on its intended use and structural defects or imperfections, and is independent of the species and the geographical region in which it is produced. However, allowable values of a certain grade of lumber, belonging to an individual species, may be different from those of other species, due to variations in specific gravity and mechanical properties. For example, a nominal 2×8 lumber of grade 1 and species group Douglas Fir–Larch has an allowable bending stress dissimilar to that of grade 1, 2×8 lumber of species group Southern Pine.

Building codes around the country require that only grade-marked lumber (or graded lumber) be used for all structural applications. A mark or *grade stamp,* which is seen in all structural or framing lumber, provides the grade of that piece of lumber. The grade is estimated based on inspection of physical characteristics that affect the mechanical properties of lumber.

As noted, the grade (structural) of a piece of lumber (stress graded) is based on two aspects: anticipated end use—called size or use classification—and defects. This means that the grade of a piece of lumber depends on the responses to two hypothetical questions: (1) What is the type of load the member is expected to carry, and (2) What is the nature and extent of imperfections? The response to the first question is interpreted in the Softwood Lumber Standards via size classification rules.

The size or use classification is based on the anticipated end use of a piece of lumber considering only its size (cross-sectional dimensions), even though no restrictions exist on the actual use in any construction. For example, because of its square cross section, 6×6 lumber (meaning a nominal cross section 6 in. in thickness and 6 in. in width) is suitable for use as a post or column and is thus classified (and graded) under the Post and Timber category. But it should be noted that this square piece of lumber may very well be used as a beam or beam-column, the choice being left to the designer.

5.7.1 Dimension, Decking, and Timbers

In general, according to size classification rules, framing lumber is divided into three classes:

- Dimension lumber
- Decking
- Timbers

Dimension lumber (Table 5.5 and Fig. 5.9) refers to any piece of rectangular cross section, 2–4 in. nominal in its least dimension (thickness), graded primarily for strength in bending edgewise or flatwise but also frequently used where tensile or compressive strength is important. For example, lumber of nominal size 2×4, 2×8, 4×4, and so on, belong to this category. This class of lumber is designated for use as joists, planks, rafters, studs, and so on.

TABLE 5.5 SIZE (USE) CLASSIFICATION OF STRESS-GRADED SAWN LUMBER

Size or use classification	Subcategory	Nominal thickness (in.)	Nominal width (in.)	Example	Uses
Dimension	Structural light framing and light framing	2–4	≥2	2×2, 2×4, 4×4, 4×12, 4×16	Joists, rafters, studs, truss members
	Studs				
	Structural joists and planks				
Decking		2–4	4–12	2×4, 2×8, 4×8	Floor decking, roof decking, solid wall
Timbers	Beams and stringers	≥5	>(2 + thickness)	6×10, 6×12 8×12	Beams, headers
	Posts and timbers	≥5	≤(2 + thickness)	6×6, 6×8, 8×8	Columns, posts

Dimension lumber is subdivided into four use categories:

- Light framing
- Structural light framing
- Studs
- Structural joists and planks

Structural light framing and light framing members are 2–4 in. wide, studs are 2–6 in. wide, and structural joists and planks are 5 in. and wider. Stud-grade lumber is used

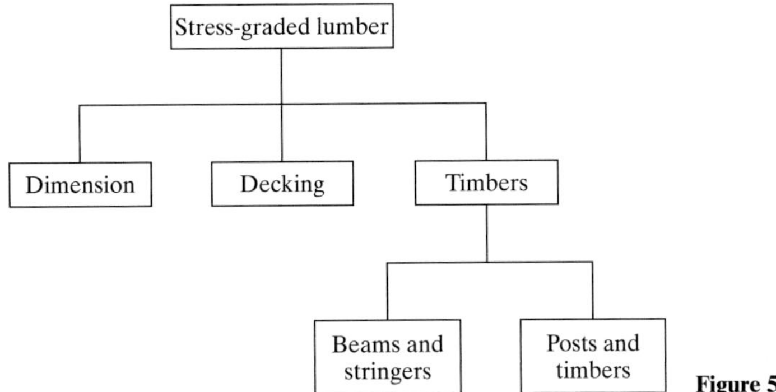

Figure 5.9 Size classification.

as studs in load-bearing walls, and joists and planks are fit for use as floor joists, rafters, headers, beams, and truss members.

A special dimension category, decking, refers to lumber 2–4 in. thick and 4–12 in. wide, tongue-and-grooved or grooved for spline on the narrow face, and intended for use in roof, floor, and wall construction. It is graded for application in the flatwise direction with loads applied on the wide face. Timber is both a general classification for the larger sizes of framing lumber and the name of a specific grade and size.

There are two basic grade groups within the Timber size classification:

- Beams and Stringers
- Posts and Timbers

The Beams and Stringers classification includes pieces of rectangular cross section, 5 in. and thicker, and width more than 2 in. greater than the thickness. It is graded for strength in bending when loaded on the narrow face, and is used primarily when materials larger than dimension lumber are required to carry flexural loads (for example, 6 × 10, 8 × 12, 8 × 24, and so on). The Posts and Timbers classification represents lumber that is approximately square in cross section, 5 × 5 in. nominal dimensions and larger (width not more than 2 in. greater than the thickness), and graded primarily for use as columns or beam-columns. This subgroup is meant for use in construction where materials larger than square dimension lumber are required. Sizes such as 6 × 6, 8 × 8, and 6 × 8 belong in the Post and Timbers category.

Dimension is the principal stress-graded lumber item available in a lumberyard, and is used primarily in light construction for joists, rafters, and studs. It is found in nominal 2-, 4-, 6-, 8-, 10- and 12-in. widths, and 8–18-ft lengths in multiples of 2 ft. (Lengths greater than 18 ft are not commonly available in many lumberyards.) In the non–stress-graded category, the shape and size of the lumber, combined with some grading requirements, are used to provide a measure of structural integrity. Board, lumber less than 2 in. in nominal thickness, is the most important non–stress-graded softwood lumber. This type of lumber is used for siding, cornices, shelving, and paneling.

5.8 MOISTURE CONTENT CLASSIFICATION AND GRADE STAMP

Softwood lumber is manufactured in length multiples of 1 ft; in practice, however, multiples of 2 ft (in even numbers such as 6, 8, 10, and 12 ft) are the most common. In the United States, board measure is in common use for lumber. A board foot, which is a volume measurement, represents a piece of lumber that is 1 ft long, 1 ft wide, and 1 in. thick (based on nominal dimensions for width and thickness, and actual dimensions for length). Thus a nominal 2 × 6 in., lumber that is 24 ft long has a volume of 24 board feet (also see Section 5.6).

5.8.1 Moisture Content Classification

Softwood lumber is surfaced in green or dry condition depending on the size classification. Dimension and board lumber may be surfaced green or dry, and both green and dry standard sizes are given in Table 5.4. The sizes of green lumber are such that when the lumber shrinks, it will have a size approximately that of dry lumber. Timbers (lumber in the Beams and Stringers and Posts and Timbers categories) are usually surfaced while green, and only green sizes are given.

The terms "green" and "dry" express the level of moisture content in a piece of lumber during the surfacing operation. Graded softwood lumber has three moisture content classifications:

- Surfaced dry (S-Dry)
- Surfaced green (S-Grn)
- 15 percent maximum moisture content (or MC15)

Softwood lumber that is dried to 19 percent moisture content or less during manufacture is identified as S-Dry, which is shown on the grade stamp. On an average, the actual moisture content in these lumber pieces may be around 15 percent. Lumber that is kiln dried to a maximum moisture content of 15 percent is called MC15 lumber and has that mark on the grade stamp. When the moisture content is in excess of 19 percent, the grade stamp shows the identification mark S-Grn.

If the moisture content is critical in any construction or during service, the moisture content grade should be specified. Note that the grade stamp marks, S-Grn, S-Dry, and MC15 (Fig. 5.10) do not represent the moisture content existing in any piece of lumber, but express the moisture content during the lumber manufacture.

5.8.2 Grading

As explained in Section 5.7, most lumber is graded based on uniform grading rules. The grading is carried out under the supervision of inspection bureaus or grading agencies, some of which also author grading rules for the species in the geographic regions they represent. But to provide for a uniform grading procedure and rules, many of the rules and standards follow the American Softwood Lumber Standards, PS20-70.

Stress-grading rules incorporate a sorting or grading criterion, a set of allowable values (mechanical properties for engineering design), and a grade name. The

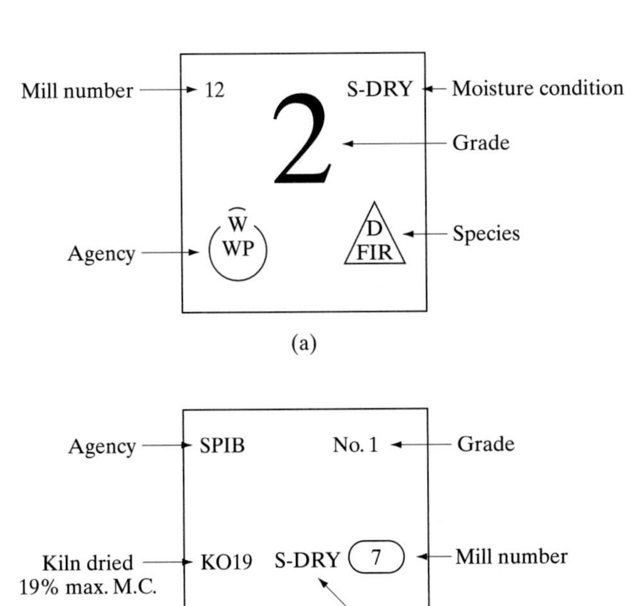

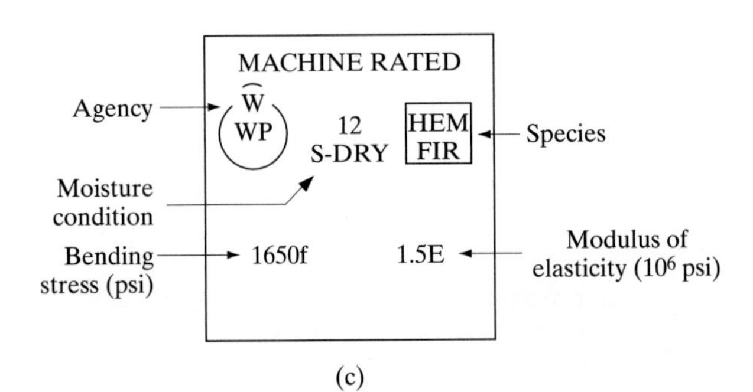

Figure 5.10 Typical grade stamps: (a, b) visual grading; (c) machine grading.

grading or sorting is accomplished visually or through nondestructive measurements (also called *E-Rating*), such as the bending stiffness. Up to six allowable properties are associated with a stress grade (Fig. 5.11):

- Modulus of elasticity
- Tensile stress parallel to grain
- Compressive stress parallel to grain
- Compressive stress perpendicular to grain
- Shear stress parallel to grain
- Bending stress

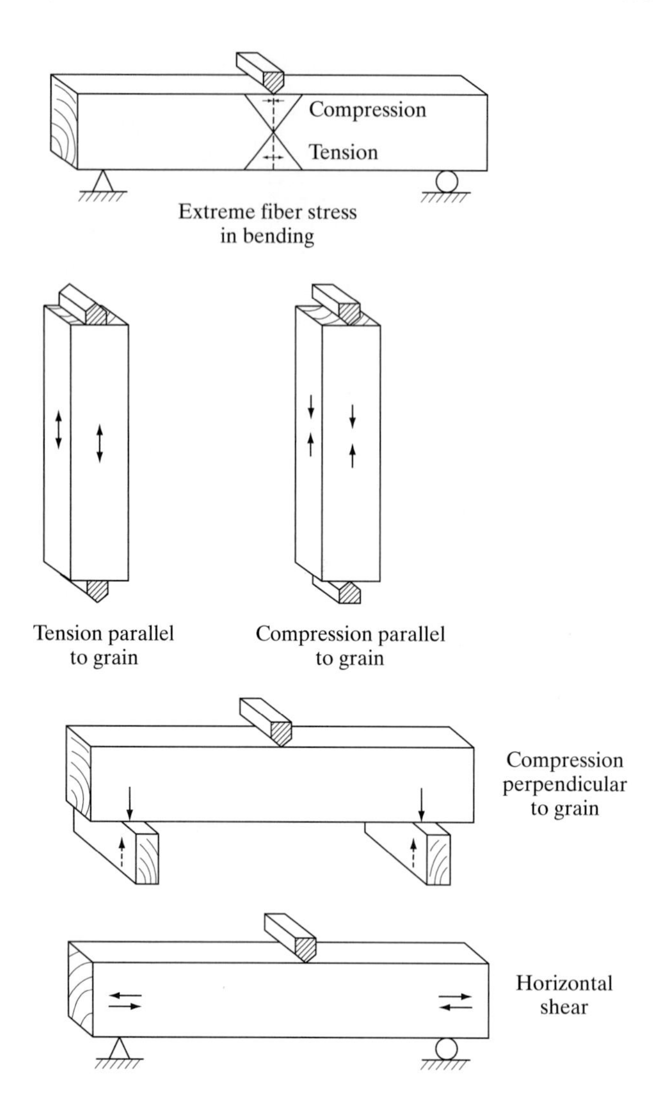

Figure 5.11 Types of stress.

All lumber manufactured for use in light construction or other structural applications (in which the sizes are determined based on the mechanical properties of the lumber) is stress graded and grade stamped to that effect. This also means lumber that does not have a grade stamp (stress grade), or that in which the portion containing the grade stamp has been cut off, should not be used in light construction.

Grade Stamp. The grade name or number (number grade or word grade) of a piece of lumber is shown in its grade stamp. The typical stress grades for visually graded lumber are Select Structural, No. 1, No. 2, and so on. In addition to the grade (or grade name), a grade stamp also provides other information, such as species name and moisture content classification.

Typical grade stamps of visually graded lumber are shown in Fig. 5.10. The grade stamp in part (a) contains information on species (D Fir), grade (No. 2), moisture content classification (S-Dry), agency (WWP), and mill number (12). Visual grading is the oldest stress-grading method, and is based on the premise that mechanical properties of lumber differ from those of clear wood specimens of the same species (because there are many growth characteristics that can be seen and judged by the eye and that affect these properties). *Clear wood* is defined as lumber that does not contain (growth) defective characteristics such as knots, cross grain, checks, and splits. The characteristics used to sort lumber into separate stress grades are density (number of annual rings per inch and percent of summerwood), decay, slope of grain, knots, and cracks.

In the nondestructive system of grading or E-rating, a stress-rating equipment measures the stiffness of the material and sorts it into various modulus of elasticity classes. (Lumber graded in this way is called MSR lumber.) Following the assignment of E values, each piece must also meet certain visual requirements. The modulus of elasticity values are then used to predict the bending strength or modulus of rupture and the tensile strength of the lumber. The higher the E value, the higher are the modulus of rupture and tensile strength. At present, only small lumber (2×4 nominal) is available as E-rated. When specifying MSR lumber, the bending stress (allowable) and modulus of elasticity values should be specified.

The grade stamp of E-rated lumber differs from that of the visually graded in the grade name designation. In visually graded lumber, as shown, the grade stamp shows the grade of the piece of lumber, whereas in E-rated lumber, the modulus of elasticity (in million psi) and allowable bending stress (in psi) are shown. Figure 5.10(c) shows a typical grade stamp of E-rated lumber, where 1.5E represents the modulus of elasticity and 1650f represents the allowable value in bending. Note that the other information is the same as that in the grade stamp of visually graded lumber.

5.9 MECHANICAL PROPERTIES AND ALLOWABLE VALUES

Wood has unique, independent properties in the three mutually perpendicular axes: longitudinal, radial, and tangential (Fig. 5.5). The longitudinal axis is parallel to the grain, the tangential axis is perpendicular to the grain but tangent to the annual rings, and the radial axis is normal to the annual rings (and perpendicular to the grain direction).

Modulus of Elasticity. The modulus of elasticity in the longitudinal direction is the highest, and is between $1–2 \times 10^6$ psi ($6.9–13.8 \times 10^3$ MPa; at 12 percent moisture content), depending on the species. The modulus of elasticity of many hardwoods (such as Birch, Hickory, and Oak) is closer to or higher than the high range, but that of many softwoods (such as Douglas Fir, Hemlock, Redwood, and Loblolly Pine) is closer to the low range. The modulus of elasticity (both parallel and perpendicular to the grain) depends on the moisture content, and decreases (from dry

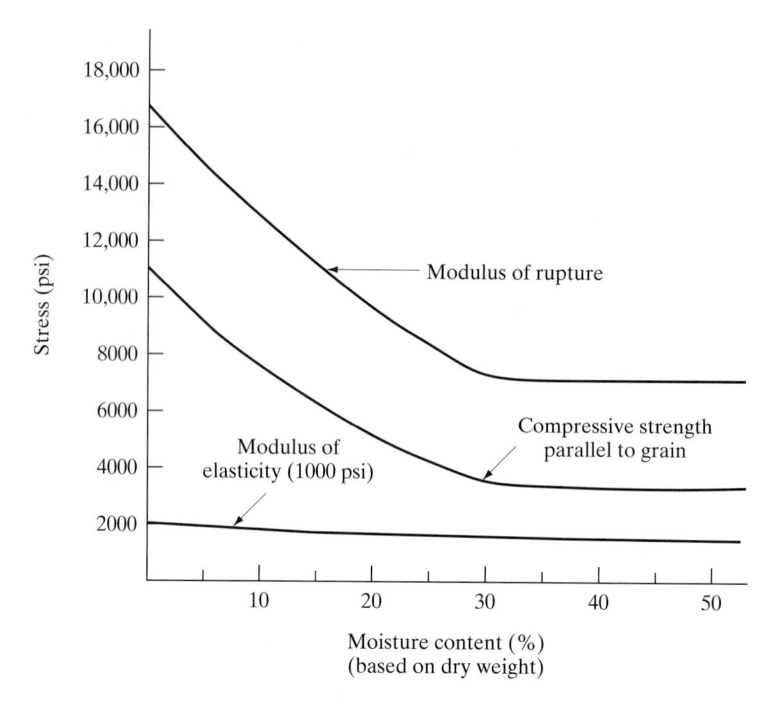

Figure 5.12 Effect of moisture content on strengths of small clear specimens of Western Hemlock.

condition) by 1–3 percent for every 1 percent increase in the moisture content (Fig. 5.12). For example, the modulus of elasticity (parallel to the grain) of Western Hemlock at 12 percent moisture content is 1.63×10^6 psi (11.2×10^3 MPa), whereas it is 1.31×10^6 psi (9×10^3 MPa) in green condition. The average modulus of elasticity of higher grade lumber (Construction, No. 1, etc.) is nearly the same as that of clear wood. In a particular species, the modulus of elasticity (as well as other mechanical properties) increases with the amount of summerwood. The modulus of elasticity in the radial direction is about 10 percent of that in the longitudinal direction, and that in the tangential direction is about one-half of that in the radial direction.

Compressive Strength. The compressive strength of lumber parallel to the grain is much higher than that perpendicular to grain. Columns, posts, and members of a truss are subjected to axial loads parallel to the grain of the wood. When a column rests on a beam, the load from the column creates compressive (bearing) stress on the beam that is perpendicular to the grain of the wood.

The compressive strength of most softwood lumber parallel to the grain is in the range 2000–4000 psi (13.8–27.6 MPa; green condition). The stress at the proportional limit is about 80 percent of this strength. The compressive strength perpendicular to the grain varies between 12–18 percent of that parallel to the grain. The compressive stress at the proportional limit, for loading perpendicular to the grain,

TABLE 5.6 AVERAGE MECHANICAL PROPERTIES OF SOME COMMON SOFTWOODS[a]

Species	Modulus of rupture (psi)	Modulus of elasticity (psi × 10^6)	Compression parallel to grain (psi)	Compression perpendicular to grain (psi)	Tension parallel to grain (psi)	Tension perpendicular to grain (psi)	Shear parallel to grain (psi)
Northern White Cedar	6500	0.8	4000	310	—	240	850
Western Red Cedar	7500	1.1	4500	460	8600	220	900
Douglas Fir	12,500	1.5–1.8	6200–7200	750	20,000	350	1100–1500
Hemlock	6500–12,000	1.2–1.6	5500–7200	550–850	17,000	350	1000–1500
Western Larch	13,000	1.85	7500	930	21,000	430	1300
Ponderosa Pine	9400	1.30	5300	580	11,000	420	1100
Virginia Pine	13,000	1.52	6700	910	17,500	380	1300
Redwood	8000–10,000	1.1–1.3	5200–6200	500–700	12,000	250	950–1100
Sitka Spruce	10,000	1.5	5600	580	11,000	370	1100

[a]Based on clear wood tests at 12 percent moisture content.

is about 12–25 percent of that for loading parallel to the grain. In dry condition, lumber can possess significantly higher compressive strength (Table 5.6). For example, the compressive strength (parallel to the grain) of Douglas Fir in green condition is about 3700 psi (25.5 MPa), while that at 12 percent moisture content is about 7000 psi (48.2 MPa). Seasoning increases the compressive strength. Typically, the compressive strength (parallel and perpendicular to the grain) increases an average 4–6 percent for every 1 percent decrease in moisture content (Fig. 5.12). Species such as Oak, Maple, Douglas Fir, Southern Yellow Pine, and Western Larch possess very high compressive strengths.

Bending Strength. Lumber has very good bending strength; because of this, and its lightness, it remains one of the important flexural materials in light construction. It is used for beams, joists, rafters, headers, and other members that are subjected to bending moments.

If you have taken a course in the strength of materials, you may recall that stress f, resulting from bending moment M, can be calculated from the flexural formula

$$f = \pm\frac{My}{I}$$

where f is measured at distance y from the neutral axis, and I is the moment of inertia of the cross-section about the bending axis. The maximum stress, or stress at the extreme fiber, can be found as

$$f = \pm\frac{Mc}{I}$$

where c is the distance from the neutral axis to the extreme fiber.

Although these two equations are commonly used to determine bending stresses in lumber beams, it should be kept in mind that they are based on the assumption that the material behaves elastically (Fig. 5.13). The maximum bending stress so computed is also called the modulus of rupture (MOR), which is a measure of the tensile or the compressive capacity of the material, depending on the type of failure.

But lumber does not exhibit truly elastic behavior up to failure. Although the stress-strain diagram is linear at small loads, the latter part of the diagram represents inelastic behavior (Fig. 5.13). Because of this the MOR value calculated from the flexural formula is intermediate between the tensile and compressive strengths of

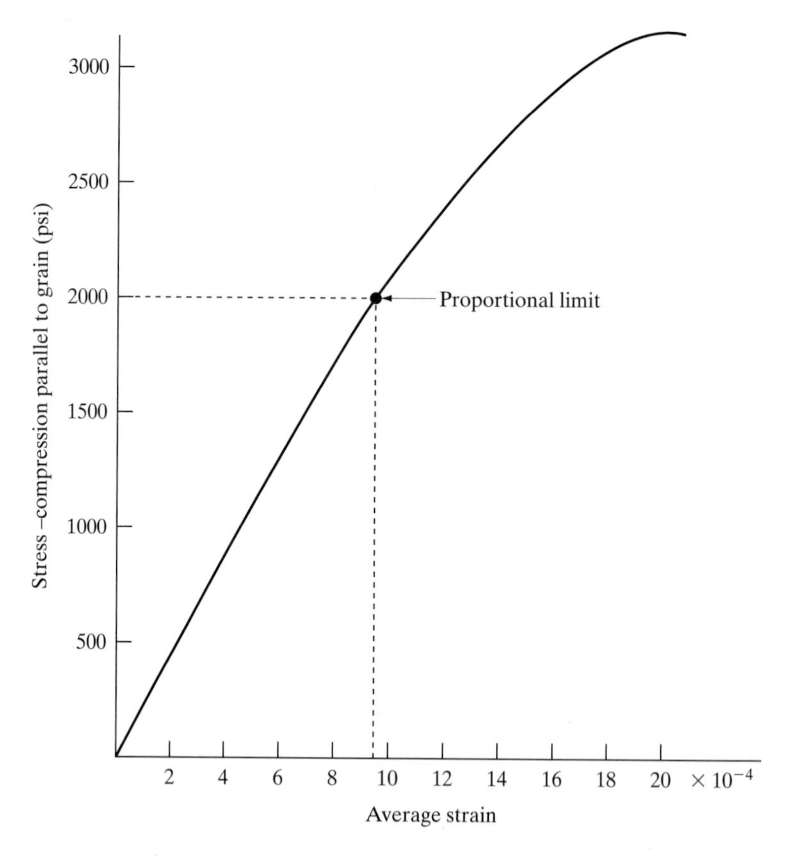

Figure 5.13 Typical stress-strain diagram of lumber (compression parallel to grain).

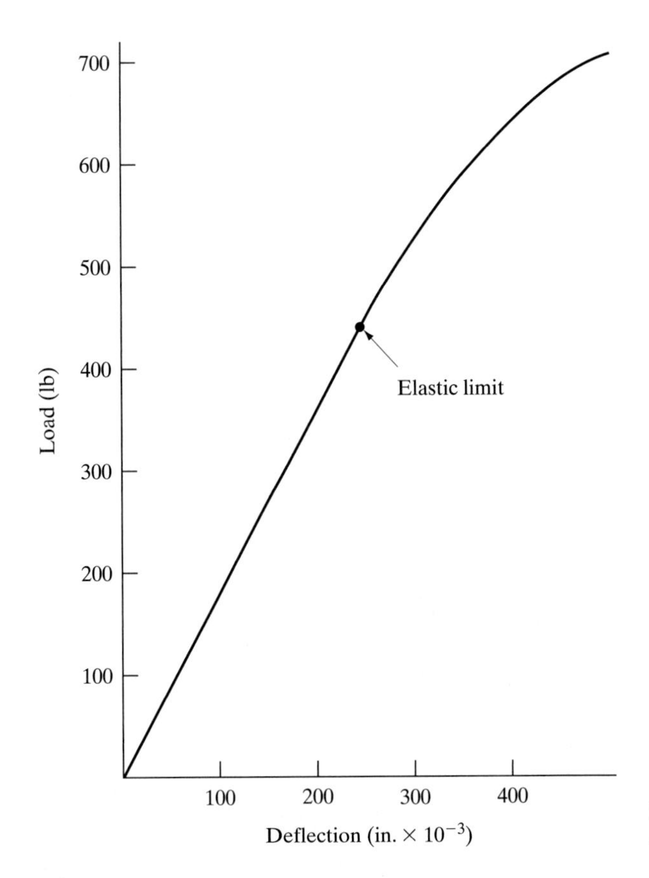

Figure 5.14 Typical load deflection diagram for 2 × 2-in. clear wood beam.

wood, and does not truly represent the extreme fiber stress. The discrepancy between the actual unit stress and the calculated stress (using the flexural formula) is thus due to inelastic behavior and also to the shifting of the neutral axis (Fig. 5.14). It should thus be emphasized that although the MOR and flexural formula are used to calculate bending properties of lumber, these measurements are not a valuable index of the quality of wood.

Failure of lumber beams is generally characterized by wrinkling or crushing of fibers in the compression zone, followed by a final failure, marked by splitting or snapping of fibers in the tension zone. Green lumber exhibits compression failure (with or without tension failure), whereas dry lumber generally fails suddenly in tension, accompanied by a loud noise.

The MOR value of most softwoods (clear wood) is greater than 6000 psi (41.3 MPa; Table 5.6). The higher the specific gravity and the lower the moisture content, the larger the MOR value. For example, Douglas Fir, with a specific gravity of about 0.48 (at 12 percent moisture content) has a MOR value of about 12,500 psi (86.1 MPa), whereas Eastern Hemlock, with a specific gravity of 0.4 (at 12 percent moisture content) has a MOR value of about 9000 psi (62 MPa). When the moisture

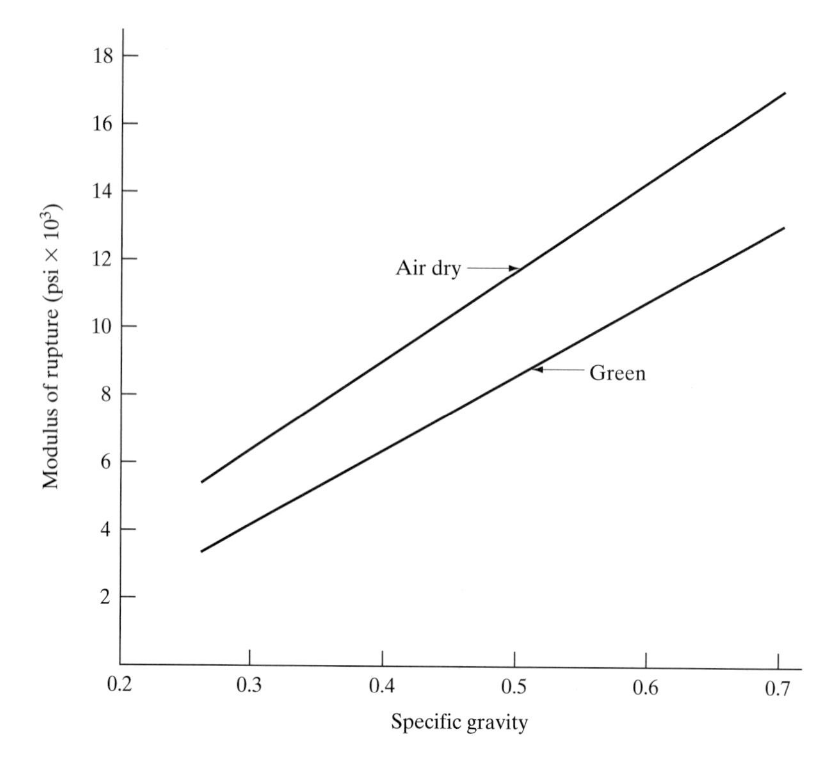

Figure 5.15 Increase in bending strength with increase in specific gravity (average values obtained for a large number of hardwoods and softwoods).

content increases to about 20 percent, the corresponding values are 7700 and 6400 psi (53 and 44 MPa), respectively (Fig. 5.15).

Hardwoods generally have higher bending strengths than most softwoods. Black Locust, Hickory, and White Oak are some of the strong hardwoods (in bending), and Longleaf Pine, Douglas Fir, and Tamarack are some of the strong softwoods.

Tensile Strength. When a clear wood specimen is subjected to tensile forces parallel to the grain, it is found to have the greatest of all strength characteristics. Tensile strength parallel to the grain is about two to four times the compressive strength parallel to the grain (Table 5.6). Failure in a tension specimen is characterized by transverse rupturing of the cell walls. Knots greatly reduce the tensile strength (both parallel and perpendicular to the grain); however, it seems to be less affected by moisture content than are other mechanical properties.

The tensile strength of wood perpendicular to the grain is very small. Failure is characterized by the separation of cells and fibers in longitudinal planes. Defects such as knots, shakes, and checks further reduce tensile strength perpendicular to the grain. The tensile strength of clear softwoods, perpendicular to the grain, varies between 180–450 psi, whereas that of strong hardwoods is higher than 500 psi. Since these values are very low, the allowable tension value perpendicular to the grain in ordinary lumber is taken as zero; loads that cause tensile stresses perpendicular to the grain should not be applied to lumber.

Shear Strength. A flexural member is always subjected to shear forces. The resulting horizontal shear stress at the neutral axis of a wood beam may cause shear failure. The shear strength of lumber is small, generally in the range 700–1500 psi (4.8–10.3 MPa). Most hardwoods have higher shear strengths than most softwoods. Defects such as knots and shakes decrease the area under shear, and the shear strength of lumber with these defects is lower than that of clear wood.

Summarizing the preceding discussion, mechanical properties of lumber are affected significantly by two properties:

- Moisture content
- Specific gravity

Most mechanical properties increase linearly with specific gravity and decrease with increase in moisture content. Beyond the fiber saturation point, the mechanical properties remain independent of changes in moisture content. Knots weaken the lumber, and at sections containing knots, most mechanical properties are lower than those in clear wood. Knots displace the clear wood and wood grain; this discontinuity of wood fibers leads to stress concentration, and checking often occurs around knots during the drying operation. The effects of knots are more on axial tensile strength, less on bending strength, and very little on axial compressive strength.

5.9.1 Effect of Slope of Grain

Slope of grain, as previously defined, refers to the angle between the direction of fibers and the direction of the edge of a piece of lumber. Wood has different properties parallel and perpendicular to the grain. Strength properties in any direction between parallel and perpendicular to the grain direction can be approximated using a *Hankinson-type formula*. The strength of wood at an angle ϕ to the fiber direction can be found as

$$N = \frac{PQ}{P \sin^n \phi + Q \cos^n \phi}$$

or

$$N = \frac{P}{(P/Q) \sin^n \phi + \cos^n \phi}$$

where N is the strength property at an angle ϕ to the fiber direction, Q the strength perpendicular to the grain, P the strength parallel to the grain, and n an empirical constant, the value of which depends on the type of stress (or load). For example, the value of n ranges between 1.5–2 for tensile strength and bending strength (Table 5.7). Note that the ratio Q/P gives the relative strength value between perpendicular and parallel to the grain, and generally lies between 0.04–0.1 for most loads.

TABLE 5.7 EFFECT OF GRAIN ORIENTATION ON STRENGTH PROPERTIES

Property	n	Q/P	Comments
Tensile strength	1.5–2	0.04–0.07	
Compressive strength	2–2.5	0.03–0.4	n is generally taken as 2.0
Modulus of elasticity	2	0.04–0.12	
Bending strength	1.5–2	0.04–0.1	n is generally taken as 0.5

5.9.2 Strength Ratio and In-grade Testing

All structural lumber is stress graded, which establishes a grade stamp for every piece of lumber. The hypothetical strength properties of lumber belonging to any stress grade of an individual species (that is, lumber with some measurable defects), are established from the measured mechanical properties of clear wood specimens of that species, followed by the application of a limiting factor called the strength ratio. The strength properties so derived are adjusted further to include a safety factor and compensate for duration of load and other service-related factors, which then provides the allowable (design) properties of the grade.

All allowable properties of lumber belonging to Beams and Stringers and Posts and Timbers categories given in current design standards have been derived from results of tests done on small clear specimens. A comprehensive reevaluation of the mechanical properties based on tests done on full-sized members—termed an *in-grade testing* program—is the basis of recommended allowable properties of lumber belonging to the Dimension category.

The strength properties of lumber, such as MOR, compressive strength parallel to the grain, and modulus of elasticity, have been discussed. The *strength ratio* is the hypothetical ratio of the strength of a piece of lumber with visible strength-reducing growth characteristics to its strength if those characteristics were absent. The true strength ratio of a piece of lumber is never known and can only be estimated. The assigned strength ratio (based on a defect), ranging from 0–100 percent, serves as a predictor of lumber strength.

For example, to account for the weakening effect of knots, a knot is treated as a hole that reduces the cross-section of the lumber. In a beam containing an edge knot, the strength ratio is equal to the ratio of the MOR value of the beam with reduced cross-section to that of the full beam (Fig. 5.16). This can be calculated from the flexural formula as

$$\text{strength ratio} = \left(\frac{h-k}{h}\right)^2 = \left(1 - \frac{k}{h}\right)^2$$

where k is the height of the knot and h is the beam depth.

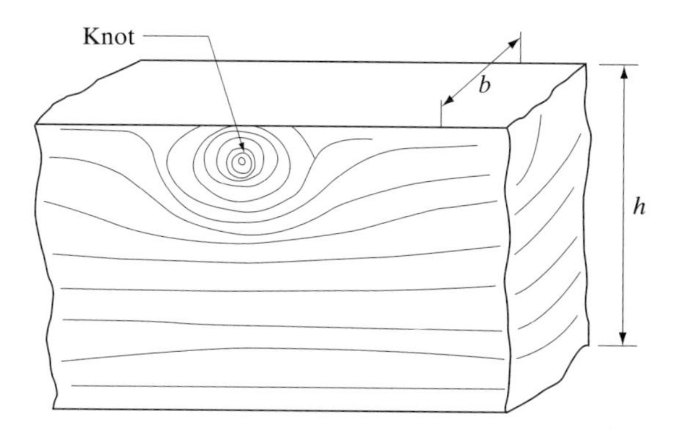

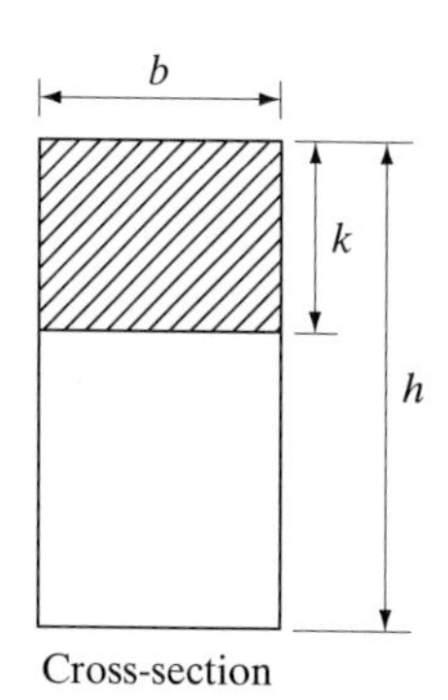

Figure 5.16 Edge knot in beam.

The strength ratios for all knots, shakes, checks, and splits are derived using concepts similar to the one just discussed, and are given in ASTM D245. When several characteristics that reduce the strength properties are simultaneously present, only the characteristic that gives the lowest strength ratio is used to derive the estimated strength of that piece of lumber.

To account for the variations between pieces of lumber in each grade, the strength ratio is applied to a near-minimum clear strength value. As mentioned, the allowable value in each grade is then obtained by adjusting this assumed strength value of the grade to account for the safety factor, duration of load, and other factors. These allowable properties are listed in the National Design Specifications for Wood Construction.

5.9.3 Examples

Example 5.3

Find the size classification of the following lumber:

(a) Nominal 2 × 4
(b) Nominal 2 × 8
(c) Nominal 4 × 16
(d) Nominal 6 × 6
(e) Nominal 6 × 16

Solution. (Refer to Table 5.5):

(a) Size 2 × 4: thickness not larger than 4 in.; Dimension
(b) Size 2 × 8: thickness not larger than 4 in.; Dimension
(c) Size 4 × 16: thickness not larger than 4 in.; Dimension
(d) Size 6 × 6: thickness larger than 4 in.; square; Post and Timber
(e) Size 6 × 16: thickness larger than 4 in.; rectangular; Beams and Stringers

Example 5.4

Find the cross-sectional properties of 4 × 16 nominal-size lumber header beam.

Solution. (Refer to Table 5.4.)

Nominal size: 4 × 16
Dressed size: 3.5 × 15.25

$$\text{Cross-sectional area} = 3.5 \times 15.25 = 53.4 \text{ in.}^2$$

$$\text{Moment of inertia} = \frac{3.5 \times (15.25)^3}{12} = 1034.4 \text{ in.}^2$$

$$\text{Section modulus} = \frac{1034.4}{15.25} \times 2 = 135.7 \text{ in.}^2$$

Example 5.5

Determine the cross-sectional properties of glulam beam (Table 5.8) built with six laminations of 2 × 6 Douglas Fir lumber.

Solution

$$\text{Number of laminations} = 6$$

$$\text{Width of beam} = 5.125 \text{ in.}$$

$$\text{Beam height} = 6 \times 1.5 = 9 \text{ in.}$$

$$\text{Cross-sectional area} = 5.125 \times 9 = 46.1 \text{ in.}^2$$

$$\text{Moment of inertia} = \frac{5.125 \times (9)^3}{12} = 311.3 \text{ in.}^4$$

TABLE 5.8 WIDTH OF GLULAM MEMBERS

Nominal width of laminating lumber (in.)	Net finished width of glulam (in.)	
	Southern Pine	Western species
3	—	2.25
4	3	3.125
6	5	5.125
8	6.75	6.75
10	8.5	8.75
12	10.5	10.75

Example 5.6

In a piece of lumber the slope of the grain is measured as 1 in 20. Assume the perpendicular-to-parallel strength ratio to be 0.1, and using a value of 1.5 for n, determine the strength of the lumber.

Solution. For a slope of 1 in 20,

$$\phi = 2.86$$
$$\tan \phi = 0.05$$
$$\sin \phi = 0.05$$
$$\cos \phi = 0.998$$
$$Q/P = 0.1$$

$$\text{Strength of lumber} = \frac{P}{\dfrac{\text{Sin}^{1.5}\phi}{0.1} + \cos^{1.5}\phi}$$

$$= \frac{P}{\dfrac{0.05^{1.5}}{0.1} + (0.998)^{1.5}}$$

where P is the strength parallel to the grain. Thus there is a 10 percent reduction due to a slope of grain of 1 in 20.

Example 5.7

For a knot of depth equal to one-fourth the depth of the beam, estimate the MOR value if that of clear wood is 10,000 psi.

Solution. The strength ratio for $k/h = 0.25$ is

$$(1 - 0.25)^2 = 0.5625$$

If the clear wood strength (MOR) is equal to 10,000 psi,

$$\text{estimated strength of lumber} = 0.5625(10{,}000)$$
$$= 5625 \text{ psi}$$

5.10 WOOD PRODUCTS

Although the traditional application of wood is in the form of lumber, a large number of wood products have come into the market in recent years and are being used extensively in construction. The most common wood products are glulam (which stands for "glue-laminated timber"), panel products, and manufactured structural components (Fig. 5.17). Properties and applications of glulam are discussed in the following section. Manufactured structural components are those built using lumber and panel products (other than glulam). These are described briefly in

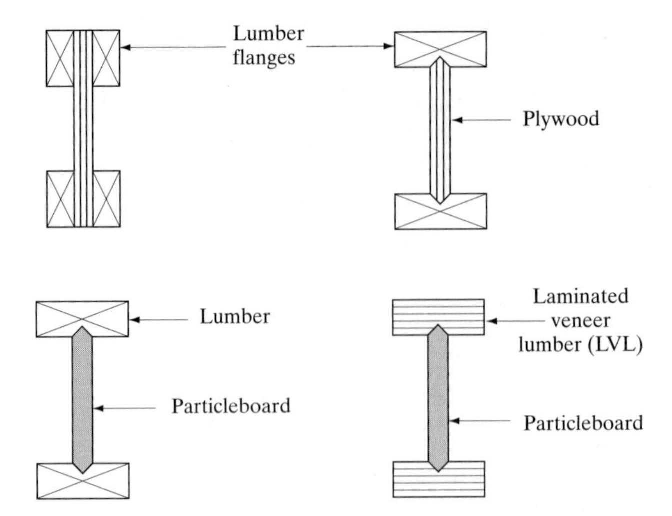

Figure 5.17 Manufactured I beams.

Section 5.10.4. General features of panel products are explained below, but some special characteristics are described in Sections 5.10.2 and 5.10.3.

Panel Products. There are two types of panels: veneered and nonveneered. Veneered panels, also called plywood, are made from thin sheets of wood, whereas nonveneered panels are manufactured from wood particles or fibers. Both types of panels are used for structural applications such as floor and wall panels, and for nonstructural applications such as furniture and cabinets. These panel materials have also been used as web material for built-up wood I beams (because of their outstanding shear characteristics). Structural applications of nonveneered panels have become more common since around 1980. Nonveneered panels are divided into two major categories:

- Particleboards
- Fiberboards

Particleboard is a generic term that identifies various types of panels manufactured from discrete wood particles, as distinguished from fibers, combined with a synthetic resin or glue, and bonded together under heat and pressure in a hot press in which an entire interparticle bond is created. It is classified into three types: low-density (density less than 37 pcf (590 kg/m^3) and specific gravity 0.59), medium-density (density between 37–50 pcf (590–800 kg/m^3) and specific gravity between 0.59–0.80), and high-density particleboard (density greater than 50 pcf (800 kg/m^3) and specific gravity 0.80).

Fiberboard is a common term for a homogeneous panel made from wood fibers (as opposed to particles) that has not been consolidated under heat and pressure as a separate stage in manufacture. Fiberboard panels have a density between 10–31 pcf (160–500 kg/m^3), and specific gravity between 0.16–0.50. Manufacturing details, properties, and applications of these panel products are described in Sections 5.10.3 and 5.10.4.

5.10.1 Glulam

"Glulam," or glue-laminated timber, introduced in Europe in the late nineteenth century, consists of sawn lumber laminations bonded with an adhesive so that the grain of all laminations runs parallel with the long direction (Fig. 5.18). One of the early uses of this material in the United States was a glulam arch erected in 1934 at the USDA Forest Products Lab in Madison, Wisconsin. Glulam is now manufactured to follow the Standard ANSI/AITC A190.1—1993, and developed and sold as a structural product. (ANSI is the American National Standards Institute, and AITC the American Institute of Timber Construction.)

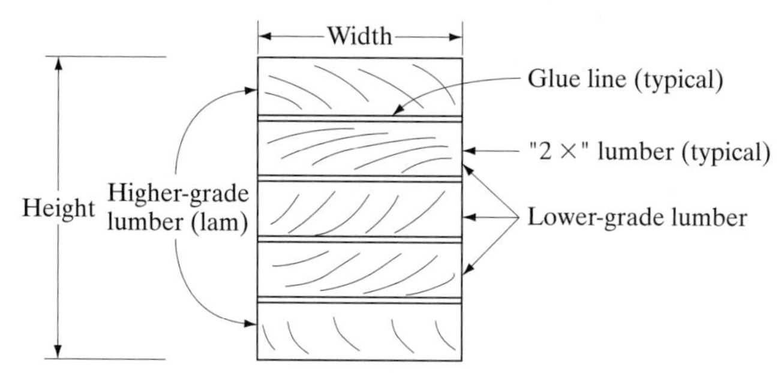

Figure 5.18 Cross section of glulam.

Glulam can be manufactured in a variety of shapes and sizes. Curved, tapered, circular, and spiral-shaped members have been used (Fig. 5.19). Curved arches have been made that span more than 300 ft (91 m); and glulam domes exceeding 500 ft (152 m) in diameter have been built.

Glulam is manufactured using lumber with nominal thicknesses of 1 and 2 in. (3/4 and $1\frac{1}{2}$ in. dressed sizes). The "1 ×" lumber is generally used to form curved members and the "2 ×" lumber to manufacture straight members. Straight beams can be designed and manufactured with horizontal laminations (load applied perpendicular to the wide face of laminations) or vertical laminations (load applied parallel to the wide face of laminations); see Fig. 5.20. Glulam can be fabricated in various widths, as shown in Table 5.8.

Glulam is usually manufactured using lumber having moisture content in the range 10–16 percent, and comes with camber or upward deflection. When fabricated with lumber of moisture content less than 12 percent and used under conditions where the moisture content remains less than 12 percent, glulam is practically free from shrinkage and swelling. Dry condition is assumed when the moisture content during service does not exceed 16 percent. Wet condition is assumed when the moisture content during service exceeds 16 percent. However, glulam is rarely used in wet condition.

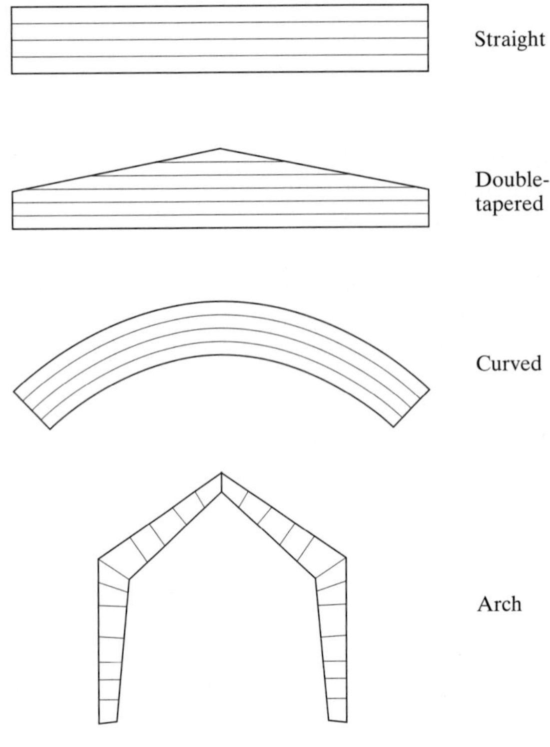

Figure 5.19 Shapes of glulam.

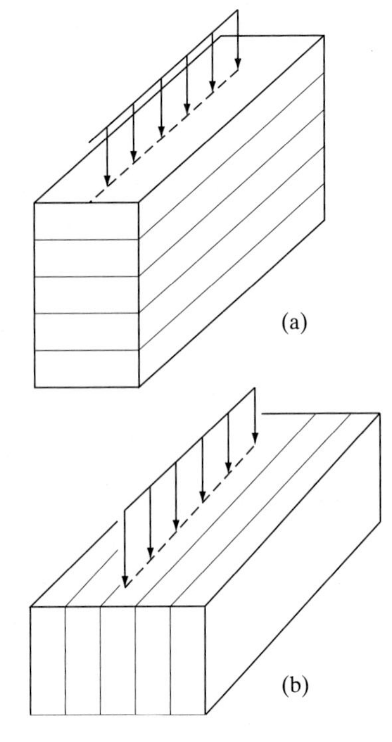

Figure 5.20 Loading direction: (a) loaded perpendicular to wide face of laminations; (b) loaded parallel to wide face of laminations.

Glulam is stronger in the longitudinal direction and weaker in the transverse direction. Manufacturers place high-grade lumber near the surface (top and bottom), and use low-grade lumber for the center (near the neutral plane). All joints (between lumber pieces in a lamination) must be of the scarf or at least of the finger type. Strength-reducing joints and knots are staggered.

The process of lamination allows control over the location of materials of different quality within the member cross section. By placing the strongest material in the regions of greatest stresses (near the top and bottom in the case of a flexural member), member performance can be improved. Laminating also allows the dispersion of lumber defects throughout the length of the member (Fig. 5.21).

Glulam is commonly used as a replacement for sawn lumber when higher-sized lumber is unavailable. In practice, sawn lumber beams of nominal size greater than 6×18 (length over 25 ft) are difficult to obtain. In these situations, glulam or other prefabricated members can be used (Fig. 5.22). There are additional advantages that come with the use of glulam. Allowable strength properties of glulam are generally superior to those of sawn lumber. Glulam is a dimensionally stable material with average moisture content of 12 percent, compared with about 30 percent for sawn

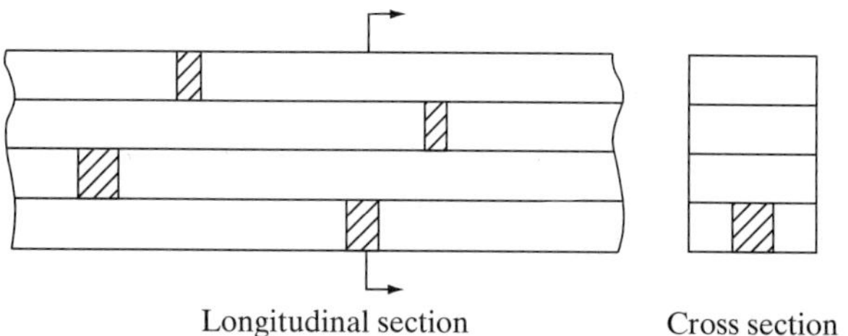

Figure 5.21 Staggered knots in glulam.

Figure 5.22 A glulam beam.

lumber (green). Adhesives used in glulam are not combustible and do not lose strength under heat. Because of all these properties, glulam is commonly used for purlins, joists, headers, beams, and truss members. It is also used in the design of pedestrian and highway bridges.

5.10.2 Plywood

Plywood is a panel product with an odd number of layers of veneer or plies, glued together so that the grains of adjacent layers are perpendicular to one another. Outer layers and all odd-numbered layers generally have the grain direction

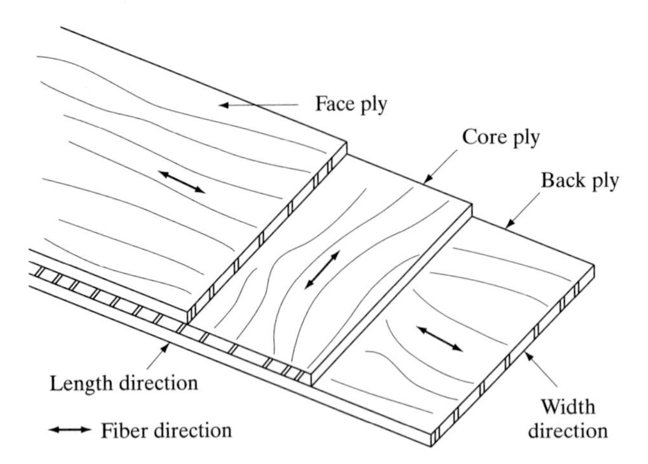

Figure 5.23 Plywood construction.

oriented parallel to the long direction of the panel (Fig. 5.23). Plywood comes in a common size of 4 × 8 ft.

Veneers. A veneer (or ply) is a thin sheet of wood, peeled from a log, from which plywood is made (Fig. 5.24). Early evidence of the practice of veneering comes from mural paintings on walls of tombs in Egypt. The art of overlaying and inlaying woodwork with decorative veneers from rare and imported wood was well established in Egypt around 2000 B.C. Up to about 150 years ago, veneers were usually applied to a solid base.

Nowadays, to conserve timber and the dwindling supply of good-quality wood, the base for the outer veneers is itself made up of several layers of wood, placed one upon the other, such that the direction of the grain in each ply alternates in direction at right angles. Ordinarily, no single ply or veneer will exceed 5/16 in. (7.9 mm) in thickness. Plywood *core* refers to all plies or layers between the face ply and the back ply.

Veneers are manufactured in various grades, depending on defects such as knots, holes, discoloration, and grain orientation. Grades N and A are the highest, and grade D is the lowest. Grade N has a smooth surface with a natural finish and is free of natural defects. Grade A has smooth paintable surface and a limited number of repairs. Grade B has a large number of patches or repairs and minor splits. Grade C, often used as the face on flooring and ceiling sheets, has knot holes and other defects. When these defects have been repaired, patched, or plugged, the grade becomes C-plugged. Grade D, which is commonly used as the back face and core of higher-grade plywood, has much larger knot holes and splits. It is used primarily in the production of interior-grade plywood. The durability characteristics of the species from which the veneers are produced should also be considered in the selection of a veneer for plywood.

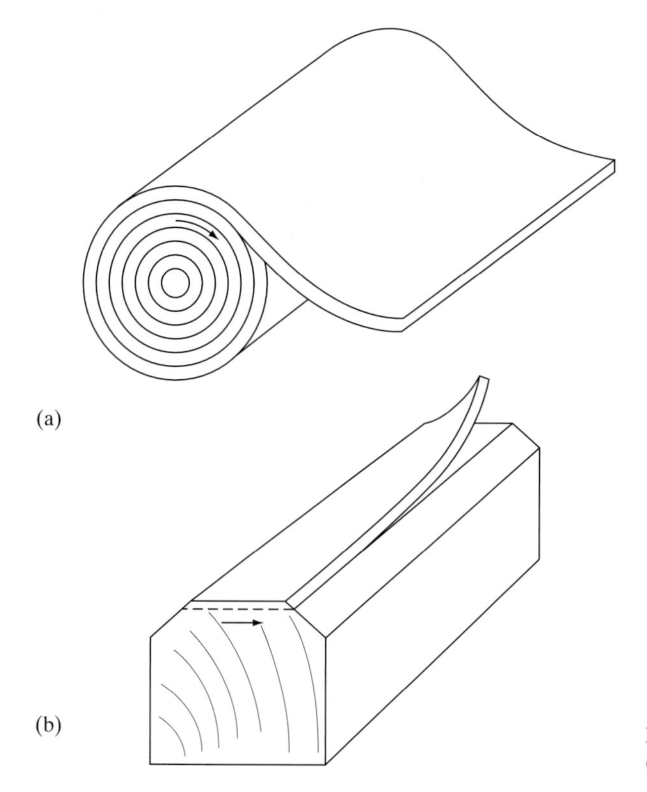

Figure 5.24 Veneer production: (a) rotary cutting; (b) slicing.

Manufacture. The plywood industry dates back to about 1905, and World War II greatly accelerated plywood technology (for use in small naval military aircraft). The manufacture of plywood involves placing heated lumber blocks (or bolts) into a lathe, peeling the blocks, clipping the resulting sheets to size, and stacking the clipped sheets. (Note that prior heating—steaming or soaking in hot water—softens the wood and the knots, making it easier to peel or cut.) These sheets are then dried to a target moisture content (2–4 percent). Adhesives are applied on individual sheets and panels are assembled. The panels are first consolidated on a cold press and later loaded into a hot press, where heat and pressure (as high as 200 psi or 1.4 MPa) are applied on the panels. These panels are then set aside for the adhesive to cure, following which they are trimmed. After sanding and patching, the panels are strapped in bundles for shipping.

Virtually all plywood manufactured in the United States is bonded using phenol–formaldehyde (PF) or urea–formaldehyde (UF) adhesive. Softwood plywood for interior and exterior applications (Interior grade or interior type, and Exterior grade or exterior type), and exterior-grade hardwood plywood are manufactured using PF adhesive. Interior-grade hardwood plywood is manufactured with UF adhesive.

Plywood is manufactured seasoned. Drying of veneers is essential to hot-press gluing. Gluing adds about 5 percent extra moisture. After manufacture, plywood has only about 6–10 percent moisture; any subsequent moisture absorption can cause it to warp and buckle. Like all wood materials, plywood (and other panel products) picks up and loses moisture with changes in the environmental humidity. These changes in moisture content lead to dimensional changes: linear and thickness swelling and shrinkage. Changes in size lead to altered strength and stiffness and affect the performance under load. Generally, dimensional changes at relative humidity below 80 percent are insignificant. Above this relative humidity, properties decrease rapidly.

Types. Plywood is divided into two types, based on the type of wood used:

- Softwood plywood
- Hardwood plywood

Softwood plywood is manufactured using softwood logs, most commonly Douglas Fir and Southern Pine. It is used principally as a construction material, in floor panels, roof panels, wall sheathing, siding, and concrete forms, and in manufactured structural components. Most of the softwood plywood (also called structural plywood) produced in the United States is manufactured to meet the requirements of United States Products Standard PS 1-83 for Construction and Industrial Plywood (Fig. 5.25). Hardwood plywood, which is manufactured from a variety of hardwood logs, such as Oak, Maple, and Birch, is principally used for paneling, industrial parts, and furniture.

Plywood is manufactured in three exposure classification classes:

- Exterior type
- Interior type
- Exposure-1 type

Exterior plywood has the exterior glue and the lowest veneer grade is Grade C. Interior plywood is manufactured using the interior glue and may include D-grade veneer. When interior plywood is manufactured with the exterior glue, it is called Exposure-1 plywood. (Note that grade A is the highest, and grade D is the lowest veneer grade.)

Softwood plywood (and other structural paneling) has a grade stamp that shows the *span rating* for the panel. The span rating (Fig. 5.25) consists of two numbers arranged as a fraction. The larger number is the allowable span length when used as roof sheathing, and the smaller number is the allowable span length when used as a floor panel. In some panels, the span rating is given as a single number, which is the spacing of floor supports.

A strip of structural plywood placed between supports will not carry more load than a piece of softwood lumber of the same width and thickness as the plywood. However, plywood has strength in bending in either direction, and hence it will serve to carry the floor load whether laid parallel or perpendicular to the floor joists. But for the most efficient use of plywood, it should be placed with the face grain

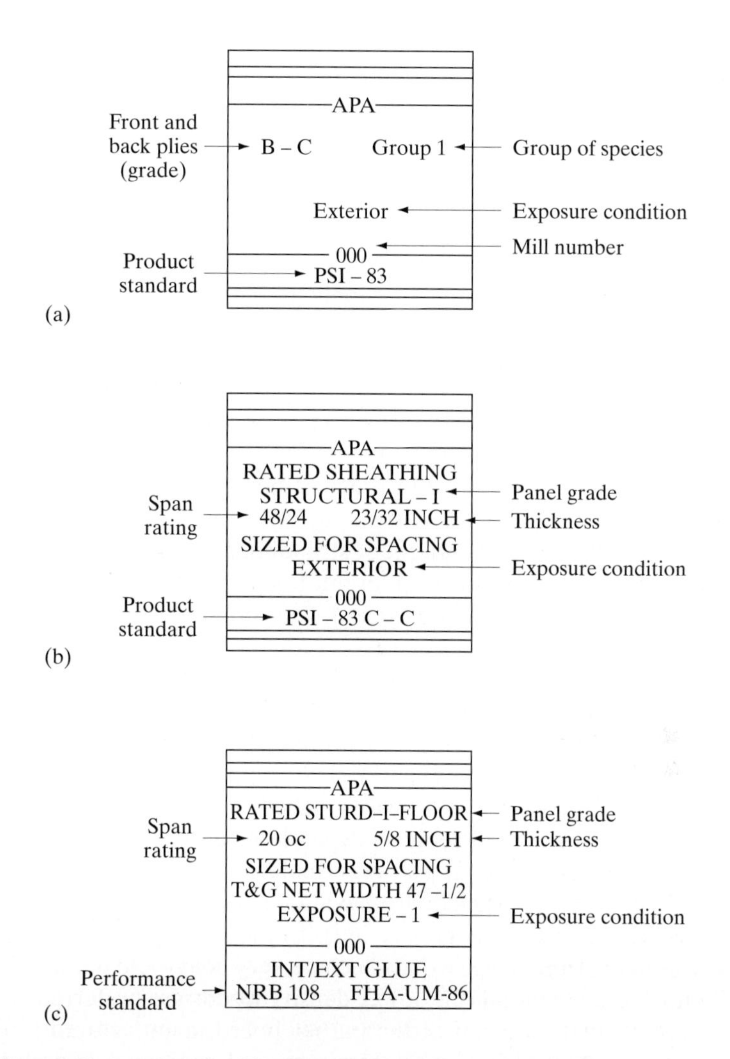

Figure 5.25 Panel grades: (a) nonstructural panel, veneered; (b) structural panel, veneered; (c) structural panel, nonveneered.

perpendicular to the joists. In addition, plywood comes in a 4 × 8 ft size (1.2 × 2.4 m), which makes it impossible to deform the plane by a force applied parallel to the plane. Thus plywood and similar panel products are excellent for use as wall and roof sheathing as well as wall siding.

5.10.3 Other Panel Products

Nonveneered panel products are divided into two categories:

- Particleboard
- Fiberboard

Particleboard. Particleboard is made from small particles of wood, and consists of chipboard, flakeboard, splinterboard, waferboard, strandboard, oriented strandboard, and other similar panel products, based on the size and type of particles and their orientation. Particles are pieces of wood smaller than veneer sheets but larger than the wood fiber. The raw material for the manufacture of particles comes from many sources, such as plywood mill waste, sawdust, roundwood (round logs), planer shavings, and wood residue (such as broken logs and branches). Residues from milling operations are commonly used as particles.

The manufacturing stages involve forming the particles by cutting the wood chips, drying to about 2–5 percent moisture content, and removal of fines, followed by the addition of resin or glue. This mixture is formed into a low-density mat which is compressed, heated, and cured. These boards are cooled prior to stacking, and most are sanded before shipping.

The most common resin used in the manufacture of particleboard is urea–formaldehyde, and boards containing UF should be used for inside applications only. Waferboard and oriented strandboard are manufactured with phenol–formaldehyde, and can be used for interior or exterior applications. These two types of panels (oriented strandboard and waferboard) come under a common category called flakeboard.

Flakeboard, introduced in the 1950s, identifies a panel manufactured from specially produced flakes, generally from low-density or low-quality species such as Aspen and Pine. Waferboard is produced exclusively from Aspen. *Wood wool (excelsior)* consists of long, curly, slender strands of wood used as an aggregate component for some particleboard—called excelsior board.

Properties of particleboard depend on its density; most have an average density of 40–50 pcf. When used as a raw material, Hickory (which has a specific gravity of 0.72) generates particleboard having a density of about 64 pcf, whereas Douglas Fir (which has a specific gravity of 0.48) produces particleboard of density 43 pcf. Boards made from dense woods become very heavy and difficult to handle and ship. Therefore, as a general rule, lower-density woods are preferred.

More than one-half of the particleboard manufactured in the United States goes into the manufacture of furniture and cabinets (as architectural veneered panels). Some particleboards are also used as underlayment (the layer below the floor covering) and in mobile homes. Flakeboard (oriented strandboard and waferboard) is used primarily in engineered construction as a replacement for plywood.

Waferboard and Strandboard. As mentioned, the term flakeboard refers to a panel produced with flakes. Flakes are small wood particles, resembling a small piece of veneer, with fiber direction essentially in the plane of the flake. They are a little thinner than wafers (0.25 to 0.5 mm) and somewhat longer than they are wide. Waferboard is a panel made from wood wafers. A wood wafer is a nearly rectangular particle of wood of thickness in the range 0.5–1 mm and length of at least 30 mm ($1\frac{3}{16}$ in.). Waferboard has nearly equal properties in all directions parallel to the plane of the panel.

Strandboard is made using wood strands, which are similar to wafers except that they have a larger length to width ratio (about 2:1). Oriented strandboard (OSB) is usually made in three layers, with strands alternating 90 degrees in orientation. Alignment of strands is done by means of an electrical field, or mechanically by vibrating the particles through fins. The alignment of strands makes this panel stronger and stiffer than waferboard.

All flakeboards (strandboard, oriented strandboard, and waferboard) have made major inroads into the construction industry, due primarily to lower cost, thanks to low-quality wood species, which otherwise have no appreciable structural properties. The use of exterior glue and the elaborate laboratory testing programs of the American Plywood Association and others have made it possible to use a span rating stamp, similar to that on structural plywood, on these panels. As a result, oriented strandboard and waferboard are being widely accepted as substitutes for softwood plywood in building construction. They are used as roof sheathing, floor sheathing, siding, and in other light residential applications.

Waferboard has a modulus of rupture between 2500–3000 psi (17.2–20.7 MPa; Table 5.9) and a modulus of elasticity around 500,000 psi (3.5×10^3 MPa). It is not as stable as plywood. The lack of alignment of fibers in waferboard is the reason for its low bending properties. The swelling of particleboard, in general, is higher than that of solid wood of the same thickness. Mechanical properties of OSB are nearly the same as plywood.

TABLE 5.9 AVERAGE BENDING PROPERTIES OF PANEL PRODUCTS

Panel	Modulus of elasticity (psi)	Modulus of rupture (psi)
Plywood (parallel to face grain)	1.25×10^6	6500
Waferboard	0.5×10^6	3050
Oriented strandboard (parallel to face strand)	1.25×10^6	7000

Fiberboard. Fiberboard (or reconstituted wood board) comes in two types,

- Hardboard
- Medium-density fiberboard (MDF)

Fibers are slender threadlike elements—of length less than 1/3 in. (8.5 mm), but generally of 1/25 in. (1 mm)—resulting from chemical or mechanical defiberization, or both. A number of raw materials, such as coarse residues from other forest products, bagasse (fiber residue from sugarcane), wastepaper, and pulp chips, can be used in the manufacture of fibers. These raw materials are broken down into fibers through thermomechanical treatment. Then the fibers are interfelted under controlled conditions of hot pressing. This will cause rebonding of the lignin, which in addition to the binding agents added (synthetic and natural resins, paraffin, and asphalt) will produce a bonded panel product.

Hardboard has a specific gravity in the range 0.5–1.45, and density of 31–90 pcf (500 to 1450 kg/m^3); and medium-density fiberboard has a specific gravity of 0.5–0.88, and density of 31–55 pcf (500–880 kg/m^3). In addition to these two products, fibers are also used in the manufacture of insulation board and laminated paperboard.

Medium-density Fiberboard and Hardboard. Medium-density fiberboard, developed in the 1960s, is similar to hardboard and particleboard and has a density in the range 31–55 pcf (500–880 kg/m^3). Early steps in the manufacture of MDF are similar to those for hardboard. Logs are reduced to chips, thermomechanically pulped, dried, and blended with resin (and occasionally, wax). The amount of resin used is much higher than that for hardboard. The mixture is formed into a mat that is subsequently pressed to the desired thickness and density.

MDF has a more uniform density and a smoother edge than particleboard. It is used primarily in the furniture industry, replacing solid wood, plywood, and particleboard. It is also used in the manufacture of doors, moldings, and other elements. Its smooth surface facilitates wood-grain printing and veneering.

Hardboard (also called grainless wood), developed in 1924, is a medium- to high-density wood fiber product that has a specific gravity closer to 1.0. It is manufactured in sheets of thickness in the range 1/16–1/2 in. (0.16–1.27 mm) and can be molded into a variety of shapes. Like MDF, it is also produced from thermomechanically produced pulp. Note that when the wood chips are fiberized, lignin is not dissolved, and the attrition mill, which is like a grinding machine, rubs the fibers apart. The resulting pulp contains fine fibers, bundles of fibers, and larger fiber aggregates. After drying, fibers are formed into a mat and pressed using either water (wet process) or air (dry process). Heat and pressure reactivate lignin as a binder, which makes the boards stronger.

Hardboard contains almost no moisture as it leaves the press. Its modulus of rupture varies between 2000–7000 psi (13.8–48.2 MPa), and tensile strength (parallel to the surface) ranges between 1000–3500 psi (6.9–24.2 MPa). Because it is denser, it is also harder than natural wood. Its grainless character gives equal strength characteristics in all directions. As it contains lignin, with changes in moisture content it shrinks and swells just like wood. Hardboard is used in the manufacture of furniture, cabinets, and exterior siding. It is also used as floor underlayment and as concrete formwork. High-density hardboard is also used as a web member of manufactured I beams.

5.10.4 Manufactured Components

In addition to panel products and glulam, wood is used in the manufacture of a number of prefabricated building components, such as I beams, manufactured trusses (trussed rafters), composite panel, composite wood I beams, and laminated veneer lumber. Most of these are proprietary products, and one should refer to the manufacturer's literature to understand their properties.

Composite panel or "COM-PLY" consists of veneers and other wood-based materials. It was introduced into the market around 1975 and is made with face veneers bonded to a core layer of oriented strandboard or other particleboard. It can also be manufactured with lumber core. Composite panel is interchangeable with similar-grade plywood.

Composite wood I beams are made with lumber or laminated boards for top and bottom flanges, and plywood or flakeboard as webs (Fig. 5.16). They are gaining in popularity where long-span members are required. Hardboard-webbed I beams have been in use in Sweden for more than 50 years. In the United States, plywood is commonly used for webs; oriented strandboard and waferboard are also used.

Laminated veneer lumber, or LVL (also called parallel laminated lumber), is made from veneers similar to that used in making plywood but with the grain direction in all veneers oriented parallel to one another. The thickness of veneers is between 1/8–1/10 in. (3.2–2.5 mm). LVL is made by a continuous process, in lengths of 8–60 ft (2.4–18 m), in which grains of all veneers run parallel to the beam length. Veneers are bonded together using PF adhesive.

Because of the high-quality softwood veneers that are used, and as the defects are small and randomized, mechanical properties and reliability of LVL are improved. The allowable value in flexure ranges between 2200–4200 psi (15.2–29 MPa). The modulus of elasticity varies between $1.8–2.8 \times 10^6$ psi ($12.4–14.3 \times 10^3$ MPa).

The cost of this veneer product is relatively higher than that of ordinary lumber. LVL members are used in building construction as headers over openings and main beams. They are also used in the manufacture of I beams (as flanges) and parallel-chord trusses, and as scaffold planks.

5.11 CREEP

A wood member subjected to constant load, well below the short-term failure load, may nonetheless fail if that load is sustained for enough time. Creep is the increase in strain or deformation with time under constant stress.

When a member is loaded, it deforms elastically. When the load is maintained continuously, additional deformation takes place, which is the creep or creep deformation. Due to creep, with time, wood beams sag. Creep can take place at very small loads and can continue for many years.

Creep or creep strain becomes a significant factor as the temperature or moisture content increases. At ordinary temperatures, creep is aggravated by changes in moisture content. In lumber, creep strain depends on the direction of measurement (longitudinal or transverse), and is much more severe when loaded perpendicular to the grain. At low stress levels, creep deformation stabilizes after some time, but at higher levels, the rate of creep increases with time.

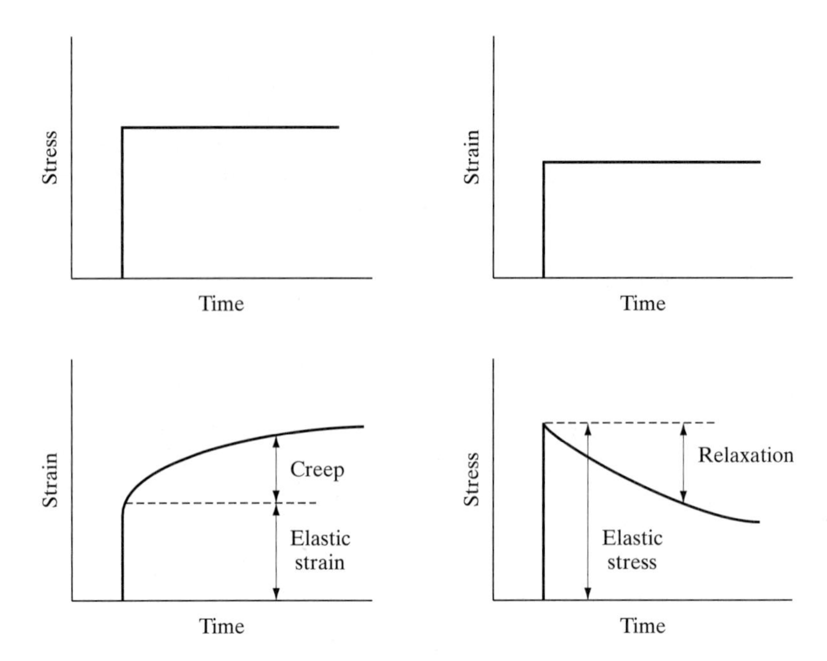

Figure 5.26 Effects of creep.

When seasoned wood remains dry during service or if unseasoned wood remains wet during service, creep deformation, at low stress levels, can be taken as 50 percent of elastic deformation. After the load is removed, most of the elastic deformation will be recovered, but the creep component will be permanent. When green lumber is allowed to dry in service, the creep component of deformation is high and can be as much as 3.5 times the elastic component.

Creep can also occur in panel products such as plywood and particleboard. High and cyclic moisture environments accelerate the creep of panel products. In laboratory tests, an increase in relative humidity from 65 to 80 percent has resulted in a creep increase (in bending) of 200–300 percent. Creep deflection of hardboard and particleboard specimens at an equilibrium moisture content of 18 percent is about three to four times that of specimens at 6 percent moisture content (measured at 40 days).

Creep has two functions: When stress is kept constant with time, creep results in an increase in deformation, as shown in Fig. 5.26. But when deformation is controlled (member being clamped), creep results in a stress loss, also called *stress relaxation.*

5.12 WOOD CONSTRUCTION

Wood is the most common construction material for dwellings. It is typically utilized in buildings of two to three stories. The advantages of wood in permanent and temporary construction are:

- Light weight
- Faster construction
- Reduced foundation load
- Economy
- Material availability
- Simpler connections
- Adaptability to modifications and remodeling
- Need for only simple tools

The method of building construction using lumber or wood products is called *wood frame construction.* The type of construction shown in Fig. 5.27 is referred to as *platform framing construction,* in which the vertical members do not continue through a floor but stop at the ceiling level at every story.

Typical wood construction consists of a wood frame composed of horizontal (or inclined) and vertical elements that form the structural components of the building (Fig. 5.28). These elements are required to transfer loads from the superstructure to the foundation, which is generally a concrete footing. The vertical components of the frame are called *studs,* and the members of the horizontal or inclined element are called *rafters* or *joists.*

A stud is one of a series of slender (wood) structural members used as supporting elements in load-bearing walls or partition walls. Nominal 2 × 4 or 2 × 6 members are commonly used as studs. They bear on a continuous horizontal element called a *sill plate* or *bottom plate*, which is bolted down to the footing. The sill plate is of nominal 2 × 4 or 2 × 6 treated lumber. At the top, the studs are connected by a continuous wood member(s) called a *top plate* or *double plate.* The top plate serves to distribute the floor and roof load to all the studs. In addition, it ties the individual wall elements together, and also works as a firestop. The studs, bottom plate, and top plate make up the load-bearing wall element in a wood frame construction.

The size of lumber used as studs depends on the number of stories and the type of wall. For load-bearing walls of one- to two-story buildings, nominal 2 × 4 in. lumber is adequate. For taller buildings and in plumbing walls (when plumbing fixtures are built into the thickness of the wall), nominal 2 × 6 or 2 × 8 may be required. In partition walls, nominal 2 × 4 lumber is sufficient. Where concentrated loads are expected, posts (size equal to the thickness of the wall) are introduced in the wall.

The spacing of studs is either 16 or 24 in. (0.4 or 0.6 m) on centers; the latter is more common. The connection to the bottom plate and the top plate is accomplished with end nailing or toe nailing. Intermediate blocking may be required to support the panel nailing and to act as a firestop. Lateral support to the frame is provided by either diagonal bracing or paneling.

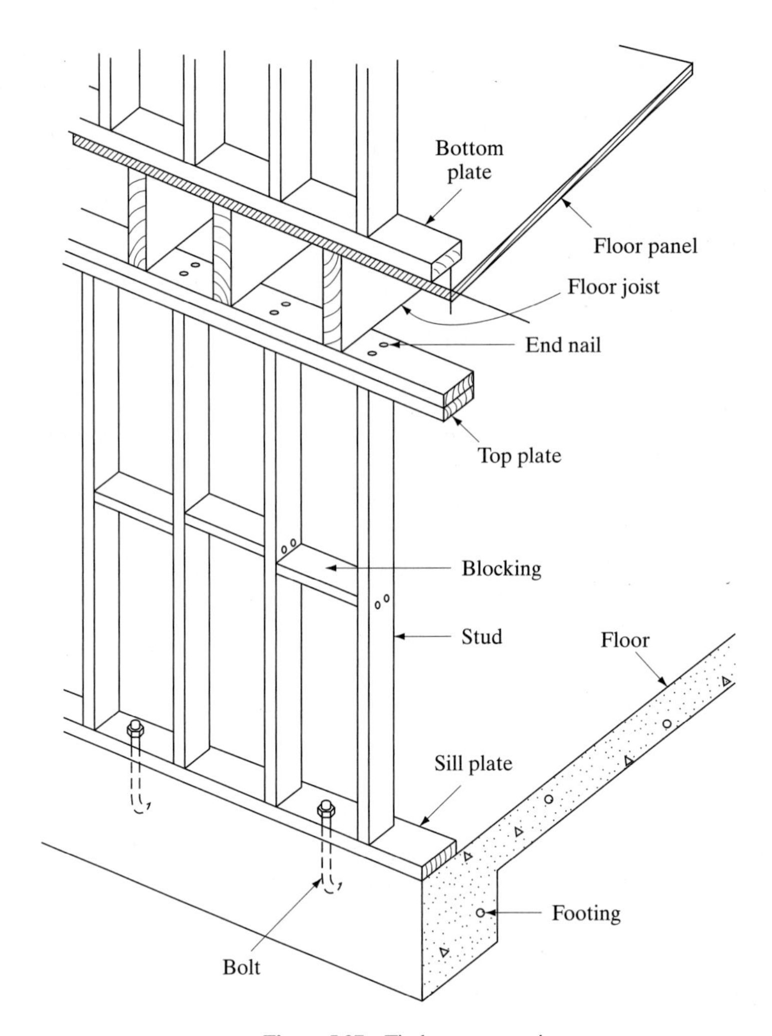

Figure 5.27 Timber construction.

The floor and roof loads are carried through the floor or roof panel to the supporting joists. A *joist* is one of a series of beams, parallel to each other, that support floor and ceiling loads. It rests on the top plate, which distributes the load to the studs. When the joists cannot be supported directly by the top plate, they are connected to the supporting element using joist hangers.

The joists are spaced, generally on 24-in. (0.6 m) centers, and are of dimension lumber (nominal thickness of 2 in.). Lateral support to the joists is from the floor panel (plywood or particleboard) or by blocking. Electrical ducts and plumbing fixtures are made to run through the depth of the joists by drilling holes through them. Drywall (or gypsum board) nailed to the bottom of the joists hides these fixtures and provides a flat ceiling.

Figure 5.28 Wood construction.

The roof joists (or rafters) are connected to the top plate in a manner similar to the floor joists. A *rafter* is one of a series of structural members of a roof that supports roof loads. The load distribution is accomplished by nailing a roof panel onto the top of the rafters. The horizontal thrust created by the inclination of the roof can be supported by ceiling joists or some other means.

Openings in the walls require slight modification to the regular stud wall construction. At both ends of an opening, double studs are generally provided. To carry the load from the superstructure over the opening, a beam is required, called a *header.* To save on labor costs, sometimes a deeper beam, of depth equal to the distance between the top of the opening and the bottom of the top plate, is used. The width of the beam is equal to the thickness of the wall. For example, a nominal 4-in.-thick wall may use a header that is (nominal) 4 in. wide, and a nominal 6-in.-thick wall requires a 6-in.-wide header. Connection between individual members is accomplished using galvanized nails. A *stringer* is a cross beam supporting load from the floor or ceiling joists. (The supporting beam on which a stair tread rests is also called a stringer.) A stringer is generally of dimension lumber of 4 in. nominal thickness.

In addition to the primary elements shown in Fig. 5.27, wooden buildings are provided with many other load-carrying elements. The conventional lumber floor

Figure 5.29 Trussed rafters.

joists can be replaced by manufactured components such as parallel chord trusses, I beams, and others. The roof can be built using roof trusses called *trussed rafters* (Fig. 5.29). The beams and headers can be of glulam or laminated veneer lumber.

5.13 TESTING

A number of laboratory tests are described in this section. Following is a list of wood properties and the corresponding test numbers.

Property	Test no.
Moisture content of wood	WOOD-1
Specific gravity of wood	WOOD-2
Compressive strength of wood	WOOD-3
Modulus of rupture of wood	WOOD-4
Modulus of rupture of particleboard	WOOD-5

Test WOOD-1: Moisture Content of Wood

PURPOSE: To determine the moisture content of green wood.

RELATED STANDARDS: ASTM D143, D4442.

DEFINITION:

- *Moisture content* is the weight of water in wood, expressed as a percentage of its oven-dry weight.

EQUIPMENT: Oven, balance.

SAMPLE: Wood in green condition cut to a size of 2 × 2 × 6 in. (50 × 50 × 150 mm).

PROCEDURE:

1. Weigh the sample in green condition: A (g).
2. Accurately measure the dimensions of the specimen.
3. Dry the sample to constant weight in the oven at 103 ± 2 °C (for approximately 24 h).
4. Weigh the oven-dry sample: B (g).
5. Calculate the moisture content as

$$\text{moisture content} = \frac{A - B}{B} \times 100$$

REPORT: Indicate the specimen species and specimen size. Round the calculated moisture content and report.

Test WOOD-2: Specific Gravity of Wood

PURPOSE: To determine the specific gravity of wood in green condition.

RELATED STANDARDS: ASTM D2395, D143.

DEFINITION:

- *Specific gravity* is the ratio between the weight of wood and the weight of an equal volume of water.

EQUIPMENT: Graduated cylinder, pointer rod, balance.

SAMPLE: Wood sample of size 2 × 2 × 6 in. (50 × 50 × 150 mm) in green condition.

PROCEDURE:

1. Get the initial weight of the specimen in green condition, W (lb).
2. Measure length, l, width, w, and thickness, t (in.).
3. Fill the graduated cylinder (about 12 in. tall) about two-thirds full with water.
4. Read the water level.
5. Immerse the specimen in the graduated cylinder, and hold it submerged with the slender pointer rod. Read the water level again.
6. The difference in water level is equal to the volume of the specimen, V (in.^{3}). *Note:* Alternatively, the volume, V, can be obtained using the actual dimensions of the specimen.

$$V = l \times w \times t$$

7. Determine the specific gravity (SG) in green condition as

$$\text{SG} = \frac{27.68(W)}{V(1 + 0.01M)}$$

where M is the moisture content in percent and V is the volume in green condition.

8. Specific gravity in oven-dry condition (and volume at test) can be calculated as

$$\text{SG} = \frac{27.68(W)}{V(1 + 0.01M)}$$

where V is the volume in oven-dry condition.

REPORT: Indicate the species. Report the specific gravity values in green and oven-dry conditions.

Test WOOD-3: Measurement of Compressive Strength of Wood Parallel to Grain

PURPOSE: To determine the parallel-to-grain compressive strength of a clear wood specimen.

RELATED STANDARD: ASTM D143.

EQUIPMENT: Universal testing machine, compressometer.

SAMPLE: Clear wood specimen of size 2 × 2 × 8 in. (50 × 50 × 760 mm).

PROCEDURE:

1. The ends of the specimen should be level.
2. Position the specimen under the crosshead of the testing machine, and apply load continuously at 0.024 in./min.
3. Measure the change in length using the compressometer [over a central gage length of 6 in. (150 mm)]. Deformation will be read to 0.0001 in. (0.002 mm).
4. Continue the loading until the proportional limit is well passed.
5. Get the deformation readings to the nearest 0.0001 in. (0.002 mm).

REPORT: Indicate the species, the moisture content, and the specific gravity. Report the clear wood compressive strength.

Test WOOD-4: Modulus of Rupture of Lumber

PURPOSE: To determine the bending strength or modulus of rupture of a clear wood sample.

RELATED STANDARD: ASTM D143.

DEFINITION:

- *Modulus of rupture* is the extreme fiber stress in bending.

EQUIPMENT: Universal testing machine, hard maple bearing block (to apply the load), dial gage.

SAMPLE: Clear wood specimen of size 2 × 2 × 30 in. (50 × 50 × 760 mm).

PROCEDURE:

1. Measure the cross-sectional dimensions accurately.
2. Position the specimen on supports with a span length of 28 in. (710 mm).
3. Apply load at the center of the specimen through the loading block at a rate of 0.1 in./min (2.5 mm/min).
4. Deflection readings should be taken to 0.001 in. (0.02 mm).
5. Measure the load and deflection at intervals.
6. Continue loading until failure.
7. Note the maximum load and type of failure.
8. Plot the load–deflection diagram.
9. Measure the slope of the load–deflection diagram: P/Δ.
10. Calculate the modulus of elasticity, E, as

$$E = \frac{(P/\Delta)L^3}{48I}$$

where L is the span length and I is the moment of inertia. Note that E has psi units when dimensions are measured in inches and P is in pounds.

11. Draw the moment diagram for the failure load.
12. Calculate the MOR value as

$$\text{MOR} = \frac{3PL}{2bd^2}$$

where P is the failure load, b is the beam width, and d the beam depth.

REPORT: Indicate the species, the specific gravity and the moisture content. Show the size of the specimen. Report MOR and E values.

Test WOOD-5: MOR of Particleboard

PURPOSE: To determine the bending capacity and modulus of rupture of a particleboard specimen.

RELATED STANDARD: ASTM D1037.

EQUIPMENT: Universal testing machine, bearing block, dial gage.

SAMPLE: 1/2-in.-thick particleboard of width 3 in. and length 14 in.

PROCEDURE:

1. The specimen is subjected to center loading on a span of 12 in.
2. Support the specimen on bearing blocks 3 in. in width and $1\frac{1}{2}$ in. in thickness.
3. Apply the load continuously through the loading block (Test Wood-4) at a rate of approximately 1/4 in./min.
4. Measure the deflection of the center of the specimen using a dial gage located at the bottom of the specimen.
5. Take deflection readings to the nearest 0.005 in. (0.10 mm).
6. Note the character of failure. Also note whether the failure was in compression or tension.
7. Calculate the MOR value as

$$\text{MOR} = \frac{18P}{bd^2}$$

where P is the failure load, b the width, and d the thickness. For $b = 3$ in. and $d = 1/2$ in.,

$$\text{MOR} = 24(P) \text{ psi}$$

8. Plot the load–deflection diagram.
9. Calculate the stress at the proportional limit and the modulus of elasticity.

$$E = 1152(P/\Delta) \text{ psi}$$

where P/Δ is the slope of the load–deflection diagram.

REPORT: Report the values of stress at the proportional limit, modulus of rupture, modulus of elasticity, and type of failure.

PROBLEMS

1. Explain the difference in the grade stamps of visually graded and machine-rated lumber.
2. The cross-sectional dimensions of a 4-ft-long wood member at 30 percent moisture content is 1.5×7.5 in. and its weight is 11 lb. At oven-dry condition, its dimensions and weight are 1.35×6.8 in. and 10.2 lb, respectively. What is its specific gravity at 30 percent moisture content?
3. What is the most common use classification of the lumber used in construction?
4. A piece of lumber dries in service from 50 percent moisture content to 10 percent. What is the expected shrinkage?
5. Define fiber saturation point and equilibrium moisture content.
6. Of the two strength values of clear wood—tensile strength parallel to grain, and compressive strength parallel to grain—which is larger?
7. What is the allowable tension perpendicular to grain?
8. Explain the effects of moisture content on the strength characteristics of wood.
9. Explain the difference between nominal size and dressed size.
10. What is clear wood?
11. What are the common defects in lumber?
12. Find the section modulus of 2×6, 4×8, and 2×16.
13. What is the general value of the fiber saturation point?
14. What is the average MOR value of Douglas Fir?
15. What are the three common types of use classification?
16. Name any three wood products.
17. Explain the decay and destruction of wood.
18. Define growth ring, check, grain, and slope of grain.
19. What are the use classifications of 2×8, 4×6, 4×12, 6×8, and 6×12?
20. Explain seasoning and treatment of lumber.

6

Bituminous Materials and Mixtures

Bitumen is a solid, semisolid, or viscous cementitious material, natural or manufactured, and composed principally of various mixtures of complex hydrocarbons. The term *hydrocarbon* refers to any of a class of compounds containing only hydrogen and carbon atoms. Large hydrocarbon molecules such as those in petroleum are formed by the bonding of many hydrogen and carbon atoms. The properties of hydrocarbons depend on the number as well as the arrangement of hydrogen and carbon atoms in the molecule. All bituminous materials are completely soluble in carbon disulfide (CS_2) and are nonvolatile, nontoxic, and soften when heated. The most common bituminous materials are asphalts, tars, and pitches (Fig. 6.1).

Asphalts are cementitious materials in which the predominant constituent materials are bitumens. They are available as natural deposits or are produced from petroleum processing. The consistency of asphalts varies widely from solid to semisolid at normal temperature. *Tars,* along with pitches, are the products of distillation of materials such as wood, coal, and shale. When these materials undergo destructive distillation (subjecting the raw material to heat alone, without access to air), the resulting condensate is tar. The most widely used tars are byproducts of the distillation or carbonization of coal. Further processing yields a solid or semisolid residue known as *pitch*. Coal-tar pitch is defined as a bituminous material produced by the partial distillation of coal. Pitches are black or dark brown in color and liquefy gradually when heated.

Bitumens possess a number of properties that make them useful in the construction industry. Most important are their ability to adhere to solid particles such as aggregate and brick, and their impermeability or water resistance. These properties make a bituminous material ideal for use in pavements, as an impervious layer

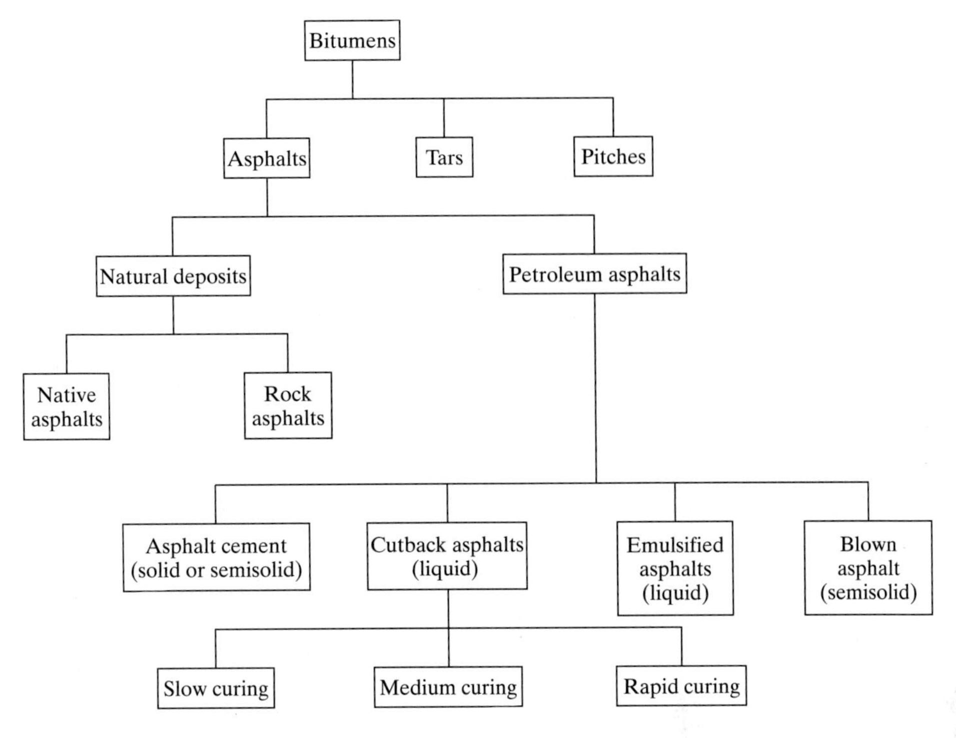

Figure 6.1 Bituminous materials.

in built-up roofing, and in the manufacture of roof shingles. Bituminous materials are normally utilized in roadways and many construction products, such as roofing felt, caulking compound, waterproofing compounds, wallboard, building paper, paints, and shingles. Bituminous compounds are also used widely to provide waterproof coatings for walls and below-grade structures.

6.1 TARS AND PITCHES

When coal is heated in coke ovens, it is reduced to coke. The gases (vaporized oils) generated during the process are collected from the oven and refined. The distillation of lighter oils produces tar, leaving a hard semisolid residue, which is pitch. As noted, tars and pitches can be produced from organic materials other than coal, but coal tar is the most common.

Tar produced by the destructive distillation of bituminous coal, called bituminous coal tar, is the most commonly used tar in pavement construction. The viscous liquid produced at the end of the distillation process, called *crude tar,* contains a large amount of water as well as some fuel oils. For it to be used in pavement construction, it should be refined. The coal refinery uses a distillation process that is similar to that of bitumen, in which the light and heavy oils are separated. The residual matter or pitch (also called base tar) is fluxed back with tar oils to yield *road tar.*

Tars are highly resistant to natural weathering action and are not as susceptible as bitumen to the dissolving action of petroleum distillates. They tend to oxidize (becoming brittle) when exposed to heat and air, somewhat more quickly than do asphalts. Overheating spoils tars more easily. Tars are more susceptible than asphalts to the changes in consistency from temperature variations. Tars have been used in the past in pavements and as a waterproofing layer in roof construction, but they have largely been replaced by petroleum asphalts.

Pitches are obtained as residual in the partial evaporation or fractional distillation of tar—primarily coal tar. They gradually liquefy when heated. Tar pitches (and other asphaltic materials) are used for built-up roofing and waterproofing of underground and aboveground construction, such as foundations, retaining walls, dams, and bridges. Aboveground structures experience great fluctuations in environmental conditions, and waterproofing compounds should be able to withstand these changes without becoming brittle or too soft.

6.2 ASPHALTS

Asphalts are the most common and most widely used bituminous materials. They are found as natural deposits or are produced from petroleum crude. The term "asphalt" refers to a black cementitious material which varies widely in consistency from solid to semisolid at normal temperatures. With heat, asphalt softens and becomes a liquid. Nearly all the asphalts manufactured in the United States are derived from crude petroleum, and are called petroleum asphalts. The amount of asphalt a crude oil contains varies from source to source. Higher-grade crudes may contain as little as 10 percent asphalt (the balance being lighter products such as gasoline, kerosene, and fuel oil), whereas lower-grade crudes have as much as 90 percent asphalt.

The natural deposits of asphalts are of two types:

- Natural rock asphalts
- Native asphalts or lake asphalts

Natural rock asphalts or *asphalt rocks* are found in many parts of the world as deposits of sandstone or limestone filled with asphalts through a geologic process. Some soft varieties of limestone contain from 6–14 percent bitumen, and a few crystalline limestones contain 2–20 percent bitumen. These types of rocks are found in some regions of Switzerland, France, and Italy (Sicily). They are also found in California, Texas, and a few other states. Remarkably straight asphalt veins several miles long are found in Asphalt Ridge, Vinta Basin, near Vernal, Utah. Asphalt deposits are also seen in Edna, about 8 miles southeast of San Luis Obispo, California, and in the Sisquoc region of Santa Barbara County, California.

Rock asphalts can be used for surfacing roads when combined with other materials. The use of asphalt rocks is very minimal in the United States due to high costs.

Bitumens also occur naturally as *native asphalts* found in lakes. A number of deposits of asphaltic materials, of wide-ranging properties, are found in Lake Trinidad, West Indies (a 110-acre lake containing about 40 percent bitumen mixed with water, fine siliceous silt, clay, and organic matter); Bermudez, Venezuela (lake asphalt containing about 65 percent bitumen); Cuba; Texas; Los Angeles (La Brea asphalt pits); and other locations.

In a few places in the Middle East (Iran to Pakistan), asphalt seeps from the earth and becomes absorbed in sand and dust. These seepages also occur at the bottom of the Dead Sea (on the seabed), and float on the surface of the water. Natural bituminous soils were collected and used to build walls as far back as the middle of the fourth millennium B.C. (in Babylonia). A combination layer of natural asphalt and clay was used to lay the bricks on regular courses. The bituminous earth was also found suitable for making baked bricks and used in pavements. Excavations of settlements in the Indus Valley dating to about 3000 B.C. have revealed the application of asphalt as a waterproof material.

Powdered rock asphalt was used with bitumen in the eighteenth century to build floors and pavements in France. Later, in the nineteenth century, the same technique was used in Switzerland, Germany, and England. A British patent was issued in 1837 for "a mastic cement applicable for paving, road making, covering buildings and other purposes." The first roadway in England was resurfaced with rock asphalt (from France) in 1869. Following the discovery of petroleum asphalt and with the growth of automotive transportation, use of asphalt paving material expanded rapidly. An asphalt pavement is built using two materials: asphalt and aggregate. There are many types of asphalt and many varieties of aggregates. Thus it is possible to build different kinds of asphalt pavements—pavements that are suitable for different applications.

6.3 PETROLEUM ASPHALTS

Refining and distillation of petroleum results in various types of asphalt or asphaltic material. Distillation is a process in which various fractions (products) are separated out of the crude by raising its temperature in stages. Gasoline is distilled at a temperature of 100–400 °F (37–204 °C), whereas asphalts are distilled at temperatures above 900 °F (482 °C).

Petroleum refining processes are divided into two groups:

- Fractional distillation
- Destructive distillation (cracking)

Fractional distillation involves the separation of crude oil into various materials without significant changes in the chemical composition of each material. The various materials are removed at successively higher temperatures (using steam or vacuum) until petroleum asphalt is obtained as the residue. Further processing of the residue yields asphalts of different grades and types.

Destructive distillation involves the application of intense heat [temperatures as high as 1100 °F (593 °C)] and high pressure [as high as 735 psi (5 MPa)], which causes chemical changes in the material. This method is used when larger amounts of lighter fractions of the materials (such as motor fuels) are required. Asphaltic material obtained from the destructive distillation process is not widely used in pavement construction, as it is more susceptible to weather changes than is asphalt produced from fractional distillation. Most asphalt used in pavements is steam-refined.

Petroleum asphalts are classified into four types (Fig. 6.2):

- Asphalt cements
- Cutback asphalts
- Emulsified asphalts
- Air-blown asphalts

Cutback asphalts are further divided into three types:

- Slow-curing
- Medium-curing
- Rapid-curing

Petroleum asphalts are used extensively in the construction of flexible pavements. They are also widely used in roof construction (asphalt shingles, rolled roofing, and built-up roofing), for insulation (electrical insulation and building paper), and for waterproofing (sprays, paints, and bituminous fabrics). Furthermore, they are utilized for canal lining and to seal subgrades against the migration of water. Joint seals generally use petroleum asphalts (with or without mineral fillers).

6.3.1 Asphalt Cement

Asphalt cement (also called *paving asphalt*) is very sticky and highly viscous, especially prepared with the quality and consistency required in the manufacture of asphalt pavements (hot-mix pavements). It is obtained after separation of lubricating oils from crude petroleum.

Asphalt cement possesses excellent binding properties and adheres very well to aggregate particles. In addition, it has superior waterproof qualities. Pavement built using paving asphalt is waterproof and resistant to many types of chemical attack.

Asphalt cement is a highly viscous material; consequently, when used in pavement construction it is necessary to heat both the asphalt cement and the aggregates prior to mixing. Asphalt cements have penetration readings ranging between 5 and 300. For use in pavement construction, the asphalt cement must have a minimum penetration reading of 40 and a maximum of 300 (at a load of 100 g in 5 seconds at 77 °F).

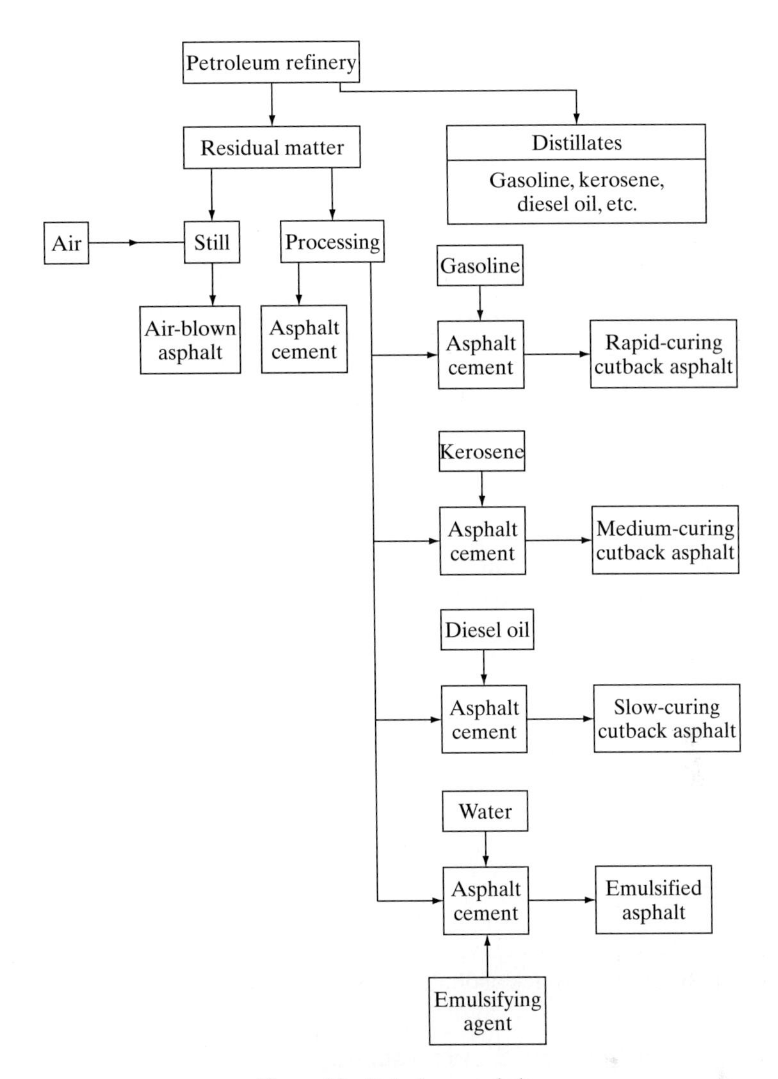

Figure 6.2 Petroleum asphalts.

6.3.2 Cutback Asphalts

Two types of asphalt are available in liquid form: one is cutback asphalt and another is emulsified asphalt. *Cutback asphalt* is asphalt cement that is liquefied by blending with petroleum solvents (called *diluents*). The diluents evaporate upon exposure to the atmosphere, leaving asphalt cement behind. This type of asphalt is suitable for direct application, requiring little or no heating before use.

Cutback asphalts are constituted from a base asphalt of selected hardness or viscosity by dissolving in a solvent of high, medium, or low volatility. The resulting asphalts have distinctive properties that make them useful in different applications.

As noted, when the solvent evaporates, asphalts of different hardness or viscosity are left behind on the aggregates. The curing rate and the characteristics of the residual asphalt depend on the type of cutback asphalt: rapid-curing, medium-curing, or slow-curing, referring to the volatility of the diluents.

Rapid-curing cutback asphalt is asphalt cement blended with naphtha or gasoline-type diluents of high volatility. The diluents in rapid-curing asphalt evaporate quickly. It is used primarily in mixed-in-place pavement mixes and surface treatments. Medium-curing cutback asphalt is obtained by blending asphalt cement with kerosene-type diluents of medium volatility. It hardens faster than the slow-curing and slower than the rapid-curing type; and is used for cold-laid pavement bases (plant mixed and mixed in place) and surface treatments. Slow-curing cutback asphalt is asphalt cement blended with oils of low volatility, such as diesel oil. Its viscosity value is low and it hardens very slowly. It is used for cold-laid pavement bases.

The curing rate and characteristics of the residual asphalt (after the evaporation of the solvent) depend on the type of cutback asphalt. These two factors should be considered in selecting one type of cutback asphalt over the others. The type of aggregate used may also influence the selection. For example, an aggregate deficient in finer particles may require harder asphalt, while a softer asphalt may be used with *dense-graded* aggregate (in which the voids between compacted aggregate particles are relatively small).

Cutback asphalts are also used in highway construction for priming the road surface prior to surface treatment, for seal coating, and in the base course for cold application.

One of the disadvantages of cutback asphalts is that the volatile distillate may evaporate in the air and add to air pollution. The amount of emission depends on the type, grade, and amount of cutback asphalt, the type of construction, and the atmospheric conditions. The rapid-curing type produces the largest amount of air pollution, and the slow-curing type produces the lowest. Another disadvantage is that the distillate in the cutback asphalt poses fire and toxicity hazards.

6.3.3 Emulsified and Blown Asphalt

A relatively stable suspension of one liquid, in a minute form, dispersed throughout another liquid, in which it is nonsoluble, is called an emulsion. Paints, ice cream, and mayonnaise are common examples of emulsion. *Emulsified asphalt* is composed of asphalt cement and water that contains a small amount of an emulsifying agent. Soaps, water-soluble chemicals, and fine clay are used as emulsifying agents. The water forms the continuous phase of the emulsion, and minute globules of asphalt form the discontinuous phase. The water content is generally on the order of 12–15 percent. The particles of asphalt remain in suspension as long as the water does not evaporate or the emulsifier does not break down.

Emulsions have been available since 1903, and used extensively since the 1920s. Prior to this, most flexible pavements had been solely hot plant mixtures using asphalt cement. Some emulsions that are presently available have better adhesion to aggregates that are high in silica and quartz, whereas others have better bond on limestone aggregates.

A total of 12 grades of emulsified asphalt are covered in ASTM D977. These can be used for spray application, as slurry seal, and in subbase, base, and surface courses of pavements. The advantage of emulsified asphalt is that it can be applied during damp weather and with aggregates that are cold or hot. In addition, the quantity of materials applied is controlled automatically because any excess (water) will drain into the subgrade. The use of emulsified asphalt also eliminates the use of fuel required to heat and dry the aggregates in the asphalt cement mixtures, and reduces energy requirements through reduction or elimination of petroleum distillates used in cutback asphalts.

One of the disadvantages of emulsified asphalt is that because the asphalt is suspended in water, it is susceptible to being washed off a road surface by rainwater if not sufficiently cured. The unbroken emulsions may be carried into streams, causing ground contamination.

Blown asphalt is obtained by blowing air through the semisolid residue obtained during the later stages of the distillation process. It is relatively firm in consistency compared to other asphalts and is the stiffest asphalt made. Blown asphalts are reasonably ductile. They are highly weather resistant and can withstand temperature changes very well. But extensive blowing reduces the ductility, and hence fully blown asphalts are not used in pavement construction. Blown asphalts possess unique properties that make them suitable for use in the manufacture of roofing materials, waterproof paints, and joint fillers (for concrete pavements), and for undersealing old, rigid pavements. They are also used in the production of mineral rubber, a rubberlike material derived from petroleum.

6.4 PROPERTIES OF ASPHALT

Properties of asphalt or asphaltic material that are of great importance in pavement design and construction are:

- Adhesion
- Consistency
- Specific gravity
- Durability
- Rate of curing
- Ductility
- Aging and hardening
- Resistance to reaction with water
- Temperature susceptibility

Some of these properties are explained in the following sections.

6.4.1 Consistency

Asphalts can exist in liquid, semisolid, or solid form. Consistency measures the degree of fluidity or plasticity of asphalt cement at any particular temperature. The consistency of an asphaltic material may vary from solid to liquid, depending on the temperature of the material, and because of this, there is no single technique or instrument that will effectively measure the consistency of all types of asphalts. At any particular temperature, the consistency indicates the grade of the material.

Asphaltic materials having the same consistency at a standard temperature may show different consistency readings at elevated temperatures. The effect of temperature on consistency is not the same for all asphaltic materials. For example, consistency of blown asphalt is less affected by temperature than that of asphalt cement.

Viscosity. At temperatures between 160–300 °F (70–150 °C), asphalt behaves as a viscous liquid. Below 160 °F (70 °C), the behavior can be described as linearly viscoelastic. Below about −4 °F (−20 °C), asphalt becomes a weak and brittle elastic solid.

In a liquid state, the consistency of asphalt is established by measuring the viscosity of the material. Viscosity is a measure of the resistance to flow and is the fundamental consistency measurement in absolute units. It is expressed as the ratio of shear stress to shear rate; at temperatures above 300 °F (150 °C), the ratio is constant for asphalt cement. When the temperature decreases, as noted, the asphalt cement behaves like a viscoelastic semisolid material, and the ratio of shear stress to shear rate fluctuates.

The deformation of liquids and semisolids under load is dependent on the load as well as the duration of load: the longer the duration, the greater the deformation. Differences in deformation between liquids are due to the differences in their viscosities. In materials with high viscosity, deformation is low. For example, diesel oil is nearly 10 times as viscous as water and has lower deformation. Asphalts are more than 10,000 times as viscous as diesel oil. Cutback asphalts, due to the solvents in them, have lower viscosity than that of the original asphalt. Since the range of viscosities of bituminous materials is large, no single instrument or method can be suitable for measuring all. Viscosity of asphalt cements is measured using a viscometer.

Absolute viscosity is a measurement method in which *poise* is the basic measuring unit. It makes use of partial vacuum to induce flow in the viscometer. *Kinematic viscosity* is a method of measuring viscosity using *stoke* as the basic measuring unit, which is a measure of the flow behavior or resistance to flow under gravity.

Kinematic viscosity is defined as the ratio between the viscosity and the density of the liquid, expressed in mm^2/s or m^2/s. Stoke is kinematic viscosity in cm^2/s; the centistoke (cSt), which is 1 mm^2/s, is customarily used.

Poise is a centimeter–gram–second (cgs) unit of absolute viscosity, equal to the viscosity of a fluid in which a stress of 1 dyne per square centimeter is required to maintain a difference in velocity of 1 centimeter per second between two parallel planes in the fluid that lie in the direction of flow and are separated by a distance of 1 centimeter.

Absolute viscosity, which is also called *coefficient of viscosity,* is often used to measure the resistance to flow of a liquid. It measures the ratio between the applied shear stress and the rate of shear. The cgs unit of absolute viscosity is 1 g/(cm · s), which is the poise (P). The SI unit of viscosity is 1 Pa · s which is equivalent to 10 P. The viscosity of water is about 1 centipoise (1×10^6 Pa · s), and that of light motor oil is in 200–500 centipoise ($200\text{–}500 \times 10^6$ Pa · s). The viscosity of molasses is over 100,000 centipoise ($100{,}000 \times 10^6$ Pa · s). Stoke is a unit of kinematic viscosity equal to the viscosity of a fluid in poise divided by the density of the fluid in grams per cubic centimeter. The density is mass per unit volume of a liquid and the cgs unit of density is 1 g/cm^3.

As noted, viscosity of asphalts is measured using viscometers (ASTM D2170). A number of reverse-flow viscometers are available for use, such as the Cannon-Fenske, Zeitfuchs cross-arm, and Lantz-Zeitfuchs viscometers. In all of these, the liquid flows into a timing bulb not previously wetted by the sample, allowing the measurement of the time. The kinematic viscosity is then calculated by multiplying the time (for outward flow) in seconds by the viscometer calibration factor.

The coefficient of viscosity or absolute viscosity of asphalts is measured using vacuum capillary viscometers (ASTM D2171). A number of capillary tube viscometers are available for use, such as the Cannon-Manning, Asphalt Institute, and modified Kopper vacuum capillary viscometers.

A capillary tube viscometer consists of a calibrated glass tube that measures the flow of asphalt. It is mounted in a temperature-controlled water bath that is preheated to 140 °F (60 °C). A sample of asphalt cement heated to the same temperature is poured into it. The preheated cement is too viscous to flow readily through the narrow opening in the tube, and to aid the flow, a partial vacuum is applied to the small end of the tube. The liquid asphalt, with the aid of the vacuum, is drawn through the tube. The flow rate is determined by carefully timing the flow from one mark to the other. The measured time is converted to poise by multiplying the flow time by the viscometer calibration factor.

Penetration. The viscosity or consistency of solid and semisolid bituminous materials can also be measured empirically by the *penetration test* (ASTM D5). For many years, the only method that was commonly used to record consistency of asphaltic materials was the penetration test (invented by H. C. Bowen in 1889).

Penetration is the consistency of a bituminous material expressed as the distance in tenths of a millimeter that a standard needle penetrates a sample of material vertically under standard conditions of loading, time, and temperature (Fig. 6.3). A higher value of penetration indicates a softer consistency. The load is the total moving weight of the needle plus the attachments (100 g and 200 g, as required for the condition of the test). The needle is left to penetrate the sample for 5 seconds. The penetration is given as the distance (in units of 0.1 mm) that the needle penetrates. For example, if the needle penetrates a distance of 2.2 cm, the penetration is 220.

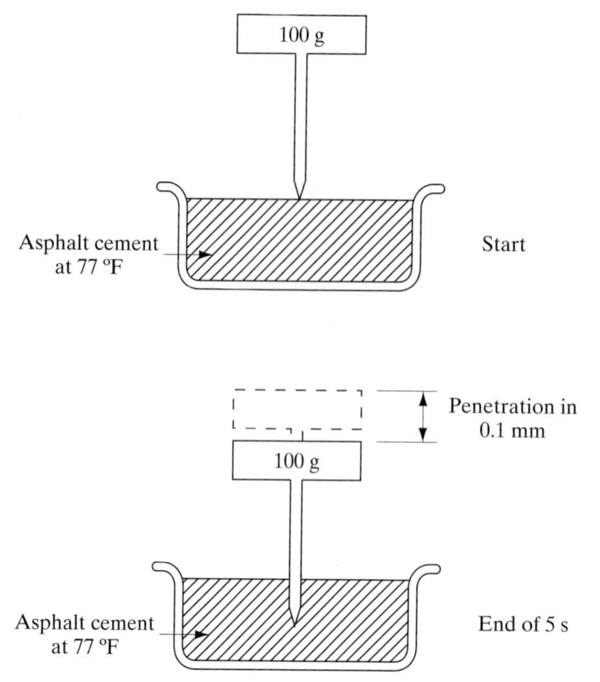

Figure 6.3 Penetration test.

The penetration test provides an indirect measurement of consistency, and the more fundamental tests previously described are better measures. Asphalt cements have an upper limit on penetration of 300 (under a load of 100 g for 5 seconds at 77 °F); samples with a penetration reading above 300 should be considered a liquid asphalt (such as cutback asphalt). Samples with penetration readings below 10 are solid asphalts. Penetration tests cannot be used to determine the viscosity of that material, and are also unsuitable for tars due to the high surface tension and large amounts of carbon in these materials.

The addition of natural rubber in very small quantities (less than 5 percent) to asphalts increases the viscosity and decreases the penetration. In addition, rubber renders the asphalt less susceptible to temperature fluctuations. Due to these changes in properties, asphalt pavement with added rubber becomes more elastic and durable.

6.4.2 Specific Gravity

Specific gravity and density of semisolid bituminous materials can be determined using the procedure described in ASTM D70. Specific gravity is normally measured using a pycnometer. Since the volume of a bituminous material changes with temperature (with the change from semisolid to liquid state), the specific gravity should be expressed at a given temperature.

The specific gravity of asphalt is defined as the ratio of the mass of the given volume of the material at 77 °F (25 °C) or at 60 °F (15.6 °C) to that of an equal volume of water at the same temperature. Petroleum asphalts have specific gravity values close to unity (0.95–1.05), and for road tars it varies between 1.08–1.24. Specific gravity decreases with increasing temperature. For example, asphalt cement has a specific gravity of 1.0 at a temperature of 60 °F (15 °C), 0.9187 at 300 °F (149 °C), and 1.0176 at 10 °F (−12 °C).

A knowledge of specific gravity is essential to determine the percentage of voids in a compacted material; its measurement is required to convert the volume measurement of asphalt to the units of mass.

6.4.3 Durability

Durability can be defined as the property that permits a pavement material to withstand the detrimental effects of moisture, air, and temperature. The performance and durability of an asphaltic pavement are affected by a number of factors, including mix design, properties of aggregates, and workmanship, as well as the properties of the asphalt. With so many factors controlling durability, its measurement is rather difficult. Nonetheless, there are laboratory tests to determine the durability of asphalt: the thin-film oven test (ASTM D1754) and rolling thin-film oven test (ASTM D2872) determine the effects of heat and air on asphaltic materials. These tests offer a measure of the changes that take place in asphalt over time.

Oxidation is the chemical reaction that takes place when asphalt is exposed to the oxygen in air. When asphalt oxidizes, the hydrogen in the asphalt combines with oxygen in the air and is removed as water molecules. The loss of hydrogen increases the carbon/hydrogen ratio, leading to hardness in the material and loss of ductility and adhesion. This reaction occurs most readily at higher temperatures and in thin asphalt films. Oxidation causes gradual hardening and loss of plasticity, but not all asphalts harden at the same rate. Oxidation in pavements can be minimized by keeping the amount of air voids low and coating the particles thickly.

Volatilization is the process by which lighter hydrocarbons from the asphalts are evaporated. As in oxidation, this process results in the loss of plasticity in asphalt. Volatilization is also affected by the temperature. The higher the temperature, the greater the rate of volatilization. The rates of oxidation and volatilization nearly double for each 18 °F (10 °C) rise in temperature.

When a sample of asphalt is heated and then allowed to cool, its molecules will be rearranged to form a gel-like structure, causing it to harden with time. The rate of hardening decreases with time and is insignificant beyond 1 year. This process happens even when oxidation and volatilization are prevented from occurring. This process of natural hardening of asphalt is called *age hardening.*

6.4.4 Rate of Curing

The process of evaporation of solvents from cutback asphalts, and the attendant thickening of the material, is called *curing*. It can also be described as the change in consistency of an asphaltic material due to the progressive loss of diluents by evaporation. The *rate of curing,* or the time required for a cutback asphalt to harden (from its original liquid consistency) and develop a consistency that is satisfactory for the function as a binder in pavements is an important property of cutback asphalts.

The rate of curing is influenced by the volatility or evaporation rate of the solvent, the amount of solvent, and penetration (or viscosity) of the asphalt base. In addition, the curing rate is affected by the temperature of the environment, the surface area of the pavement, and the wind velocity. The less solvent, the faster the rate of evaporation. The curing time increases with the penetration (or increase in softness) of the base asphalt.

In emulsified asphalts, the rate of evaporation of water depends on the weather conditions, such as humidity and temperature. With favorable weather conditions, the water evaporates quickly and curing progresses rapidly.

6.4.5 Resistance to Action of Water

Asphaltic materials designed for pavements should be able to withstand the effects of water. The durability of the pavement is greatly affected by the ability of asphalt to adhere to aggregate particles in the presence of water. Loss of bond in the presence of water may lead to pavement deterioration. Generally, pavements built using paving asphalts are waterproof and resistant to many types of chemical action. Complete drying of the aggregate particles prior to mixing improves pavement performance.

6.4.6 Ductility and Adhesion

Ductility is an important property of asphaltic materials. Brittle materials do not perform well in pavement construction. But the exact value of ductility (if it can be measured) is not as important as the existence of ductility characteristics. The ductility of asphalt samples can be tested and compared using the procedure described in ASTM D113.

In laboratory measurements, ductility is taken as the distance to which an asphalt sample will elongate before breaking when the ends of a briquette specimen are pulled apart at a specified speed and temperature. The test is typically conducted at a temperature of 77 °F (25 °C) and a speed of 5 cm/min. The ductility thus measured provides a measure of the tensile properties of the material. Experience has shown that asphalts with low ductility show the most pitting and cracking or disintegration.

Ductility is sometimes used as an indirect gage of adhesion and cohesion of asphaltic material. *Adhesion* is the ability to stick to aggregate particles in the pavement; *cohesion* is the ability to hold the particles firmly in place. The former refers to the molecular force that exists in the area of contact between unlike bodies (as in adhesive tape), whereas the latter stands for the molecular force that acts to unite the particles (cohesive organization). Ductile materials with sufficient bonding characteristics, such as asphalt and tar, exhibit excellent adhesive and cohesive properties.

But the requirements for these properties vary from one application to another. For example, compounds that are used as dampproofing or waterproofing materials in aboveground structures should possess greater ductility and higher tensile strength than those in underground structures exposed to moderate temperature fluctuations.

To guarantee good adhesion of asphalt in pavement construction, it is necessary that the aggregate particles be clean, dry, and free from dust. The presence of moisture may make it difficult for the asphaltic binder to displace moisture from the surface and adhere to the particles.

6.4.7 Temperature Susceptibility

All asphalts are thermoplastic; they become harder (more viscous) with decrease in temperature and softer (less viscous) with increase in temperature. The temperature susceptibility of asphalt varies from one source to another, even when the asphalts are of identical grade (Fig. 6.4): Asphalts A and B plotted in the figure have the same viscosity grade but are from different petroleum sources. At temperature T_1, both have the same viscosity, but at other temperatures the viscosities are different. At temperatures higher than T_1, asphalt B is less viscous—or more fluid—than asphalt A.

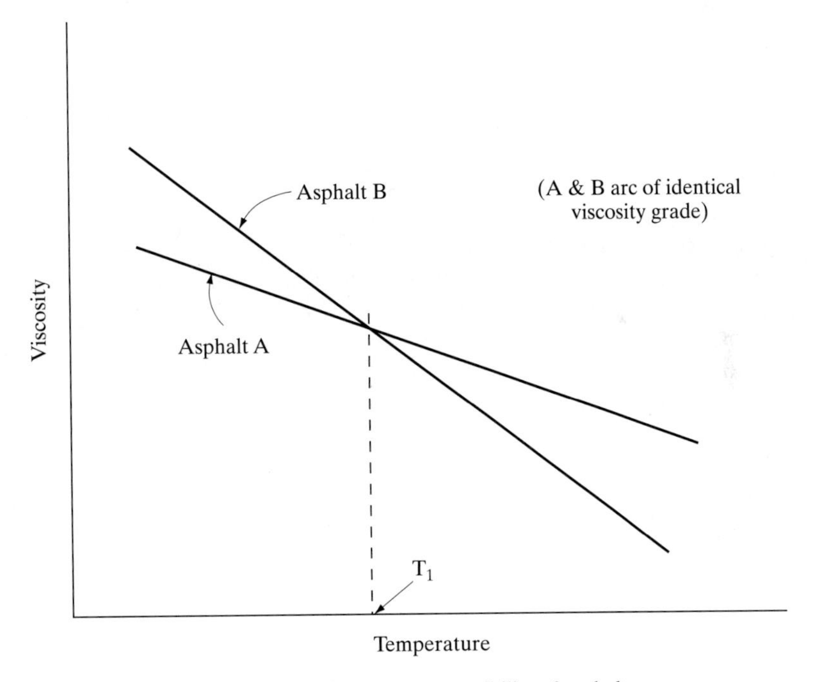

Figure 6.4 Temperature susceptibility of asphalts.

As a result, asphalt B requires a lower temperature than asphalt A to make it fluid enough to properly coat all aggregate particles. The temperature susceptibility of asphalts is the characteristic that makes them fluid enough at elevated temperatures to coat particles, and for the movement of particles during compaction. At normal temperatures, the asphalt must remain elastic to bind or hold the particles together.

As in all viscoelastic materials, the deformation behavior of asphalt depends on the rate of loading and temperature. The change in stiffness of asphalt from changes in temperature, which is also a characteristic of temperature susceptibility, is due to change in viscosity. An asphalt cement sample showing a greater rate of change of viscosity or consistency with temperature is more susceptible to temperature deformations than a sample with a lower rate. In other words, the sample with the higher rate of change in viscosity with temperature is more liable to rutting at high temperatures and to fracture at low temperatures.

At low temperatures and short-duration loading, the behavior of asphalt remains elastic, and at high temperatures and long-duration loading, the behavior is viscoelastic or viscous, as shown in Fig. 6.5. For short-duration loads, the graph remains horizontal, with stiffness approximately equal to modulus of elasticity. At intermediate loading, the behavior is viscoelastic, and the deformation is both elastic and viscous. Under long-term loading, the behavior resembles that of a purely viscous fluid (Newtonian), and the stiffness decreases approximately linearly. A bending beam rheometer is used to measure the creep and determine the stiffness of

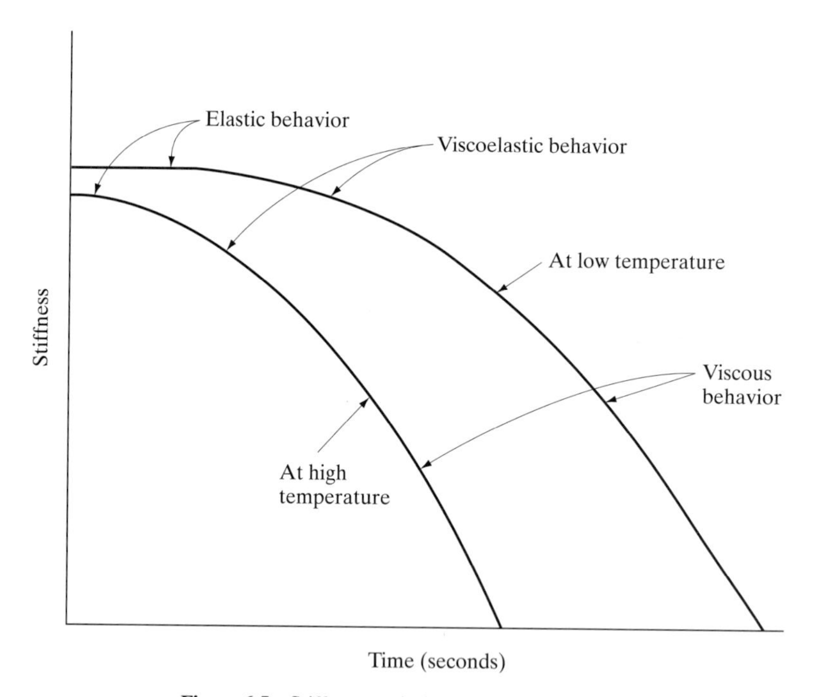

Figure 6.5 Stiffness variation of asphalt cement.

asphalt binder at low temperatures. At high temperatures, a dynamic shear rheometer measures the rheological and elastic properties of asphalt.

6.4.8 Hardening and Aging

The increase in the viscosity of asphalt caused by heating a thin film is shown in Fig. 6.6. The initial viscosity of the asphalt is significantly lower than that after the sample is heated. The rate of hardening with temperature varies among asphalts.

The properties of asphalt binder prior to use differ significantly from those in asphalt concrete pavement. Subjecting the asphalt to widely varying temperatures at different times during mixing and placing forces changes in the structure and composition of asphalt molecules. Volatilization of hydrocarbon elements during construction, and oxidation (reaction with oxygen from the environment during service) makes the asphalt binder harder and brittle. Oxidation more readily occurs at higher temperatures and in thin films. During mixing, asphalt is both at high temperatures and in thin film, as in the coating of particles, and this makes the construction stage the most severe for oxidation. The hardening of the binder during service is from

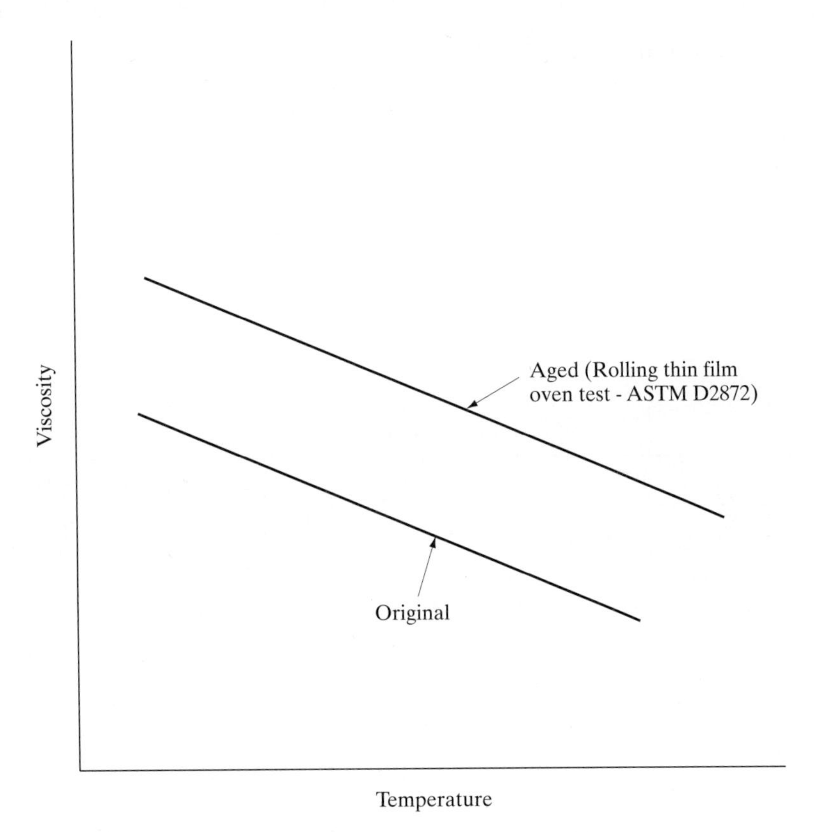

Figure 6.6 Hardening of asphalt after exposure to high temperatures.

oxidation and polymerization, and can be minimized by keeping the number of connecting pores or air voids low and coating the particles thickly.

6.5 ASPHALT GRADES

Asphalt cements are graded according to four different systems:

- Viscosity grading system
- Viscosity-after-aging grading system
- Penetration grading system
- Performance-based grading system

Traditionally, asphalt cements have been graded by measuring the penetration of a standard needle into the sample or by viscosity readings. In these systems, different grades of asphalt, having different physical property requirements, are all measured at the same temperature. However, the correlation between the properties of asphalt at the high temperatures used in these grading systems, and the in-situ performance of asphalt in service (at low ambient temperatures) is poor.

6.5.1 Viscosity and Penetration Grading

The viscosity grading system (ASTM D3381) prescribes limiting values on the viscosity of asphalt (as supplied or unaged) at two temperatures: 140 °F (60 °C) and 275 °F (135 °F). The lower temperature represents the maximum temperature the pavement is likely to experience during service, and the higher temperature approximates the mixing and laydown temperature. The five viscosity grades are AC-2.5, AC-5, AC-10, AC-20, and AC-40 (Table 6.1). AC-2.5 asphalt has a viscosity of

TABLE 6.1 REQUIREMENTS FOR ASPHALT CEMENT, VISCOSITY GRADED AT 140 °F (16 °C)

	Viscosity grade				
Test	AC-2.5	AC-5	AC-10	AC-20	AC-40
Viscosity, 140 °F (60 °C), P	250 ± 50	500 ± 100	1000 ± 200	2000 ± 400	4000 ± 800
Viscosity, 275 °F (135 °C), min, cSt	80	110	150	210	300
Penetration, 77 °F (25 °C), 100 g, 5 s, min	200	120	70	40	20
Flash point, Cleveland open cup, min, °F (°C)	325 (163)	350 (177)	425 (219)	450 (232)	450 (232)
Solubility in trichloroethylene, min, %	99.0	99.0	99.0	99.0	99.0
Tests on residue from thin-film oven test:					
Viscosity, 140 °F (60 °C), max, P	1250	2500	5000	10000	20000
Ductility, 77 °F (25 °C), 5 cm/min, min, cm	100[a]	100	50	20	10

Note: Grading based on original asphalt.

[a]If ductility is less than 100, material will be accepted if ductility at 60 °F (15.5 °C) is 100 minimum at a pull rate of 5 cm/min.

Source: ASTM D3381

250 poise at 140 °F, and is referred to as "soft" asphalt, and AC-40 asphalt has a viscosity of 4000 poise, and is known as "hard" asphalt. The higher the poise, the more viscous the asphalt.

Asphalt cements of grades AC-5 and AC-10 are used in asphalt concrete pavements in regions where the mean annual temperatures are less than 45 °F (7 °C). When the mean annual temperature exceeds 75 °F (24 °C), asphalt cements of grades AC-20 or AC-40 are recommended. In regions of mean annual temperature between 45 °F (7 °C) and 75 °F (24 °C), asphalt cements of grades AC-10 or AC-20 is appropriate.

In the viscosity-after-aging grading system, asphalt cements are graded according to their viscosity readings after laboratory aging, so as to have an estimate on the viscosity characteristics after the asphalt is placed in the pavement. To simulate the aging of asphalt before mixing in the laboratory, it is subjected to a standard aging-exposure test. The asphalt residue that remains after the test is then graded based on its viscosity measurements, as shown in Table 6.2. The designation AR stands for *aged residue*. The AR-1000 grade, which has a viscosity of 1000 poise, is referred to as "soft" asphalt, and the AR-16,000 grade, with a viscosity of 16,000 poise, is referred to as "hard" asphalt.

Asphalt cements can also be graded based on penetration measurement, which is the depth of penetration of a standard needle in five seconds, measured in tenths of a millimeter, using the penetration test described previously. If the needle penetrates 9.8 mm, or 98 tenths of a millimeter, the penetration value is 98. The

TABLE 6.2 REQUIREMENTS FOR ASPHALT CEMENT VISCOSITY GRADED AT 140 °F (60 °C)

Tests on residue from rolling thin-film oven test[a]	Viscosity grade				
	AR-1000	AR-2000	AR-4000	AR-8000	AR-16000
Viscosity, 140 °F (60 °C), P	1000 ± 250	2000 ± 500	4000 ± 1000	8000 ± 2000	16000 ± 4000
Viscosity, 275 °F (135 °C), min, cSt	140	200	275	400	550
Penetration, 77 °F (25 °C), 100 g, 5 s, min	65	40	25	20	20
Percent of original penetration, 77 °F (25 °C), min	—	40	45	50	52
Ductility, 77 °F (25 °C), 5 cm/min, min, cm	100[b]	100[b]	75	75	75
Tests on original asphalt:					
Flash point, Cleveland open cup, min, °F (°C)	400 (205)	425 (219)	440 (227)	450 (232)	460 (238)
Solubility in trichloroethylene, min, %	99.0	99.0	99.0	99.0	99.0

Note: Grading based on residue from rolling thin-film oven test.

[a]Thin-film oven test may be used but the rolling thin-film oven test shall be the referee method.

[b]If ductility is less than 100, material will be accepted if ductility at 60 °F (15.5 °C) is 100 minimum at a pull rate of 5 cm/min.

Source: ASTM D3381

TABLE 6.3 REQUIREMENTS FOR ASPHALT CEMENT FOR USE IN PAVEMENT CONSTRUCTION

	Penetration grade									
	40–50		60–70		85–100		120–150		200–300	
	min	max	min	max	min	max	min	max	min	max
Penetration at 77 °F (25 °C) 100 g, 5 s	40	50	60	70	85	100	120	150	200	300
Flash point, °F (Cleveland open cup)	450	—	450	—	450	—	425	—	350	—
Ductility at 77 °F (25 °C) 5 cm/min, cm	100	—	100	—	100	—	100	—	100[a]	—
Solubility in trichloroethylene, %	99.0	—	99.0	—	99.0	—	99.0	—	99.0	—
Retained penetration after thin-film oven test, %	55+	—	55+	—	47+	—	42+	—	37+	—
Ductility at 77 °F (25 °C) 5 cm/min, cm after thin-film oven test	—	—	50	—	75	—	100	—	100[a]	—

[a]If ductility at 77 °F (25 °C) is less than 100 cm, material will be accepted if ductility at 60 °F (15.5 °C) is 100 cm minimum at the pull rate of 5 cm/min.

Source: ASTM D946

penetration grading system is based on the results of the penetration test, and consists of five grades: 40–50, 60–70, 85–100, 120–150, and 200–300 (ASTM D946). The grade numbers, shown in Table 6.3, indicate the penetration readings; for example, the penetration of the standard needle into a sample of grade 120–150 is from 120–150 tenths of a millimeter. Grade 40–50 is "hard" asphalt and grade 200–300 is "soft" asphalt. It can be noted from Tables 6.1 and 6.2 that the viscosity grading system for asphalt cements has penetration requirements in addition to viscosity limits. For example, grade AC-40 has a minimum penetration requirements of 20, and AC-2.5 has a minimum penetration requirement of 200.

6.5.2 Performance-based Grading

Unlike traditional specifications, which require a fixed temperature during testing and vary the requirements for different grades, the new performance-based grading system requires that tests be performed at the critical pavement temperature—which is different for different grades, depending on the temperature during service—with the criteria fixed or the same for all grades. This system was developed as part of the Superior Performing Asphalt Pavements Program (*Superpave*), through the five-year U.S. Strategic Highway Research Program (SHRP) begun in 1987. The term "asphalt binder" is used in place of "asphalt" or "asphalt cement," as small quantities of modifiers are often added to asphalt cements to improve their performance during service. The research studies carried out in the program concluded that asphalt pavements often fail or crack due to three major reasons:

1. Pavement deformations or rutting at high temperatures, as asphalt softens and the binder loses elasticity

2. Fatigue cracks resulting from high pavement loads or aging
3. Low-temperature cracks, as asphalt becomes brittle and the pavement shrinks in cold weather

The new grading system is designed to provide measures of the performance-related properties of asphalt binder that can be related, more rationally, to the in-situ performance of the material in pavements, especially in terms of its ability to resist these three types of failure. Four characterization tests (the bending beam rheometer, rotational viscometer, dynamic shear rheometer, and direct tension tests) are used to characterize the asphalt binder at temperatures that represent the upper, middle, and lower ranges of service temperatures (discussed in test BITU-5, at the end of the chapter).

Three pavement design temperatures are required to detail the asphalt binder specifications: a maximum, an intermediate, and a minimum. The maximum and minimum pavement temperatures at a specific geographical location in the United States are generated with the help of algorithms contained in the performance-system-grading software, and are based on weather information from 7500 weather stations. The maximum temperature is the highest successive seven-day average maximum pavement temperature, and the minimum is the lowest temperature expected over the service life of the pavement. The intermediate design temperature is the average of the maximum and minimum design temperatures plus 4 °C (7.2 °F). Laboratory tests to evaluate the potential for rutting are carried out at the maximum design temperature, and thermal cracking tests are done at the minimum design temperature plus 10 °C (18 °F).

Performance grades have two numbers, the first representing the high range, or maximum, and the second the low range, or minimum, of the service temperature (Table 6.4). The high temperatures vary from 46 to 82 (°C) by increments of 6 °C, and the low temperatures from −10 °C to −46 °C, also in increments of 6 °C. For example, an asphalt binder of performance grade PG 64 − 34 is suitable for application when the maximum temperature is between 39 and 64 °C and the minimum not lower than −34 °C.

TABLE 6.4 PERFORMANCE GRADES OF ASPHALT BINDER

High temperature grades	Low temperature grades
PG 46	−34, −40, −46
PG 52	−10, −16, −22, −28, −34, −40, −46
PG 58	−16, −22, −28, −34, −40
PG 64	−10, −16, −22, −28, −34, −40
PG 70	−10, −16, −22, −28, −34, −40
PG 76	−10, −16, −22, −28, −34
PG 82	−10, −16, −22, −28, −34

Modifiers are added to asphalt cements to enhance their resistance to permanent deformation (rutting), cracking at low temperatures or from fatigue, or oxidation, so as to meet the specifications of the performance-based grading system. They are essential in applications involving very high and very low temperatures, and are generally added by the supplier.

6.5.3 Cutback Asphalt Grades

Cutback asphalts are graded based on their kinematic viscosity at 140 °F (60 °C) in centistoke (ASTM D2026, D2027, and D2028). All three types of cutback asphalt (slow curing, medium curing, and rapid curing) have four grades: 70, 250, 800, and 3000, depending on the kinematic viscosity values. Grade 70 has a minimum kinematic viscosity value (at 140 °F) of 70 cSt and a maximum of 140 cSt. Grade 250 has minimum and maximum values of 250 and 500 cSt, respectively. Grade 800 has minimum and maximum values of 800 and 1600 cSt, whereas grade 3000 has 3000 and 6000 cSt, respectively.

In addition to these four grades, the medium-curing type has an additional grade 30, which has minimum and maximum kinematic viscosities of 30 and 60 cSt. The difference in these grades is attributed to the amount of diluents; the smaller the amount of diluent, the higher the grade. The choice of a particular grade or type of cutback asphalt depends on several factors, including the environmental conditions relative to the application. A pavement that is under a hot sun most of the time during service may require rapid-curing asphalt to provide a harder asphalt base and stable pavement. On the other hand, to obtain similar pavement characteristics in a cold region, slow-curing or medium-curing asphalt may be preferred.

6.6 ASPHALT CONCRETE

The term "concrete" generally refers to a solid mass made by cementing together aggregate particles. Portland cement concrete is made by mixing together portland cement and water with coarse and fine aggregates. Asphalt concrete is made by mixing hot asphalt cement with heated aggregates and compacting the mix to form a uniform dense mass (emulsified asphalt can be used instead of asphalt cement).

Asphalt concrete is used for the construction of asphalt pavement, which is also called *flexible pavement,* due to its ability to conform to settlement of the foundation. Asphalt concrete pavements are used for highways, airports, parking lots, and driveways.

In asphalt concrete, the framework of aggregates is held in place by the binding action of the asphalt cement (Fig. 6.7). The function of asphalt cement is to hold the particles in place. Selected aggregates are compacted so that the external loads are distributed by point-to-point contact of the aggregate particles. This mechanism of load transfer is called *aggregate interlock* (or *intergranular pressure* or *internal friction*).

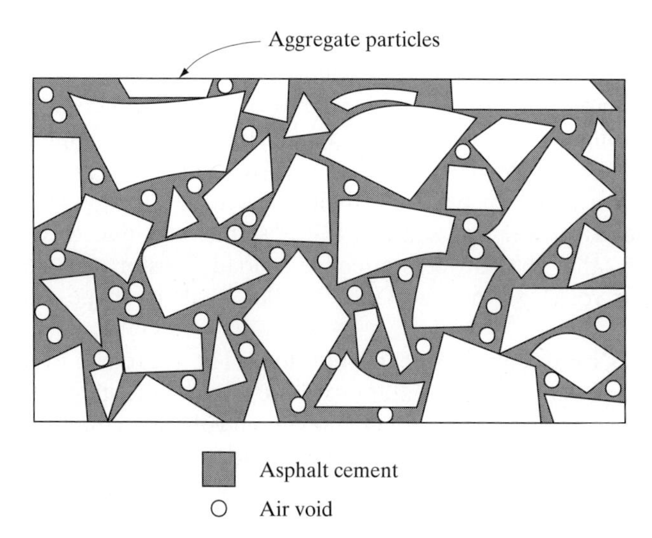

Figure 6.7 Aggregate interlock mechanism.

The voids or spaces between the particles are partially filled with asphalt cement, leaving sufficient air pockets, which account for roughly 2 to 6 percent of the concrete volume. The amount of asphalt cement should be such that the concrete has more than 2 percent air voids. Too much cement (or too few air voids) allows the particles to float, displacing the particles. Too little cement or binder causes imperfect bonding, and potholes are produced in the pavement.

6.6.1 Aggregates

In asphalt concrete pavement, aggregates may constitute about 70–75 percent by volume or 90–95 percent by weight. The external loads are distributed through point-to-point contact (aggregate interlock). Aggregates also provide resistance to abrasion and skid resistance for vehicular traffic. The functions of the asphalt are to bind the particles and act as a waterproofing ingredient. Excessive amounts of the binder tend to lubricate the particles and lower the stability of the pavement. Asphalt also allows the particles to rebound upon removal of the load, but the major share of the resistance to pavement deformation is from the aggregate framework.

A number of factors, such as quality, gradation, shape, stiffness, and quantity of aggregates, determine the effectiveness of this load transfer and the stability of pavement. Stability is the resistance to deformation under load. It is this property that enables a pavement to withstand traffic without sustaining permanent deformation.

Internal friction or aggregate interlock represents the resistance to movement of particles past one another under the action of applied load. The resistance is offered by the interlocking of particles and the surface friction. Crushed stone aggregate has very good internal friction. Gravel is relatively low in internal friction because of the round shape and smooth surface of the particles. Blast-furnace slag, which is a by-product in the manufacture of iron (consisting primarily of silicates and aluminosilicates of lime), has very high internal friction.

It was indicated that an average of about 95 percent of asphalt concrete by weight or 75 percent by volume is made up of aggregate particles. The asphalt binder is about 5 percent by weight. As the aggregates occupy a very large portion of asphalt concrete, the performance of pavement is influenced by the aggregate properties. Rounded particles (gravel and sand) are less stable, and due to the tendency to slide over each other they are less effective in load transfer. Angular particles (crushed stone or rock) are more stable, thanks to their shape and texture.

Soft aggregate particles can break and wear under load. The aggregate mass should be as impermeable as possible to keep water out of the mix and the subgrade. Dense-graded aggregates can be compacted to fill the maximum volume in the mass, producing the maximum surface contact and the least void space.

Particles should always remain on the surface to yield a skid-resistant surface. The surface texture of the pavement should be rough to grip the tires when the surface is wet. When the asphalt is too soft or there is too much asphalt in the pavement, the aggregate particles are forced down into the pavement. As a consequence, the surface becomes smooth and slick, resulting in a weak top layer.

The amount of asphalt cement should be such that the cement binds or holds all the particles but is not enough to lubricate them so as to cause slippage. Too much asphalt content in the mix results in a downward movement of aggregate particles or upward movement of the binder to the top surface, which is called *bleeding* or *flushing*. Placement of aggregate chips on a coating of asphalt cement (seal coat) will control bleeding. (It can be recalled from discussions in Chapter 3 that bleeding in concrete is the upward movement of softer particles: cement, water, and fine sand).

Aggregates for asphalt concrete consist of fine aggregate or a mixture of fine and coarse aggregate, with or without a mineral filler. The mineral filler is a finely divided mineral product at least 70 percent of which will pass a 75-μm (No. 200) sieve. Pulverized limestone is the filler most commonly used; other fillers include stone dust (rock dust), hydrated lime, portland cement, and fly ash.

Grading. Coarse aggregate is a graded aggregate made up of particles that are retained on a No. 4 sieve. Fine aggregate is a graded aggregate composed of particles that almost entirely pass a No. 4 sieve. The coarse and fine aggregates used for asphalt concrete can be crushed stone, crushed slag, crushed gravel, natural or manufactured sand prepared from stone, crushed blast-furnace slag, gravel, or a combination thereof. The standard requirements for these two aggregates are given in ASTM D692 and D1073 (Table 6.5).

Well-graded (or dense-graded) aggregate is uniformly graded from the maximum size down to the filler, with the object of obtaining an asphalt mix that has a controlled void content or high stability. *Macadam* is a coarse aggregate of uniform size, usually of stone, slag, or gravel.

TABLE 6.5 GRADING REQUIREMENTS FOR AGGREGATES

	Amounts finer than each laboratory sieve (square openings), weight %			
Sieve size	Grading No. 1	Grading No. 2	Grading No. 3	Grading No. 4
3/8-in. (9.5-mm)	100	—	—	100
No. 4 (4.75-mm)	95 to 100	100	100	80 to 100
No. 8 (2.36-mm)	70 to 100	75 to 100	95 to 100	65 to 100
No. 16 (1.18-mm)	40 to 80	50 to 74	85 to 100	40 to 80
No. 30 (600-μm)	20 to 65	28 to 52	65 to 90	20 to 65
No. 50 (300-μm)	7 to 40	8 to 30	30 to 60	7 to 40
No. 100 (150-μm)	2 to 20	0 to 12	5 to 25	2 to 20
No. 200 (75-μm)	0 to 10	0 to 5	0 to 5	0 to 10

Source: ASTM D1073

Three types of gradation are commonly used in pavement construction:

- Open graded
- Intermediate graded
- Dense graded

Open-graded aggregate is an aggregate containing little or no fine aggregate, or one in which the void space in the compacted aggregate is relatively large (Fig. 6.8). It contains less than 10 percent passing a No. 10 (2-mm) sieve and less than 2 percent passing a No. 200 (75-μm) sieve. The nominal maximum size is generally equal to or less than $2\frac{1}{2}$ in. (63.5 mm).

Open-graded aggregate is used to make open-graded asphalt concrete, which provides good skid resistance and high permeability so as to permit good surface drainage. A high void content (which can be as large as 20 percent) allows the surface water to be removed very quickly, giving a rough surface texture; this quick removal of moisture from the surface minimizes the danger of hydroplaning and improves skid resistance during wet season.

Intermediate-graded mixes contain a larger percentage of sand than open-graded, usually 8–18 percent passing a No. 8 sieve (2.36 mm). It is well-graded from coarse to fine, and may contain crushed stone or gravel. Both open-graded and intermediate-graded base layers require a surface course (or seal coat) to make the surface impermeable.

Dense-graded aggregate may be composed of crushed stone, slag, crushed gravel, or a mixture thereof. It is well-graded, made up of coarse to fine material, but the amount of fines passing a No. 200 sieve (75 m) should not exceed 5 percent. Increase in fine fraction increases the asphalt requirement and slows the curing rate. Use of dense-graded aggregate provides a dense and impermeable layer and does not normally require surface treatment or a seal coat.

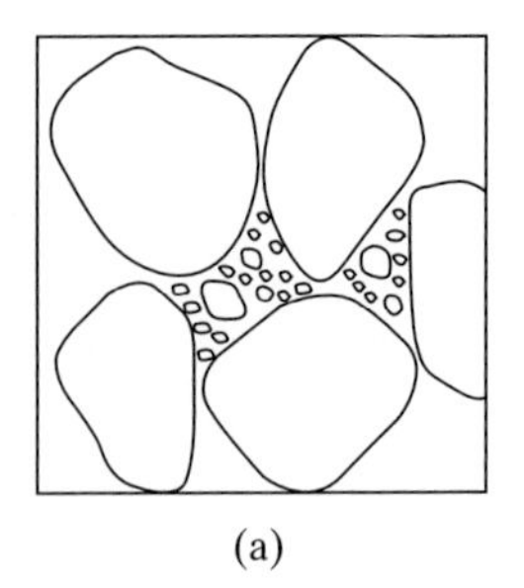

(a)

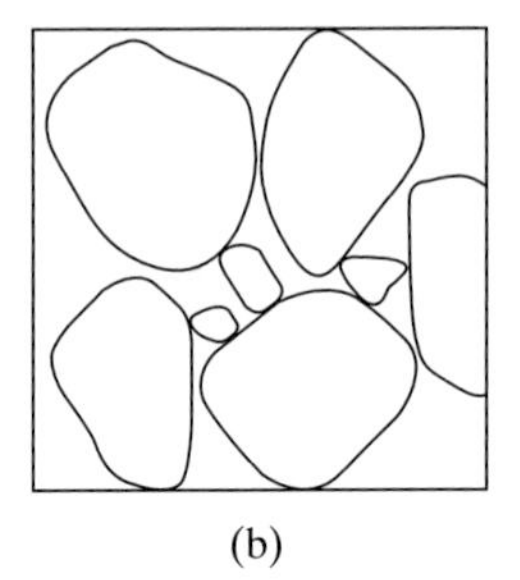

(b)

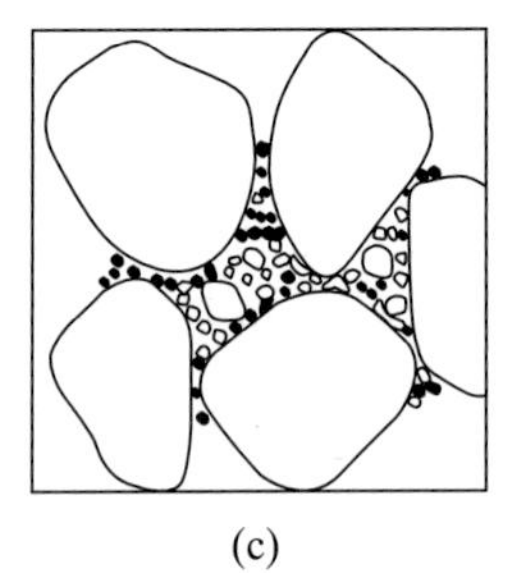

(c)

Figure 6.8 Types of gradation: (a) dense-graded; (b) open-graded; (c) too much fines.

The composition of bituminous mixtures should follow the recommendations in ASTM D3515. The fine and coarse aggregates are proportioned so that the combined mix falls within the specified grading requirements. Although the nominal maximum size of aggregate can be as high as $2\frac{1}{2}$ in. (63.5 mm), it is typically 3/4 in. (19 mm) for most pavements, and more than 90 percent of aggregate particles are retained on a No. 200 sieve.

As discussed, rock dust (mineral filler), which is a by-product of mining operations, is used to reduce the amount of voids. The finer particles also assist in smooth placement of the mix and prevent segregation.

6.6.2 Types of Asphalt Concrete

Hot-mixed asphalt (HMA) is a mixture of hot asphalt and aggregates produced at a batch or drum mixing facility that must be spread and compacted while hot. Both aggregates and asphalt cement are heated prior to mixing to drive off moisture from the particles and make the asphalt cement sufficiently fluid. After heating, all the raw materials are mixed in the plant, and the hot mixture is transported to the paving site and spread on a loosely compacted layer to a uniform, even surface with the help of a paving machine. While the mixture is hot it is compacted by heavy, motor-driven rollers to produce a smooth, well-compacted paving course.

Two types of asphaltic concretes are:

- Hot-mixed, hot-laid bituminous mixtures
- Cold-mixed, cold-laid bituminous mixtures

A hot-mixed, hot-laid bituminous mixture is fine aggregate, or a mixture of coarse and fine aggregate, with or without mineral filler, uniformly mixed with asphalt cement, tar, or emulsified asphalt. The aggregates and the bitumen are heated prior to mixing. The asphalt cement is heated to a temperature not exceeding 350 °F (176.6 °C) and the emulsified asphalt to a temperature not exceeding 180 °F (82.2 °C). The viscosity grade or penetration grade of the bitumen depends on the type of construction, climatic conditions, and the amount and nature of traffic. As the aggregates are thoroughly dried prior to mixing, stripping of asphalt (removal from the pavement) will not take place in hot-mixed, hot-laid asphalt pavements.

A cold-mixed, cold-laid bituminous paving mixture is coarse aggregate, fine aggregate, or a mixture of coarse and fine aggregates, with or without mineral filler, uniformly mixed with liquid bitumen (emulsified asphalt, cutback asphalt, or tar) and laid at or near ambient temperature. Note that the mixing is done at normal temperature and drying of aggregates is not necessary except when the particles have surface moisture. To improve bonding, commercial additives are needed in this type of asphalt concrete. In winter months, some heating of both the aggregates and the asphalt may be required.

A bituminous surface or base course produced by mixing aggregates and cutback asphalt, emulsified asphalt, or tar at the job site is called *road mixed* or *mixed in place*. A mixture of aggregates and cutback asphalt, emulsified asphalt, or tar prepared at a central mixing plant and spread and compacted at the job site at or near ambient temperature is called *plant-mixed, cold-laid.*

The road-mixed or mixed-in-place asphalt is prepared at the job site using travel plants, motor graders, or other road mixing equipment. The aggregates may consist of open- or dense-graded aggregates, sand, or sandy soil. The liquid asphalt is generally heated to a temperature that will provide the necessary viscosity. This type of mix is of lower quality than the hot-mixed, hot-laid concrete.

A number of additives are used in the manufacture of asphalt concrete. Asphalt rubber (crumb rubber), which is a recycled product from old tires, is used in hot-mixed asphalt pavements to improve the binding property of aggregates. About 1–5 percent rubber can, in addition, increase the viscosity and the softening point. Issues of worker safety and the long-term effects of rubber on the environment remain to be addressed. Polymers (manufactured rubber) are added to asphalt to produce polymer-modified asphalt. Most commonly, ethyl vinyl acetate (EVA) is added for better dispersion of materials. This compound is also found to increase ductility and adhesiveness. Other polymers used in pavement construction are latex, silicone, and epoxies.

Glasphalt is the term used to describe asphalt that is partly replaced by glass. Sulfur is added to asphalt concrete to provide higher stiffness at elevated temperatures. Fibers (asbestos or polypropylene) are added to improve fatigue resistance and decrease thermal cracking. Carbon-black pellets are added as a reinforcing material to increase the pavement durability. Sewage-sludge ash and roofing-shingle waste have also been recycled in asphalt concrete. Adding hydrated lime to hot mixes increases the strength and stability of the mix while making them more water resistant. Lime also reduces swell, water absorption, and stripping, and allows for faster compaction.

6.7 ASPHALT PAVEMENT

Asphalt (flexible) pavement has sufficiently low bending resistance to maintain intimate contact with the underlying structure or base, yet has the required stability to support traffic loads. Its function is similar to that of a mattress that provides good contact with the human body, yet supports the load without a failure. A number of pavements (such as crushed stone, gravel, and all bituminous types not supported on a rigid base) fall into this category.

Most flexible pavements are constructed with an asphaltic concrete surface layer placed over a treated or untreated base layer (or course) and an untreated subbase layer (Fig. 6.9). The advantage of asphalt concrete pavement is that it is able to adjust more readily to differential settlement likely to occur when the roadway is constructed on relatively flexible (or variable quality) soil or ground. In addition, this type of pavement can be repaired and recycled quickly and easily.

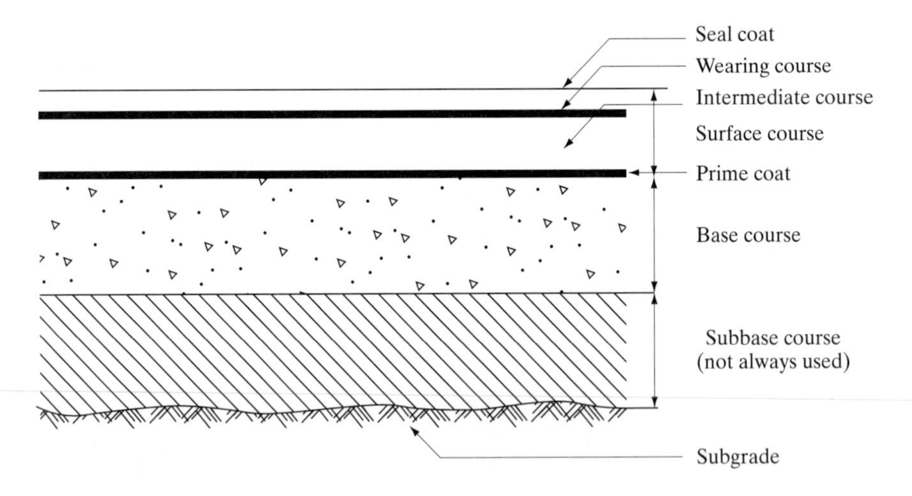

Figure 6.9 Typical flexible pavement cross-section.

But the drawbacks of a flexible pavement are that it requires a higher level of maintenance than portland cement concrete pavement (called *rigid pavement*), as well as periodic surface treatments due to hardening of asphalt with age. It was explained earlier that asphalt tends to become hard and brittle with age and under load. The hardening is principally due to oxidation (combining with oxygen in atmosphere), which is more severe at higher temperature and in thin films. It can be minimized by keeping the number of air voids in the mixture low and by coating the particles thickly.

Asphaltic concrete pavement is designed with a proper mixture of aggregate particles and asphalt cement. The amount of asphalt cement is crucial to a good performance rating. Adding more asphalt than required for void control results in less than the minimum amount of air space; similarly, less than required creates a high amount of air space. A large percentage of air voids (more than 6 percent) may cause early hardening of asphalt from weathering (oxidation and volatilization), leading to failure. A low asphalt content may produce a brittle pavement.

The amount of asphalt cement generally lies between $3\frac{1}{2}$ and 12 percent of the total weight of the materials. When the maximum aggregate size is reduced, the percent of asphalt cement increases. For example, aggregates of maximum size 3/8 in. (10 mm) may require asphalt cement in the range 5–12 percent, whereas 1-in. (25-mm) maximum aggregate has a limit of 3–9 percent asphalt cement.

6.7.1 Elements of Flexible Pavement

A flexible pavement consists of four elements:

- Subgrade
- Subbase course
- Base course
- Surface course

A subgrade (also called *prepared roadbed*) consists of natural or imported soil, such as in an embankment, improved by stabilization. Stabilization is done using various techniques and materials, such as mechanical and chemical stabilization. When the native soil is found unsuitable to carry the loads, it will be stabilized. The most common stabilization method is the addition of portland cement. The subgrade acts as the foundation for the road and should possess good *resilient modulus,* the ability to carry normal loads under repeated application. It is a measure of the elasticity of roadbed soil or other pavement material. Additional details on stabilization are addressed in the following section.

A subbase is a layer between the subgrade and the base course. It is made from materials generally superior to that of subgrade. When the subgrade material possesses properties that makes it suitable as a subbase, the subbase layer is omitted.

The material for a subbase consists of particles of various sizes. Up to a maximum of 12 percent is made up of particles passing sieve No. 200. Lime-treated subbase is aggregate base mixed with lime. Aggregate subbase, which may consist of more than a single layer, is the most economical form. When the normal pavement design process results in a thin layer of subbase, the subbase can be eliminated in place of a thick base layer.

A base course (or base layer) lies directly beneath the wearing surface. It can be constructed as an asphalt concrete base (called asphalt base course) or as an untreated aggregate base. Aggregate may be one or more of crushed stone, crushed slag, gravel, sand, or combinations of these materials. The base layer can also be constructed using cement-treated (base) aggregates.

In addition to portland cement, a combination of portland cement, pozzolan material, lime, and other cementing agents can be combined with aggregates or soil for construction of the base course and subbase layer. The choice of types depends on the relative cost, availability of materials, type of native material, environmental conditions, and traffic. The base layer is generally identified by the type of material used:

- Aggregate
- Asphalt cement
- Asphalt-treated permeable
- Cement-treated
- Cement-treated permeable
- Lean concrete

Cement-treated permeable base is a highly open-graded mixture of coarse aggregate, portland cement, and water, formed to provide drainage of the pavement section and give structural support. Asphalt-treated permeable base course consists of a mixture of coarsely graded aggregate and asphalt cement. The permeable base courses are commonly selected to render drainage of the pavement section, as shown in Fig. 6.10. In a permeable base course, up to about 80 percent of the aggregate particles are of size equal to or larger than 3/8 in. (10 mm).

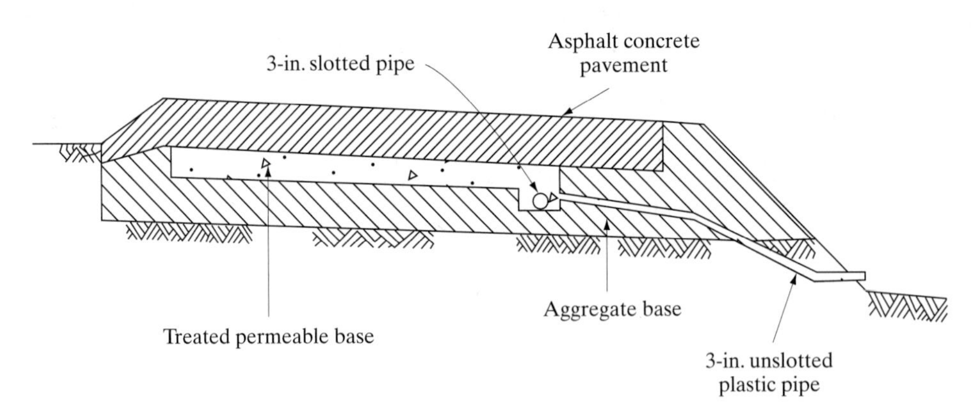

Figure 6.10 Treated permeable base pavement.

A full-depth asphalt concrete is a pavement constructed with asphalt mixture for all courses above the subgrade. This type of pavement has little or no aggregate base or subbase. Full-depth asphalt concrete provides a faster method of building flexible pavements; it is generally thinner than a typical layered pavement and thus requires less soil excavation.

The surface course (also called the wearing course) and the base course may consist of two or more layers, somewhat different in composition from each other, that are placed in separate construction operations. The surface course must be capable of withstanding wear and abrasion forces from vehicles and be able to offer sufficient stability to prevent the surface from shoving or rutting. In addition, it should prevent the migration of excessive amounts of surface water into the base course and subgrade. To improve skid resistance and minimize hydroplaning effects during high-speed traffic, some wearing courses are topped with a highly drainable open-graded friction course. The thickness of the surface course varies from 1 in. (25 mm) in the case of bituminous surface treatment on low-traffic roads, to 6 in. (150 mm) or more for heavy-traffic highways.

White topping is a term normally applied for the layer of concrete placed over an old asphalt base, and is typically employed in areas where repeated concentrated loads cause the asphalt to shove and rut in a relatively short period of time. It can also be used in places where the surface has deteriorated to the point of needing replacement. The process involves milling down of deteriorated asphalt pavement for a depth of 2–5 in. (50–125 mm), and then overlaying with ordinary or fiber-reinforced concrete.

6.7.2 Stabilization

Soil and aggregate stabilization involves the use of specially graded granular materials or of strengthening agents such as lime, asphalt, portland cement, or salt, to permit the use of low-quality or in-situ materials as subgrade, base course, or surface course of low-volume roads. Granulated base materials consist of dense-graded aggregates mostly 1 in. (25 mm) or less in size. The amount of fines is more than that allowed in a typical base course. About 40–60 percent of particles must pass sieve No. 4 (4.75 mm), for increased strength and to allow surface drainage. An impermeable layer is preferred for a surface course to force the surface water to run off rather than get into the base course.

With highly plastic, fine-grained soils, lime up to 8 percent by weight, in the form of quicklime or hydrated lime, can be used to increase the strength. The amount of lime in fine-grained, low-plasticity soils is about 4 percent. Mixing lime followed by compaction reduces the plasticity and the swelling potential of clayey soils and increases the load carrying capacity. The improvement in strength is the result of a slow reaction between the silica and alumina in clay and the lime, forming compounds with binding characteristics.

When aggregates are mixed with salts such as calcium chloride (up to 2 percent by weight), the strength is significantly improved. Calcium chloride is a hygroscopic material, which means that it has the ability to attract and retain moisture. Salts help in expediting the compaction of soil-aggregate mixtures by slowing down the rate of evaporation of moisture during compaction and increasing the density at a given compaction. Absorption of water from the atmosphere aids in maintaining moisture content, especially in wearing surfaces, keeping them from drying out and maintaining their stability (preventing raveling). Moreover, dust is reduced. Rural roads having high amounts of fines can be treated with salts to form sound, smooth, and impermeable surfaces.

Portland cement can be used for soil-cement base courses, to improve the quality of granular soils or granular base aggregates, and for reconstruction of roads. When used as a base material, soil cement must be surfaced with a wearing course of asphalt concrete. The amount of cement is established from compaction tests, and is normally in the range of 5–9 percent by weight for coarse-grained soils and 9–15 percent for fine-grained soils. Soil treated with cement must be cured. The seven-day strength of cement-modified materials is between 250–300 psi (1.73–2.07 MPa).

6.8 SPRAY APPLICATIONS

There are four common spray applications to pavements:

- Surface treatments and seal coats
- Fog seal
- Prime and tack coats
- Slurry seal

Asphalt surface treatments and seal coats are spray applications of asphalt followed by the application of a skin of aggregates (stone chips or fine aggregate) embedded by rolling. A single application of a mixture of asphalt and aggregate is called seal coat. A surface treatment may consist of single or multiple applications of asphalt-aggregate mixture placed one on the other. In the case of a single application, the thickness of the resulting layer (generally not exceeding 1/2 in. or 12.5 mm) corresponds to the nominal aggregate size. When multiple applications are used, the thickness of the layer is only slightly larger than the thickness of the finish course and may be as high as 2.5 in. (63.5 mm).

Seal coats and surface treatments provide a waterproofing layer on a base course and improve the skid resistance of the existing pavement. Moreover, they extend the service life of a weathered surface, and are cost-effective. They can be used on an existing surface that is cracked or oxidized to improve the texture and waterproofing, or on a new surface.

Aggregates for seal coat and surface treatment consist essentially of one-size material, and the maximum size is generally not more than twice the smallest size. For example, for 1/2-in. (12.5 mm) maximum size aggregate, about 95 percent of the particles pass a 1/2-in. (12.5 mm) sieve and 30 percent pass a 3/8-in. (10 mm) sieve. Nearly 90 percent of aggregate particles are within 1/2- and 1/4-in. (12.5 and

6.3 mm) sieve sizes. If the size difference is large, the larger particles may be dislodged by the traffic, and the smaller particles lead to "fat" spots. The shape of the particles should be angular.

Fog seal is a light application of a slow-setting emulsified asphalt (diluted with additional water) without the mineral aggregates. The purposes of fog seal are to provide an impermeable layer to the pavement and prevent movement of aggregate particles downward.

A *prime coat* is the application of a liquid asphalt to an untreated foundation layer or subgrade of stabilized soil, gravel, or water-bound macadam. Its function is to coat and bind loose particles and to provide adhesion between the base and the wearing surfaces. The prime coat also helps to consolidate the surface on which the new treatment is to be placed and prevents the upward migration of capillary water. A well-compacted and finished soil-cement base coarse is given a coat of asphalt emulsion as a curing membrane to prevent the drying of the base.

Liquid asphalts and road tars of low viscosity can be used for the prime coat. The application should be to a clean surface that is dry or slightly damp and when the ambient temperature is above 55 °F (13 °C).

A *tack coat* is a thin coat of bituminous material applied to an existing surface (asphalt concrete or portland cement concrete) to provide bond between the new construction (overlay) and the existing surface. Emulsified asphalt, diluted with additional water, is commonly used. The surface on which the tack coat is applied can be bituminous, portland cement concrete, brick, or stone. Tack coat is the first step in the construction of a new asphalt wearing surface and demands a clean and dry or slightly damp surface. It should be noted that both prime and tack coats are not wearing surfaces themselves.

Slurry seal is a mixture of a slow-setting emulsified asphalt, fine aggregate, mineral filler, and water, applied to the pavement surface without heat. The slurry (binder and other components) penetrates cracks and seals them. It also gives a new surface to the pavement. The application rate can vary between 4–12 lb/yd^3.

Slurry seal is used primarily as a maintenance material for light-traffic roadways, mainly to seal the surface cracks on asphalt concrete pavements. Its application helps to fill voids, maintain a minimum wearing surface, and correct surface erosion conditions. Slurry seal is also employed to provide a new wearing surface.

Aggregates for slurry seal may consist of sand, slag, and crushed fines. The amount of asphalt emulsion to be blended with the aggregates is estimated from laboratory mix design. Water is added as necessary to obtain a fluid, homogeneous consistency. Mineral fillers such as portland cement, lime, and fly ash are often added to the slurry.

A *chip seal* is a high-viscosity emulsified asphalt surface coat incorporating rolled-in rock screenings (chips) over an asphalt concrete pavement (Fig. 6.11).

An *overlay* is a layer of asphalt concrete placed on an existing flexible or rigid pavement to restore quality, improve load-carrying capacity, minimize maintenance costs (for highways and vehicles), reduce skid resistance, and extend the service life of the pavement. A nonwoven bonded-fiber synthetic fabric (pavement reinforcing

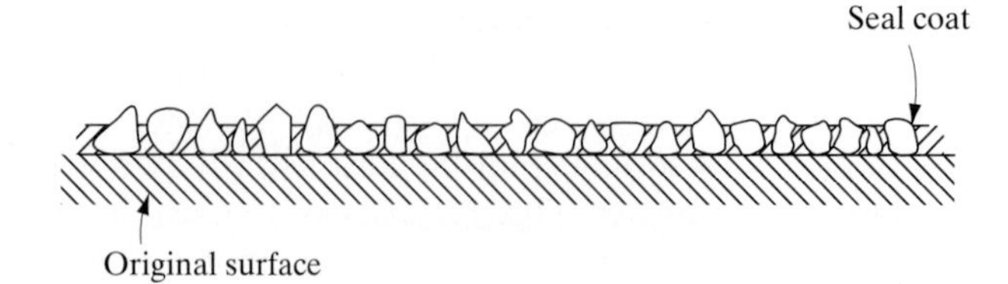

Figure 6.11 Seal coat with crushed stone/gravel cover.

fabric) is placed as an interlayer in asphalt concrete overlays to minimize infiltration of surface water. The fabric is also helpful in retarding reflection cracking, however it does not add to the strength of the pavement.

A properly designed overlay can correct a number of deficiencies, such as excessive permanent distress and inadequate ride quality. Using overlays, old roads can be improved in a shorter time at less cost. When excessive permeability, surface roughness, and low skid resistance are noticed, overlays may be required.

6.9 TESTING

A number of laboratory tests are described in this section. Following is a list of asphalt properties and the corresponding test numbers.

Property	Test no.
Penetration of asphalt	BITU-1
Ductility of asphalt	BITU-2
Kinematic viscosity of liquid asphalt	BITU-3
Coefficient of viscosity of asphalt	BITU-4
Performance-based grading system tests for asphalt binders	BITU-5

Test BITU-1: Penetration of Asphalt Materials

PURPOSE: To determine the consistency of an asphaltic material using a penetration test.

RELATED STANDARD: ASTM D5

DEFINITION:

- *Penetration* is the consistency expressed as the distance in tenths of a millimeter that a standard needle (100 g of total weight) penetrates a sample of the material vertically at 77 °F for 5 s.

EQUIPMENT: Standard needle [50 mm (2 in.) in length and 1.0 mm (0.0394 in.) in diameter], standard flat-bottomed sample container of metal or glass [55 mm diameter and 35 mm depth (or 3 oz) for penetration below 200, and 70 mm diameter and 45 mm depth (or 6 oz) for penetration between 200–350] water bath (capacity 10 liter), timing device, thermometer, transfer dish.

SAMPLE: Semisolid or solid bituminous sample.

PROCEDURE:

1. Heat the sample [temperature $\leq$ (90 °C + softening point of the sample)] until it is sufficiently fluid to pour.
2. Pour the sample into the sample container to a depth 10 mm greater than the depth to which the needle is expected to penetrate when cooled to 77 °F.
3. Allow the sample to cool ($1\text{–}1\frac{1}{2}$ h for small container and $1\frac{1}{2}\text{–}2$ h for large container).
4. Place the container in the transfer dish and place it in the water bath at 77 °F (25 °C). Allow the small container to remain in the water bath for $1\text{–}1\frac{1}{2}$ h and the large container for $1\frac{1}{2}\text{–}2$ h.
5. Place the 50-g weight above the needle (total load is 100 g).
6. Cover the container with water from the water bath, and place the transfer dish on the stand of the penetrometer.
7. Position the needle by lowering it slowly until its tip just makes contact with the surface of the sample.
8. Note the reading of the penetrometer dial.
9. Quietly release the needleholder for 5 s.
10. Adjust the instrument to measure the distance penetrated in tenths of a millimeter.
11. Make at least three measurements. (Return the sample and transfer dish to the constant temperature bath between the measurements.)

REPORT: Report to the nearest whole unit the average of three penetrations whose values do not differ by more than 2 for penetrations less than 50, 4 for penetrations less than 150, 6 for penetrations less than 250, and 8 for penetrations less than 350.

Test BITU-2: Ductility of Asphalt

PURPOSE: To determine the ductility of an asphalt specimen using a briquette test.

RELATED STANDARD: ASTM D113.

DEFINITION:

- *Ductility* is the distance to which the material will elongate before breaking when two ends of a briquette specimen are pulled apart at a specified speed (5 cm/min) and at a specified temperature (77 °F).

EQUIPMENT: Briquette mold, water bath, testing machine (specimen is continuously immersed in water), thermometer.

SAMPLE: Asphalt specimen.

PROCEDURE:

1. Assemble the mold on a brass plate.
2. Coat the plate and interior surface of the mold (side pieces only) with a thin layer of a mixture of glycerin and dextrin, talc, or china clay (kaolin) to prevent sticking.
3. Heat the sample until it is sufficiently fluid to pour.
4. Strain the sample through a No. 50 (300 μm) sieve.
5. Pour the sample into the mold until it is more than level full.
6. Let the sample cool to room temperature for 30–40 min.
7. Place it in the water bath (at 77 °F) for 30 min.
8. Cut off the excess bitumen with a hot straight-edged putty knife to make the mold just level full.
9. Place the brass plate and mold in the water bath (77 °F) for a period of about 90 min.
10. Remove the briquette from the plate, detach the side pieces, and test immediately.
11. Attach the rings at each end of the clips to the pins or hooks of the testing machine, and pull the end clips apart at a speed of 5 cm/min until the briquette ruptures (thread has no cross-sectional area).
12. Measure the distance in centimeters through which the clips have been pulled.

REPORT: Report the ductility in centimeters.

Test BITU-3: Kinematic Viscosity of Liquid Asphalt

PURPOSE: To determine the kinematic viscosity of cutback or other liquid asphalt using reverse-flow viscometers.

RELATED STANDARD: ASTM D2170.

DEFINITIONS:

- *Kinematic viscosity* is the ratio between the viscosity and the density of a liquid. The standard measuring unit is the centistoke (equal to 1 mm^2/s).
- *Density* is the mass per unit volume of a liquid. The standard unit is 1 g/cm^3.

EQUIPMENT: Reverse-flow viscometer, thermometer, bath, timer.

SAMPLE: Sample of liquid asphalt that is thoroughly mixed. If the sample is too viscous, place it in the tightly sealed container in a bath or oven maintained at about 145 °F (63 °C) until it becomes sufficiently liquid for stirring.

PROCEDURE:

1. Maintain the bath at a temperature of about 140 °F (60 °C).
2. Preheat the clean, dry viscometer.
3. Charge the viscometer (following the recommendations of the manufacturer). Allow the charged viscometer to remain in the bath long enough to reach the test temperature (between 10–30 min).
4. Start the flow of asphalt in the viscometer.
5. Measure (to within 0.1 s) the time required for the leading edge of the meniscus to pass from the first timing mark to the second (more than 60 s).
6. Clean and dry the viscometer.
7. Calculate the kinematic viscosity to three significant figures using the following equation:

$$\text{Kinematic viscosity, cSt} = C \times t$$

where C is the calibration constant of the viscometer in cSt/s (ASTM D2170) and t is the efflux time in seconds.

REPORT: Report the test temperature and kinematic viscosity in centistoke.

Test BITU-4: Viscosity of Asphalt Cement Using Capillary Viscometer

PURPOSE: To determine the viscosity of asphalt cement using capillary viscometer.

RELATED STANDARD: ASTM D2171.

DEFINITIONS:

- *Viscosity* is the resistance to flow of a liquid.
- *Coefficient of viscosity* (also called *absolute viscosity*) is measured as the ratio between the applied shear stress and the rate of shear. The cgs unit, 1 g/cm · s, is called *poise* (P).
- When the ratio between the shear stress and the rate of shear is not constant, the liquid is called *non-Newtonian*.

EQUIPMENT: Capillary-type viscometer, thermometer, bath, vacuum system, timer.

SAMPLE: A minimum sample of 20 mL of asphalt cement in a suitable container and heated to about 275 °F (135 °C). Constant stirring may be required to prevent local overheating and to release entrapped air.

PROCEDURE:

1. Maintain the bath at 140 °F (60 °C).
2. Select a clean and dry viscometer that will give a flow time greater than 60 s, and preheat to about 275 °F (135 °C).
3. Charge the viscometer by pouring the prepared sample to within ± 2 mm of fill line E (refer to the viscometer).
4. Place the charged viscometer in an oven or bath maintained at about 275 °F (135 °C) for a period of about 10 min.
5. Remove the viscometer from the oven or bath, and insert it in a holder. Position the viscometer vertically in the bath so that the uppermost timing mark is at least 20 mm below the surface of the bath liquid.
6. Establish a 300 Hg vacuum below atmospheric pressure in the vacuum system and connect the vacuum system to the viscometer with the toggle valve or stopcock closed. Maintain the vacuum for 30 min.
7. At the end of 30 min., start the flow of asphalt in the viscometer by opening the toggle valve or stopcock.
8. Measure, within 0.1 s, the time required for the leading edge of the meniscus to pass between successive pairs of timing marks.
9. Clean and dry the viscometer.
10. Calculate the viscosity using the following equation:

$$\text{Viscosity, } P = K \times t$$

where K is the selected calibration factor in P/s and t is the flow time in seconds.

REPORT: Report the test temperature and vacuum with the calculated viscosity value in poise.

Test BITU-5: Performance-based Grading System Tests for Asphalt Binders

PURPOSE: To characterize asphalt binders for their suitability for application in pavements, and to obtain the properties for grading based on the performance-grading system.

EQUIPMENT: Physical properties of asphalt are measured from four tests: rotational viscometer, dynamic shear rheometer, bending beam rheometer, and direct tension.

SAMPLE: Some physical properties of the binder are measured on samples that have been laboratory-aged to simulate their aged condition in a pavement. A pressure-aging vessel is used to simulate, in the laboratory, the aging condition of the pavement. Some properties are also measured on binder samples aged in the rolling thin-film oven test, to simulate oxidation hardening that occurs during mixing and placing.

PROCEDURE: The rotational viscometer measures the stiffness of asphalt at 135 °C (275 °F), when it behaves mostly as a viscous liquid. It is a rotational coaxial cylinder viscometer that measures viscosity by the torque required to rotate, at a constant speed, a spindle submerged in a sample of hot binder. Another unit measures the temperature. The test is conducted on unaged samples, and this property allows the user to ensure that the binder can be pumped during manufacture. The viscosity is read from the viscometer in Pa · s (Pascal seconds) as the average of three readings at one minute intervals, and the specifications require a viscosity of less than 3 Pa · s.

The dynamic shear rheometer test is used to measure the viscoelastic properties of the binder. It measures the complex shear modulus (G-star) and the phase angle, by subjecting a small sample (8-mm diameter and 2-mm thick, or 25-mm diameter and 1-mm thick) sandwiched between two parallel plates—one fixed and other rotating—to oscillatory shear stresses. The complex shear modulus measures the total resistance to deformations—composed of elastic and viscous components—when exposed to a repeated pulse of shear stress. The phase angle is used as an indicator of the relative amounts of recoverable and nonrecoverable deformations. The two quantities (G-star and phase angle) are determined by measuring the strain response of the specimen to a fixed torque. The test temperature is selected according to the grade of the binder. The ratio between complex shear modulus and sine of the phase angle is used as a means of controlling the asphalt stiffness. By controlling the stiffness at high temperatures, the binder can be expected to have adequate resistance to rutting at high temperatures. By limiting the stiffness at intermediate temperatures, the binder is expected to provide sufficient resistance to fatigue cracking.

The bending beam rheometer test is used to measure the low temperature stiffness properties of the asphalt binder. The test measures the creep stiffness and the creep rate (logarithmic) by recording the response of a small binder beam specimen to a creep load at low temperatures. A prismatic beam of asphalt binder is subjected to a constant load at its midpoint. The instrument consists of a loading frame, a controlled temperature bath that can maintain any temperature between −40 °C (−40 °F) and 25 °C (77 °F), and an automated data collection system. The creep stiffness is calculated by knowing the load applied and the deflection at any time during the test, as the ratio between the maximum stress (calculated from the load and the beam dimensions) and the maximum strain (from the deflection readings and the beam dimensions) for the specified time. Binder specifications place limits on creep stiffness and creep rate, depending on climate. A binder with a low creep rate has higher resistance to cracking in cold weather. Similarly, a binder with a higher creep rate is more resistant to low temperature cracking. Some polymer-modified asphalts, however, may show large creep stiffness at low temperatures, but their crack resistance remains high due to their ductility (ability to stretch without fracture).

The ductility, measured by the direct tension test, provides information on the ability to resist cracking at low temperatures. The equipment consists of a displacement-controlled tensile testing machine with gripping system, a temperature-controlled chamber, and a data acquisition system. The sample—a small dog-bone–shaped specimen—is pulled up to failure at a constant rate of deformation (1 mm/min.) at a test temperature selected based on the binder grade. An extensometer measures the elongation using which tensile failure strain is recorded. The strain at failure provides information on the ability of the binder to resist cracking at low temperatures.

PROBLEMS

1. What are the three forms of bitumen commonly used in construction?
2. How is tar manufactured?
3. What are the various forms of asphalt?
4. What are the uses of asphalt?
5. What are the four classifications of petroleum asphalt?
6. Explain the uses of cutback asphalt.
7. Define viscosity, penetration, and specific gravity of asphalt cement.
8. What are the various grades of asphalt?
9. Discuss the requirements of aggregates for use in asphalt concrete.
10. Explain the various elements of a flexible pavement.

7

Iron and Steel

Historical records and ancient sites show worldwide use of iron and steel during the second and third millennia B.C. Many historians think that the original center of steelmaking can be placed in India: the Hyderabad province and Tiruchirapally are considered to have been the centers of production of steel more than 3000 years ago. Iron ore was heated with charcoal, and the spongy metal thus obtained was drawn and worked by hammering to make wrought iron. Small pieces of wrought iron and carbonaceous matter were heated at high temperature in crucibles lined with clay. Resulting impure steel was heated and hammered to remove impurities, and was marketed in the form of round cakes a few inches in diameter, known as *wootz* steel.

The technique of ironworking also existed in Asia Minor. Philistines (sea people from whose name comes the word "Palestine") were skilled ironworkers. The availability of iron plow tips made it possible to till heavy soil in Palestine. The use of iron also improved the productivity of workers. Ingots of steel from India were taken to Damascus, where Syrian ironworkers made them into sword blades. This popularized Damascan (damascene) knives and swords.

Knowledge of iron smelting became common around 1200 B.C. The Chinese are known to have used iron plows around 500 B.C. Iron bars were used as currency around 400 B.C. in Europe, employed as a barter medium for cattle, land, and so on. The iron pillar of New Delhi [16 in. (40.6 cm) in diameter and about 16 ft (4.8 m) in height], which is nearly 1500 years old, is thought to have been made with wrought iron. One of the early uses of cast iron was in the making of church bells and guns; iron was also used in transportation. One of the Chinese monks who visited India around A.D. 400 described seeing suspension bridges held up by iron chains.

The first iron produced in the United States was at Saugus, Massachusetts, in 1645, by employing ironworkers brought over from England. The iron was used to make plows, tools, swords and, of course, nails. By about 1770, some 30,000 tons of iron was being produced in the United States annually. A machine to make cold-cut nails was invented in 1777. Introduction of rail transportation opened up huge new markets for the iron and steel industry.

During the 1860s, an Englishman, Henry Bessemer, invented a new method of converting iron to steel. Adaptation of this process set American steel interests in head-to-head competition with British steel. The first steel rails were produced commercially in Johnstown, Pennsylvania, in 1867. Steel rails helped the nation move forward industrially.

New steel structures and products began to proliferate. Production of high-tensile steel wire made it possible to build suspension bridges. Demand for steel for naval use increased the number of steel plants. The development of universal mills that could roll wide-flange structural sections during the beginning of the twentieth century revolutionized the building industry. Around the same time the auto industry started to expand, further increasing the demand for steel. The post–World War II era continued the large-scale production of steel that helped rebuild war-torn countries. Today, steel is produced in rebuilt or new plants that are totally modern, efficient, and use computerized equipment.

Steel is now used in countless products, from the heaviest machines and structures to paper clips and watch springs. Using steel, sheets, bars, angles, wide-flange beams, heavy structural products, joists, decks, screws, bolts, bearings, metal buildings, and other construction products are manufactured. Steel is one of the most versatile of human-made construction materials.

7.1 IRON

The basic constituent of steel is iron. Iron is widely available all over the world, but only in combination with other elements. Next to aluminum, iron is the most abundant metallic material in the earth's crust (about 4–5 percent). It is found in the form of ores as oxides, carbonates, silicates, and sulfides.

The most important iron-bearing minerals or iron ores are *limonite* (brown iron ore, $Fe_2O_3 \cdot nH_2O$) *hematite* (red iron ore, Fe_2O_3) and *magnetite* (magnetic oxide of iron, Fe_3O_4). Hematite, which is anhydrous ferric oxide, the most commonly used ore, contains about 70 percent pure iron. Its specific gravity varies between 4.5–5.3.

The mined ore is crushed to small particles of size 1 in. (25 mm) or smaller. These particles are further reduced to fine powder, which is later converted to pellets or sinters. *Sintering* is a process of application of heat that results in the conversion of fine ore into hard and porous lumps [size 0.4–2 in. (10–50 mm)]. *Pelletizing* is a process of forming balls [0.4–0.8 in. (10–20 mm) in diameter] in the presence of moisture and additives such as bentonite or lime. (Bentonite is a fine-grained plastic clay that is water absorbent.) When powdered limestone is incorporated in the pellets, *fluxed pellets* result.

Blast Furnaces. Iron is presently produced in blast furnaces, developed in Europe around 1400. At that time, the hot metal tapped in the furnace and cast in sand was designated a "pig" if the weight was less than 112 lb (50.7 kg). From this, the term "pig iron" was derived, which represents iron produced in a blast furnace.

A blast furnace is a tall, circular shaft that increases in cross-sectional area from the top to the base (Fig. 7.1). The inside of the furnace is lined with firebricks or carbon bricks. A tapping hole is located at a height of about 2–3 ft (0.6–0.9 m) from the furnace bottom. The main function of a blast furnace is to reduce the ore to metal, followed by separation of the metal from the impurities.

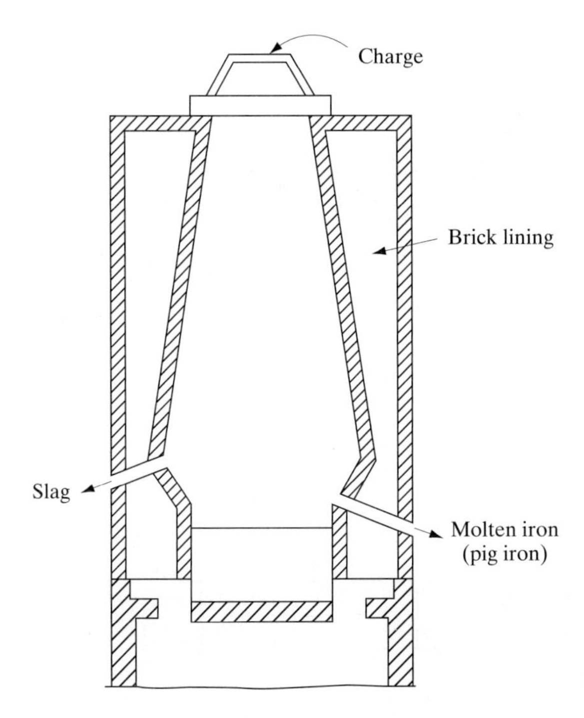

Figure 7.1 Schematic view of blast furnace.

The iron ore in the form of pellets or sinters is charged into the furnace with coke and limestone (Fig. 7.2). Coke is a carbonaceous solid made from coal (bituminous), petroleum, or other raw materials by thermal decomposition, and is typically produced by heating a mixture of coal in the absence of air (in coke ovens). It is charged into the furnace as pieces [0.8–3.2 in. (20–81 mm) in size]. Limestone is added to act as a flux that holds the silica and alumina impurities of the ore and coke. Dolomite with or without limestone can also be used as a flux.

A powerful air blast through the bottom raises the temperature sufficiently to burn coke, melt iron, and burn off oxygen. As the charge comes to the bottom of the furnace, the temperature increases to about 3000 °F (1650 °C), which is enough to melt the iron. The molten iron, which has a high carbon content, is collected at the

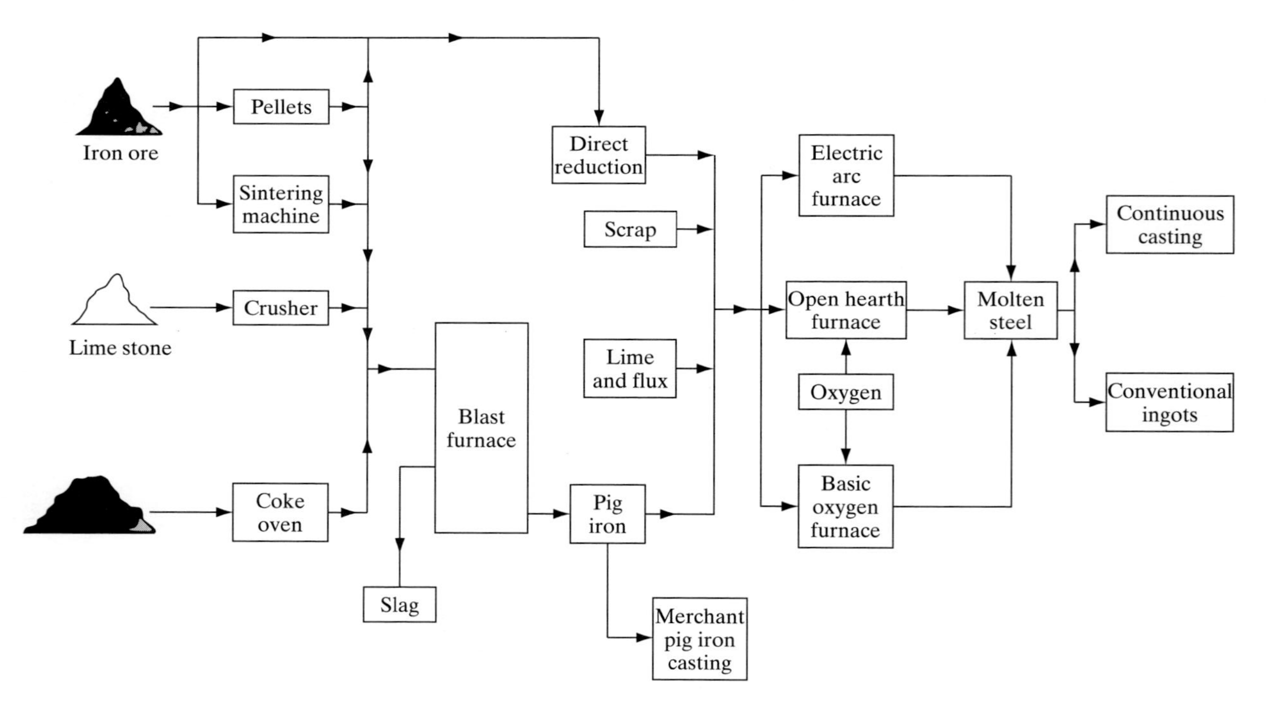

Figure 7.2 Steelmaking process.

bottom of the furnace every few hours. It flows into hot metal cars, which transport it to the steelmaking furnace.

A nonmetallic product, consisting essentially of silicates and aluminosilicates of calcium and other bases, is produced simultaneously with the molten metal. This product, called slag (or fluid slag), floats to the top and is also collected at the base of the furnace. It is diverted to slag pits or pots for further processing. Pulverized or crushed slag is used in the manufacture of special cements.

As noted, pig iron (tapped every 4–6 hours) runs to hot metal cars for conveyance to the steelmaking location. When pig iron castings are required, it is run into a pig-casting machine. At present, most pig iron is made into steel; the balance goes to make cast iron.

Pig iron is not pure iron; it is metal nearly saturated with carbon (taken from the coke). In addition, pig iron contains manganese, silicon, and other materials. The type and constituents of pig iron depend on the composition of the ore.

7.1.1 Cast Iron and Wrought Iron

Iron products come in three commercial forms (Table 7.1):

- Wrought iron
- Steel
- Cast iron

Wrought iron has the least, and cast iron the greatest amount of carbon. Increase in the amount of carbon decreases the melting point of the metal. In fact, carbon exerts the most significant effects on the microstructure and properties of iron products.

TABLE 7.1 SOME PROPERTIES OF IRON-BEARING METALS

Material	Carbon content (%)	Silicon (%)	Manganese (%)	Melting point [°F (°C)]	Tensile strength [ksi (MPa)]	Tensile modulus [ksi (GPa)]
Pig iron	3.5–4.5	1–2	0.25–1	3040 (1670)		
Pure iron	0.01–0.02	<0.01	0.01–0.02	2795 (1535)	49 (335)	
Wrought iron	0–0.1	0.1–0.2	<0.1		45–55 (310–380)	29×10^3 (200)
Mild steel	<0.25	0.05–0.25	<0.68	varies with carbon content	64 (450)	3×10^3 (207)
High-carbon steel	1.4	<0.8	<1.5	varies with carbon content	130 (900)	30×10^3 (207)
Cast iron	5.0	1.25	1–2.5	2084 (1140)	16 (110)	$15–22 \times 10^3$ (103–152)

The upper theoretical limit of carbon in steel is 1.7 percent; in structural steel the carbon content is generally less than 0.25 percent (by weight).

Wrought Iron. Wrought iron is manufactured by melting and refining iron to a high degree of purity. The molten metal is then poured into a ladle and mixed with hot slag. The fluxing action of the slag causes a spongy mass to form which is processed by rolling and pressing. This is wrought iron. It is the only iron-bearing material containing slag.

Wrought iron can be described as a low-carbon steel (less than 0.1 percent carbon by weight) containing a small amount of slag, usually less than 3 percent. The slag is distributed throughout the metal and appears as long fibrous elements, which distinguish this metal from steel with the same carbon content. Wrought iron contains small amounts of manganese (less than 0.1 percent) and silicon (0.2 percent).

The mechanical properties of wrought iron are nearly the same as those of pure iron, and its ductility is somewhat lower than that of steel. Its tensile strength along the grain varies between 45–55 ksi (310–380 MPa), and across the grain (transverse to the direction of rolling), the tensile strength is lower.

Wrought iron can be cold worked, forged, and welded like steel. Forging is working a metal to a predetermined shape by one or more processes such as hammering, pressing, and rolling at a temperature above the recrystallization temperature. Cold working is the process of working at a temperature that does not alter the structural changes caused by the work or that is below the recrystallization temperature. Wrought iron can be molded easily and has good resistance to corrosion. It is used to make pipes, corrugated sheets, grills, bars, chains, and other products.

Cast Iron. Cast iron is manufactured by reheating pig iron (in a cupola) and blending it with other materials of known composition. Alternate layers of pig iron (with or without scrap steel) and coke are charged into the furnace. Limestone is added to flux the ash from the coke. Heat necessary, for the smelting is supplied by the combustion of coke and air supplied by the blast. The function of the cupola is to purify iron and produce a more uniform product. When sufficient metal is accumulated at the bottom of the furnace, it is tapped.

Ordinary cast iron is called *gray cast iron,* owing to the color of fracture. When the fracture surface has a silvery white metallic color, it is called *white cast iron.* The amount and form of carbon affect the strength, hardness, brittleness, and stiffness of cast iron. Gray cast iron contains free carbon (graphite flakes), which makes the metal weak and soft. It is the most widely used cast iron. In white cast iron, carbon is combined chemically with iron in the form of cementite (iron carbide, Fe_3C), which makes this metal strong, hard, and brittle. (Note that pure iron is a soft and ductile metal; the addition of carbon to iron increases its hardness and strength, but lowers the ductility.) Cast iron has high compressive strength, but its tensile strength is low. Until about 1800, columns in buildings were built using only cast iron. Later, cast iron beams were developed. Wrought iron was used in bridges and multistory buildings during the 1850s. But by the end of the nineteenth century, steel dominated bridge and building construction. Presently, a number of types of steel are produced that find application in nearly all phases of the construction industry.

7.2 STEEL

Steel is a combination of iron and carbon. The carbon content may range between about 0.01–1 percent. The addition of carbon not only hardens the metal but also imparts other distinct properties. Steel in addition contains varying amounts of manganese (less than 1.6 percent), phosphorus, sulfur, and silicon (less than 0.6 percent), together with some 20 other alloys. The alloys are added to molten steel to produce steel of different characteristics, such as hardness, tensile strength, and toughness.

Steel is produced in a basic oxygen furnace (BOF), open hearth furnace, or electric arc furnace. Ingredients for the BOF and the open hearth furnace are the same: one or more of molten pig iron, iron ore, mill scale, scrap metal (for example, junked cars and steel cans), and limestone. Iron ore controls the carbon content and limestone acts as a fluxing agent. The principal ingredient for the electric arc furnace is scrap metal.

A charge consisting of iron-bearing materials is introduced into the furnace (Fig. 7.2). In the BOF, high-purity oxygen is blown at high velocity through the upright furnace to promote rapid oxidation, which is the combination of oxygen with nonferrous elements, especially carbon. This reaction generates enough thermal energy to raise the temperature to the required levels without requiring external fuel to be supplied to the furnace. The oxides either leave the furnace as gases or combine with the slag.

The slag-producing materials (limestone or lime) are added after the oxygen blow. When the molten liquid has reached the specified composition, the furnace is tapped into a ladle. During the tapping, alloying elements are added. Basic oxygen furnaces, largely controlled by computers, can produce a heat of steel in about 45 minutes. At present, BOF is the chief method of producing steel.

An open hearth furnace (so named because the charge is open to the surface of flames) utilizes the same raw materials (charge) as in the BOF. The charge is melted through liquid fuel and gas with preheated air for combustion. The flames reduce the amount of unwanted elements such as carbon, manganese, silicon, phosphorus, and sulfur. The oxidizing agent is iron ore.

An electric arc furnace is used to produce regular carbon steel grades. These furnaces also allow close control of temperature and refining conditions necessary in the manufacture of some special steels. The charge consists essentially of scrap metal; some iron ore and lime are added during the melting process. High-temperature electric arcing generates the high temperatures required for melting and refining the steel.

Modern electric arc furnaces work on 100 percent scrap, but some use a small amount of cold or molten pig iron. Before the scrap metal is placed in storage, it is screened for contaminants and nonmetallic parts. Desired carbon content in the molten metal is achieved by the addition of coke. Lime is used to remove impurities in the steelmaking.

Various alloying elements are added during tapping of steel. Manganese gives strength and toughness to steel; in structural steel, manganese content is about 1.35 percent. Silicon provides strength and hardening properties, and it is also an oxidizing agent.

Aluminum is added to control deoxidation and for fine graining of steel. Chromium is added along with nickel to produce stainless steel. Nickel also gives toughness to the metal, and chromium steels can withstand abrasion and shock very well. Molybdenum is added to improve tensile and hardening properties. The corrosion resistance of steel containing sufficient levels of copper, chromium, nickel, and silicon is superior to that of steels without these elements. Copper increases the tensile properties and hardness but decreases the ductility, and carbon–manganese levels are important to determine the weldability of steel.

7.2.1 Steel Products

The molten steel from the furnace is conveyed through ladles to ingots or goes through a continuous slab caster (Fig. 7.2). The continuous casting converts a heat of molten steel to semifinished products of varying widths, lengths, and thicknesses. The semifinished products come in three types (Fig. 7.3):

- Blooms
- Slabs
- Billets

Blooms are generally square or rectangular in cross section (area greater than 36 in.^{2} or 232 cm^2), while slabs are flat and wide. Billets are, generally, square lengths of steel 4–6 in. (10–15 cm) in cross-section, which are shipped to rod mills for rolling. Steel rods are produced on high-speed continuous rod mills from preheated billets.

The process of pouring molten steel into ingot molds is called *teeming*. The molten steel is allowed to cool in the ingot molds (usually of cast iron), and then the molding is removed. The steel ingots, which weigh between about 10–100,000 lb (4530–45,300 kg), are later heated in special furnaces called *soaking pits* until they attain uniform temperature throughout. Reheated ingots travel to either blooming or slabbing mills, where they are squeezed down to a smaller size so that they can be handled on a finishing mill. Then they are rolled into semifinished products. This conventional casting procedure is time consuming and costly, due to the number of stages involved, the additional spaces required for casting, stripping, and reheating, and the transportation and handling costs.

In the continuous casting approach, molten steel is poured between a pair of water-cooled flanged rolls so as to solidify it continuously into a thin-rolled strip or

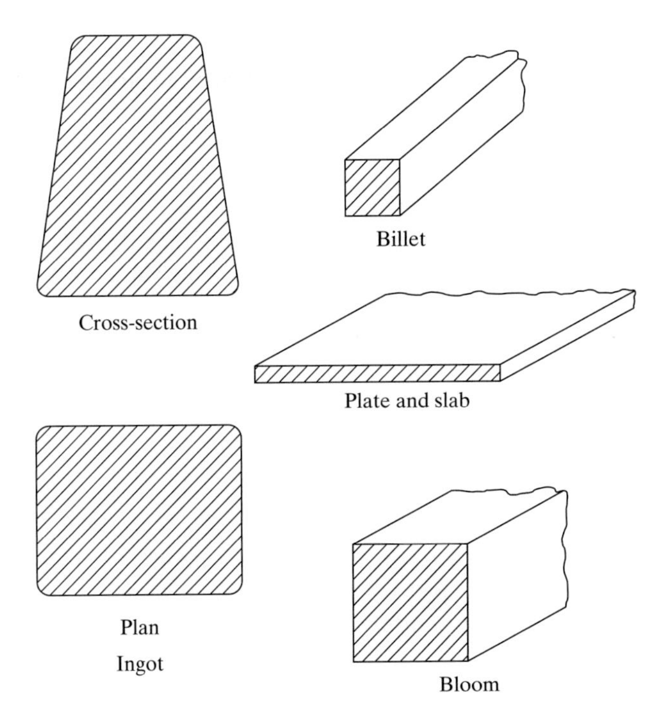

Figure 7.3 Steel products.

sheet. The intermediate stages of first casting the hot steel into slabs or ingots and later gradually rolling the reheated metal into the desired shapes are missing in this operation.

Universal mills are used for rolling wide-flange shapes. Both wide-flange and H-pile sections are generally hot-rolled products. To save time in rolling, structural shapes are also cast continuously, which, along with low carbon levels, contributes to improved surface quality. In the continuous-casting approach (using universal roughing and edging mills), steel is cast as a beam blank in a shape that approximates the net shape of the beam. The intermediate milling stage is eliminated in this process. Beams up to 24 in. (61 cm) deep are presently cast using this method.

Conventionally, structural shapes and H-piles are produced on structural and shape mills. Standard mills form various shapes by passing reheated (to a uniform temperature) blooms or billets through the mill rolls. Hot rolling involves passing the material between two rolls where the gap between the rolls is lower in height than the material entering. The metal passes repeatedly back and forth through the same rolls to reduce it to the desired thickness and shape. Plain barrel rolls are used for flat products, and grooved rolls are for structural shapes, rounds, and squares.

A cubic foot of steel weighs about 490 lb. The weights of the structural shapes are calculated from their theoretical dimensions and rounded off to the nearest pound.

Bars are hot-rolled from billets. (Billet mills reduce blooms to billets.) In the bar mill, reheat furnaces bring the billets to rolling temperature. The rolls exert pressure horizontally and vertically to produce twist-free bars. During each pass through the rolls, the metal elongates and is reduced further in cross section. Grooves on the surface of the roller produce the desired bar shape and size. Steel wires and rods are made similarly in rod mills.

Structural tubes are made from large steel coils which are uncoiled at the beginning of a long process line and pulled through a series of forming stands, which shape the flat steel into a round tube. After welding, the tube goes through an air-cooling section, then a water-cooling section, and finally into a sizing section, which finalizes the round shape into other shapes. The tube, which has the desired shape, is cut into the required lengths and discharged onto a bundling table, stacked in the required configuration, and banded for shipment.

7.3 PROPERTIES OF STEEL

Steels usually contain less than 1 percent carbon by weight; structural steel has less than 0.25 percent of carbon. Manganese is the principal alloying element and is added when the steel is in molten condition, in amounts up to about 1.6 percent.

Carbon exerts the most significant effects on microstructure and properties of steel. Increase in carbon content increases the hardness, strength, and abrasion resistance, but decreases the ductility, toughness, and impact resistance (Fig. 7.4). Ductility, as measured by the percentage of elongation during the tension test, decreases drastically with increase in carbon content. Toughness, as measured by the area under the stress–strain diagram, decreases rapidly with carbon content exceeding 0.4 percent.

Tensile strength and yield point of steel are maximum when carbon content is about 1 percent. High-strength low-alloy structural steel (ASTM A572) is expected to have a tensile strength of 60 ksi when the carbon content is 0.2 percent, and 80 ksi when the carbon content is 0.26 percent. With heat treatment, it is possible to change the properties of steel without changing the chemical composition. Ductility is also affected by heat treatment.

Some of the carbon exists within the crystals of iron (in solution), and most of the balance forms a chemical compound (interstitial or intermetallic compound) with the iron as Fe_3C (cementite).

Cementite is the hardest and most brittle component of steel and is found in high-carbon steel, and has 6.67 percent carbon and 93.33 percent iron. In structural steel carbon exists mostly within the crystals of iron. Ferrite iron (also called alpha iron) is the stable form of iron at temperature below 1670 °F (910 °C), and dissolves the carbon in interstitial solid solution to a maximum of 0.08 percent. Ferrite is the solid solution of carbon in alpha iron and contributes to the cold-working properties of steel. Alpha iron is the softest compound in steel and is very ductile.

At 2065 °F (1130 °C), iron undergoes a polymorphic change to gamma iron, which is capable of dissolving carbon a maximum of 2 percent. *Austentite* is the solid

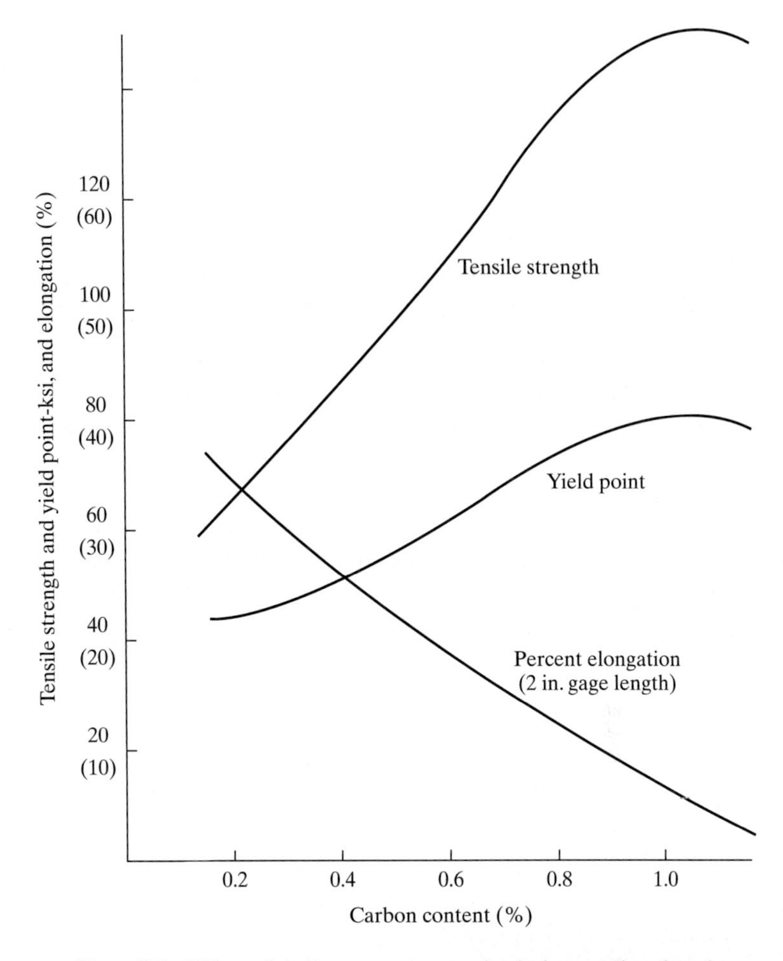

Figure 7.4 Effects of carbon content on mechanical properties of steel.

solution of carbon in gamma iron. It is soft and ductile. At about 2535 °F (1390 °C), a further polymorphic change takes place and delta iron is produced. It has magnetic properties.

In general, properties of steel are greatly affected by three factors:

- Chemical composition
- Heat treatment
- Mechanical work

Carbon and alloying elements affect both physical properties (such as weldability and corrosion) and mechanical properties (such as yield strength, tensile strength, and ductility).

Mechanical Properties. Figure 7.5 shows typical stress–strain diagrams of steel in tension. All steels show an initial elastic phase followed by yielding or strain hardening. The elongation or strain at failure depends on the type of steel. The modulus of elasticity of steel, which is the slope of the initial straight-line portion of the diagram, is constant, equal to 29,000 ksi (200 GPa) for all types of ordinary mild steel.

Working the steel to get desired shapes and sizes affects the stress–strain diagram. Hot working is mechanical deformation at temperatures slightly above the transformation temperature (recrystallization temperature), and cold working is mechanical deformation at temperatures below the transformation temperature. Cold working increases the strength properties but decreases the ductility.

The yield point is a stress at which there is a marked increase in strain without an increase in stress. Some steels exhibit a sharp knee or discontinuity at the end of the elastic state. Stress corresponding to the top of the knee is the yield point. In steels that do not show marked yielding during the tension test, yield strength is measured as stress corresponding to a known strain or an offset (proof stress). In the offset method, a line is drawn parallel to the initial straight-line part of the stress–strain diagram at the prescribed offset value. The stress

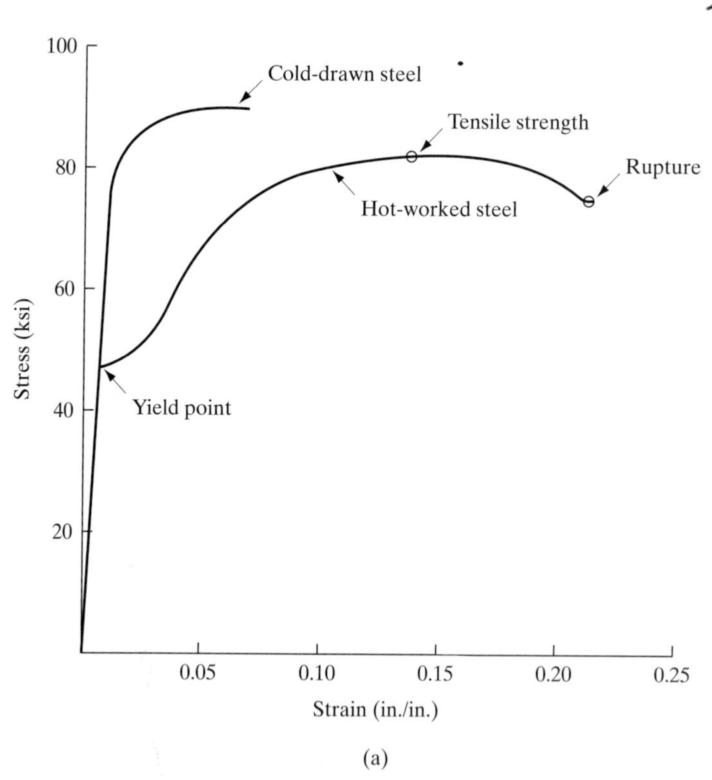

Figure 7.5 Typical stress–strain diagram of steel.

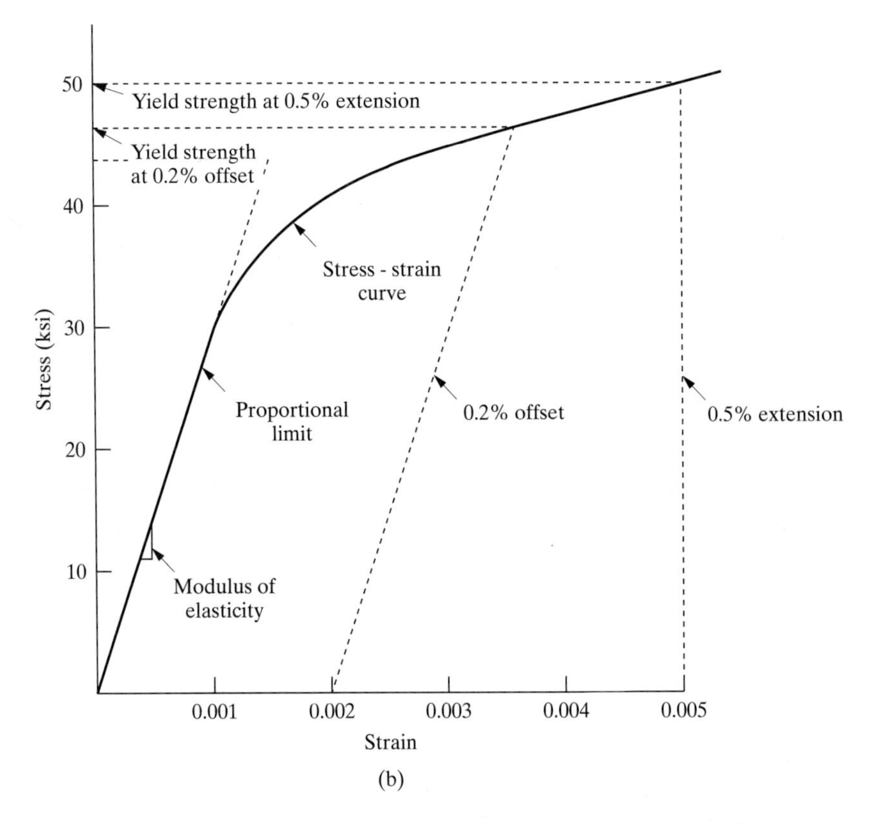

Figure 7.5 (*Continued*) Typical stress–strain diagram of yield strength determination.

corresponding to the intersection of this line with the stress–strain diagram is called yield strength.

Notch toughness is a measure that is commonly used to calculate steel's behavior in a ductile or brittle transition range. It is influenced by many factors, such as chemical composition, hot and cold working temperatures, internal cleanliness and method of fabrication, carbon content, oxygen level, and grain size.

Properties of steel are greatly affected by elevated temperatures (Fig. 7.6). Very high temperatures [exceeding about 900 °F (480 °C)] may not only cause loss of cross-section but may also result in metallurgical changes and severe deformation. Steel exposed to very high temperatures will have heavy scale, pitting, and surface erosion. Exposure to severe fire of nominal duration will destroy the ability of steel members to sustain loads. The modulus of elasticity of steel decreases from 29,000 ksi (200 GPa) at 70 °F to about 25,000 psi (172 GPa) at 900 °F. Beyond this temperature, the modulus of elasticity decreases rapidly at higher temperatures.

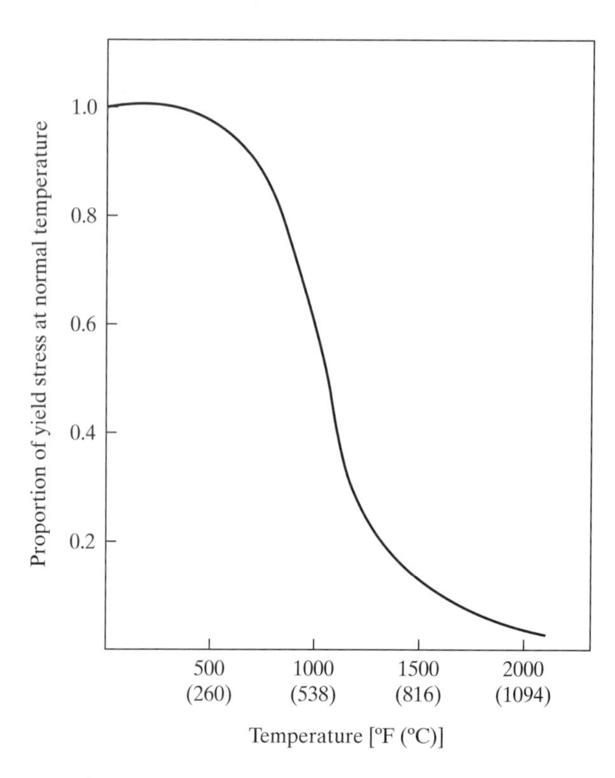

Figure 7.6 Effect of temperature on yield stress of steel (typical).

7.4 STRUCTURAL STEEL

Structural steel shapes are rolled flanged sections having one dimension of the cross-section equal to or greater than 3 in. (7.5 cm). They are manufactured in many shapes and grades (Fig. 7.7). These products are used as columns, beams, and bracing members in buildings; for trusses; as bridge girders; and in prefabricated structures and similar construction.

W shapes are doubly symmetric wide-flange shapes whose inside flange surfaces are substantially parallel (Fig. 7.8). The profile of a W shape of a given nominal depth and weight available from different manufacturers is essentially the same, except for the size of fillets between the flange and the web. These shapes can be used as beams or columns.

HP shapes are wide-flange shapes whose flanges and webs are of the same nominal thickness, and whose depth and width are essentially the same. They are used as bearing piles. *S shapes* are doubly symmetric shapes whose inside flange surfaces have approximately 16.67 percent slope (or 2 in 12 in.). They are used primarily as beams or girders in building construction. *M shapes* are doubly symmetric shapes which cannot be classified as W, S, or HP shapes.

C shapes are channels produced with inside flange surfaces having approximately 16.67 percent slope (or 2 in 12 in.). *MC shapes* are channels that cannot be

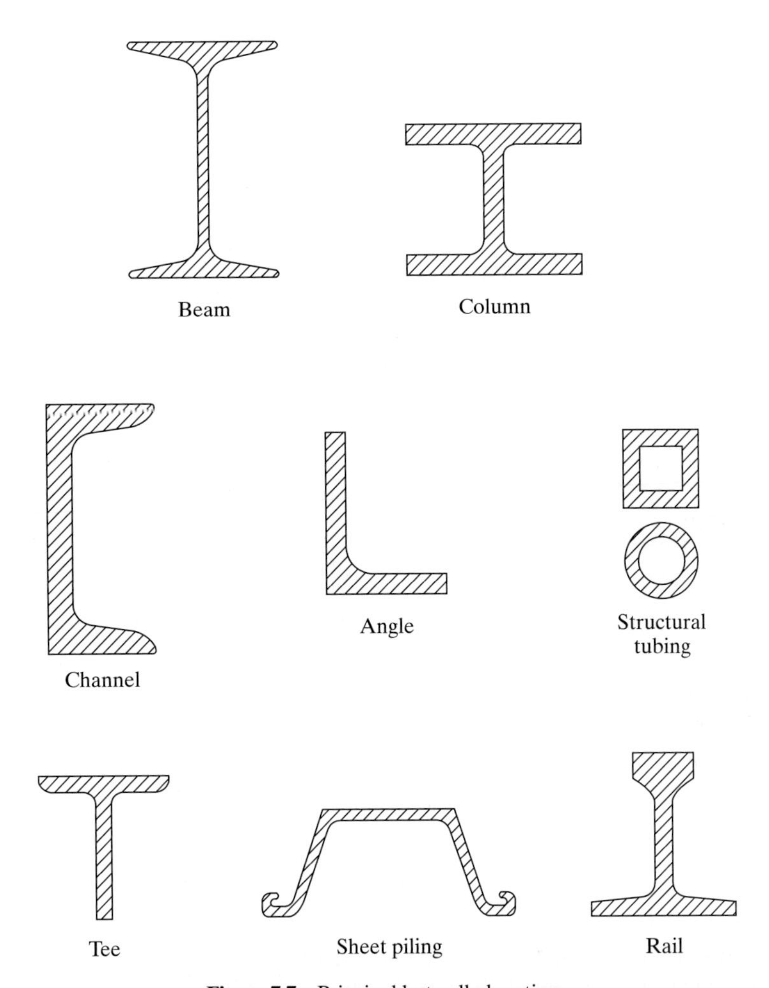

Figure 7.7 Principal hot-rolled sections.

classified as C shapes. Both C and MC shapes are available only from a limited number of producers. *L shapes* are equal-leg or unequal-leg angles.

A structural steel shape is designated by a letter followed by two numbers separated by a "×" sign. An example is "W8 × 67." The letter (W) identifies the shape. The first number (8) is the nominal depth of the cross-section in inches. The second number (67) is the weight of the member in pounds per linear foot. The nominal depth is not always equal to the actual depth of the member; the actual depth can be equal to, more than, or less than the nominal depth. For example, the actual depth of W8 × 67 is 9 in. and that of W8 × 10 is 7.89 in. Cross-sectional properties of these structural shapes (such as flange thickness and width, web thickness and width, cross-sectional area, and moment of inertia) are listed in tables in the *Manual of Steel Construction* of the American Institute of Steel Construction (Chicago). The

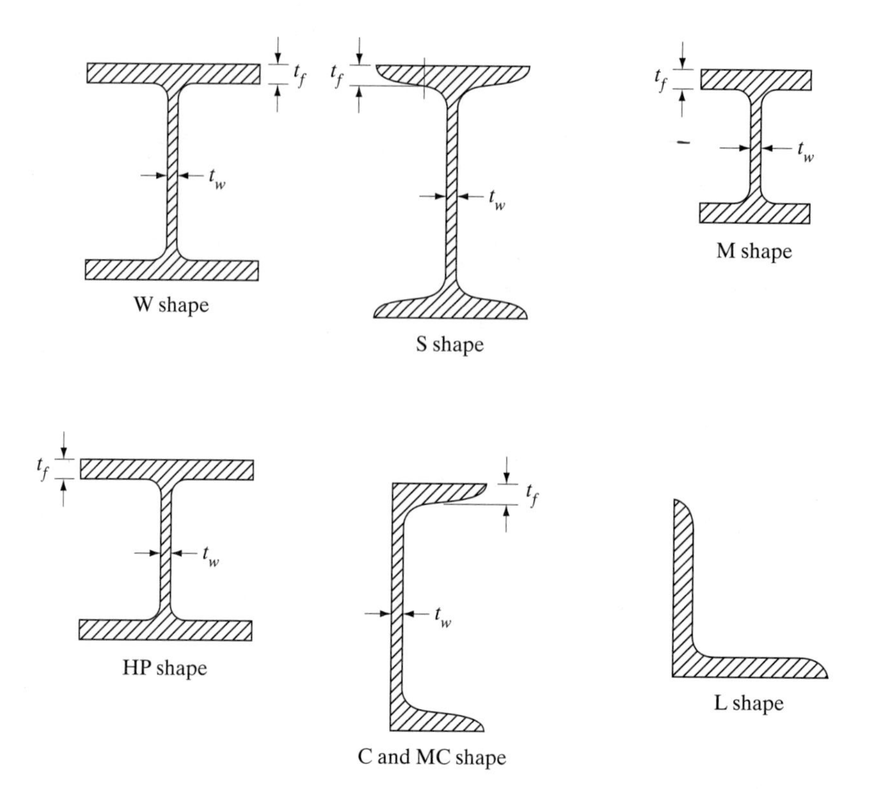

Figure 7.8 Structural shapes (t_f and t_w are flange and web thicknesses, respectively).

flange thickness listed in these tables for S, M, C, and MC shapes is the average flange thickness. In calculating the theoretical weights, fillets and roundings are included for all shapes except angles.

7.4.1 Structural Grades

Structural steel (hot-rolled structural shapes, plates, and bars) is produced in seven grades: A36, A529, A441, A572, A242, A588, and A514. Chemical and tensile requirements for these grades are shown in Table 7.2. A36/A57250 dual grade is a steel formulated to meet overlapping chemical and mechanical specifications for both ASTM A36 and ASTM A572 grade 50 steel. Weldability for this special steel is the same as that of grade 36 steel.

A36 grade, which is called all-purpose carbon steel (or structural carbon steel), is the most widely used steel in building and bridge construction. A529 steel, called structural carbon steel, has a 42-ksi minimum yield point. Grades A441 and A572 are called high-strength low-alloy structural steel; Grade A441 steel is discontinued from ASTM as of 1989. Grades A242 and A588 arc atmospheric corrosion-resistant

TABLE 7.2 PROPERTIES OF STRUCTURAL STEEL

ASTM designation	Type	Grade	Chemical requirements (%)			Tensile requirements		Availability
			Carbon (max)	Manganese (max)	Copper (min)	Tensile strength [ksi (MPa)]	Yield point min [ksi (MPa)]	
A36	Structural carbon steel	36	0.26	—	0.20	58–80 (400–550)	36 (250)	All plates, shapes, and bars
A572	High-strength low-alloy steel of structural quality	42	0.21	1.35	0.20	60 (415)	42 (290)	All shapes, sheet piling, and tees
		50	0.23	1.35	0.20	65 (450)	50 (345)	All shapes, sheet piling, and tees
		60	0.26	1.35	0.20	75 (520)	60 (415)	Limited shapes, all sheet piling, and tees
		65	0.26	1.35	0.20	80 (550)	65 (450)	Limited shapes and all tees
A529	Structural steel with 42 ksi min yield point	42	0.27	1.2	0.20	60–85 (415–485)	42 (290)	Selected shapes, plates, and bars of 1/2 in. and less in thickness
A441	High-strength low-alloy structural steel	(Discontinued as of 1989; replaced by A572)						
A242	High-strength low-alloy structural steel (corrosion resistant)	42–50	0.15	1.0	0.20	63–70 (435–480)	42–50 (290–345)	Limited shapes, plates, and bars
A588	High-strength low-alloy structural steel with 50 ksi min yield point (corrosion resistant)	50	0.17–0.19	0.5–1.25	0.2–0.5	70 (485) 63–70 (435–485)	50 (345) 42–50 (290–345)	All shapes Plates and bars
A514	High-yield-strength quenched and tempered alloy steel	90–100	0.12–0.21	0.4–1.10	0.15–0.50	100–130 (690–895)	90–110 (290–690)	Plates

high-strength low-alloy structural steel. Grade A514, quenched and tempered alloy structural steel, is available only as plates.

The properties of carbon steel are due primarily to the carbon content. The amounts of alloying elements, except manganese, are low. Alloy steel is any type of steel in which the amounts of alloying elements exceed certain limits, and it owes its distinct properties to elements other than carbon. High-strength low-alloy steel has improved mechanical properties compared to those of low-carbon steel.

As noted, A36 is the most commonly used structural steel grade (Fig. 7.9). It has been used successfully in buildings, transmission towers, bridges, and many other structures in various temperature conditions. Past experience has shown that this grade of steel can offer satisfactory performance in low temperatures. Brittle fracture is not usually experienced in these structures even at below-normal temperatures.

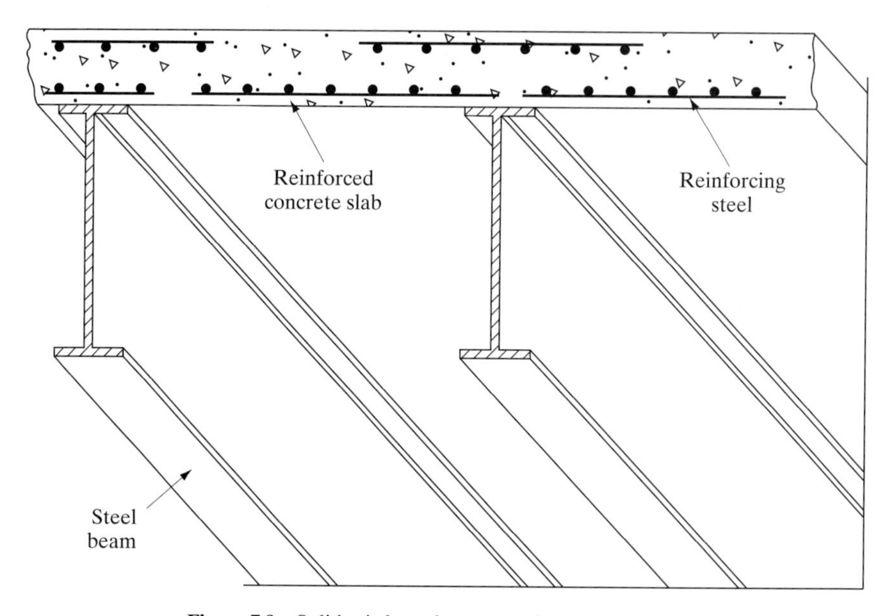

Figure 7.9 Solid reinforced concrete slab on steel beam.

Design using high-strength steel results in lighter sections and may prove economical, especially for tension members and beams in continuous and composite construction. Grades A588 and A242 can be used in bare (uncoated) condition in most atmospheres. When these two steels are exposed to normal atmospheric conditions, a tightly adherent oxide will form on the surface, which protects the steel from further atmospheric corrosion. But these two grades of steel can also be coated, and the coating life is typically longer than that for other steels. The use of coated steel results in lower maintenance costs over the life-span of the structure, although the initial material cost is high.

Failure Types. Typical failures in steel construction may be due to yielding of a member, failure of connections, or buckling of a member. Connection failure is the most common cause of collapse of steel buildings. Buckling of columns rarely has resulted in structural failure. Hot-rolled steel shapes used in the majority of steel buildings are not prone to brittle failure. When fracture occurs by cleavage at a nominal tensile stress that is below yield stress, it is called brittle fracture.

Increased strain rates—or structures that are loaded at a fast rate—tend to increase the possibility of brittle fracture. Cold working and residual tensile stress (such as those caused by welding) may increase the likelihood of brittle fracture. Welding may also leave notches or flaws in the parent metal, which may promote brittle fracture.

With decrease in temperature, most of the strength properties of structural steel (yield strength, tensile strength, modulus of elasticity, and fatigue strength) increase. But the ductility of the material, or the total elongation before fracture, decreases. Furthermore, there is a temperature below which structural steel, when subjected to tensile loads, may fracture by cleavage with little or no plastic deformation.

Investigations of earthquake damage have revealed that buildings of structural steel have performed excellently and better than any other type of construction in protecting life and preventing economic loss. This is due to physical and mechanical properties of steel:

- Light weight (compared to similar structures made of reinforced concrete)
- Ductility
- High tensile strength

In addition, damaged steel buildings can be repaired relatively easily. All structures (steel, reinforced concrete, and masonry) rely on steel (structural or reinforcing) to supply the ductility and toughness needed to resist severe earthquakes.

In general, structural steel used in bridges, buildings, and towers must possess adequate tensile strength, high ductility, excellent toughness, and sufficient elasticity (Fig. 7.10). Weldability is important for steel used in a majority of building projects, and this property deteriorates with increase in the carbon content. Structural carbon steel, or A36 steel, has good ductility and is weldable. In fact, all ASTM grades of structural steel are weldable, and their carbon content is limited to about 0.25 percent. In steel structures exposed to severe environment, adequate corrosion resistance is required. The major advantage of high-strength steel (A572 Grade 50) is in seismic design. This steel is manufactured so as to perform better—in terms of yielding—in seismic conditions. Currently the price difference between A36 and A572 (Grade 50) steels is marginal, and the use of the latter may offer cost savings due to reduction in the weight that comes with the higher-strength material.

Figure 7.10 A high-rise steel building.

Steel bridges and buildings are assembled by connecting the individual elements of various shapes and sizes using bolts or welding or both (Fig. 7.11). Connections between various members in a structure are of many types: simple, eccentric, tension, shear, shear and tension, and moment-resisting connections. When a beam transfers only the shear forces to the supporting columns or girders, the connection is a simple or shear connection; if, in addition, it transfers the couple (or flange forces), the connection is moment-resisting. In high-rise buildings, columns are made continuous using splices. A column may be fixed or pin-connected to its footing. A steel beam may resist all the superimposed loads by itself, or it can be made to perform like a composite beam, along with a concrete floor or roof slab (using shear connectors), wherein the stresses are shared by both steel and concrete.

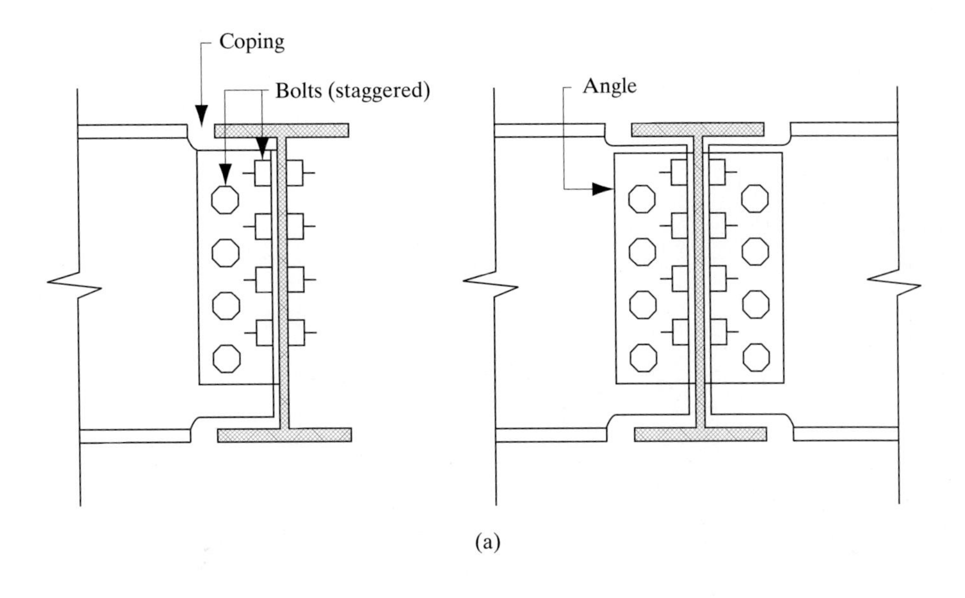

(a)

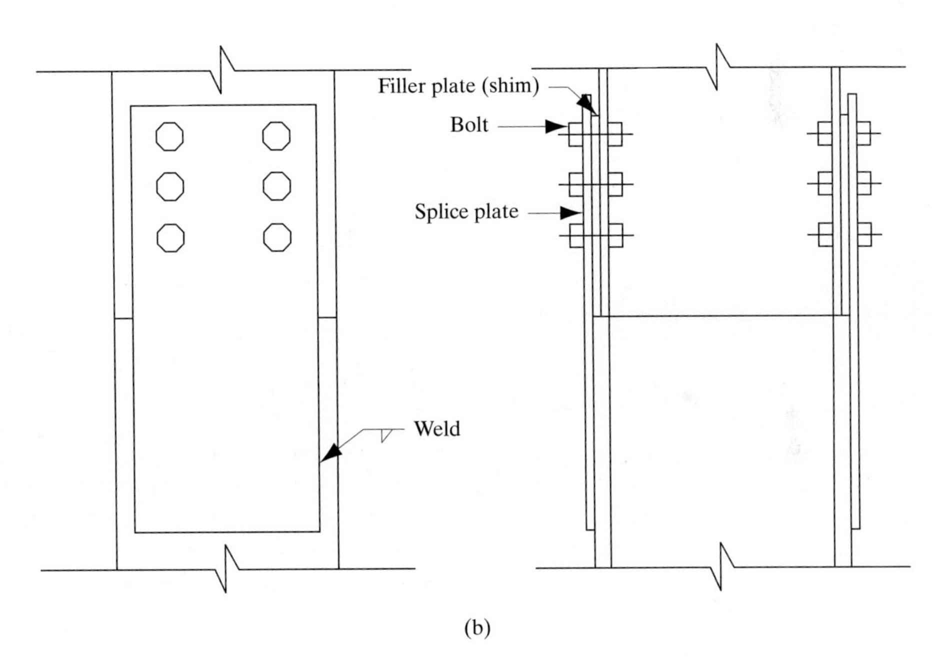

(b)

Figure 7.11 Typical connection details in structural steel:
(a) Double web angle shear connection (bolted to both the beam and girder webs).
(b) A typical column splice.

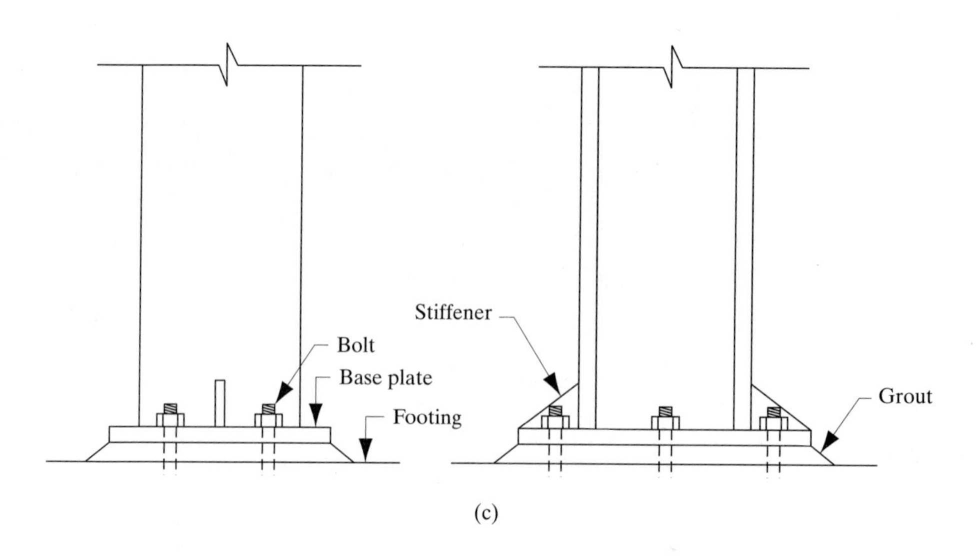

(c)

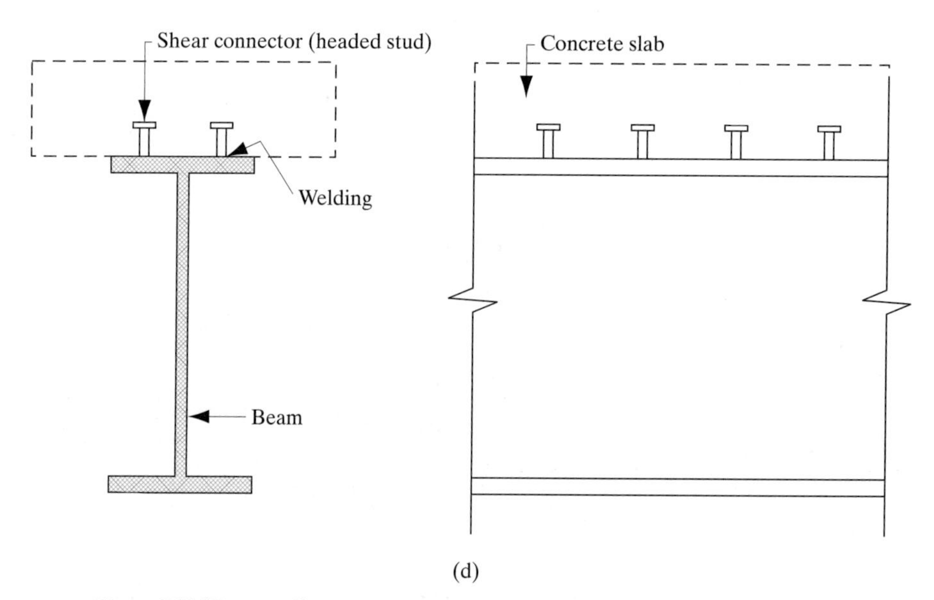

(d)

Figure 7.11 (*Continued*)
(c) Moment base connection (triangular stiffener reduces bending in the base plate and increases the fillet weld length for shear).
(d) Typical shear connector that causes the steel beam and the concrete slab to act as an integral unit (composite beam).

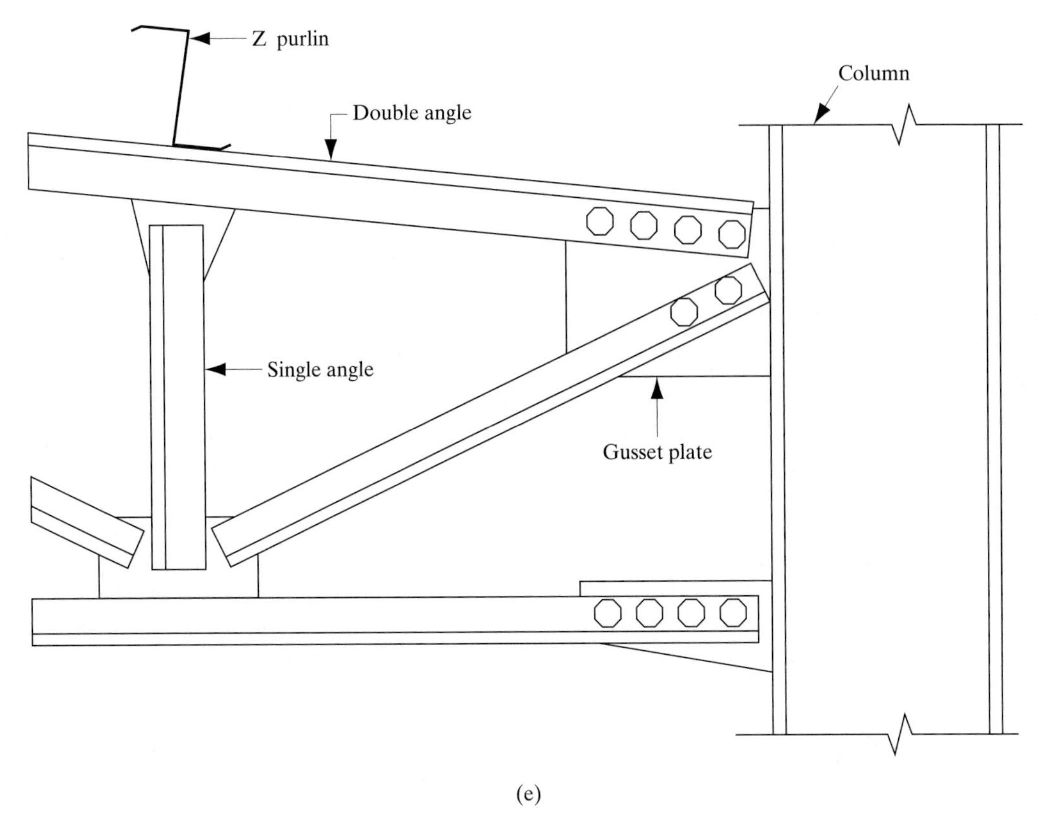

Figure 7.11 (*Continued*)
(e) Connection between a steel truss (using angles) and steel column.

7.5 REINFORCING STEEL

Concrete, the most common construction material all over the world, has low tensile strength and modulus of rupture, albeit its compressive strength is high, and can be as much as 12,000 psi (83 MPa). To use concrete in places where the applied loads induce tensile or bending stresses—such as in beams, slabs, walls, and beam-columns—the tensile capacity of the cross-section should be improved. Adding steel as reinforcement within the concrete cross-section enhances the bending capacity as well as the resistance to cracking from tensile forces. In addition, this composite material—steel carrying the tensile forces, and concrete the compressive forces—possesses higher ductility and greater safety factor. The cost of steel reinforcement in a reinforced concrete structure is about 50–70 percent of the total cost, including the cost of formwork and construction.

Reinforced concrete can thus be treated as a composite material, made from two distinctly different materials: steel, which has high modulus of elasticity and tensile strength, and concrete, which has superior compressive and bond strengths. If a reinforced concrete section is to perform according to the theoretical model—a composite—it is essential that the steel reinforcement remains bonded to concrete throughout the service life. Adequate bond is ensured by using *rebars* (reinforcing bars) with surface deformations or lugs, called deformed bars. The bond capacity in these bars is primarily from the surface friction and roughness.

Reinforcing steel is manufactured in three forms:

- Plain bars
- Deformed bars
- Plain and deformed wire fabric

A deformed bar (Fig. 7.12) is intended for use in reinforced concrete construction; the surface has lugs (protrusions or deformations) that inhibit longitudinal movement (or slip) of the bar relative to the concrete surrounding it. Deformed bars are used in almost all types of reinforced concrete construction: slabs, beams, columns, walls, footings, foundations, dams, and others (Fig. 7.13). They are also used as reinforcement in reinforced masonry construction. A plain bar is a round steel bar without surface deformations, and a wire fabric is a full steel in which the wires pass each other at right angles, and one set of elements is parallel to the fabric axis. Fabrics with welded joints, or welded wire fabrics, are used as nominal reinforcement in floor slabs, roof slabs, slabs-on-grade, and pipes (Fig. 7.14).

Figure 7.12 Deformed bars.

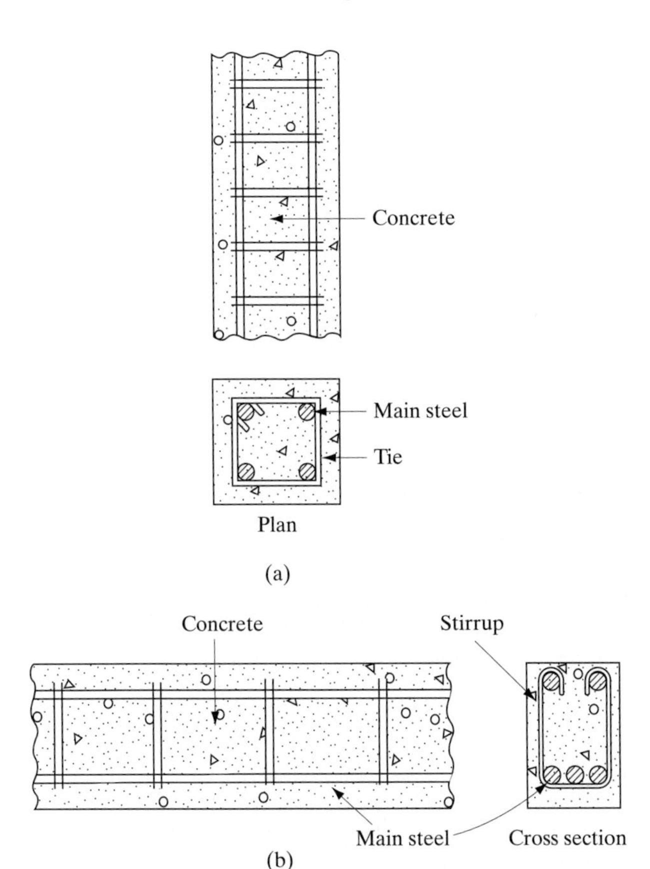

Figure 7.13 Use of reinforcing steel in (a) concrete column and (b) beam.

The nominal dimension (cross-sectional), or nominal diameter of a deformed bar is equivalent to that of a plain round bar having the same weight per foot as the deformed bar. Deformed bars are identified by their bar numbers which are based on the number of eighths of an inch included in the nominal diameter of a bar. For example, a No. 3 bar has a nominal diameter of 3/8 or 0.375 in., and No. 18 bar of 18/8 or 2.26 in. The minimum average spacing of surface deformations, on each side of the bar, is equal to seven-tenths of the nominal diameter. Thus, No. 3 bar has an average lug spacing of 0.262 in., and No. 18 bar, of 1.58 in. Numbers 3 to 11 bars are available in all grades, which refer to the expected yield strength of the steel in ksi.

7.5.1 Grades and Types

Reinforcing steel is rolled from properly identified heats of mold-cast or strand-cast steel, using one or more of the three primary processes:

- Open-hearth
- Basic oxygen
- Electric-arc

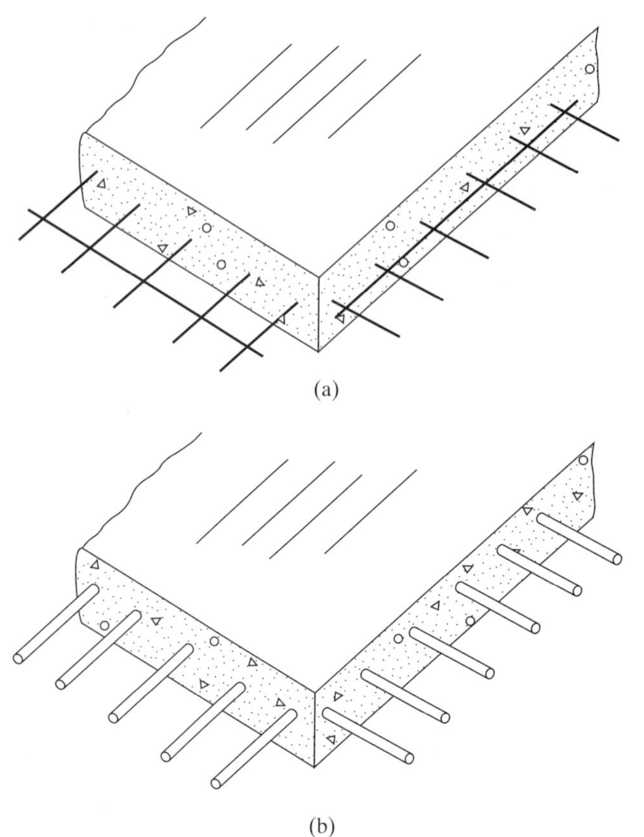

Figure 7.14 Use of (a) welded wire fabric and (b) reinforcing bars in ground-supported slab.

The primary melting of steel may incorporate separate degassing or refining, and this may be followed by secondary melting (such as electroslag and vacuum arc remelting).

Deformed or plain bars are produced in four principal (minimum) yield levels:

- 40,000 psi (300 MPa)
- 50,000 psi (350 MPa)
- 60,000 psi (400 MPa)
- 75,000 psi (500 MPa)

These are designated as Grade 40 (300), Grade 50 (350), Grade 60 (400), and Grade 75 (500), where grade refers to the minimum yield point or yield strength in ksi (or MPa). The yield point, which is determined by the drop of the beam or halt in the gage length of the testing machine, is the first stress, less than the maximum obtainable stress, at which an increase in strain occurs without increase in stress. The stress-strain diagram of materials with a definite yield point is characterized by a sharp knee or discontinuity. When the material does not have a well-defined yield point, the yield strength is established, which is the stress at which the material exhibits a specified limiting deviation from the proportionality of stress to strain, expressed in terms of strain, percent offset, or total extension under load.

Reinforcing steel comes in four ASTM types:

- Deformed and plain billet-steel bars, ASTM A615
- Deformed and plain rail-steel bars, ASTM A616
- Deformed and plain axle-steel bars, ASTM A617
- Low-alloy steel deformed bars, ASTM A706

The tensile properties and grade availability of these bars are shown in Table 7.3.

A615 steel is manufactured in three grades: Grade 40, Grade 60, and Grade 75. When the material does not have a well-defined yield point, the yield strength is determined at an extension of 0.005 in./in. (or 0.5 percent) for Grades 40 and 60, and of 0.0035 in./in. (0.35 percent) for Grade 75. All A615 bars (except plain round bars) that are tagged for grades are identified by a distinguishing set of marks, legibly rolled onto the surface on one side of the bar, to denote the following:

- Point of origin (Letter or symbol—mill designation)
- Size designation (Bar number)
- Type of steel (Letter S)
- Minimum yield-strength designation

The yield designation may include the number 60 or a single continuous longitudinal line—offset from the center of the bar—for Grade 60 steel, or the number 75 or two continuous longitudinal lines for Grade 75 steel. No yield designation marks are provided for Grade 40 bars.

A616 (rail steel) plain or deformed bars are produced in two minimum yield levels: 50,000 psi (350 MPa) and 60,000 psi (400 MPa), designated as Grade 50 (350) and Grade 60 (400), respectively. When the material does not have a well-defined yield point, the yield strength is determined at an extension of 0.005 in./in. of gage length (or 0.5 percent). The bars have the same set of marks as A615 bars, and a rail symbol indicates that the bar was produced from rail steel.

A617 (axle steel) plain and deformed bars are produced in two minimum yield levels: 40,000 psi (300 MPa) and 60,000 psi (400 MPa), designated as Grade 40 (300) and Grade 60 (400), respectively. When the material does not have a well-defined yield point, the yield strength is determined at an extension of 0.005 in./in. of gage length (or 0.5 percent). The bars have the same set of marks as those in A615 steel, and the letter "A" indicates that the bar was produced from axle steel.

A706 (low-alloy steel) deformed bars are produced in a single grade level of 60,000 psi (400 MPa), designated as Grade 60 (400). When the material does not exhibit a well-defined yield point, the yield strength is obtained at an extension of 0.0035 in./in. (or 0.35 percent). The same set of marks as those in A615 bars are rolled on the surface, and the letter "W" indicates that the bar was produced to satisfy the requirements of A706. The yield designation includes either the number 60 (4) or a single continuous longitudinal line, offset from the bar center.

TABLE 7.3 TENSILE REQUIREMENTS OF REINFORCING BARS

In pound units

ASTM Designation	ASTM Type	Grade	Tensile strength, psi (min)	Yield strength, psi (min)	Elongation in 8 in. (%)	Bar no.
A615	Billet steel, plain and deformed	40	70,000	40,000	11–12	3–6
		60	90,000	60,000	7–9	3–18
		75	100,000	75,000	6	11–18
A616	Rail steel, plain and deformed	50	80,000	50,000	5–7	3–11
		60	90,000	60,000	4–6	3–11
A617	Axle steel, plain and deformed	40	70,000	40,000	7–12	3–11
		60	60,000	90,000	7–8	3–11
A706	Low alloy steel, deformed	60	80,000[b]	60,000[a]	10–14	3–18

[a]Maximum yield strength = 78,000 psi.

[b]Tensile strength should be equal to or greater than 1.25 times the yield strength.

In SI units

ASTM Designation	ASTM Type	Grade	Tensile strength, MPa (min)	Yield strength, MPa (min)	Elongation in 8 in. (%)	Bar no.
A615	Billet steel, plain and deformed	390	500	300	11–12	10–20
		400	600	400	7–9	10–55
		500	700	500	6	35–55
A616	Rail steel, plain and deformed	350	550	350	5–7	10–35
		400	600	400	4–6	10–35
A617	Axle steel, plain and deformed	300	500	300	7–12	10–35
		400	600	400	7–8	10–35
A706	Low-alloy steel, deformed	400	550[b]	400[a]	10–14	10–55

[a]Maximum yield strength = 540 MPa.

[b]Tensile strength should be equal to or greater than 1.25 times the yield strength.

Weldability in A706 bars is accomplished by imposing two limits on chemical composition: one on the individual chemical elements and the other on the carbon equivalent (C.E.). The carbon equivalent is a measure of the chemical elements affecting the weldability of the material and can be calculated as:

$$\text{C.E.} = \%\text{C} + \frac{\%\text{Mn}}{6} + \frac{\%\text{Cu}}{40} + \frac{\%\text{Ni}}{20} + \frac{\%\text{Cr}}{10} - \frac{\%\text{Mo}}{50} - \frac{\%\text{V}}{10}$$

TABLE 7.4 CHEMICAL REQUIREMENTS OF REINFORCING BARS

ASTM Designation	Percentage (max)				
	Carbon	Manganese	Phosphorus	Sulfur	Silicon
A615	—	—	0.06	—	—
A616	—	—	0.06	—	—
A617	—	—	—	—	—
A706[a]	0.3	1.5	0.035	0.045	0.5

[a]This type of steel is intended for welding, and is manufactured with restrictions on chemical composition.

where C is carbon, Mn is manganese, Cu is copper, Ni is nickel, Cr is chromium, Mo is molybdenum, and V is vanadium. The carbon equivalent limit is 0.55 percent. In addition, there is a maximum limit of 0.3 percent for carbon and of 1.5 percent for manganese, as shown in Table 7.4. But it should be noted that weldable bars are not really required for reinforced concrete construction, and there is an increasing use of proprietary mechanical connectors for butt-splicing of bars.

Reinforcing bars should be free from detrimental surface imperfections. Rust, seams, surface irregularities, or mill scale is permitted provided the weight, dimensions, cross-sectional area, and tensile properties of a hand wire-brushed specimen satisfy the requirements.

The bars can be furnished in coils; coil stock is preferred for use with automatic stirrup and tie bending machines. Coil stocks are also favored by manufacturers of reinforced concrete pipes. Rebar manufacturers produce steel in units called *heats.* A heat of steel, depending on the mill, may range from 30–200 tons.

7.5.2 Handling

Bars are supplied in bundles, secured by wires or bands, each bundle containing bars of only one size, length, and grade. The bundles are limited to weights that can be conveniently handled by common equipment, approximately 3000 lb (1360 kg). To identify the bars, each bundle has a tag made of rope fiber or metal that identifies the order details—name, number, and so on—and the size, length, and grade of bars.

Considerable savings in cost can be realized if the bundles can be hoisted directly from the truck to the area on the project site where the bars are to be placed. When the bundles are unloaded in a storage area, the location should provide easy access by trucks for rehandling of the bars. During storage, lumber pieces must be placed under the bars to keep them free from mud and water, for the surface condition of bars affects their bond characteristics in concrete. The presence of heavy scale, rust deposits, oil, or mud may lower the bond strength significantly.

Scale occurs at the time the bars are rolled in the mill, and results from the cooling of the hot metal. Loose scale should be removed before placement. Normal rust stains increase the normal roughness of the surface and thus improve bond. But when rust is severe, it affects the height of deformations or the weight of the bar, and is harmful. Generally, bars with normal mill scale or rust or a combination are accepted in "as is" condition, without cleaning or brushing. Mud coating on bars should be washed off. Oil or grease should be removed with the help of a torch or by wiping with solvents before the placement of concrete.

7.6 WELDED WIRE FABRIC

Welded wire fabric is a prefabricated reinforcing material available in rolls or sheets for use in slabs and pavements. The common types of such fabric are shown in Table 7.5. The rolls are of width 5–7 ft (1.5–2.1 m) and length 150–200 ft (45–60 m). The sheets come in widths of 5–10 ft (1.5–3 m) and lengths of 10–20 ft (3–6 m).

7.6.1 Grades

Welded wire fabric is available in grades and types shown in Table 7.5. ASTM A82, plain steel wire for concrete reinforcement, is cold-drawn steel wire, as-drawn or galvanized, available in sizes not less than 0.08 in. (2.03 mm) nominal diameter. ASTM A185 covers specifications for welded wire fabric made with A82 plain wires, fabricated into sheets or rolls by electric resistance welding. Wire fabric consists of a series of longitudinal and transverse wires arranged substantially at right angles and welded together at points of intersection. The welded joints can withstand normal shipping and handling without breakage, but the presence of broken welds does not constitute cause for rejection unless the total number per sheet exceeds 1 percent of the total number of joints. Similarly, rust, surface seams, or surface irregularities will not be cause for rejection provided the minimum dimensions, cross-sectional area, and tensile properties of a hand wire-brushed test specimen satisfy the requirements.

TABLE 7.5 SPECIFICATIONS FOR WIRE AND WELDED WIRE FABRIC

ASTM Designation	Type	Yield strength[a] [ksi (MPa)]	Tensile strength [ksi (MPa)]	Size range
A82	Cold-drawn steel wire (plain)	56, 65 (385, 450)	70, 75 (485, 515)	W 0.5–W 31
A185	Welded steel wire fabric (plain) using A82 wires	56, 65 (385, 450)	70, 75 (485, 515)	—
A496	Deformed steel wire	70 (485)	80 (550)	D-1–D-31
A497	Welded deformed steel wire fabric using A496 wires	70 (485)	80 (550)	—

[a]Yield strength is determined at an extension of 0.005 in./in. of gage length.

The minimum tensile strength of A82 wires (material for welded wire fabric) is 70,000 psi (485 MPa) for sizes smaller than W1.2 and 75,000 psi (515 MPa) for other sizes, and the yield strength is 56,000 psi (385 MPa) for sizes smaller than W1.2 and 65,000 psi (450 MPa) for other sizes.

ASTM A496 provides specifications for cold-worked, deformed steel wire for use, as such or in fabricated form, as concrete reinforcement, and is available in sizes having nominal cross-sectional area not less than 0.01 in.^{2} (6.45 mm^2) and not greater than 0.31 in.^{2} (200 mm^2). The minimum tensile strength of wires (when used for welded wire fabric) is 80,000 psi (550 MPa) and the yield strength is 70,000 psi (485 MPa). The requirements of welded wire fabric using this type of wire are covered in ASTM A497.

Specifications for fabricated deformed steel bar mats, or sheets, are covered in ASTM A184. The mats consist of two layers of bars that are assembled at right angles by clipping or welding at the intersections. They should be fabricated in a manner that prevents the dislodgment of members during handling, shipping, placing, and concreting. Welds at the intersections should provide attachment at exterior intersections, and at not less than alternate interior intersections. Clips for clipped mats are formed mechanically prior to or during the fabrication and assembling of the mats.

Bars of Grades 40, 50, and 60 (300, 350, and 400) used in the manufacture of clipped mats should conform to A615, A616, A617, or A706 specifications; bars of Grades 40 (300) in welded mats must conform to A615 specifications. Bars of Grade 60 (400) used in the manufacture of welded mats are to conform to either A615 or A706 specifications.

7.6.2 Sizes

As indicated, wires intended for use in welded wire fabrics come in two forms:

- Plain wires
- Deformed wires

They are distinguished by a system that identifies the form of wire and its cross-sectional area. The letter "W" is for smooth (or plain) wire, and the letter "D" identifies deformed wire. These letters are followed by a number to represent the cross-sectional area of the wire in hundredths of a square inch. For example, W4.0 is plain wire with a cross-sectional area of 0.04 in.^{2}, and D4.0 is deformed wire of the same area.

Plain wire fabrics develop the anchorage in concrete at the welded intersections, whereas in deformed wire fabrics the development of anchorage is through the surface deformations and at the welded intersections. The physical properties of welded wire fabric are shown in Table 7.6.

Welded wire fabric is designated by two numbers and two letter–number combinations. An example is "6 × 8—W2.9 × W2.9." The first number (6) gives the spacing in inches of the longitudinal wires (Fig. 7.15). The second number (8) is the spacing in inches of the transverse wires. The first letter of the letter–number combination (W) shows the type of longitudinal wire (plain wire), and the second letter (W) shows the

TABLE 7.6 PROPERTIES OF WELDED WIRE FABRICS

Designation	Nominal diameter (in.)	Area of wire (in.^{2})	Area (in.^{2}/ft) Longitudinal	Area (in.^{2}/ft) Transverse	Weight (lb/100 ft^2)
6 × 6—W1.4 × W1.4	0.135	0.014	0.028	0.028	21
6 × 6—W2.0 × W2.0	0.159	0.02	0.04	0.04	29
6 × 6—W2.9 × W2.9	0.192	0.029	0.058	0.058	42
6 × 6—W4.0 × W4.0	0.225	0.04	0.08	0.08	58
4 × 4—W1.4 × W1.4	0.135	0.014	0.042	0.042	31
4 × 4—W2.0 × W2.0	0.159	0.02	0.06	0.06	43

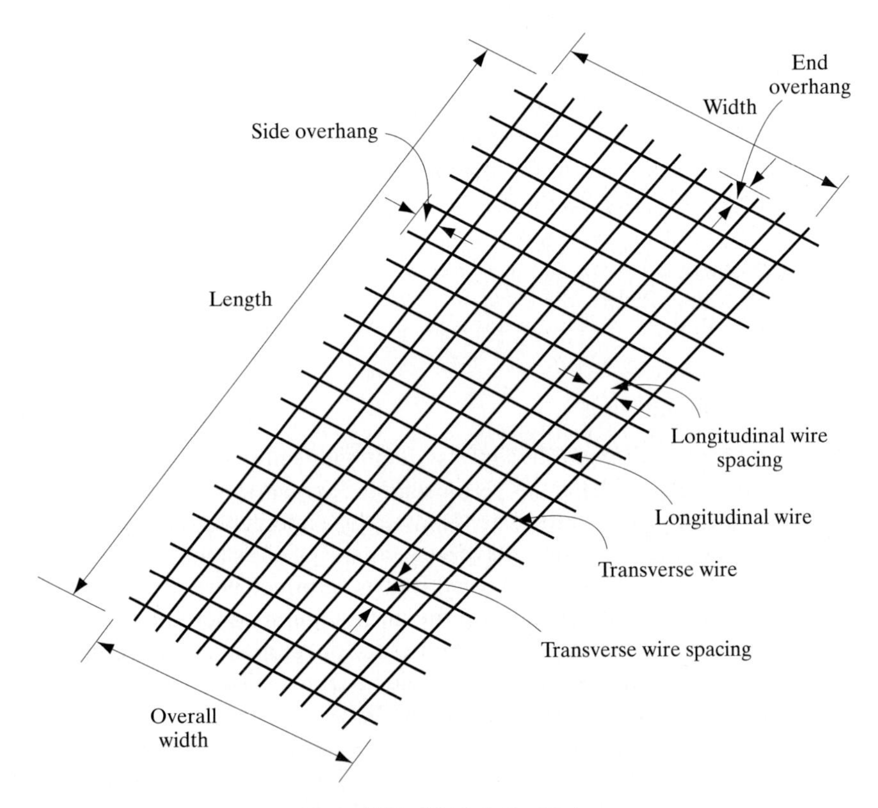

Figure 7.15 Welded wire fabric.

type of transverse wire (plain wire). The first number in the combination (2.9) gives the area of longitudinal wire in hundredths of a square inch, and the second number (2.9) represents the area of transverse wire in hundredths of a square inch.

The longitudinal wire spacing varies from 2–12 in. (5–30 cm). Spacings of 4, 6, 8, 9, and 12 in. are common in building construction. Transverse wire spacing is normally 4, 6, 8, 12, or 16 in.

7.6.3 Uses

Welded wire fabric is used as concrete reinforcement for crack control in residential slabs, driveways, sidewalks, and slabs for light construction. In a slab on grade (or ground-supported concrete slab), cracking is due primarily to three causes:

- Drying shrinkage
- Change in temperature and moisture
- Weak subgrade or soil

When the slab cracks, the cracked surfaces become jagged. If the width of the crack is sufficiently small [less than about 1/16 in. (1.5 mm)], the jagged faces are interlocked, which helps to transfer the loads. This mechanism is called aggregate interlock. On the other hand, if the crack gets wider, the interlock decreases.

When properly positioned within the cross-section of the slab, welded wire fabric acts to control cracks by keeping the cracked section of the slab closely knit together. This will allow for effective transfer of the load (through the crack) and also makes the crack less noticeable. Smaller cracks minimize the movement of water into and through the slab. The use of reinforcement in a slab also reduces the number of cracks.

For proper crack control, welded wire fabric (or similar reinforcement) should be placed in the middle one-third of slabs 4–6 in. (10–15 cm) thick [or about 2 in. (5 cm) below the top surface]. Some authorities recommend placing the steel at one-third depth from the top of the slab.

The primary objective of placing reinforcement in slabs on grade is crack control. In plain concrete slabs, crack control is accomplished by placing control joints at spacings less than 15 ft (4.5 m). When these slabs are reinforced with wire fabric, the control joints can be farther apart.

The most common types of fabric used are 6 × 6—W1.4 × W1.4 (10 gage) and 6 × 6—W2.9 × W2.9 (6 gage). The most common methods of placing welded wire fabric are (1) by using chairs (or concrete blocks), and (2) by placing concrete in two layers and placing the reinforcement on the first layer. The chairs are supports for steel reinforcement, and are made of steel or plastic. The spacing of chairs or blocks depends on the reinforcement size and spacing; the common practice is to place them 2–3 ft (0.6–0.9 m) apart. When the concrete is placed in two layers, the first course is generally to middepth (or slightly more); the reinforcement is then spread on this layer. The top layer of concrete is placed before the lower layer starts to harden.

7.7 EPOXY-COATED REINFORCING STEEL

The reinforcing bars and welded wire fabrics described in the preceding section are also available with a protective coating of epoxy (applied by the electrostatic spray method). The film thickness after curing is 5–12 mil (0.13–0.30 mm), and the coating should be free of holes, voids, cracks, and deficient areas discernible to the unaided eye (ASTM A775).

When compared to other corrosion-protection systems or no protection at all, epoxy-coated reinforcement offers low life-cycle costs and has proven to be a very cost-effective corrosion-protection method. These bars are used as reinforcement in the construction of bridges, parking garages, and sea-front structures.

Ordinary steel can be coated with epoxy either before or after it is fabricated. The most common method is to coat the steel in straight lengths of 40–60 ft (12–18 m), and then cut or bend the bars as specified. A few custom-coating facilities have the capacity to coat reinforcing bars and other steel shapes after they are fabricated. In this process, typically, individual bars are hung from a conveyor system and moved through the coating assembly. Welded wire fabrics are coated in this manner. Applying the coating using the traditional single-bar process is more cost effective, but has the disadvantage of potential damage during construction or fabrication. Coating after fabrication is more labor intensive and costly, but eliminates damages caused by bending.

The ideal solution to prevent potential damage to the coating is to schedule the delivery of the coated bars as close to the placing schedule as possible. If the storage is expected to be more than two months, the bars should be protected from weather by draping opaque plastic sheeting over the bundle, while still allowing for air circulation around the bars to minimize condensation under the sheeting. To prevent damage during concreting, only rubber or nonmetallic vibrator heads should be used.

Epoxy coating typically adds about 25 percent to the installed cost of the reinforcement, or about a 2 percent increase in the total structural cost of the project.

7.8 TESTING

The following procedure describes the test to determine the tensile properties of steel using round specimens.

Test STL-1: Tension Test of Steel

PURPOSE: To draw the stress–strain diagram of steel, and determine yield, tensile strength, and percent elongation.

RELATED STANDARDS: ASTM A370, E8.

DEFINITIONS:

- *Yield point* is the first stress at which an increase in strain occurs without an increase in stress.

- *Yield strength* is the stress at which the material exhibits a specified limiting deviation from the proportionality of stress to strain.

EQUIPMENT: Tension testing machine, wedge grips, micrometer, extensometer.

SAMPLE: Standard 0.5-in. (12.5-mm) round tension test sample with 2-in. (5-cm) gage length.

PROCEDURE:

1. Measure the diameter at the center of the gage length to the nearest 0.001 in. (0.025 mm).
2. Mark the gage length with ink. Gage points should be approximately equidistant from the center of the length of the reduced section.
3. Grip the specimen at sections outside the gage length.
4. Apply the load at a constant rate.
5. Measure extensions at predetermined load intervals.
6. When the increase in load stops or hesitates, record this load. The corresponding stress is the yield point.
7. When no yielding is noticed, determine yield strength, using the offset (0.2 percent offset) or percent extension method, from the stress–strain diagram.
8. Plot the stress–strain diagram.
9. Calculate the tensile strength by dividing the maximum load during the test by the original cross-sectional area.
10. Calculate the elongation over the gage length and percent increase.

REPORT: Report the tensile strength, yield strength (or point) and the method of determination, and percent elongation.

PROBLEMS

1. What are the three commercial forms of iron products?
2. Explain the differences between wrought iron and cast iron.
3. Explain the effects of carbon content on the strength and ductility of steel.
4. Define yield point, offset, tensile strength, and modulus of elasticity.
5. What is the yield strength of Grade 50 steel?
6. What are the advantages of using high-grade steel?
7. What are the three forms of reinforcing steel?
8. What is the purpose of using reinforcing steel in concrete?
9. Show typical reinforcing details in a slab.
10. What is the cross-sectional area of a No. 11 bar?

8

Plastics and Soils

The most common civil engineering materials were described in previous chapters. Concrete, steel, wood (and wood products), masonry, and asphalt make up the majority of structural materials. A few other materials important to civil and construction engineers are discussed in this chapter.

8.1 PLASTICS

The term "plastic" is used to describe any of a group of synthetic or natural organic materials that can be shaped when soft and then hardened. The basic component of organic plastics is a resin that can flow when softened. Although the resin used in the manufacture of a plastic product can be synthetic, partially synthetic, or natural, the majority of plastics are made from synthetic chemicals. Note that all three types of resins are organic, which means that these chemicals are similar to other organic compounds, such as sugars and dyes. Organic compounds, by definition, contain carbon.

The component molecules of resins used in the manufacture of plastics are very large; they are made up of relatively simple repeating units. A repeating unit, designated by the suffix "-mer," may exist in a single unit (called a monomer), in two units (called a dimer), or in many units (called a polymer).

Thus, a monomer can be defined as a unit consisting of relatively few hydrocarbon atoms joined to other units to form a polymer. Similarly, a polymer can be defined as a molecule made up of repeating groups of monomers. It may also be described as a large molecule comprised of hundreds or thousands of atoms, formed by a continuous chain of small molecules which are monomers. For example, polyvinyl chloride is a polymer made from a monomer called vinyl chloride. The most common natural polymers are wood and rubber.

The name "polymer" comes from the Greek words *poly,* meaning "many," and *meros,* meaning "part." Of the polymer materials (also called macromolecules) in engineering, plastics form the dominant group. Plastics are produced using petroleum or natural gas as raw materials. The process of building up small molecules in a long chain to form a large molecule is called polymerization. This is accomplished under heat, heat and pressure, or in the presence of catalysts.

The structure of plastics is similar to that of other petrochemical products or hydrocarbons. Hydrocarbons are substances containing only the chemical elements carbon and hydrogen. Note that petroleum and natural gas are complex mixtures of hydrocarbons formed naturally in the earth. Methane, for example, which is the main component of natural gas, is a hydrocarbon. The molecular structure of methane consists of a single carbon atom forming a single molecular bonding with each of the four hydrogen atoms.

A large number of polymers are hydrocarbons. A synthetic hydrocarbon polymer such as polyethylene is composed of hydrocarbon chains consisting of hundreds or thousands of carbon atoms as compared to a simple molecular structure of methane. Here the word "synthetic," which commonly means "not genuine" or "artificial" (as in synthetic stone, synthetic diamond), is used to describe a product formed by chemical reaction as opposed to that of natural origin.

Polymers, in the form of plastics, rubbers, and fibers, play an important role in everyday life. They are used in thousands of products, ranging from window curtains and computer hardware to water pipes and entry doors. In fact, it is nearly impossible to live without the help of this human-made material, which we learned to make during the second half of the nineteenth century.

8.1.1 Types of Plastics

Plastics (or polymer resins) fall into two classifications:

- Thermoplastics
- Thermosets

A *thermoplastic* material is a polymer that softens and melts gradually when heated and can be reshaped when still warm. Thermoplastics have linear molecular chains (without cross-linkage) that move in relation to one another when heated or cooled. The most common thermoplastics are polyethylene, polypropylene, polystyrene, and polyvinyl chloride (Table 8.1). The majority of plastics in use today are thermoplastics. Polyethylene and polyvinyl chloride comprise nearly 50 percent of the total plastic production in the United States.

A *thermosetting* material is a polymer that cannot be reshaped after manufacture. The molecular structure of thermosets is similar to that of thermoplastics before molding. But during the hardening process these chains are cross-linked to

TABLE 8.1 COMMON PLASTICS

Thermoplastics	Thermosets
Polyethylene	Epoxide (epoxy)
Polyvinyl chloride (PVC)	Phenol–formaldehyde
Polypropylene	Melamine–formaldehyde (phenolic)
Polystyrene	Unsaturated polyester
Polyamide (nylon)	Polyurethane
Polymethyl methacrylate (acrylic)	

form an interconnected network of chains that are not free to move. The most popular thermosets are unsaturated polyester, epoxides, phenol–formaldehyde, and polyurethane.

Thermoplastic resins do not undergo permanent change on reheating. They soften and melt when heated, and resolidify upon cooling, like wax or butter. The molded shapes are retained upon cooling. Any object made with thermoplastics can be remolded into a new shape. Thermoplastics creep considerably, more than thermosets, particularly at higher temperatures. They are generally isotropic and show consistent properties and thus can be used for light structural applications.

Thermosetting resins first soften on heating, and on continuous heating harden or set to become relatively rigid, like an egg. When reheated, they do not soften and cannot be remolded. When reinforced, they show higher strength and less tendency to creep than thermoplastics, and can also be used in light structural applications.

Molecules of both thermoplastic and thermosetting resins are high polymers with long and continuous hydrogen-carbon–atom chains for a molecular framework. But (as noted) the setting process of thermosets binds the molecular chains together, making them larger and still more complex.

Thermoplastic products are manufactured using a number of methods; the most common in mass production are extrusion and molding. Machinery for the extrusion process consists of a heated barrel with a screw turning inside it. Resin in the form of granules or powder is fed into the barrel, and upon heating, is forced through the screw in much the same way that toothpaste is squeezed from a tube. The fluid comes out through a die or mold that forms the shape of the extruded plastic. PVC and polyethylene products of indefinite length (such as pipes, tubes, rods, and edgings) are commonly manufactured using this process.

Thermoplastic products can also be produced by molding. In the injection molding process, the resin granules are fed into a heated barrel. The fluidized material is then forced through a nozzle at high pressure (by a ram) and later led into the cavities of the mold. Cold water circulates in the mold, and when the fluid sets, it takes up the required shape. Thermoplastics such as polyethylene, polystyrene, and nylon are suitable for injection molding.

Thermosetting products are generally made by a combination of compression molding, transfer molding, or injection molding. In compression molding, the shaping of the plastic is accomplished by the application of heat and pressure in a suitable mold. In this method, which can be used for both thermoplastics and thermosets, but is best for the latter, the mold is opened and after placement of a charge inside it is then closed and heated. The hot viscous mass flows into all parts of the mold as pressure is applied. The mold remains closed until the part is fully formed. A thermoplastic softens, flows, fills the mold, and remains soft until the mold is cooled. A thermosetting material first softens and then hardens or cures to a solid infusible mass. It may then be ejected from the mold while hot, whereas thermoplastic material requires that the mold be cooled before the part can be removed. The injection molding method for thermosets is similar to that of thermoplastics. In the transfer molding method, thermosetting powder is put in the bottom half of a heated mold. As the material melts and begins to set, the upper half of the mold drops and forms it into the desired shape under both heat and pressure.

Light cross-linking of certain linear polymers produces an additional class of plastics: rubberlike materials called *elastomers* (also called *synthetic rubbers*). This material has a high level of elasticity (the capacity to recover large deformations) and is similar to natural rubber.

The effect of the length of application of elevated temperature on thermoplastics and thermosets is illustrated in Fig. 8.1. Continuous heating increases the softening point of thermosets and makes them hard. But it should be noted that some plastic materials show characteristics of both thermoplastics and thermosets.

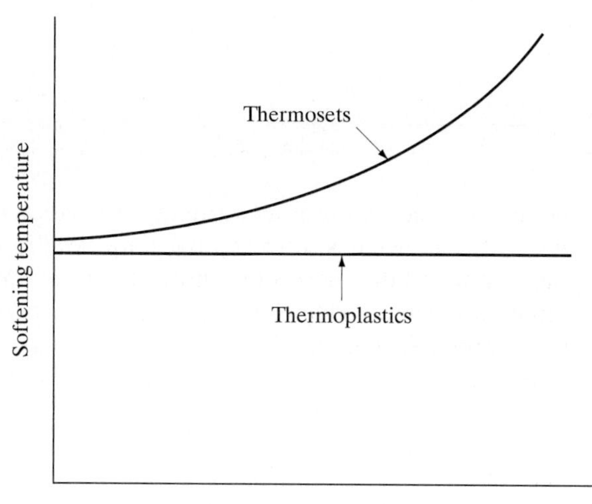

Figure 8.1 Effect of length of time at elevated temperature on plastics.

8.1.2 Properties of Plastics

The mechanical properties of plastics are affected by several factors:

- Rate of loading
- Temperature
- Environmental conditions

The behavior of plastics can be described as partly elastic and partly viscous. This viscoelastic behavior is both temperature and time dependent. Plastics that are hard and strong at ordinary temperatures may be weak and soft at higher temperatures. Some plastics (such as polystyrene) that are strong and tough under compressive loads may be brittle under tensile loads (Fig. 8.2).

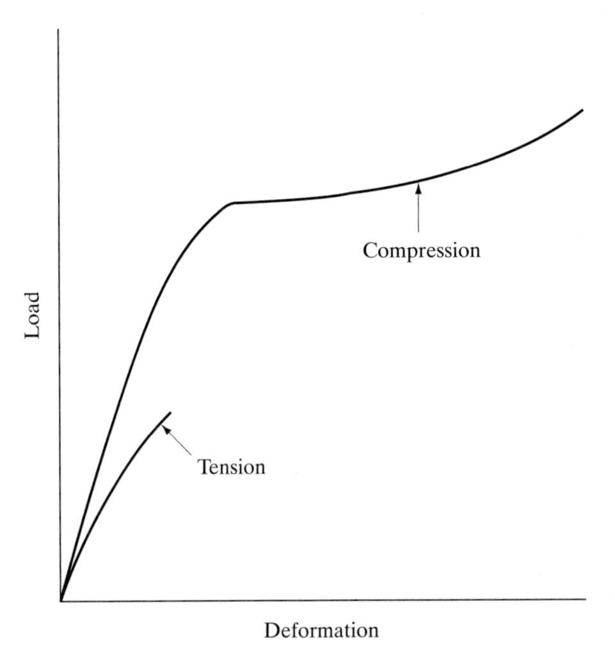

Figure 8.2 Load-deformation diagrams of polystyrene.

The mechanical properties are not the same for all varieties of plastics; they depend on the length and shape of the molecules of individual polymers. Typical stress–strain diagrams of thermoplastics and thermosets are shown in Fig. 8.3. Most plastics exhibit limited elastic range and a low modulus of elasticity. Some plastics are brittle and weak; others are relatively strong and flexible. In the latter type of plastics, after the occurrence of necking down or yielding, elongation continues for about 100 to 200 percent strain (percent elongation).

For most plastics, the modulus of elasticity in tension is not the same as that in compression, and is generally less than 1×10^6 psi (6.89 GPa). This low modulus gives rise to unsafe deformations and extensive deflections when plastics are used in structural or load-bearing applications. Some of the properties of selected

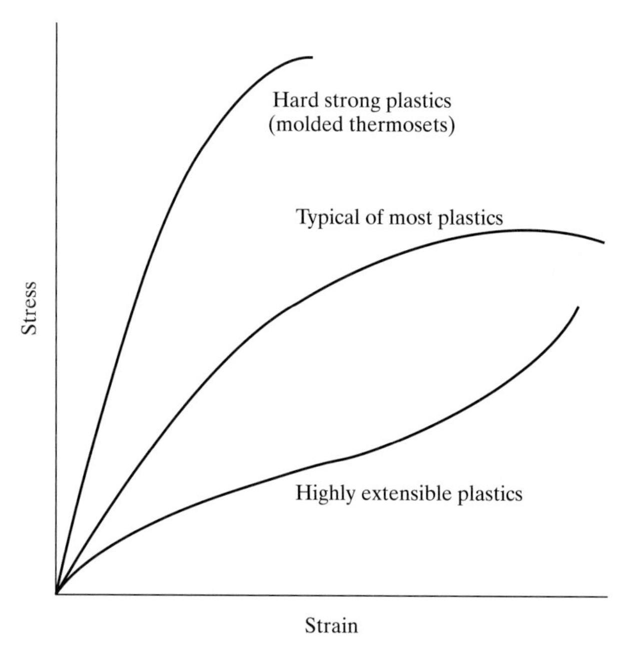

Figure 8.3 Typical load-deformation diagrams of plastics.

thermoplastics and thermosets are listed in Table 8.2 and discussed briefly in the following paragraphs.

Polystyrene, one of the most common thermoplastics, was discovered in 1831, but industrial production of the resin did not start until 1930. It has a specific gravity of 1.05, and its tensile strength is low, typically less than 7000 psi (48.5 MPa); its modulus of elasticity is also low. It has good water resistance and dimensional stability and is inexpensive. But it possesses poor heat resistance and limited weather resistance. Moreover, polystyrene is brittle and lacks toughness.

Pure polystyrene, which is crystal clear, is used for storage containers and cups. Polystyrene and expanded polystyrene (made by heating it with a foaming agent) are used in tiles, packaging, containers, foams, and housewares. Foamed polystyrene panels are used in poured-in-place concrete wall systems and as insulation boards for foundation walls, masonry and wood walls, and roofing.

Polyvinyl chloride (PVC) is a commonly used thermoplastic material with a specific gravity of about 1.39. The industrial production of PVC began in Germany around 1925. It is made from a monomer, vinyl chloride. Its modulus of elasticity is about 0.5×10^6 psi (3.6 GPa), and its tensile strength is nearly the same as that of polystyrene. Its coefficient of thermal expansion is less than 0.3×10^{-4} per °F; PVC is also an excellent insulator. Popular applications of PVC are in raincoats and shower curtains. It is used extensively in floor tiles, electric cables, flexible sheeting, hoses, pipes, luggage, moldings, expansion joint fillers, decorative wall coverings, and other products.

TABLE 8.2 PROPERTIES OF SELECTED PLASTIC

Material	Type	Specific gravity	Tensile strength [ksi (MPa)]	Modulus of elasticity [ksi (GPa)]	Compressive strength [ksi (MPa)]	Coefficient of expansion [per °F × 10^{-6} (per °C × 10^{-6})]
Polyethylene	Thermoplastic	0.92	1.8–4.0 (12.4–27.5)	19–160 (0.13–1.1)	0.4–2.4 (2.8–16.5)	100–120 (180–215)
PVC	Thermoplastic	1.3–1.4	11.5–8.0 (10.3–55.2)	1–800 (0.007–5.5)	0.1–13 (0.7–89.6)	30 (54)
Polypropylene	Thermoplastic	0.91	4.0–5.5 (27.6–40)	120–175 (0.8–1.2)	8.5–10 (58.6–69)	170 (308)
Polystyrene	Thermoplastic	1.05	4.0–7.0 (27.6–27.5)	20–50 (1.4–3.5)	1.3 (7.1)	38 (68.5)
Nylon	Thermoplastic	1.14	7–12 (48.3–82.8)	260–410 (1.8–2.8)	7–13. (48.3–89.6)	55 (90)
Unsaturated polyeser	Thermoset	1.6–2.2	4 (27.6)	300–1500 (2.1–10.31)	20.–25. (138–173)	42 (75)
Phenol–formaldehyde	Thermoset	1.3–1.9	4–8 (27.6–55.2)	1000–1500 (6.9–10.3)	20–25 (138–172)	45 (81)
Polyurethane	Thermoset	1.2	5 (34.5)			32 (58)
Acrylic	Thermoplastic	1.19	8–10 (55.2–68.9)	420 (2.9)		40 (72)

Polyethylene is also a common thermoplastic, with a specific gravity of about 0.92. It is a tough, weather-resisting plastic and is durable. The resistance to moisture of polyethylene is very good. In addition, polyethylene has excellent electrical properties and favorable chemical resistance. Its tensile strength is very low, generally less than 2000 psi (13.8 MPa). Its modulus of elasticity is also very low, and does not exceed 0.16×10^6 psi (1.1 GPa); the standard value of modulus of polyethylene is 0.019×10^6 psi (0.13 GPa). Its coefficient of linear expansion is 1×10^{-4} per °F. The common application of polyethylene is as polyethylene bags. It is used in the manufacture of films, sheets (clear or opaque), moldings, pipings, electrical conduits, tanks, foams, and bottles. Polyethylene films and sheets are used as dampproof courses, membranes, and curing membranes.

Polypropylene is another thermoplastic that has a low specific gravity (the same as polyethylene) and low modulus of elasticity. In fact, polypropylene is the lightest of all thermoplastics. Its coefficient of thermal expansion is comparable to that of polyethelene, but it has a higher softening point and is shinier. It has good heat resistance but degrades under exposure to sunlight. Thanks to good abrasion resistance and hardness, polypropylene is used for pipes, moldings, sheets, geomembranes, and so on.

Acrylics are thermoplastic materials that are crystal clear in natural form but are commonly manufactured in a wide range of colors and shades. They exhibit good weathering resistance and ease of forming. Their tensile strength, compared to other thermoplastics, is relatively high, about 10,000 psi (69 MPa), but their modulus of elasticity is comparable with that of others. They are used in light fittings, skylights, screen doors, and so on. They are also used in paints and adhesives.

Polyester is a thermosetting resin that has a specific gravity ranging between 1.6 and 2.2. Its modulus of elasticity is fairly low, generally around 0.3×10^6 psi (2 GPa). But glass-reinforced polyester has higher modulus, as much as 2×10^6 psi (13.8 GPa). Its coefficient of thermal expansion is about 0.4×10^{-4} per °F. Polyester is used in the manufacture of fiberglass products and composite materials.

Phenol–formaldehyde is a common thermosetting resin with a high specific gravity (1.9). Its tensile strength varies between 4000 and 8000 psi (27.5 and 55 MPa), and its coefficient of linear expansion is one of the lowest among plastics. As a molded material, this plastic (also called Bakelite) is used in the manufacture of lavatory seats, electrical fittings and equipment, and decorative laminates.

8.1.3 General Properties

Plastics are dimensionally unstable materials. The dimensional stability of a material depends on its stiffness or elastic modulus. Plastics are less elastic than steel or concrete. They are subject to creep, which is the increase in deformation, under load, with time. The rate of creep in plastics is much higher than the rate in concrete. The creep deformation increases with the stress level. Because of this, the modulus of elasticity of plastics decreases with the increase in load and its duration. As a result, plastics become dimensionally unstable under load. But it should be noted that most plastics are dimensionally unstable not only because of creep but also because of the tendency to shrink. Thermosetting plastics shrink appreciably in comparison to thermoplastics.

Generally, plastics do not transmit vibrations very well. They tend to absorb part of the vibrational energy, which is called damping capacity. The high damping capacity of plastics make them suitable in applications where vibration is encountered.

Plastics retain their color in indoor applications. In outdoor applications, however, colorfastness depends on weather conditions. Ultraviolet radiation in sunlight is a significant factor in the weathering of plastics. Weathering can take several forms: fading, yellowing, and roughening of the surface. Changes in temperature (heat) and humidity (moisture) affect the color and accelerate the aging of plastics.

Carbon in plastics has a tendency to combine with oxygen from the air to form CO_2. This tendency is more pronounced at high temperatures. Thus all plastics have an upper limit on service temperature, beyond which they will decompose. Organic compounds in general are partially soluble in similar compounds, a property that also applies to plastics.

Fire resistance is an important property in buildings. Most plastics are flammable; however, there are large differences between various plastics in terms of flammability. Polyethylene burns readily when fired by a flame and extinguishes itself when the flame is removed. The addition of fire-resistant materials in plastics may not provide a permanent resistance. Moreover, some plastics may give out heavy toxic smoke in a fire. Such plastics should not be used in exit areas in buildings, such as around stairways and corridors.

Most plastics have low water-absorption properties. They are particularly suitable as impervious membrane layers to prevent the movement of water. Examples of these applications are in foundations and in storage, sanitary, and water supply installations. In applications requiring permanent or temporary solution to water movement (such as in foundation slabs and for concrete curing), plastics are an ideal choice.

The coefficient of thermal expansion of plastics is higher than that of other materials, and is 5 to 10 times that of steel or concrete. This means that allowance should be made in plastics construction for thermal expansion and contraction of the member (such as providing construction joints, slotted holes, and so on).

The thermal insulation property of most plastics is superior to that of most other materials. A 1-in. (2.5-cm) polyurethane foam provides as much thermal insulation as 20 in. (50 cm) of clay brick. Polyurethane foam and expanded polystyrene are commonly used as thermal insulation material in sandwich construction and in buildings. However, the acoustical insulation of plastics is poor. This means that improvement in energy consumption may be offset by poor sound or noise protection.

Poisson's ratio is a measure of the reduction in cross-section that accompanies longitudinal elongation. For most brittle plastic materials, Poisson's ratio is around 0.3. For flexible plastics, its value is close to 0.4.

8.2 MODIFIED PLASTICS

Many plastics have their basic properties modified by the addition of substances such as plasticizers, antioxidants, fillers, and colorants (pigments). Plasticizers soften the plastic and make it easier for shaping. Antioxidants prevent degradation by light and heat, fillers provide toughness and strength, and colorants give color to plastics. Fibers are the most common type of fillers added for changing the mechanical properties of plastics.

To improve the strength properties of thermosetting or thermoplastic resins they are reinforced with fibers. A wide variety of materials can be employed as reinforcement or filler, including glass, asbestos, jute, and mineral fibers. In applications requiring very high tensile strength, graphite (carbon), aramid (Kevlar), glass, or boron fibers are used, in which case the ensuing material—a composite—is called *fiber-reinforced plastic* (FRP).

Fiber-reinforced plastics were originally developed for the aerospace and defense industry. In addition to increasing stiffness and strength, fibers may reduce shrinkage, improve abrasion resistance, increase toughness, and provide dimensional stability to plastic resins. These lightweight, high-strength, and corrosion resistant composites are used primarily in applications requiring low-density, high-strength materials. However, the elastic modulus of FRPs are low, and they are highly anisotropic as well as brittle. In civil engineering applications they are used for pedestrian bridges, bridge repair, and as prestressing tendons.

Glass fibers consist primarily of silica, SiO_2, and some metallic oxides—modifying elements—and are generally produced by mechanically drawing molten glass through a small orifice in a platinum plate. E-glass, named for its electrical properties, accounts for the majority of glass fiber production and is the most widely used reinforcement for composites. Another form of glass fiber, S-glass, has 30 percent higher tensile strength than E-glass but is more expensive. The tensile strength of glass fibers ranges between 250–350 ksi (1720–2410 MPa). They are transparent (the colors of the resin are not changed), isotropic, and have superior resistance to high temperatures. As they are relatively inexpensive and possess favorable properties, glass fibers are the most widely used reinforcement in plastics. Glass-epoxy and glass-polyester composites are used extensively in applications ranging from fishing rods to storage tanks, automobiles, and airplanes.

Graphite or carbon fiber is produced by subjecting organic precursor fibers (such as polyacrylonitrile, pitch, or rayon) to a sequence of heat treatments, converting the precursor to carbon. Graphite fibers are 99 percent carbon, whereas carbon fibers have around 95 percent carbon. The tensile strength of graphite fibers may be as high as 500 ksi (3450 MPa); they also have high modulus and their use in plastics imparts sufficient strength and stiffness to the composite. They are chemically inert, dimensionally stable, and possess high electrical and thermal conductivity along the fiber axis. But one of the disadvantages of carbon fiber is its color (black), which will change the color of the plastic made with it. Graphite and carbon fibers are most widely used in aerospace structures and in products such as tennis rackets and golf clubs. Graphite-epoxy and carbon-epoxy are the more common composites.

Aramid fiber (also called Kevlar) was originally developed for radial tires, and has a density about half that of glass fibers. Boron fibers are produced as a composite, and consist of a boron coating on a substrate of tungsten or carbon. They have much higher strength, density, and stiffness than graphite. Boron-epoxy and boron-aluminum composites are used in the aerospace industry.

In addition to polymers, metals and ceramics are also used as matrix materials in composites. However, polymers are the most widely used, with epoxies and polyesters being the principal matrix materials. Lightweight metals such as aluminum, titanium, and magnesium are used in high temperature applications. They offer, in addition, higher strength, stiffness, and ductility than polymers. Ceramic materials such as silicon carbide and silicon nitride can be used for the matrix in materials required to resist very high temperatures, above 3000 °F (1650 °C). But their tensile strength is low and they are brittle. Silicon carbide fibers are also used with metal or ceramic composites.

As repair materials and for retrofitting, FRPs are typically installed by a technique known as wet lay-up, which consists of forming layers of flexible sheets or fabrics of the fiber material on uncured two-part resin. The resin functions as both the polymer matrix and as the adhesive to bond the system to concrete. The advantage of FRPs in construction is that they can be applied to surfaces of any type or shape.

The fibers are the main load-bearing elements, and the polymer protects the fibers from damage and ensures that the fibers remain aligned. In addition, the polymer allows for the load distribution between individual fibers. The selection of fibers is based on strength, stiffness, and durability; that of resin on method of manufacture and ambient conditions.

FRP members (structural shapes) for bridge construction are commonly created through pultrusion, a process in which the reinforcing fibers are pulled continuously through a resin bath at a speed of up to 5 ft/min, and then molded and cured. This method creates products of constant cross-section and a high degree of reinforcement orientation. Miyun Bridge in Beijing, China, opened in 1983; No-Name Creek Bridge in Russel County, Kansas, opened in 1996; and Bonds Mill Lift Bridge, built in 1995 over a canal connecting the Severn and Thames Rivers in England are some examples that show the increasing popularity of FRP shapes in building and highway construction.

In summary, most ordinary plastics have low strength and very low modulus, and due to this inherent weakness they are not employed in load-bearing construction. The principal advantages of plastics lie in their low specific gravity, moldability, aesthetic qualities, low cost, electrical insulation, thermal insulation, simplicity in fabrication, and easy adaptability. But the major disadvantages are that they are difficult to repair, have high expansion potential, creep considerably, and possess extremely low fire resistance.

8.3 USES OF PLASTICS

Plastics have revolutionized a number of industries, such as automobile, electrical, and electronic. Lower prices have contributed to the universal application of plastics. Polyethylene, polypropylene, polystyrene, and PVC represent over 85 percent of world plastics consumption. In civil engineering applications, plastics possess a number of advantages. The most important of these are:

- Plastics have a favorable strength/weight ratio.
- Because of their lightness, plastics present ease in handling, transportation, storage, and assembly.
- Plastics can be molded to any shape and pattern.
- Most plastics are generally maintenance free and have good corrosion resistance.
- Plastic products come in variety of colors and textures.
- Plastics can be manufactured with consistent quality. They are easy to manufacture into components.
- Plastics are very economical.

Based on these advantages, it can be said that plastics are very suitable in the manufacture of prefabricated components for both indoor and outdoor applications.

Moldings, floor and wall coverings, windows, skylights, and countertops exemplify some of the common uses of plastics. Success in the use of plastics depends on several factors, such as fire resistance requirements, thermal and acoustical needs, dimensional stability, and weathering tolerances.

Dimensional-stability problems of plastics have been discussed. To attain sufficient dimensional stability in civil engineering structures, large components of plastics are needed in load-bearing construction. This may prove uneconomical and impractical. Because of low stiffness and strength, plastics are not generally used for load-bearing construction.

However, a new type of reinforcement using fiber-reinforced plastics is being tried for concrete. Aramid fiber–reinforced rods have been used for a prestressed concrete berth in Japan. A 210-ft (64-m) cable-stayed bridge has been built in Scotland in which the deck and tower are constructed from interlocking cellular glass-reinforced plastic protrusions that are glued together. Composite plastic and steel piles have been used to help solve problems of durability and corrosion in marine environments. Steel pipes encased in recycled plastics are found to be immune to attack by marine borers, which destroy timber piles.

Manufactured plastics have good surface finish and may not require maintenance. Because of this and due to their availability in a large range of colors and textures, plastics are commonly employed for floor and wall tiles and countertops.

The excellent water resistance property of plastics was also discussed. Polyethylene, PVC, and a few other plastics are used extensively as dampproof membranes (plastic sheeting) to prevent the rise of moisture from soil through concrete floors. Plasticized PVC waterbars are used in construction and expansion joints. Plastics are commonly used as dampproof courses or dampproof layers in brick and block wall construction.

With very good thermal characteristics, plastics are the most popular thermal insulation material. Polyurethane, polystyrene, and vinyl resins are used extensively as wall and ceiling insulation material in both cast-in-place (as foamed insulation) and precast constructions (as rigid insulation).

Summarizing, plastics can be used in applications that demand moisture prevention, thermal insulation, lightness, electrical resistance, passage of light, good wearing surface, and moldability. However, the most important factor in the universal adaptation of polymer materials is that they are relatively inexpensive.

8.4 SOILS

The earth's crust, which is 6–9 miles (9.6–14.4 km) thick, is made up of rocks and unconsolidated sediments or overburden, termed *soil* (Fig. 8.4). In engineering geology, soil is the unconsolidated material in the earth's crust, and is defined as all the material above the bedrock and made up of mineral particles (as in sand and clay), organic material (found in the topsoil and marsh deposits), air, and water. In engineering, soil is an aggregate of earth materials that will disaggregate or decompose when subjected to mechanical forces or under water.

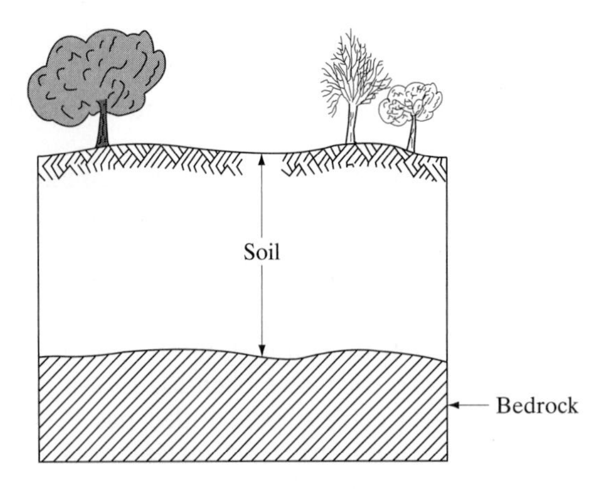

Figure 8.4 Section of the earth's crust.

Soil is the natural product of the weathering and mechanical disintegration of rocks that form the earth's crust, such as limestone, sandstone, shale, basalt, and granite. The process of weathering, which has been going on for millions of years and continues, has several forms: physical, chemical, and organic. Mechanical disintegration occurs during the transportation of rock fragments by water, ice, or wind.

The mineral particles in soil result from physical and chemical weathering. Physical weathering is the mechanical breakdown of rocks through processes that cause pressure and temperature changes within them. The action of water, wind, ice, plant or animal life, and other weathering agents break away the mineral particles from the bedrock, which are then transported by wind, ice, water, or gravity. If water is available within rocks and the temperature drops below 0 °C, ice forms and the resulting increase in volume can produce tremendous forces, causing internal cracks. Ice and frost are probably the primary agents for the breakup of jointed bedrock outcrops in mountainous areas. The combination of moisture and salts creates salt crystals that can cause stresses in the rock, breaking it apart. The type of breakdown and the length of transport have a substantial influence on the character of soils. During the transportation, the size of the particles is further reduced, and the shape and texture are altered. The type of soil formed from physical weathering is called granular soil.

When water flows through the bedrock, some of the dissolved mineral compounds are leached out or decomposed, and this process is called chemical weathering. Dissolved mineral ions in ground water are largely the result of chemical weathering of rocks. Hydration, hydrolysis, solution, and oxidation are the chief common chemical weathering reactions. Hydration is a process whereby a chemical combines with water to form a hydrated mineral. Hydrolysis is the reaction between a mineral and water to produce a new mineral. Some compounds may dissolve in water, and this process is called solution. Oxidation involves the ionic combination of an element with oxygen. The soil formed from chemical weathering is called clay, which consists of mineral crystals with properties distinctly different from the parent rock. A brief description of common types of rock is given in the following section.

8.4.1 Types and Properties of Rocks

Characteristics of a soil depend on the parent rocks from which the soil is derived. Rocks are mixtures of several minerals, and may be one of the following:

- Igneous
- Sedimentary
- Metamorphic

A mineral is a naturally occurring chemical element or compound that has a definite crystalline structure and distinctive physical properties. The most common rock minerals are feldspars, quartz, kaolinite, muscovite, and calcite. Feldspars are the most important group of rock-forming minerals and are abundant in the earth's crust.

Igneous rocks are formed by the cooling and hardening of molten magma. About 95 percent of the earth's crust is made up of igneous rocks. Some of the important igneous rocks are granites, basalts, pumice, scoria, and rhyolite. The mineral composition of granite is principally feldspars [potassium aluminum silicate, $K(Al)Si_3O_8$, sodium aluminum silicate, $Na(Al)Si_3O_8$, or calcium aluminum silicate, $Ca(Al)Si_3O_8$] and quartz (silicon dioxide, SiO_2).

Rocks formed by the accumulation or deposit of transported fragments, followed by consolidation, are sedimentary rocks. Transportation is by water, ice, and wind. Only about 5 percent of the earth's crust is made up of sedimentary and metamorphic rocks. Of the exposed rocks (land surface), nearly 75 percent are sedimentary rocks. The more common sedimentary rocks are sandstones, limestones, and shales. Quartz is the most abundant mineral in sedimentary rocks. Limestone contains primarily calcite mineral (calcium carbonate, $CaCO_3$). Shale is formed from the hardening of silts and clays. Sandstone is formed from sand (quartz) under pressure.

Rocks that are formed by alteration or metamorphosis of sedimentary or igneous rocks from heat, pressure, or both are called metamorphic rocks. Schist, gneiss, slate, and marble are some examples of metamorphic rocks.

The stress-strain relationship of rocks is influenced by the mineralogy, banding, bedding planes, and other structural characteristics. Four common types of deformation behavior are shown in Fig. 8.5. Part (a) represents a rock material that is dense, massive, and uniform. The stress-strain or load-deformation relationship is elastic, and the material suffers neither progressive failure (of individual minerals) nor densification. Part (b) shows typical behavior of rocks that are uniform and massive with some pore space that causes inelastic deformation as the material densifies. Failure results when all of the mineral components fail as a unit. Rocks that show ductile or progressive failure, in which the weakest mineral fails first and then the failure surface progresses through the material, exhibit load-deformation behavior as shown in part (c) or (d).

8.4.2 Types of Soil

Based on the grain size, mineral soils are grouped into two broad classifications: coarse-grained and fine-grained; and into four types: gravel, sand, silt, and clay, as

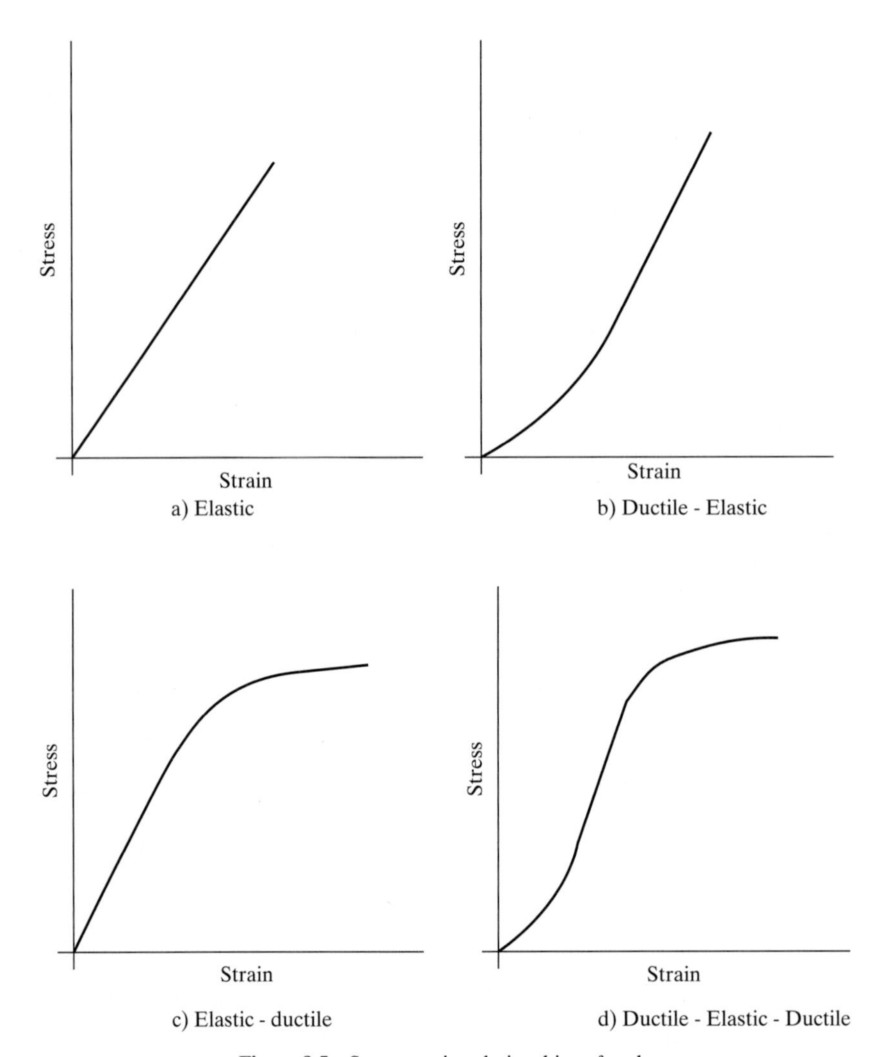

Figure 8.5 Stress-strain relationships of rocks.

shown in Table 8.3. Particles of size larger than 3 in. (75 mm) are called *cobbles* and those bigger than 8 in. (200 mm) are termed *boulders*.

Gravel consists of large particles, of size ranging between 3/16–3 in. (4.75–75 mm), derived from the weathering or mechanical disintegration of parent rock. The mineral composition of gravel is similar to that of the parent rock. Quartz (silica) is the predominant mineral in gravel, and also in sand. Sand consists of particles resistant to weathering, and is defined as a mass of unconsolidated mineral grains of nearly uniform size that can be detected by the unaided eye. The composition of sand varies widely, from pure silica (siliceous sand) to calcium carbonate (calcareous

TABLE 8.3 SOIL CLASSIFICATION AND CHARACTER

	Grain			
	Size	Shape	Group	Type
Gravel	Over 3/16 in. (4.75 mm) (retained on No. 4 sieve)	Spherical or cubical	Coarse-grained	Granular or cohesionless
Sand	From 3/16 in. (4.75 mm) to smallest visible to eye (0.075 mm)	Spherical or cubical	Coarse-grained	Granular or cohesionless
Silt	Not visible to eye (0.002 mm to 0.075 mm)—or, passing No. 200 sieve	Spherical or cubical	Fine-grained	Granular or cohesionless
Clay	Smaller than silt—or, passing No. 200 sieve	Flat or plate-shaped	Fine-grained	Cohesive

sand) to a complex mixture of many minerals. Quartz, however, is the dominant mineral in a majority of sand. Boulders, gravel, and sand do not possess the interparticle bond that exists in clays and silts, and can be excavated with a shovel. However, when moist, as on the shore, sand particles are held together by the water held from the capillary attraction between the grains.

Clay deposits consist of clay minerals and result from the chemical weathering or decomposition of minerals of igneous or sedimentary rocks (feldspars and mica), followed by consolidation. Clay can be defined as a fine-grained soil that is sticky or plastic when wet, is coherent when dry, and becomes stone-like if baked at a high temperature. Clay particles differ widely in composition, but silica is generally the primary ingredient, and alumina the second. The main groups of clay minerals are kaolinite, illite, and smectite, which are principally the oxides of silica and alumina (mostly hydrated aluminum silicate), packed in a manner that shows a plate-like structure. Varying amounts of water, lime, iron oxide, magnesia, alkalis, and organic matter, left due to decomposition of plants and animal life, are also found in clay. When the amount of lime is moe than 5 percent the clay is called *marl,* and if mixed with silt, it is called *loam*.

Silt results from the grinding action between pieces of rock of different sizes. It can be described as a type of soil that is intermediate between fine sand and clay, or a fine-grained soil suspended in or deposited by water. Silt contains primarily powdered quartz or silica. The particles are round and smooth and can be easily molded and crushed with the fingers. If one takes a sample of moist silt in the palm of the hand and shakes it horizontally, the moisture from within the sample rises to the surface, and will recede when pressed with the fingers. Silt deposits are commonly found in plains, as alluvial and marine deposits, and are generally well-graded and stable. Silts also exist combined with sands and clays in estuarine and deltaic sediments.

Clay particles stick to the fingers when wet. When squeezed between the fingers, wet soft clay exudes. Clay particles bond to each other, in a lump, or are cohesive—whereas sand, silt, and gravel exist as individual grains. Thus clay possesses plasticity when wet, and it is this property that gives a clayey soil the ability to be remolded and deformed without breaking apart. All varieties of clay are plastic; silts also exhibit plasticity, but to a lesser degree. Moist sand and silt particles seem to stick together, and this apparent cohesion is due to the moisture film around the grains and disappears when the soil is brought back to dry condition.

Large quantities of silt and clay are found in sedimentary deposits of lakes and near shorelines or river exits. Clay particles are the principal source of cohesion in largely cohesive soils. Even when present in small amounts, clay tends to dominate the soil behavior. The term loam is used to describe a loose-textured mixture of sand, clay, and silt. For example, soil containing 60 percent sand, 30 percent clay, and 10 percent silt is called sandy loam. Organic matter in soil is derived from plant life (for example, root growth). Even modest amounts of organic material may significantly affect compressibility and other properties of soils. Soil composed principally of partially decomposed vegetable matter is called *peat.* It is highly compressible and has high water content, making it unsuitable as a foundation material.

Cohesive and Cohesionless Soils. Plasticity is the ability of a soil to undergo unrecoverable deformation at constant volume without cracking or crumbling, and is due to the presence of clay minerals and organic matter. Based on plasticity, soils are divided into two types:

- Cohesive soils
- Cohesionless soils

A cohesive soil is any soil whose grains stick together upon wetting and subsequent drying, so that some force is required to separate them in the dry state, as in clay lumps. This type of soil becomes plastic when wet, and is very compressible. With sufficient amounts of moisture, a cohesive soil may even act like a viscous fluid. Changes in the water content are followed by volume changes in the soil. However, two cohesive soils of apparently the same physical structure or mineral composition may not possess the same properties.

A cohesionless soil is any soil in which the grains fall apart upon drying but may stick together when wet. Sand, which needs to be wet for building a sandcastle, is an example of cohesionless soil. The attraction between grains in this type of soils is due to surface tension forces, called apparent cohesion, in which the water film surrounding the particles pulls the particles and holds them together. A cohesionless soil does not become plastic in any water content, and is only slightly compressible. However it is more permeable for the movement of water than is a cohesive soil.

8.4.3 Soil Classification Systems

In the Unified Soil Classification System (USC), soil is classified as coarse-grained if more than 50 percent of the sample is retained on a No. 200 sieve (0.075 mm), and it is fine-grained if more than 50 percent passes a No. 200 sieve (Table 8.4).

Coarse-grained soil is gravel if more than half of the coarse fraction is retained on a No. 4 sieve (4.75 mm), and is sand if more than half the coarse fraction is between No. 4 and No. 200 sieves. Gravel is further subdivided as clean gravel and gravel with fine, based on the amount of fines. Similarly, sand is divided into two categories, sand with fine and clean sand, depending on the amount of fines. Clays and silts are fine-grained soils. Soil containing a large proportion of fibrous organic matter is called peaty soil. Coarse-grained and fine-grained soils are further subdivided into several types and are given two-letter designation. The first letter indicates the main group and the second modifies the first, as shown in the following:

Symbol	Description
First letter	
G	Gravel
S	Sand
M	Silt
C	Clay
O	Organic
PT	Peat
Second letter	
W	Well-graded
P	Poorly graded
M	Silty fines
C	Clayey fines
H	High plasticity
L	Low plasticity

For example, the symbol GC represents clayey gravel or gravel-sand-clay mixture, and CL stands for inorganic clays of low to medium plasticity, gravelly clay, sandy clay, silty clay, or lean clays.

In highway engineering, soils are classified according to the AASHTO system, which divides soil into seven main groups, A-1 to A-7, based generally on the desirability of the soil as a subgrade for highway construction. The grain-size distribution and plasticity values are the criteria used for classification. A soil sample in which 35 percent or less passes a No. 200 sieve (0.075 mm) is called granular material, and when more than 35 percent passes through this sieve, it is called silt-clay material. Granular materials are further divided into three subgroups: A-1, A-2, and A-3, and silt-clay material into four subgroups: A-4, A-5, A-6, and A-7.

TABLE 8.4 SOIL CLASSIFICATION CHART

Criteria for assigning group symbols and group names using laboratory tests[A]				Soil classification	
				Group symbol	Group name[B]
Coarse-grained soils (more than 50 percent gravel[F] retained on No. 200 sieve)	Gravels (more than 50 percent of coarse fraction retained on No. 4 sieve)	Clean gravels (less than 5 percent fines[C])	$Cu \geq 4$ and $1 \leq Cc \leq 3$[E]	GW	Well-graded gravel[F]
			$Cu < 4$ and/or $1 > Cc > 3$[E]	GP	Poorly graded gravel[F]
		Gravels with fines (more than 12 percent fines[C])	Fines classify as ML or MH	GM	Silty gravel[F,G,H]
			Fines classify as CL or CH	GC	Clayey gravel[F,G,H]
	Sands (50 percent or more of coarse fraction passes No. 4 sieve)	Clean sands (less than 5 percent fines[D])	$Cu \geq 6$ and $1 \leq Cc \leq 3$[E]	SW	Well-graded sand
			$Cu < 6$ and/or $1 > Cc > 3$[E]	SP	Poorly graded sand[I]
		Sands with fines (more than 12 peercent fines[D])	Fines classify as ML or MH	SM	Silty sand[G,H,I]
			Fines classify as CL or CH	SC	Clayey sand[G,H,I]
Fine-grained soils (50 percent or more passes the No. 200 sieve)	Silts and clays (liquid limit less than 50)	inorganic	PI > 7 and plots on or above "A" line[J]	CL	Lean clay[K,L,M]
			PI < 4 or plots below "A" line[J]	ML	Silt[K,L,M]
		organic	(Liquid limit—oven dried) / (Liquid limit—not dried) < 0.75	OL	Organic clay[K,L,M,N] Organic silt[K,L,M,O]
	Silts and clays (liquid limit 50 or more)	inorganic	PI plots on or above "A" line	CH	Fat clay[K,L,M]
			PI plots below "A" line	MH	Elastic silt[K,L,M]
		organic	(Liquid limit—oven dried) / (Liquid limit—not dried) < 0.75	OH	Organic clay[K,L,M,P] Organic silt[K,L,M,Q]
Highly organic soils		Primarily organic matter, dark in color, and organic odor		PT	Peat

[A]Based on the material passing the 3-in. (75-mm) sieve.
[B]If field sample contains cobbles or boulders, or both, add "with cobbles or boulders, or both" to group name.
[C]Gravels with 5–12 percent fines require dual symbols:
GW-GM well-graded gravel with silt
GW-GC well-graded gravel with clay
GP-GM poorly graded gravel with silt
GP-GC poorly graded gravel with clay
[D]Sands with 5–12 percent fines require dual symbols:
SW-SM well-graded sand with silt
SW-SC well-graded sand with clay
SP-SM poorly graded sand with silt
SP-SC poorly graded sand with clay

[E] $Cu = D_{60}/D_{10} \quad Cc = \dfrac{(D_{30})^2}{D_{10} \times D_{60}}$

[F]If soil contains ≥15 percent sand, add "with sand" to group name.
[G]If fines classify as CL-ML, use dual symbol GC-GM, or SC-SM.
[H]If fines are organic, add "with organic fines" to group name.
[I]If soil contains ≥15 percent gravel, add "with gravel" to group name.
[J]If Atterberg limits plot in hatched area, soil is a CL-ML, silty clay.
[K]If soil contains 15–29 percent plus No. 200, add "with sand" or "with gravel," whichever is predominant.
[L]If soil contains ≥30 percent plus No. 200, predominantly sand, add "sandy" to group name.
[M]If soil contains ≥30 percent plus No. 200, predominantly gravel, add "gravelly" to group name.
[N]PI ≥ 4 and plots on or above "A" line.
[O]PI < 4 or plots below "A" line.
[P]PI plots on or above "A" line.
[Q]PI plots below "A" line.

(*Source*: ASTM D2487, Reprinted with permission)

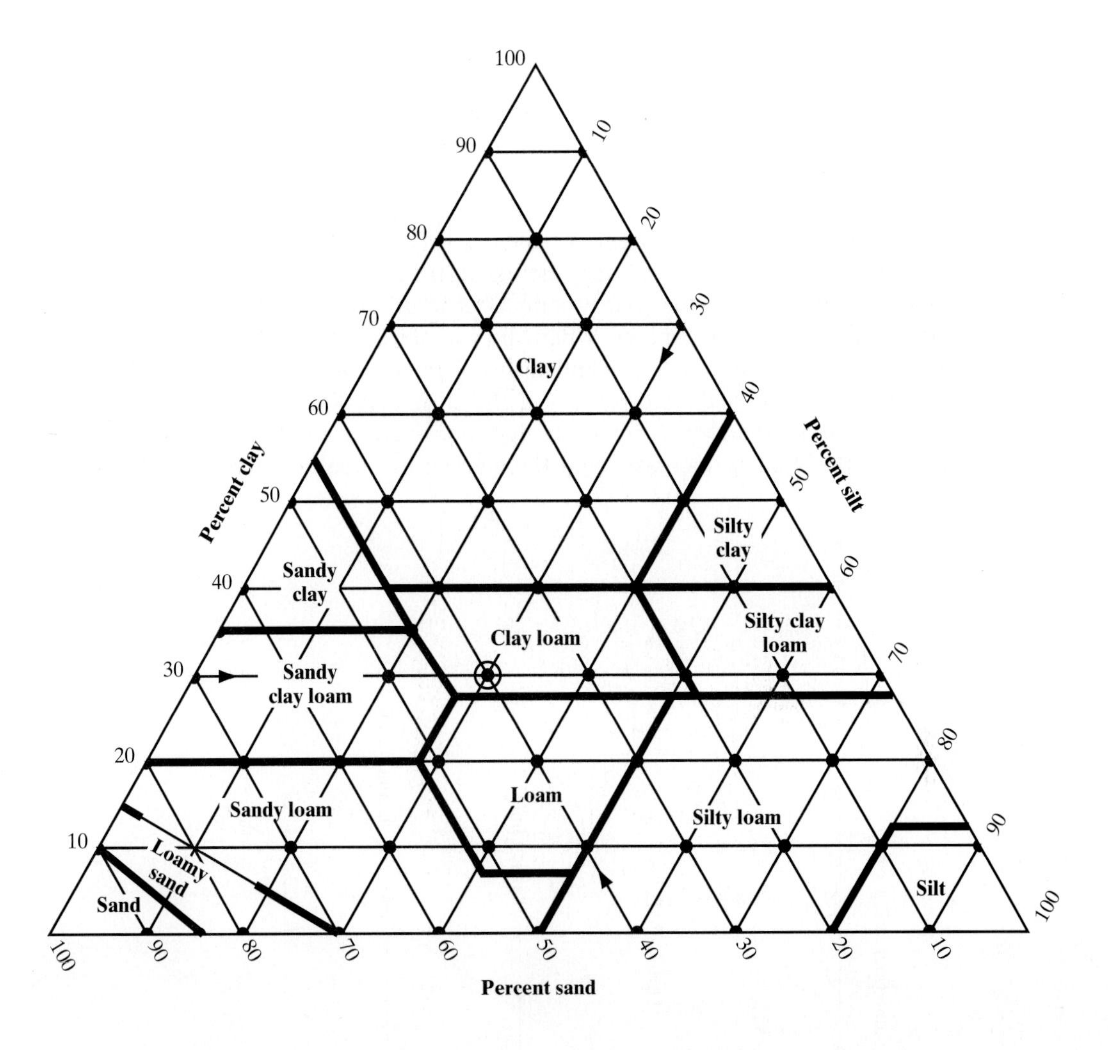

(Example: A soil sample made up of 30% silt, 30% clay, and 40% sand is clay loam)

Figure 8.6 Textural chart for classification of soils (USDA).

Most soil deposits consist of various amounts of clay, silt and sand. Once the percentages of these ingredients present in a soil sample are established, a general classification can be established from Fig. 8.6.

Phase-volume Relationships Soil grain is composed of a variety of different crystals. Materials formed from very fine crystals have higher strengths than materials produced from very large crystals. Materials made up of interlocking crystals have similar strength in all directions, while the strength of materials composed of oriented or layered crystals depends on the loading direction. Soil consists of three phases of materials: solids (containing just the grains), liquids (consisting of water or other liquids in some of the void space between the grains), and gases or air (filling the remaining void space). Many mechanical properties of a soil depend on its phase relationship.

In the case of fine-grained material, clay or silt, the mechanical properties vary as a function of moisture content or plasticity (Fig. 8.7). As water is added to a fine-grained material, the water content increases, but the volume remains constant because the water is filling the void system. After saturation, the total volume increases with increase in water content. When the sample is saturated and the volume is smallest, the moisture content is called the *shrinkage limit*. The transition between the semisolid and solid states occurs at the shrinkage limit. In any water content equal to or more than the shrinkage limit, the soil is saturated. With further increase in water, the volume increases and the sample eventually behaves like a plastic

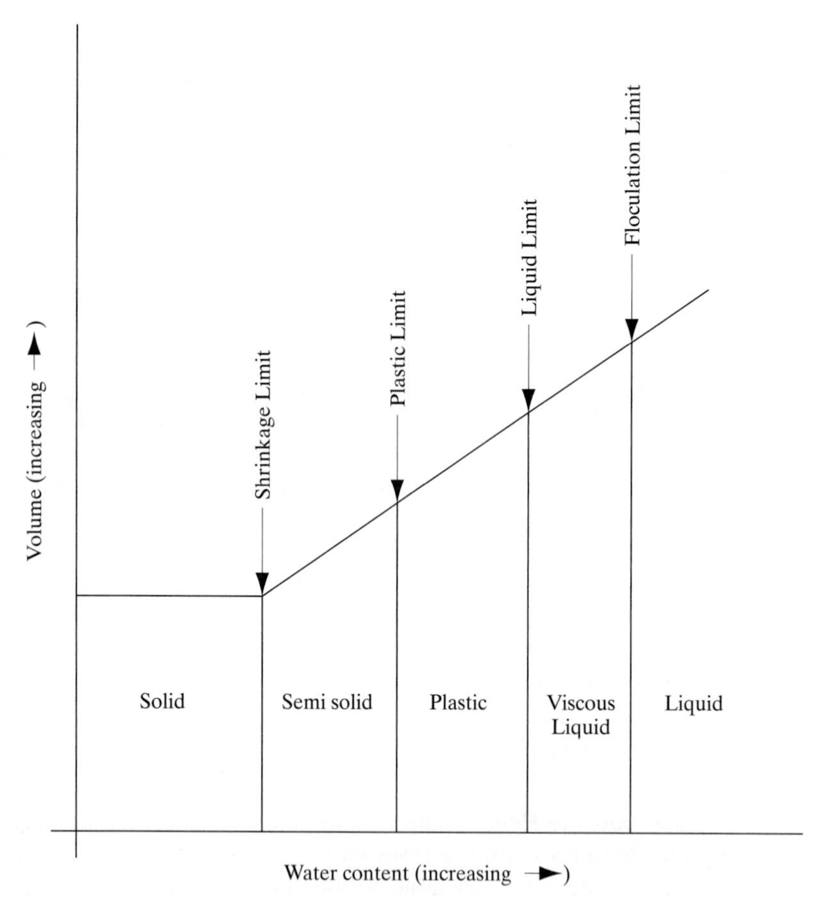

Figure 8.7 Phase–volume relationships of fine-grained soils.

material. The water content at which the sample becomes a true plastic material is called the *plastic limit.* If water is added to a plastic material, it behaves as a viscous fluid. The upper limit of water content at which a solid exhibits plastic behavior is called the *liquid limit.* It is also defined as the water content of the sample when it has a shear strength of 1 g/cm^2. The range in water content between the liquid limit and the plastic limit is called the *plasticity index.* A material with low plasticity index is called low-plastic material, and that with a high plasticity index is high-plastic material. When the water content is increased further, the particles are separated by water and the sample behaves as a true liquid, dropping the shear strength to zero. This water content is called the *floculation limit.*

8.5 STRENGTH PROPERTIES OF SOILS

Notable engineering properties of soil are shear strength, permeability, and compressibility. The shear strength depends on the type, density, moisture content, and loading history of the soil. It is needed in estimating the stability of earth dams and embankments, in assessing the lateral pressure on retaining walls, and in the calculation of bearing capacity. Permeability is a measure of the ease with which water flows through the soil, and is required in the design of earth dams and subsurface drainage systems. The settlement behavior of soil is assessed from its compressibility. The magnitude and rate of settlement vary with the type and stratification of soil.

Size and grading have a significant influence on the engineering properties of granular soils. Increase in particle size generally improves the strength. Soil with well-graded particles is normally stronger than a uniformly graded deposit. Densely packed granular soil is almost incompressible, whereas loosely packed deposits are compressible. Moisture within the soil affects the strength; when the foundation level is below the water table, rather than above, greater settlement can be expected.

In general, coarse-grained soils have higher shear strength, more permeability, and less compressibility than fine-grained soils. The nature and percentage of fine fraction (particles less than 0.075 mm) affect these properties in coarse-grained soil. The moisture condition, loading history, and amounts of clay material control the engineering properties of fine-grained soils. Dry silt or clay has higher strength and lower compressibility than saturated silt and clay. Silt and clay subjected to high stresses in the past have higher strength and lower compressibility than silt or clay that has not been subjected to such.

8.6 CONSTITUENT PROPERTIES OF SOILS

As indicated, soil is a combination of very dissimilar materials in three phases: solid, liquid, and gas. The strength and behavior of soil is a function of the group action of the soil grains and the interaction between the three phases. Properties of a soil in any one phase may be entirely different from those at other phases.

Water is held in the void space by gravitational forces, capillary action, and as absorbed water. Gravitational water is free to move under the force of gravity, and is the water that will drain from the soil. Capillary water is the water that is held above the water table by the capillaries (small pores) in the soil. This water can be removed from the soil only after the water table is lowered. Fine-grained soils have a margin of capillaries extending 3–4 ft. above the water table, whereas coarse-grained soils have only a few inches. Absorbed water (hygroscopic moisture) is the moisture remaining in the grains after the gravitational and capillary water is removed. It is moisture that is held by the soil grains in the form of very thin films.

In construction engineering, soil is used either as a foundation or construction material. For use as a foundation material, the in-situ properties are important; investigations are carried out to determine the characteristics of the soil for assessing the load-carrying, or bearing capacity, and to estimate the foundation settlement, ground water movement, and lateral stability. These aspects are required in the design, construction, and assessment of highway pavements, bridge piers and abutments, retaining walls, embankments, dams, tunnels, and foundations.

As a construction material (such as for the building of earth dams, road base, grading, and backfill), remolded properties of soil are important. The constituent properties that affect soil performance both as a foundation and construction material, discussed in the following paragraphs, are listed here:

- Particle-size distribution or composition
- Grain size and shape
- Water content
- Density (unit weight)
- Specific gravity
- Void ratio
- Crystal size and shape

Important engineering properties that have already been discussed are:

- Permeability
- Compressibility
- Shear strength

8.6.1 Particle Size

The particle or grain size of a soil sample is taken as the size of the smallest hole or sieve through which the grain will pass. It represents a nominal diameter of soil particles.

Soil particles range in size from about 3/8 in. (10 mm) to 0.00001 mm or 0.01 μm. Some particles are visible to the unaided eye, some are microscopic, and the rest are submicroscopic in size. Particles of cohesionless soil typically are visible to the unaided eye, and those of cohesive soil are microscopic and submicroscopic.

Physical separation of a sample into two or more fractions, each containing particles of a certain size only, is termed *fractionalization* or *gradation*. Using *mechanical analysis,* or *sieve analysis,* we can determine the percentage of particles of various sizes in a soil sample. The procedure involves taking a representative sample of the soil and shaking it through a stack of standard sieves, with openings of known sizes, arranged in decreasing sizes from top to bottom.

Grain-size analysis provides data that helps to determine the suitability of a particular soil for a given construction project, and to estimate the permeability and capillarity. Soil permeability is a measure of the ease with which water will flow through a soil, and capillarity is associated with the retention and rise of water above the water table (further explained in the section on moisture content). Both these properties are connected to the gradation.

Sieve analysis yields the percentages of particles retained in each sieve, which can be used to plot the *grain-size distribution,* or *gradation curve.* The horizontal axis of the gradation curve shows the particle sizes, plotted from left to right in increasing magnitude (Fig. 8.8). The ordinate is the percentage finer than (or percentage coarser than) individual sieve sizes. A continuously sloping curve (A) represents a specimen with an even distribution of all sizes. A vertical or horizontal drop in the curve (B and C) shows a sample with a heavy concentration of materials in limited sections of the total range.

Curve A in Fig. 8.8 represents a sample with good gradation. This type of soil is relatively stable, can be readily compacted, and is resistant to erosion. The bearing capacity and the shear resistance of this soil are high. The sample represented by curve B depicts poor or uneven gradation. This sample is lacking in particles of sizes

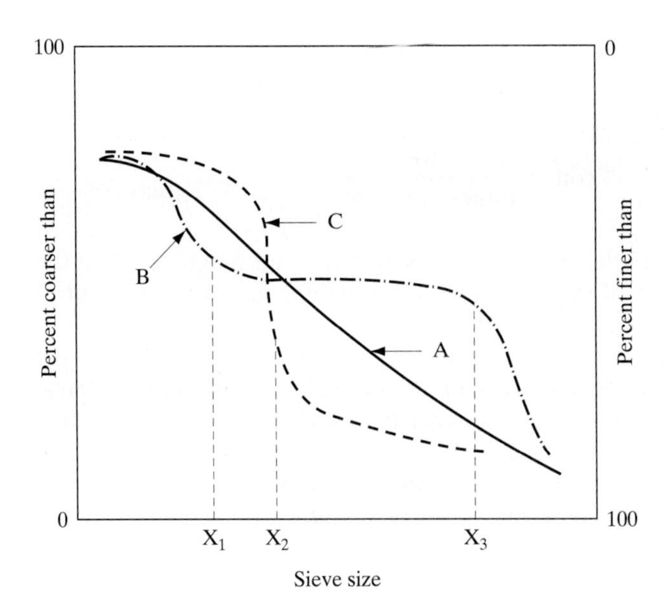

Figure 8.8 Particle-size distribution curves.

between sieve X_1 and X_3. It is primarily made up of fines; coarser particles make up a very small fraction of the total sample. The sample of curve C is composed largely of particles of sieve size X_2. It contains very few fines and coarser particles. Compaction of soils represented by curves B and C is difficult.

8.6.2 Unit Weight

If we take a look at a sample of soil, it can be seen that it is made up of grains or particles, moisture, and voids. The voids or pores are the open spaces between grains. Grains are particles of varying sizes, shapes, and mineral compositions. Moisture exists in the voids and makes the soil wet, damp, or dry. When the voids do not contain water, they are filled with air. When all the voids are filled with water, the soil is said to be saturated. If the soil is dried to constant weight in an oven (to drive off water from the voids), the weight is called dry weight.

The unit weight of soil is the weight of soil divided by the total soil volume, which is the solid volume plus the volume of voids. The weight of soil may be the weight of solids only, the weight of solids plus the weight of pore water, or the submerged weight of solids (Fig. 8.9).

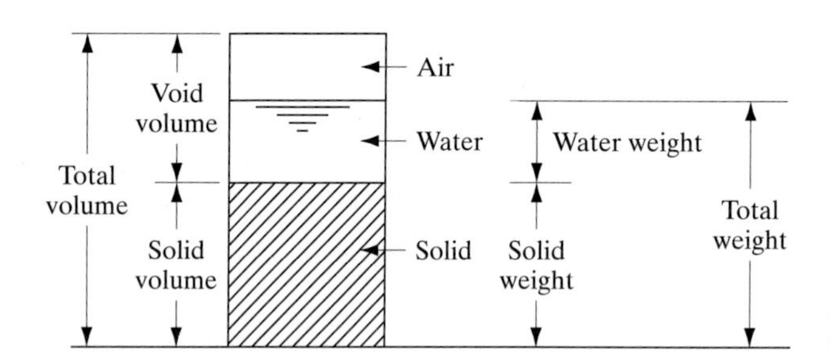

Figure 8.9 Weight–volume relationship of soil mass.

$$\text{Unit weight of soil} = \frac{\text{soil weight}}{\text{total soil volume}}$$

When the soil weight includes the weight of solids only, the unit weight is called the unit dry weight. It is an index of the density of soil, and depends on the concentration of solids in a given volume.

The dry unit weight of granular materials ranges between 80–130 pcf (1280–2080 kg/m^3). Clay soil has a dry unit weight between 50–105 pcf (800–1680 kg/m^3). The dry unit weight of mixed soils can be anywhere between 60–130 pcf (960–2080 kg/m^3). Organic matter decreases the unit dry weight. In organic silts and clays, the dry unit weight can be as low as 30 pcf (480 kg/m^3). When the soil is wet, the unit weight (wet unit weight) is higher. The submerged unit weight is the least of the three unit weights.

8.6.3 Moisture Content

The moisture content or water content is the ratio of the weight of water to the weight of solids in the soil and is usually expressed as a percentage:

$$\text{Moisture content} = \frac{\text{weight of water}}{\text{weight of solids}} \times 100$$

The water content is determined by weighing the soil sample and then drying it in an oven at a temperature of about 230 °F (110 °C; for a constant weight, normally 24 hours) and reweighing. Some marine and lake clays have a moisture content of 300–400 percent, and can be as high as 800 percent. Typically, fine-grained soils have a higher moisture content than coarse-grained soils. In moist sands, the water content is less than 60 percent. The water content of the top layer of any soil deposit may not be the same as that of the bottom layer. Soils that appear dry may, in fact, contain water of 2–3 percent.

Water content is among the most frequently determined characteristics of soil. It is one of the more useful soil properties and is a good indicator of the shear strength of clays (saturated). When water is added to a dry soil, each particle is covered with a film of adsorbed water. Increasing the water will increase the thickness of this film, which allows the particles to slide easily past one another. Thus the behavior of a soil depends on the amount of water in the soil.

Water content plays an important role in the compaction of soils, especially of fine grained soils. Compaction is a process of placing soil in a dense state. It is done to decrease settlement, increase shear strength, and decrease permeability. Earth dams, retaining walls, highways, airports, foundations, and embankments may require compacted soil.

8.6.4 Degree of Saturation

The degree of saturation (S) is the ratio of the volume of water to the total volume of voidspace:

$$\text{Degree of saturation, } S = \frac{\text{volume of water}}{\text{volume of voids}} \times 100$$

or,

$$= \frac{\text{weight of water}}{w_w \times \text{volume of voids}} \times 100$$

where w_w is the density of water. The volume of voids can be calculated as

$$\text{Volume of voids} = \text{soil volume} - \frac{W_s}{\text{SG} \times w_w}$$

where W_s is the solid weight or dry weight of soil and SG is the specific gravity of soil.

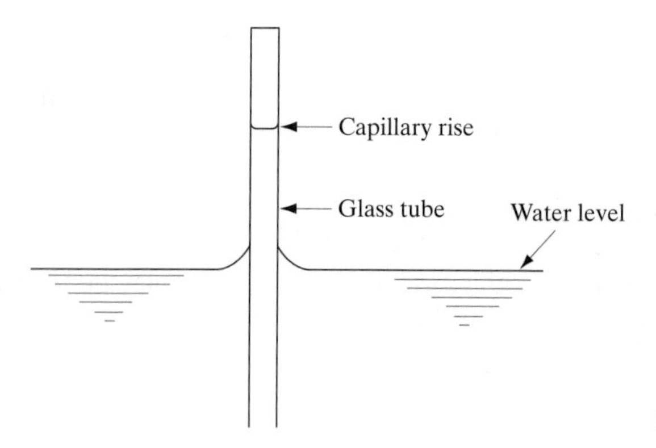

Figure 8.10 Capillary rise.

Like moisture content, the degree of saturation is expressed as a percentage. If the soil is dry, the degree of saturation is zero, and if the pores are full of water, the soil is saturated, and $S = 100$ percent.

Natural deposits of soil below the water table have a 100 percent degree of saturation. Due to capillary action, fine-grained soil above the water table may also be saturated. Capillary action refers to the rise of water in a capillary tube above the surrounding water level (Fig. 8.10). The height of the rise depends on the diameter of the tube; the smaller the tube, the higher the rise inside it. In fine-grained soils, under certain conditions, water is retained through capillary action. Moisture on concrete floors at basement or ground level may be present through capillary action.

The degree of saturation affects the permeability of the soil. In addition, it has an effect on the shear strength and compressibility of the soil.

8.6.5 Void Ratio

Void ratio (e) is the ratio of the volume of voids to the volume of solids:

$$\text{Void ratio, } e = \frac{\text{volume of voids}}{\text{volume of solids}}$$

or,

$$= \frac{\text{SG}(w_w)V - W_s}{W_s}$$

where SG is the specific gravity of soil, w_w the unit weight of water, W_s the solid weight or dry weight of soil, and V the total volume. Typical ranges of values of the void ratio for various type of soils are shown in Table 8.5. Generally, granular soils have a void ratio below 0.6 and cohesive soils have values greater than 0.7.

Void ratio represents the denseness of a soil mass. Permeability, shear strength, and other properties are related to the void ratio. Void ratio is a more important characteristic of the soil than water content. In a given saturated soil, the void ratio is proportional to the water content.

TABLE 8.5 AVERAGE PROPERTIES OF SOILS

Soil	Specific gravity	Void ratio	Density or unit weight pcf (g/cm^3)
Gravel	2.5–2.8	0.25–0.33	90–130 (1.4–2.1)
Sand	2.6–2.7	0.30–0.54	100–125 (1.6–2.0)
Silt	2.64–2.66	0.35–0.85	100–135 (1.6–2.2)
Clay	2.55–2.75	0.42–0.96	60–130 (1.0–2.1)

8.6.6 Porosity

Porosity (n) is a property that also measures the void content in a soil sample, but is used more frequently by agricultural engineers. It is defined as the ratio between the volume of voids and the soil volume, expressed as a percentage:

$$\text{Porosity, } n = \frac{\text{volume of voids}}{\text{soil volume}} \times 100$$

or

$$= 1 - \frac{W_s}{\text{SG}(w_w)V}$$

From comparing the definitions of porosity and void ratio, the following relationship can be derived:

$$e = \frac{n}{1 - n}$$

and

$$n = \frac{e}{1 + e}$$

Porosity, similarly to void ratio, also measures the denseness of the soil. When a soil compresses or swells, the void volume changes, but the volume of solids remains the same. By comparing the equations for the void ratio and the porosity, it can be verified that the void ratio is a more convenient measure of volumetric changes in the soil than the porosity. But when calculating the seepage through a soil, porosity is a more convenient measure.

8.6.7 Particle Shape

In addition to the characteristics already discussed, the properties of coarse-grained soils depend on the shape of the particles. Depending on the length of transportation history, the particles can be angular, subangular, subrounded, rounded, and well-rounded (Fig. 8.11 and Table 8.6).

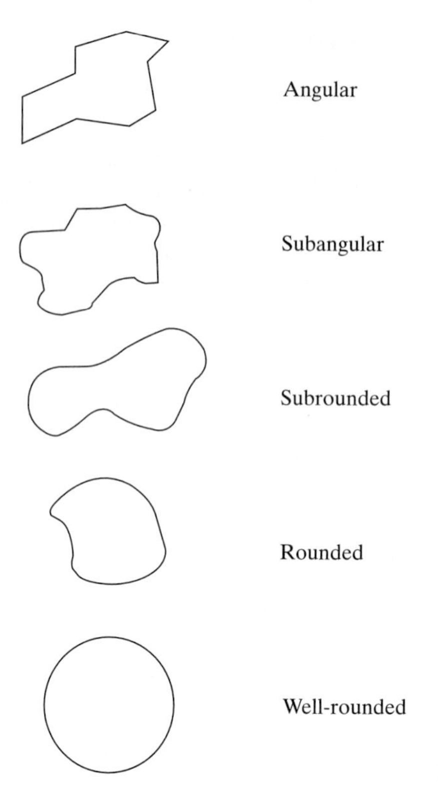

Figure 8.11 Particle shapes.

Angular particles allow more interlock between the grains than do rounded particles, and thus offer greater frictional resistance. Particle shape also influences the density, compressibility, and shear strength of the soil. Typically, coarse-grained soils with angular particles have greater strength and bearing capacity than granular soils with rounded particles.

8.6.8 Specific Gravity

Specific gravity (SG) is the ratio of the weight in air of a given volume of solid particles to the weight in air of an equal volume of distilled water. This property is used in relating the weight to the volume and in calculating the unit weight:

$$\text{SG} = \frac{\text{solid weight}}{(\text{volume of solids}) \times (\text{density of water})}$$

The specific gravity of soils depends on the mineralogical composition. Some minerals, such as mica and dolomite, have high specific gravity (2.7–3.2); a few others, such as gypsum and bentonite, have low specific gravity (2.13–2.3). Ottawa sand has a specific gravity of 2.67. The specific gravity of clays may range between

TABLE 8.6 SOIL DESCRIPTION AND CRITERIA

Basis	Description	Criteria
Angularity (coarse-grained)	Angular	Sharp edges, plane sides, unpolished surfaces
	Subangular	Rounded edges, plane sides, unpolished surfaces
	Subrounded	Well-rounded edges, plane sides, well-rounded corners
	Rounded	No edges, smoothly curved sides
Shape	Flat	(Width/Thickness) greater than 3
	Elongated	(Length/Width) greater than 3
	Flat and elongated	(Width/Thickness) greater than 3 and (Length/Width) greater than 3
Moisture condition	Dry	Dry to the touch, dusty, no moisture
	Moist	Damp, no visible water
	Wet	Visible free water or submerged
Consistency	Very soft	Thumb penetration into soil more than 1 in. (25 mm)
	Soft	Thumb penetration into soil about 1 in. (25 mm)
	Firm	Thumb indentation into soil about 0.25 in. (6 mm)
	Hard	Indentation into soil by thumbnail only
	Very hard	No indentation by thumbnail
Plasticity	Nonplastic	A 1/8-in. (3-mm) thread cannot be rolled at any water content
	Low	A 1/8-in. (3-mm) thread can barely be rolled
	Medium	A 1/8-in. (3-mm) thread can easily be rolled, but cannot be rerolled after reaching the plastic limit
	High	A 1/8-in. (3-mm) thread can be rerolled several times after reaching the plastic limit

2.2–2.75. Porous soils and soils with organic content may have a specific gravity less than 2.0. The specific gravity of most soils lies in the range 2.5–2.85.

Engineering properties of soil, such as compressibility and shear strength, depend on several factors and cannot be explained without a basic understanding of soil mechanics. These principles can be learned in a course on soil mechanics.

8.6.9 Uses of Soil

Well-graded sand, gravel, and sand–gravel mixtures are pervious when compacted. They possess excellent shear strength and negligible compressibility (following compaction). These types of soils constitute good construction material and can be used in foundations and roadways. Silty gravels and clayey gravels are excellent materials for rolled earth dams. They can also be used for road bases. Silts and clays with or without sands are impervious when compacted but possess poor shear strength. They have average compressibility characteristics and can be used for canal sections.

References and Further Reading

Abraham, H. (1960–1962). *Asphalts and Allied Substances,* Vols. I–V. D. Van Nostrand, Princeton, N.J.

ACI Committee 524 (1993). Guide to portland cement plastering. *ACI Materials Journal,* 90(1), 69–93.

Adams, J. T. (1979). *The Complete Concrete, Masonry and Brick Handbook.* Arco Publishing, New York.

Adaska, W. S. (1991). Soil–cement: a material with many applications. *Concrete International,* 13(1), 49–52.

Al-Adeeb, A. M., and Al-Mudhaf, H. A. (1992). Performance of masonry walls: case study in Kuwait. *Journal of Materials Engineering,* 4(1), 77–90.

American Iron and Steel Institute (1991). *The Performance of Steel Buildings in Past Earthquakes.* AISI, Washington, D.C.

——— (n.d.). *Steel Processing Flow Lines.* AISI, Washington, D.C.

American Society of Civil Engineers (1982). *Wood Structures: A Guide and Commentary.* ASCE, New York.

——— (1986). *Compilations of ASTM Standard Definitions.* ASTM, Philadelphia.

——— (1991). *Annual Book of ASTM Standards,* Vol. 1.03, *Steel: Plate, Sheet, Strip, Wire.* ASTM, Philadelphia.

——— (1991). *Annual Book of ASTM Standards,* Vol. 1.04, *Steel: Structural, Reinforcing, Pressure Vessel, Railway.* ASTM, Philadelphia.

——— (1991). *Annual Book of ASTM Standards,* Vol. 4.01, *Cement; Lime; Gypsum.* ASTM, Philadelphia.

——— (1991). *Annual Book of ASTM Standards,* Vol. 4.02, *Concrete and Aggregates.* ASTM, Philadelphia.

——— (1991). *Annual Book of ASTM Standards,* Vol. 4.03, *Road and Paving Materials; Pavement Management Technologies.* ASTM, Philadelphia.

——— (1991). *Annual Book of ASTM Standards,* Vol. 4.04, *Roofing, Waterproofing and Bituminous Materials.* ASTM, Philadelphia.

——— (1991). *Annual Book of ASTM Standards,* Vol. 4.05, *Chemical Resistant Materials; Vitrified Clay, Concrete, Fiber-Cement Products; Mortars; Masonry.* ASTM, Philadelphia.

——— (1991). *Annual Book of ASTM Standards,* Vol. 4.08, *Soil and Rock; Dimension Stone; Geosynthetics.* ASTM, Philadelphia.

——— (1991). *Annual Book of ASTM Standards,* Vol. 4.09, *Wood.* ASTM, Philadelphia.

——— (1973). *The World of Plastics.* Young Readers Press, New York.

ANON. (1980). Control of cracking in concrete structures. *Concrete International,* Oct., 80, 35–76.

ANON. (1991). Proposed revisions of guide to durable concrete. *ACI Materials Journal,* Sept–Oct., 544–574.

ANON. (1993). *Fiber-Reinforced Plastics International,* 1(1).

ARNISON, J. H. (1967). *Roadwork Technology,* Vol. 2. Iliffe Books, London.

ATKINS, H. N. (1983). *Highway Materials: Soils and Concretes,* 2nd ed. Reston, Reston, Va.

BAALBAKI, W., BENMOKRANE, B., CHAALLAL, O., AND AITCIN, P. C. (1991). *ACI Materials Journal,* 88(5), 499–510.

BAKER, I. O. (1899). *A Treatise on Masonry Construction,* 9th ed. Wiley, New York.

BAUER, E. E. (1949). *Plain Concrete,* 3rd ed. McGraw-Hill, New York.

BENJAMIN, B. S. (1982). *Structural Design with Plastics.* Van Nostrand Reinhold, New York.

BERGER, B. D., AND ANDERSON, K. E. (1978). *Modern Petroleum.* Petroleum Publishing, Tulsa, Okla.

BERRY, E. E., AND MALHOTRA, V. M. (1980). Fly ash for use in concrete: a critical review. *ACI Journal,* 77(2), 59–73.

BETHLEHEM STEEL (n.d.). *Sparrows Point Plant.* Bethlehem Steel Corporation, Bethlehem, Pa.

——— (n.d.). *The Steel Making Process.* Bethlehem Steel Corporation, Bethlehem, Pa.

BISWAS, A. K. (1981). *Principles of Blast Furnace Ironmaking.* Cootha Publishing House, Brisbane, Australia.

BOWLES, J. E. (1984). *Physical and Geotechnical Properties of Soils,* 2nd ed. McGraw-Hill, New York.

BOWMAN, M. D., AND BETANCOURT, M. (1991). Reuse of A325 and A490 high strength bolts. *Engineering Journal,* 28(3), 110–118.

BRABSTON. W. N., AND ROLLINGS, R. S. (1991). Design of cement stabilized base course for airfield pavements. *Concrete International,* 13(12), 19–23.

BRUNSKILL, R., AND CLIFTON-TAYLOR, A. (1977). *English Brickwork.* A Hyperion Book. Ward Lock, London.

CAPPER, P. L., AND CASSIE, W. F. (1969). *The Mechanics of Engineering Soils.* E. and F. N. Spon, London.

CHAALLAL, O., BENMOKRANE, B., AND BALLIVY, G. (1992). Drying shrinkage strains: experimental versus codes. *ACI Materials Journal,* 89(3), 263–266.

CHILINGARIAN, G. V., AND YEN, T. F. (eds.) (1978). *Bitumens, Asphalts and Tar Sands.* Elsevier, New York.

Cohen, M. D., Olek, J., and Mathew, B. (1991). Silica fume improves expansive-cement concrete. *Concrete International,* 13(3), 31–37.

Committee on Engineering Materials, ASTM (1958). *Engineering Materials.* ASTM, Pitman, New York.

Concrete Reinforcing Steel Institute (1986). *Manual of Standard Practice.* CRSI, Schaumburg, Ill.

——— (1986). *Placing Reinforcing Bars.* CRSI, Schaumburg, Ill.

——— (1992). *The Tallest Reinforced Concrete Building in Seismic Zone IV.* CRSI, Schaumburg, Ill.

Cordon, W. A. (1979). *Properties, Evaluation and Control of Engineering Materials.* McGraw-Hill, New York.

Craig, R. F. (1983). *Soil Mechanics.* Van Nostrand Reinhold, Wokingham, Berkshire, England.

Dallaire, G. (1983). Superplasticizer concrete takes off in Dallas. *Civil Engineering,* Mar., 39–41.

Davey, N. (1961). *A History of Building Materials.* Phoenix House, London.

Davis, H. E., Troxell, G. E., and Hauck, G. F. W. (1982). *The Testing of Engineering Materials,* 4th ed. McGraw-Hill, New York.

Davis, H. E., Troxell, G. E., and Wiskocil, C. T. (1964). *The Testing and Inspection of Engineering Materials.* McGraw-Hill, New York.

Day, R. (1979). *The Practical Handbook of Concrete and Masonry.* Arco Publishing, New York.

DeCamp, L. S. (1972). *Great Cities of the Ancient World.* Doubleday, New York.

Dennis, W. H. (1967). *Foundations of Iron and Steel Metallurgy.* Elsevier, Barking, Essex, England.

Derucher, K. N., and Korfiatis, G. P. (1988). *Materials for Civil and Highway Engineers.* Prentice Hall, Englewood Cliffs, N.J.

Dolar-Mantauani, L. (1983). *Handbook of Concrete Aggregates.* Noyes Publications, Park Ridge, N.J.

Dubovoy, V. S., and Ribar, J. W. (1990). *Masonry Cement Mortar: A Laboratory Investigation.* Research and Development Bulletin RD0095, Portland Cement Association, Skokie, Ill.

Duncan, C. I. (1992). *Soils and Foundations for Architects and Engineers.* Van Nostrand Reinhold, New York.

Engineering Equipment Users Association (1973). *The Uses of Plastic Materials in Buildings.* Handbook, EEUA, London.

Fletcher, B. (1954). *A History of Architecture on the Comparative Method,* 16th ed. Raphael Tuck, England.

Flinn, R. A., and Trojan, P. K. (1975). *Engineering Materials and Their Applications.* Houghton Mifflin, Boston.

Garber, N. J., and Hoel, L. A. (1988). *Traffic and Highway Engineering.* West Publishing, St. Paul, Minn.

Garraty, J. A., and Gay, P. (eds.) (1981). *The Columbia History of the World.* Harper & Row, New York.

GEBLER, S. H., AND KLIEGER, P. (1986). *Effect of Fly Ash on Some of the Physical Properties of Concrete.* Portland Cement Association, Skokie, Ill.

GIACCIO, G., ROCCO, C., VIOLINI, D., ZAPPITELLI, J., AND ZERBINO, R. (1992). High-strength concretes incorporating different coarse aggregates. *ACI Materials Journal,* 89(3), 242–246.

GUSTAFSON, D. P., AND FELDER, A. L. (1991). Questions and answers on ASTM A706 reinforcing bars. *Concrete International,* 13(7), 54–57.

HALL, C. (1989). *Polymer Materials,* 2nd ed. Wiley, New York.

HAQUE, M. N., AND KAWAMURA, M. (1992). Carbonation and chloride-induced corrosion of reinforcement in fly ash concretes. *ACI Materials Journal,* 89(1), 41–48.

HEINZ, R. (1993). Plastic piling. *Civil Engineering,* 63(4), 63–65.

HODSON, R., AND KUSHNER, D. D. (1992). Ingredients, texture, and integrally colored concrete. *Concrete International,* Sept., 21–25.

HOUGH, B. K. (1957). *Basic Soils Engineering.* Ronald Press, New York.

HOYLE, R. J., AND WOESTE, F. E. (1989). *Wood Technology in the Design of Structures,* 5th ed. Iowa University Press, Ames, Iowa.

JACKSON, N., AND DHIR, R. K. (1988). *Structural Engineering Materials.* Hemisphere, New York.

JOHNSON, A. (1992). Steel weighs in. *Civil Engineering,* July, 69–71.

KANTRO, D. L. (1981). *Influence of Water Reducing Admixtures on Properties of Cement Paste: A Miniature Slump Test.* Research and Development Bulletin RD079.01T, Portland Cement Association, Skokie, Ill.

KECK, R. (1991). A tower of strength. *Concrete International,* 13(3), 23–25.

KINNEY, G. F. (1957). *Engineering Properties of Plastics and Applications.* Wiley, New York.

KNOWLES, P. (1977). *Design of Structural Steelwork.* Surrey University Press, London.

KORHONEA, C. J., AND CORTEZ, E. R. (1991). Antifreeze admixtures for cold weather concreting. *Concrete International,* 13(3), 38–41.

KREBS, R. D., AND WALKER, R. D. (1971). *Highway Materials.* McGraw-Hill, New York.

LARSON, T. D. (1963). *Portland Cement and Asphalt Concretes.* McGraw-Hill, New York.

LAY, M. G. (1986). *Handbook of Road Technology,* Vol. 1. Gordon and Breach, New York.

LEA, F. M. (1970). *The Chemistry of Cement and Concrete,* 3rd ed. Edward Arnold, London.

LEVER, A. E. (1968). *The Properties and Testing of Plastics Materials,* 3rd ed. Temple Press Books, Feltham, Middlesex, England.

LOOV, R. E. (1991). Reinforced concrete at the turn of the century. *Concrete International,* Dec., 67–73.

MAJKO, R. M., AND PISTILLI, M. F. (1984). Optimizing the amount of Class C fly ash in concrete mixtures. *Cement, Concrete and Aggregates,* 6(2), 105–119.

MCCRUM, N. G., BUCKLEY, C. P., AND BUCKNALL, C. B. (1989). *Principles of Polymer Engineering.* Oxford University Press, Oxford.

MEININGER, R. C. (1988). No-fines pervious concrete for paving. *Concrete International,* 10(8), 20–29.

MILLS, A. P., HAYWARD, H. W., AND RADER, L. F. (1965). *Materials of construction,* 6th ed. Wiley, New York.

MURDOCK, L. J., AND BROOK, K. M. (1978). *Concrete Materials and Practice.* Wiley, New York.

NEVILLE, A. M. (1971). *Hardened Concrete: Physical and Mechanical Aspects.* ACI Monograph Series, American Concrete Institute, Detroit, Mich.

NEVILLE, A. M., AND BROOKS, J. J. (1987). *Concrete Technology.* Longman, Harlow, Essex, England.

NUCOR (n.d.). *The Nucor Story.* Nucor Corporation, Charlotte, N.C.

O'FLAHERTY, C. A. (1988). *Highways,* Vol. 2, 3rd ed. Edward Arnold, London.

PARE, T. P. (1991). The big threat to big steel's future. *Fortune.*

PEACEY, J. G., AND DAVENPORT, W. G. (1979). *The Iron Blast Furnace.* Pergamon Press, Oxford.

PISTILLI, M. F., WINTERSTEEN, R., AND CECHNER, R. (1984). The uniformity and influence of silica fume from a U.S. source on the properties of portland cement concrete. *Cement, Concrete, and Aggregates,* 6(2), 114–120.

PITTS, J. (1984). *A Manual of Geology for Civil Engineers.* Wiley, New York.

PLUMMER, H. C. (1962). *Bricks and Tile Engineering.* Brick Institute of America, McLean, Va.

POLLACK, H. W. (1977). *Material Science and Metallurgy,* 2nd ed. Reston, Reston, Va.

PORTLAND CEMENT ASSOCIATION (1971). *High Strength Concrete.* PCA, Skokie, Ill.

——— (1976). *Concrete Masonry Handbook for Architects, Engineers and Builders.* PCA, Skokie, Ill.

——— (1980). *Portland Cement Plaster (Stucco) Manual.* PCA, Skokie, Ill.

——— (1987). *Mortars for Masonry Walls.* (1987). PCA, Skokie, Ill.

——— (1983). *Concrete Floors on Ground,* 2nd ed. PCA, Skokie, Ill.

——— (1987). *Recommended Practices for Laying Concrete Block.* PCA, Skokie, IL.

——— (1991). *Masonry Information.* PCA, Skokie, Ill.

PRICE, W. H. (1982). Control of cracking of concrete during construction. *Concrete International,* 4(1), 40–43.

QUARMBY, A. (1974). *The Plastics Architect.* Pall Mall Press, London.

RAGSDALE, L. A., AND RAYNHAM, E. A. (1972). *Building Materials Technology.* Edward Arnold, London.

RASHEEDUZZAFAR (1992). Influence of Cement Composition on Concrete Durability. *ACI Materials Journal,* 89(6), 574–586.

REBOUL, P., AND MITCHELL, R. G. B. (1968). *Plastics in the Building Industry.* George Newnes, London.

RIBAR, J. W. (1990). *Water Permeance of Masonry: A Laboratory Study.* ASTM STP 778. American Society for Testing and Materials, Philadelphia, 200–220.

RICE, W. H. (1982). Control of cracking in concrete during construction. *Concrete International,* Jan., 40–43.

ROBERTS, W. L. (1983). *Hot Rolling of Steel.* Marcel Dekker, New York.

ROPKE, J. C. (1982). *Concrete Problems: Causes and Cures.* McGraw-Hill, New York.

SCHRADER, E. (1992). Mistakes, misconceptions, and controversial issues concerning concrete and concrete repairs. *Concrete International,* Sept., 52–56.

SENBETTA, E. (1989). Curing compound selection. *Concrete International,* Feb., 65–67.

SHAH, S. P. (ed.) (1979). *Proc. Workshop on High-Strength Concrete,* Dec. 2–4, 1979, University of Illinois at Chicago.

SHARP, J. D. (1966). *Electric Steelmaking.* Iliffe Books, London.

SHILSTONE, J. M. (1991). Performance specifications for concrete pavement. *Concrete International,* 13(2), 28–34.

SMITH, R. C. (1979). *Materials of Construction,* 3rd ed. McGraw-Hill, New York.

SOMAYAJI, S. (1990). *Structural Wood Design.* West Publishing, Minneapolis, Minn.

SOROKA, I. (1979). *Portland Cement Paste and Concrete.* Chemical Publishing, New York.

SPEARS, R. E. (1983). The 80 percent solution to inadequate curing problems. *Concrete International,* 5(4), 15–22.

SPENCE, R. J. S., AND COOK, D. J. (1983). *Building Materials in Developing Countries.* Wiley, New York.

STODOLA, P. R. (1983). Performance of fly ash in hardened concrete. *Concrete International,* 5(12), 64–68.

STOLL, T. M., AND EVSTRATOV, G. I. (1987). *Building in Hot Climate.* MIR Publishers, Moscow.

TAYLOR, W. H. (1967). *Concrete Technology and Practice.* Angus and Robertson, Sydney, Australia.

TRANSPORTATION RESEARCH BOARD (1975). *Bituminous Emulsions for Highway Pavements.* TRB, National Research Council, Washington, D.C.

——— (1984). *Asphalt Overlay Design Procedure.* TRB, National Research Council, Washington, D.C.

——— (1988). *Low Temperature Properties of Paving Asphalt Cements.* TRB, National Research Council, Washington, D.C.

TUTHILL, L. H. (1982). Alkali–silica reaction: 40 years later. *Concrete International,* 4(4), 32–36.

——— (1991). Long service life of concrete. *Concrete International,* 13(7), 15–17.

VONDRON, G. L. (1991). Applications of steel fiber reinforced concrete. *Concrete International,* 13(11), 44–49.

WALLACE, H. A., AND MARTIN, J. R. (1967). *Asphalt and Pavement Engineering.* McGraw-Hill, New York.

WATSON, D. A. (1978). *Construction Materials and Processes,* 2nd ed. McGraw-Hill, New York.

WILUN, Z., AND STARZEWSKI, K. (1972). *Soil Mechanics in Foundation Engineering.* Wiley, New York.

WIRE REINFORCEMENT INSTITUTE (1991). *Reinforcing Steel in Slabs-on-Grade.* Tech Facts TF 701. WRI, Washington, D.C.

——— (1992). *How to Specify, Order and Use Welded Wire Fabric in Light Construction.* Tech Facts TF 202. WRI, Washington, D.C.

WRIGHT, P. H., AND PAQUETTE, R. J. (1987). *Highway Engineering,* 5th ed. Wiley, New York.

ZOLLO, R. F., AND HAYS, C. D. (1991). Fiber vs. WWF as nonstructural slab reinforcement. *Concrete International,* 13(11), 50–55.

Index

AASHTO soil classification, 445
Abram's law, 129, 130
Abrasion resistance, 66, 85, 383, 402, 434, 436. *See also* Durability
Absolute viscosity. *See* Viscosity, coefficient of
Absorption, 9–10, 39–40, 49–53, 70, 74. *See also* Saturation
  aggregates, 50, 69–72, 74
  asphalt pavement, 384
  blocks, 242–44, 245
  bricks, 233, 236–37, 256, 271, 286–88
  mortar, 249, 251–55
  plastics, 436, 439
  wood, 305
Absorption, sound. *See* Sound absorption
Acoustics, 9–10, 16. *See also* types of sound properties
Acrylics, 12, 196, 430, 434–35
Adhesion, 366–67. *See also* Bond strength
Admixtures
  in asphalts, 372, 374, 379–80
  in cements, 82, 91–93, 372, 374
  in concretes, 101, 106, 132–33, 155–56, 158, 163, 175–87, 379–80
  in mortar, 253, 254
Adsorption, 298
Aesthetics, 30, 32
Aggregates, 5, 6, 14, 40–43, 47–67. *See also* Rocks
  absorption, 69–72, 74
  alkali reactions, 67, 155, 156, 184
  in asphalt, 41, 44, 65, 360
  for asphalt pavement, 381–82, 384–85
  classifications, 5, 42–47, 98
  for concretes, 61–65, 164–65, 169–70, 376–78, 381
  density, bulk, 43–47, 53–55, 68–69
  elasticity, 55–56, 139–41
  examples, 68–69
  fineness modulus, 63–65, 75, 126–28, 131
  gradation, 54, 56–63, 75–76
  moisture content, 48–55, 69, 74
  sampling, 74–75
  specific gravity, 42–45, 47–49, 51–52, 70–72, 74
  standards, 59–63, 64, 65, 66, 67, 70–76, 197, 376–77
  storage, 65, 110
  strength, 42, 55–56, 125, 374–76
  tests, 66, 67, 70–77
  workability and, 105

Air content, 209–10, 251
entrained in cement, 82, 91–93
entrained in concrete, 101, 106, 132–33, 155, 158, 163, 176, 180–81
entrained in mortar, 253, 254
entrapped, 115, 131, 132, 163
Alkali-aggregate reactions, 67, 155, 156, 184
Alumina, 88, 89, 90–91, 383
Aluminum, 7–11, 14, 112
elasticity, 25
ore, 81, 82, 88
Poisson's ratio, 22
in steel, 400
yield point and, 24
American Concrete Institute (ACI), 31, 32–33, 60
standard (211), 168, 169, 199
standard (318), 118–19, 134–35, 141–42, 162–63, 166–68
strength (332), 162
American Institute of Steel Construction (AISC), 33
standards, 407
American Institute of Timber Construction (AITC), glulam (A190.1), 333
American Iron and Steel Institute (AISI), 31, 33
American National Standards Institute (ANSI), 32
glulam (A190.1), 333
American Plywood Association, 341
American Society for Testing and Materials (ASTM), 33
aggregate (C330), 197
aggregate abrasion (C131), 66
aggregate denisty, bulk (C29), 72
aggregate gradation (C33, C144, D448), 60–64
aggregate gradation (D692, D1073), 376–77
aggregate impurities (C40–84, C142), 67
aggregate moisture content (C566), 74
aggregate sampling (C702, D75), 74
aggregate sieve analysis (C136), 75, 76
aggregate specific gravity and absorption (C127, C128), 70–72, 74
asphalt binder (D3315), 378
asphalt cement grading (C946, D3381), 370–72
asphalt ductility (D113), 366, 387
asphalt durability (D1754, D2872), 365, 369
asphalt, emulsified (D977), 361
asphalt specific gravity and density (D70), 364
asphalt viscosity, penetration test (D5, D2170, D2171), 363–64, 386, 388–89
brick (C62, C67), 33, 233, 237, 284–87
cement fineness (C115, C204), 94
cement, hydraulic strength (C778), 65, 218
cement, portland (C150), 91, 218–19
cement, portland consistency (C187), 95
cement, pozzolan (C595), 81, 186
concrete (C172), 208, 209, 217
concrete admixture (C260, C494, C618), 176, 180–82
concrete block (C90), 241–43
concrete brick (C55), 244
concrete consistency (C360), 104–5, 217
concrete density, bulk and yield (C138), 209
concrete elasticity (C469), 141, 213
concrete Poisson's ratio (C469), 141
concrete rebound number (C805), 219
concrete slump (C143), 101–4, 208, 217
concrete specimens (C192), 134, 211–13, 215–16
concrete strength (C39, C78, C496, C617), 134–38, 211, 213, 215–17
grout (C404, C476), 256
masonry, 227–28
masonry strength (E447), 272, 290
mortar (C144), 205
mortar bond strength (C952), 254
mortar flow (C230), 254, 288
mortar mixing (C270), 250–51
mortar strength (C109), 94–95, 218–19, 254, 288, 289
mortar water retentivity (C91), 255
particleboard bending and rupture (D1037), 352
plaster (C926), 204
soils (D2487), 446
steel (A36, A242, A441, A514, A529, A588), 408–10
steel (A572), 402, 408–10
steel, epoxy-coated (A775), 426
steel reinforcement (A421, A416), 193
steel reinforcement (A615, A616, A617, A706), 188, 419–20

American Society for Testing (*continued*)
steel strength (A370, E8), 426
steel wire fabric (A82, A185, A496, A497), 422–23
wood (D143), 349, 350, 351
wood moisture content (D4442), 349
wood specific gravity (D2395), 349
wood strength ratio (D245), 329
American Softwood Lumber Standards, 314, 315
grading (PS20–70), 318–20
Amorphous materials, 5, 6, 12, 36, 37
Annual Book of ASTM Standards, 233
Aramid (Kevlar), 436, 437, 439
Archimedes, 16
Asbestos, 4, 9, 436
Aspdin, Joseph, 87
Asphalt binders. *See* Cements, asphalt
Asphalt pavement, 14, 357, 360, 361, 380–86. *See also* Concrete, asphalt
asphalt cement grades, 372–74
layers, 380, 381–83, 445
mix design, 60, 381, 383
Asphalts, 4, 6, 36, 354–57. *See also* Cements, asphalt
adhesion, 366–67
air-blown, 358, 359, 361
conductivity, 9
consistency, 362–64, 367–68, 386–87
curing, 360, 365–66
cutback, 358–60, 366, 374, 388–89
ductility, 366–67, 387–88
durability, 365
elasticity, 7, 368–69
emulsified, 358, 359, 360–61, 385
hardening, 365, 369–70
manufacture, 357–58
mix design, 360
mixing, 369
oxidation, 365, 369
permeability, 366
plasticity, 365
repair material, 12, 65
soil stabilizing, 385
specific gravity, 364–65
standards, 361, 363–66, 369, 386–89
temperature, 9, 367–69
tests, 386–92
viscosity, 358, 362–64, 367–68, 388–90
volatilization, 365, 369
Atomic structure, 7–8, 10, 36–37
Axial forces, 19, 20, 21
Axial shrinkage, 305, 306, 307, 321

Barite, 43, 47
Basalt, 38, 43, 55, 66, 441
Batching concrete, 51, 107–11
Bending moment, 16, 20, 30
Bending strength, 20, 21. *See also* Rupture, modulus of
aggregates, 55
asphalt cement, 392
bricks, 238–39, 284–85
compressive strength and, 136–38
concrete, 136–38, 143, 215–16
concrete, reinforced, 190–92, 196
mortar, 254
plywood, 338–39
rocks, 55
tensile strength and, 284
test, 136–38
wood, 319–20, 321, 323–26
wood products, 343, 352–53
Bessemer, Henry, 394
Bitumen, 2, 11–12, 14–15, 41, 354–55. *See also* Asphalts; Pitches; Tars
Bleeding, 101, 106–7
Blending aggregates, 56–57
Blocks, 11, 14. *See also* Masonry
absorption, 242–44, 245
classification, 197, 239–44
durability, 245
expansion, 266
manufacture, 239, 241
moisture content, 241–42, 245, 266–67
shrinkage, 242, 245, 258, 266
standards, 241–44, 290
strength, 242–44, 245, 273
test, 290–91
Board foot, 312, 318
Bond strength, 252–53, 254–55, 416. *See also* Adhesion; Segregation
Brick Institute of America (BIA), 33
Bricks, 2–5, 11, 14, 228–30, 235–36. *See also* Clay; Masonry
absorption, 233, 236–37, 256, 266, 271, 286–88

classification, 229, 232–35
conductivity, 9, 239
density, 236
dimensions, 234
ductility, 6, 27, 28
durability, 236
efflorescence, 231, 237, 258, 272, 276–78
elasticity, 23, 239
expansion, 6, 266, 271
manufacture, 230–32
rupture, 238–39, 284–85
saturation, 233, 236–37, 286–87
shrinkage, 235
standards, 33, 233, 237, 272–74, 284–88
strength, 232–33, 237–39, 273, 284–86
temperature, 6, 9, 239, 266, 271
tests, 284–88
Bricks, concrete. *See* Blocks
Brickwork. *See* Masonry
Brittleness, 5, 6. *See also* Ductility
Building construction
masonry, 223–27, 245, 255, 258–71
plastics, 437–39
plywood, 112, 224, 331–32, 339, 342–43, 348
steel, 6, 112, 400–402, 406–8, 410–15
steel reinforcement, 11, 279–81, 410, 417, 421–22, 425
wood, 224, 306, 344–48
Bulking, 51, 69. *See also* Density, bulk

Calcining, 84
Calcite, 37, 38, 40, 441
Capillary action, 450, 454
Carbon, 5–7, 12, 37, 196. *See also* Graphite
in iron alloys, 11, 399, 402, 403
Carbonation, 154–55, 159
Caustic lime. *See* Lime
C/B ratio, 233, 236–37, 286–87
Cellulose, 294–95
Cement-aggregate ratio, 125–26, 131, 140, 147, 164–65, 169–70
Cementitious materials, 5, 14, 79
Cement rock, 80, 88
Cements, 6, 12, 14, 16, 38. *See also* Grout; Mortar
in asphalt pavement, 381, 382
classifications, 79–84, 152, 186–87, 204, 246
in concrete, 186–87
environmental concerns, 31
soil stabilizing, 207, 381
standards, 81, 91, 94–95, 186
test, 65, 218
Cements, asphalt, 358–59
aging, 371
in asphalt pavement, 382–83
classifications, 370–74, 391–92
ductility, 370, 371, 392
failure, 372–73
flash point, 370, 371
modifiers, 372, 374
solubility, 370, 371
standards, 370–72, 389
strength, 391, 392
temperature, 370–74
tests, 370–72, 373, 389–92
viscosity, 358, 362, 364, 370–72, 389–91
Cements, hydraulic/nonhydraulic, 80–84, 186–87, 204, 246
test, 65, 218
Cements, portland, 5, 14, 80–84, 87–88
in asphalt pavement, 381, 382
chemistry, 90–91
classification, 91–95, 125, 127
consistency, 95–97
elasticity, 140–41
fineness, 93–94
hardening, 96–97
hydration, 79, 82, 91–98, 105
manufacture, 88–90
in mortar, 11, 246, 252
setting, 95–96
soil stabilizing, 381, 384
standards, 91, 94–95
strength, 94–95
test, 218–19
Ceramics, 5–6, 11, 437
Chapman, C. M., 101
Chemical abbreviations, 90
Chromium, in iron alloys, 11
Civil engineering, 1–4
Clay, 16, 43, 47, 52, 455. *See also* Bricks; Ceramics
expanded, 45, 46, 197
in cement, 80, 81, 88
in soil, 441, 443, 444, 452, 455
Coal, 12, 395

Coal tar, 256, 354, 355
creosote, 308–9, 310
Coignet, Francois, 187
Composite materials, 6, 66
Compressive strength, 19–21, 30, 40, 139–40
aggregates, 42, 55–56, 125
bending strength and, 136–38
blocks, 242–44, 245, 273
bricks, 232–33, 237–38, 273, 285–86
concrete, 47, 55–56, 124–35, 198, 415–16
concrete water-to-cement ratio, 128–31
curing, 117–23, 124, 127, 131–32
elasticity and, 25, 142, 274, 275
fineness modulus, 126–28, 131
grain and, 320, 322–23
iron, 398
masonry, 56, 271–74, 275, 290–91
mix design and, 125–26, 131–33, 176–80, 184, 186
moisture content and, 124
mortar, 94–95, 128, 131, 249–51, 253–54, 289–90
plastics, 432, 434
rocks, 40, 42, 55, 56
soils, 449, 457
standards, 162–63
tensile strength and, 136
tests, 111, 124, 134–35, 142–43, 211–12, 217–19
wood, 319–20, 322–23, 331, 350–51
Concrete, air-entrained, 93, 101, 132–33
admixtures, 101, 106, 132–33, 155, 158, 163, 176, 180–81
water-to-cement ratio, 167, 168
Concrete, asphalt, 56, 374–80. *See also* Aggregates; Cements, asphalt
mix design, 38, 375–78
pavement construction, 357, 380–86
Concrete, high-strength, 129, 176, 179–80, 198–200
precast, 92, 193
Concrete, portland cement, 3, 5–9, 11–12, 14, 16, 84–87, 98–100, 197–99. *See also* Admixtures, in concretes; Aggregates; Cements, portland
abrasion resistance, 66
air content, 165–66, 209–10
alkali-silica, 67, 155, 156, 184
in asphalt pavement, 383
bleeding, 101, 106–7
bond strength, 416
carbonation, 154–55, 159, 185
cement-aggregate ratio, 125–26, 131, 140, 147, 164–65, 169–70
conductivity, 198
consistency, 101–6, 184, 217
consolidation, 131
corrosion, 93, 155, 158–60, 168, 177, 185
creep, 55, 153–54
curing, 117–23, 124, 127, 131–32, 138–39, 145
density, 46, 47, 198, 209–10
deterioration
ductility, 27, 28, 29, 187
durability, 155–60, 161, 168
elasticity, 22, 23, 25, 55, 139–43, 213–14
environmental concerns, 31
examples, 142–43, 166–75, 201
expansion, 16, 436
finishing, 67, 100, 106–7, 116–17, 145
flowing, 178, 199
forms for, 112–13
freeze-thaw cycle, 155, 157–58, 161, 168, 176, 177
fresh, 98, 100–107, 144–45
hardening, 96–97, 123–62
joints in, 146, 148–52
mix design, 110–11, 124, 161–75
mixing, 107–13, 161–62, 212–13
permeability, 85–86, 124, 155, 207
placing, 106, 113–16, 160
Poisson's ratio, 21, 22
properties and, 145–48, 152–54, 155–60, 170–71
rebound number, 219–20
rupture, 136–38, 143, 215–16, 415
segregation, 85, 106–7, 179
setting, 95–96
shrinkage, 55, 122, 144–53, 155, 179, 198
slump, 101–5, 106, 111, 160, 165, 208–9
specific gravity, 52
standards, 101–5, 118–19, 125, 134–38, 141, 162–63, 166–69, 176, 181–82
strength, 47, 55–56, 98, 117–38, 140, 142–43, 161–65, 167, 193, 195, 215–17, 415–16
sulfate attack, 155, 156–57, 184–85
stress-strain diagram, 139–42
temperature, 16, 198, 436
tests, 111, 124, 134–38, 142–43, 208–20

types, 38, 46, 47, 99, 152–53, 197–98
void content, 55
water-to-cement ratio, 106, 128–31, 147, 154, 167–69, 176, 177–80, 183
yield, 209–10
Concrete, reinforced, 6, 14, 17, 187, 190, 197
in asphalt pavement, 383
bleeding, 196
corrosion, 15, 93, 155, 158–60, 164, 168, 177, 185, 196, 421
cracking, 425
example, 201–2
fibers, 193–97, 206
moment capacity, 190–92
plastics, 425
precast, 92, 193
prestressed, 14, 192–93, 194–95
segregation, 196
shrinkage and, 147, 197
slump, 196
standards, 188, 193
steel, 188–90, 191, 196–97, 410, 416–27
strength, 190–93, 195, 196, 415
toughness, 195
Concrete, spray. *See* Shotcrete
Concrete blocks. *See* Blocks; Masonry
Conductivity. *See* types of conductivity
Conductors, electrical, 7–8, 12–13
Consistency
asphalts, 362–64
concrete, 101–6, 184, 217
soils, 457
Copper, 2, 8–11, 22, 24–25, 27
in iron alloys, 400
masonry joints, 268
Corrosion
cement and, 93
concrete, 93, 155, 158–60, 168, 177, 185
resistance to, 16, 436
steel, 408, 411
steel reinforcement, 15, 93, 155, 158–60, 164, 168, 177, 185, 196, 421
Cracking, 109, 144–46, 156–58
Creep, 55, 153–54, 198, 343–44, 435–36
Creosote, 308–9, 310
Crystalline materials, 5–6, 12, 36–38, 53, 66
Curing concrete, 117–23, 124, 127, 131–32, 138–39, 145
admixtures, 176–77
Deflection diagrams, 325, 432
Deformation
creep, 55, 153–54, 198, 343–44, 435–36
ductility and, 27, 402, 411, 420
elasticity and, 22–26
loads and, 17, 18–19, 441
strain and, 18, 20–22
toughness and, 28
viscosity and, 362
Density, 40
aggregates, 40, 44–47, 53
asphalt, 364
bricks, 236
clay, 47
concrete, 46, 47, 198
conductivity and, 8, 9
diffusivity and, 8
masonry, 271
moisture content and, 45, 299–300
specific gravity and, 15, 40, 44
wood, 299–302
wood products, 342
Density, bulk, 44, 53
aggregates, 43–47, 53–55, 68–69, 72–73
concrete, 209–10
soils, 452, 455
Diffusivity, 8–9
Direct stress. *See* Axial forces
Dolomite. *See* Limestone
Dry conditions, 50, 300, 307–8, 318–19, 325
Dry-loose weight. *See* Density, bulk
Dry-rodded weight. *See* Density, bulk
Drywall, 224, 346
Ductility, 6–7, 16, 26–29, 388. *See also* Elasticity
asphalt, 366–67, 387–88
asphalt cement, 370, 371, 392
bricks, 6, 27, 28
concrete, 27, 28, 29, 187
iron, 5, 27, 28, 29, 398
masonry, 28
metals, 17, 27, 28, 29
mortar, 27, 29
plastics, 437
rocks, 28
steel, 17, 27, 29, 402–5, 410–11
Durability, 30, 39. *See also* Abrasion resistance; Freeze-thaw cycle
asphalts, 365
blocks, 245

Durability (*continued*)
bricks, 236
concrete, 155–60
plastics, 433–34

Economics, 30, 32, 161, 198, 200
reinforcing steel, 415, 421, 426
Elasticity, 7, 16, 22–23, 306, 411
aggregates, 55–56, 139–41
asphalts, 368–69
bricks, 239
concrete, 25, 55, 139–43, 213–14
creep and, 55
elastic limit, 23, 24–25, 325
elastic recovery, 154
failure, 7, 17–18
fiber reinforcement, 196
masonry, 274, 275
metals, 25, 196, 404, 405, 415
plasticity vs., 26
plastics, 432, 433–35, 436, 439
porosity and, 40
rocks, 40, 55
shrinkage and, 55
soils, 381
steel, 25, 196, 404, 405, 415
stiffness and, 16, 22–23, 306
strength and, 25, 142, 274, 275
stress/strain, 25–26, 55, 139–42, 213–14
wood, 25, 306, 319, 321–22, 323
wood products, 341, 343
Elasticity, modulus of, 25, 40, 55, 213
Elastomers, 7, 431
Electrical conductivity, 7–8, 12–13, 434, 439
Environmental criteria, 31, 32
Epoxy, 14, 426, 430, 437, 439
Equilibrium, 17
Equilibrium moisture content (EMC), 299
Expansion coefficients
moisture, 271
thermal, 6, 12, 16, 266, 271, 433–36

Failure, 7, 17–18, 27–29, 53, 136, 441
asphalt cement, 372–73
concrete, 139
concrete, reinforced, 190–91, 201–2
rocks, 441
steel, 411–12
wood, 325, 326

Federal Highway Administration, 185
Feret, R., 163
Fiberboard. *See* Wood products
Fiber-reinforced materials. *See* Concrete, reinforced; Plastics
Fiber saturation point, 298, 301, 305
Fineness modulus, 63–65, 68, 75
strength and, 126–28, 131
workability and, 105
Finishing concrete, 67, 100, 106–7, 116–17, 145
Flexural strength. *See* Bending strength
Floating, 107
Flow, 178, 199, 253–54, 288–89
Fly ash, 14, 45, 122. *See also* Admixtures
in bricks, 11
in cement, 80, 88, 93, 98, 184
in concrete, 175, 182, 183–85
Free moisture. *See* Moisture, surface
Freeze-thaw cycle, 52–53
bricks, 233, 236
concrete, 155, 157–58, 161, 168
concrete admixtures, 163, 176, 177
mortar, 253
weathering index, 233
Freyssinet, Eugene, 192
Fuller, W. B., 56

Gage length, 19, 214, 419
Glass, 5, 6, 36, 37, 196
asphalt concrete, 380
ceramics, 5, 6
ductility, 27, 28
Poisson's ratio, 21, 22
reinforced concrete, 195–96, 197
reinforced plastics, 6, 12, 112, 435–37, 439
Glulam, 14, 330, 333–35
Gradation, 54, 56–63, 65, 105
fineness, 63–65, 75, 126–28, 131
sieve analysis, 57–63, 75–77
soils, 446–47, 449, 450–52
standards, 59–63, 64, 65, 75, 76
Gradation charts, 58, 61–63, 64, 451
Grain-size distribution. *See* Gradation
Granulating, 81
Graphite, 6, 398, 436, 437
Gravel, 40–45, 54–55, 66, 455
in soil, 441–42, 443, 455

Green condition, 299, 318, 325
Grout, 5, 84, 85, 152, 205
  in masonry, 11, 245, 256
  standards, 256
Gunite. *See* Shotcrete
Gypsum, 11, 36–38, 40, 83–84, 231, 456. *See also* Cements, portland
Gypsum board, 224, 346

Hardening
  asphalts, 365, 369–70
  cement, 96–97
  concrete, 96–97, 123–62
  steel, 404–5
Hardness, 37, 47. *See also* Abrasion resistance; Durability
Heat transfer. *See* Thermal conductivity
Heat transmission coefficient. *See* U-values
Hemicellulose, 294, 295
Highway construction. *See* Asphalt pavement
Hooke's law, 25, 55
Hydration, 79, 82, 91–93, 94, 95–98
  curing and, 118, 124
  heat of, 97–98
  workability and, 105

Igneous rocks, 35–36, 38, 45, 441
Inertia, moment of, 306, 312, 320
Insulating materials, 6, 7–10, 12–13, 46
Interlocking particle strength, 374–76, 447–48
Ions, 8
Iron, 2–3, 10, 36–38, 393–94. *See also* Steel
  in cement, 88, 89
  ductility, 398
  melting point, 397
  ores, 12, 38, 40, 43, 47, 394
  production, 394–97
  strength, 397
  types, 11, 395–98, 400, 402–3
Iron, cast, 2, 11, 397, 398–99
  ductility, 5, 28, 29, 398
  elasticity, 22, 25
  melting point, 397
  Poisson's ratio, 22
  strength, 397, 398
Iron, wrought, 11, 397–98
  ductility, 27
  melting point, 397
  Poisson's ratio, 22
  strength, 397, 398

Johnson, I. C., 88
Joints
  in concrete, 146, 148–52
  in masonry, 6, 237, 249, 258, 260–61, 265–71, 358, 439

Lambot, Jean-Louis, 187
Laminated veneer lumber (LVL). *See* Wood products
Latex. *See* Rubber
Lattices, 36
Lime
  asphalt concrete, 380
  in cement, 80–83, 88–89, 90–91
  in mortar, 11, 38, 83, 245, 248–49, 252
  types, 82–83, 246–48
Limestone, 4, 6, 12, 66, 156, 395, 441
  aggregates, 36–38, 40, 42
  in cement, 80, 82, 88
  conductivity, 9
  density, 40
  deterioration and, 67
  elasticity, 40, 55
  porosity, 40, 52
  rupture, 40
  specific gravity, 40, 42, 43, 456
  in steel, 399
  strength, 40, 42, 53, 55
Loads, 16, 17–18, 136, 325
Lumber. *See also* Wood
  classifications, 311–21, 329
  in construction, 224, 306, 344–48
  creep, 343–44
  cutting, 306, 315
  defects, 327, 328–29
  dimensions, 312–13, 316
  elasticity, 319, 320, 321–22, 323
  examples, 329–31
  grain slope, 304, 327–28
  moisture content, 318, 319, 322
  rupture, 321, 322, 323, 324–26, 328, 331, 351–52
  seasoning, 312–13
  section modulus, 312
  species groups, 296–97

Lumber (*continued*)
standards, 318–20, 329
strength, 319–27, 328–29, 331
stress-strain diagram, 324

Manual of Steel Construction, 407
Marble, 36, 38–39, 40, 88, 441. *See also* Limestone
Masonry, 6, 11, 38, 39, 223. *See also* Blocks; Bricks; Grout; Mortar; Plaster
aggregates, 45
bonds (layouts), 258, 263–65
conductivity, 274–76, 277
in construction, 223–27, 256, 261–62
construction procedure, 245, 255, 258–71
density, 271
ductility, 28
elasticity, 274, 275
examples, 282–84
joints in, 6, 237, 249, 258, 260–61, 265–71, 358, 439
mix design, 282–84
plastics in, 268, 439
standards, 227–28, 272–74, 290
strength, 56, 271–75, 290–91
temperature, 274–76, 277
tests, 272, 290–91
types, 227–45
Masonry, reinforced, 11, 279–81
grouting, 256, 261–63, 279, 281
Materials, 1–4, 11, 38–39, 82, 83
classifications, 5–29, 47
selection, 29–31, 32
standards, 31–34
Metal alloys, 6, 11, 13, 22, 25
iron, 11, 400, 402
Metals, 7–11, 13, 37, 437
elasticity, 23
Poisson's ratio, 22
yield point and, 24
Metal ties in masonry, 226, 263, 269
Metamorphic rocks, 36, 441
Mica, 38, 40, 441, 456. *See also* Vermiculite
Mineral rubber, 361, 379
Minerals, 36, 37–39, 441
Mixed aggregates, 56–57
Mixing concrete, 107–13, 161–62, 212–13. *See also* Concrete, portland cement, mix design
Mohs number, 37
Moisture content (MC), 16, 74, 349
aggregates, 48–55, 69, 74
blocks, 241–42, 245
concrete curing, 117–18, 119–22
density and, 45, 299–300
equilibrium (EMC), 299
example, 69
glulam, 333, 334
soils, 453, 457
specific gravity and, 51–52
strength and, 124
wood, 298–301, 304–8, 318, 319, 349
workability and, 105
Moisture expansion coefficient, 271
Moisture, surface, 50, 69, 74
Monier, Joseph, 187
Mortar, 2–3, 5–6, 11, 38, 84–85, 204–5, 245. *See also* Cement
absorption, 249, 251–55
air content, 251
bond strength, 252–53, 254–55
classification, 245–52, 253
curing, 255
ductility, 27, 29
durability, 30, 252
examples, 282–84
flow, 253–54, 288–89
mix design, 249, 250, 252–53, 254, 282–84
mixing, 251
permeability, 252
plasticity, 249, 252
Poisson's ratio, 21
shrinkage, 249, 252, 272
standards, 94–95, 205, 218–19, 250–51, 254–55, 288–89
strength, 94–95, 128, 131, 249–51, 253–54, 289–90
tests, 218–19, 288–90
water-cement ratio, 131, 249
workability, 252, 253, 254, 255

National Design Specifications for Wood Construction, 301, 329
National Forest Products Association (NFPA), 32
National Hardwood Lumber Association, 314
Newton's laws of motion, 17–18

Nominal/specified dimensions, 234, 312–13, 316
Non-Newtonian liquid, 390
Nylon, 12, 13, 195–96, 197, 430

Offset approach, 24–25, 404–5
Oxidation, 365, 369, 435

Particleboard. *See* Wood products
Particle-size distribution. *See* Gradation
Particle-size distribution curves. *See* Gradation charts
Penetration tests
  asphalt consistency, 358, 363–64, 371–72, 386–87
  concrete consistency, 217
Perlite, 43, 45–46, 47, 52
Permanent set approach, 24
Permeability, 15–16, 94, 155, 168
  asphalt pavement, 382–83
  asphalts, 366
  concrete, 85–86, 124, 155, 207
  mortar, 252
  plastics, 434
  soils, 449
  wood, 85
Phenol-formaldehyde (PF), 12, 14, 430, 434, 435
  plywood, 337, 342
Pitches, 13, 354, 355, 356
Placing concrete, 106, 113–16, 160
Plaster, 202, 203–4, 245, 257–58. *See also* Cement; Mortar; Stucco
Plasticity, 16, 26, 27, 28
  asphalts, 365
  soils, 444–47, 448–49, 457
Plasticizers, 82, 178–79, 199, 436
Plastics, 5–8, 11–12, 268, 431–39. *See also* Polymers
Plywood, 6, 10, 14, 335–39, 341
  in construction, 112, 224, 331–32, 339, 342–43, 348
  creep, 344
Poisson's ratio, 21–22, 141, 436
Polyester, 12, 14, 430, 434, 435
  glass-reinforced (GRP), 12, 435, 437
  in reinforced concrete, 196
Polyethylene, 12–13, 196, 430, 434–35, 439
Polymers, 6–7, 11–12, 26, 36, 428–29. *See also* Plastics
  in asphalt, 379
  wood, 295, 342
Polypropylene, 12–14, 196, 430, 434
  asphalt concrete, 380
  reinforced concrete, 195, 196–97
Polystyrene, 9, 12, 430, 432–34, 439
Polyurethane, 12, 430, 434, 436, 439
Polyvinyl chloride (PVC), 12, 13, 430, 433–34, 439
  stucco lath, 202
Pore volume, 48, 51–52
Porosity, 5, 16, 40, 52, 66
  absorption, sound and, 9–10
  absorption, water and, 39–40, 50, 52–53
  aggregates, 52–53
  conductivity and, 8
  elasticity and, 40
  freeze-thaw and, 52–53
  rocks, 40, 52
  soils, 455
  strength and, 40
Portland Cement Association (PCA), 32
Pozzolanic reaction, 67, 182–83, 186
Pozzolans, 80–81, 98. *See also* Fly ash
  in concrete, 163, 175, 181–87
Proportional limit, 23, 324, 405
Pumice, 17, 36, 43, 45, 47, 66, 441
  aggregates, 45, 46, 67, 197
  in concrete, 181
Pumping concrete, 113–14, 178, 198

Quartz, 6, 36–40, 43, 441. *See also* Sandstone
  in cement, 88, 89
Quenching, 5, 37
Quicklime. *See* Lime

Regolith. *See* Soils
Resins. *See* Plastics
Resistance. *See* types of conductivity
Rigidity, modulus of. *See* Elasticity
Rocks, 2–4, 6, 9. *See also* Aggregates; Masonry
  classification, 35–40, 441
  deformation, 441
  elasticity, 23
  failure, 441
  stress-strain diagrams, 442
Rubber, 7, 11–12, 17–19, 22–23
  in asphalt concrete, 364, 380

Rules for the Measurement and Inspection of Hardwood and Cypress Lumber, 314
Rupture, modulus of (MOR), 6, 40, 294, 324
  bricks, 238–39, 284–85
  concrete, 136–38, 143, 215–16, 415
  plywood, 341
  rocks, 40
  wood, 321, 322, 323, 324–26, 328, 331, 351–52
  wood products, 341, 342, 352–53
Rupture, plane of, 28–29, 212

Sand, 40–42, 44–45, 60–61, 65, 67
  in cement, 88, 89
  density, 51, 53, 455
  fineness modulus, 65, 128
  impurities, 67
  in mortar, 11, 252
  in soil, 441, 442–43, 444, 455
  specific gravity, 43, 45, 455
  standards, 60–61, 67
  types of, 65
  void content, 54–55, 455
Sandstone, 4, 9, 36, 39–40, 42–43, 66, 441. *See also* Quartz
  absorption, 50
  elasticity, 40, 55
  porosity, 40, 52
  strength, 40, 42, 53, 55
Saturated surface dry (SSD) conditions, 49–50
Saturation
  bricks, 233, 236–37, 286–87
  soils, 453–54
  wood, 298, 301, 305
Saylor, David, 88
Scaling, 107
Scoria, 43, 45, 46, 197, 441
Sealants
  in asphalt pavement, 384–86
  in masonry joints, 266, 268, 358
  repair materials, 12, 15
Seasoning wood, 307–8, 312–13, 318, 319
Secant modulus, 25, 26, 139, 141, 142
Sedimentary rocks, 36, 441
Segregation, 85, 106–7, 179, 196
Semiconductors, 8, 12–13
Serviceability, 30, 32
Setting concrete, 95–96
Shale, 17, 36, 38, 40, 43, 47, 441
  aggregates, 45, 46, 197
  in bricks, 230
  in cement, 88
  in concrete, 181–82
Shear deformation, 21
Shear plane/cone. *See* Rupture, plane of
Shear strength, 16, 19, 20, 25, 391
  soils, 449, 457
Shotcrete, 84, 85, 206–7. *See also* Concrete, reinforced; Mortar
Shrinkage, 55, 448
  blocks, 242, 245, 258, 266
  bricks, 235
  concrete, 55, 122, 144–53, 155, 179, 197–98
  mortar, 249, 252, 272
  plastics, 436
  shotcrete, 206
  wood, 298, 301, 305–7
Sieve analysis, 57–63, 75–77, 451
  fineness modulus, 63–65, 75
  standard, 75, 76
Silica, 67, 88, 89, 90–91, 182. *See also* Quartz
  reactive, 67, 156
Silica fume, 80, 182, 185–86
Silt, 67, 441, 443, 444, 455
Slag, 45, 46, 66, 81, 122, 395–97. *See also* Admixtures
  in cement, 80–81, 88
  in concrete, 175, 181, 187
  in iron, 398
  in steel, 399
Slate, 36, 38, 40, 43, 47, 441
  aggregates, 45
  in cement, 88
Slump
  concrete, 101–5, 106, 111, 160
  concrete test, 208–9
Soils, 6, 14, 36, 439–40, 444–57. *See also* Aggregates; Rocks
  classification, 88, 357, 441–47
  stabilized, 207, 381, 383–84, 385
Sound absorption, 9–10, 435
Sound conductivity, 9, 16, 436
Southern Pine Inspection Bureau (SPIB), 314
Specific gravity (SG), 40, 48, 52, 70, 349
  aggregates, 42–45, 47–49, 51–52, 70–72, 74
  asphalts, 364–65
  concrete, 52

density and, 15, 40, 44
moisture content and, 51–52
plastics, 434–35
soils, 455, 456–57
void content, 53–54
wood, 43, 300–302, 349–50
wood products, 342
Specific gravity (SG), bulk, 51–52, 70
Specific weight. *See* Density, bulk
Standards. *See* specific standard-setting organizations
Steel, 5, 11–12, 14–15, 393–94, 397
abrasion resistance, 402
absorption, 85
aggregates, 47
classification, 408–10
conductivity, 8, 9
in construction, 6, 112, 400–402, 406–8, 410–15
corrosion, 408, 411
deformation, 402, 411
dimensions, 406–8
ductility, 17, 27, 29, 402–3, 404–5, 410–11
elasticity, 7, 22, 23, 25, 404, 405, 411, 415
expansion, 12, 436
failure, 411–12
manufacture, 396, 399–402
melting point, 397
plastic-coated, 426, 439
Poisson's ratio, 22
standards, 402, 407, 408, 426
strength, 17, 397, 403–6, 409, 411, 426–27
stress-strain diagrams, 404, 405
stucco lath, 202
temperature, 8, 9, 12, 405–6, 410, 411, 436
test, 426–27
toughness, 402, 405, 411
weldability, 411
worked, 404–5
yield point, 23–24, 403–4, 408–9
Steel in reinforced concrete, 188–90, 191, 193, 194–96, 206
classifications, 416–21, 422–24
in construction, 410, 417, 421–22, 425
corrosion, 15, 93, 155, 158–60, 164, 168, 177, 185, 196, 421
deformation, 420
economics, 415, 421, 426
elasticity, 196
fibers, 193–97, 206
moment capacity, 191–92
plastic-coated, 426
standards, 188, 193, 419–20, 422–23, 426
strength, 188, 196, 418–20, 422–23
weldability, 420–21
wire fabric, 190, 193, 202, 416, 418, 422–26
yield point, 418–19
Steel in reinforced masonry, 261–62, 270, 280, 281
Stiffness, 16, 22–26. *See also* Ductility; Elasticity
Stone. *See* Aggregates; Rocks
Strain, 18–19, 20–22, 25–26. *See also* - Stress-strain diagrams
Strandboard. *See* Wood products
Strategic Highway Research Program (SHRP), 372
Strength, 16, 26, 27–28, 29–30, 32. *See also* types of strength
grain and, 319–23, 326, 327–28, 398
ratio, 328–29
Stress, 17–18, 19–20, 22–23, 25–26
Stress-strain diagrams, 23–26, 55, 139–42, 213–14
concrete, 139–42, 213–14
plastics, 433
rocks, 442
steel, 404, 405
wood, 324
Stucco, 85, 202–4, 224, 257–58
Suction. *See* Absorption
Sulfate attack, 91, 93, 155, 156–57
Superpave (Superior Performing Asphalts Pavements) program, 372–74

Tangent modulus, 25, 26, 139, 141
Tars, 354, 355–56
coal, 256, 308–9, 310, 354, 355
Temperature-dependent properties, 8–9. *See also* types of thermal properties
asphalts, 367–69
concrete, 157
concrete curing, 119–23, 132, 148, 176–77
concrete mixing, 109, 146
plastics, 12–14, 429–31
steel, 405–6, 410, 411

Tensile strength, 19, 20, 21. *See also* Rupture, modulus of
bending strength and, 284
bricks, 239
ceramics, 437
compressive strength and, 136
concrete, 135–36, 143, 193, 195, 216–17, 415
durability and, 53
elasticity and, 25
fibers, reinforcing, 196, 436–37
grain and, 398
iron, 397, 398
plastics, 432, 433–35, 436
porosity and, 40
reinforcing, 188, 196, 420, 422–23
steel, 17, 397, 403, 404, 409, 411, 426–27
tests, 135–36
wood products, 342
Tests
absorption (AGG-1, AGG-2), 70–72
absorption (MAS-3, MAS-4, MAS-5), 286–88
asphalt binders (BITU-5), 391–92
concrete mixing (CON-4), 212–13
consistency (BITU-1), 386–87
consistency (CON-8), 217
consistency (MAS-6), 288–89
density, bulk (AGG-3), 72–73
density, bulk, (CON-2), 209–10
ductility (BITU-2), 387–88
elasticity (CON-5), 213–14
moisture content (AGG-4), 74
moisture content (WOOD-1), 349
rebound number (CON-11), 219–20
rupture, modulus of (MAS-1), 284–85
rupture, modulus of (WOOD-4, WOOD-6), 351–53
sample capping (CON-9), 217–18
sampling (AGG-5), 74–75
sieve analyses (AGG-6, AGG-7), 75–77
slump (CON-1), 208–9
specific gravity (AGG-1, AGG-2), 70–72
specific gravity (WOOD-2), 349–50
strength (CON-3, CON-6, CON-7, CON-10), 211–12, 215–17, 218–19
strength (MAS-2, MAS-7, MAS-8), 285–86, 289–91
strength (STL-1), 426–27
strength (WOOD-3), 350–51
viscosity (BITU-3, BITU-4), 388–90
voids (AGG-3), 72–73
Thermal conductivity, 8, 9, 12–13
asphalts, 9
bricks, 9, 239
concrete, 198
masonry, 274–76, 277
steel, 8, 9
U-value, 9, 276
Thermal diffusivity, 8–9
Thermal expansion, coefficient of
blocks, 266
bricks, 6, 266, 271
concrete, 16, 436
plastics, 12, 433–35, 436
steel, 12, 436
Thermoplastics, 12, 13, 367, 429–31, 433–35
Thermosets, 12, 14, 429–31, 434, 435
Thompson, S. E., 56
Tiles. *See* Bricks
Timber. *See* Wood
Toughness, 27–28
aggregates, 66
concrete, 195
plastics, 436
steel, 402, 405, 411

Ultimate strength, 26
Unified Soil Classification (USC) System, 445, 446
Uniform Building Code masonry standard, 272–74
Unit cell, 6
U.S. Department of Agriculture (USDA) soil classification, 447
United States Products Standard for Construction and Industrial Plywood (PS1–83), 338
Unit volume, 44
Unit weight. *See* Density, bulk
Urea-formaldehyde (UF), 14, 337, 340
U-values, 9, 276

Veneered panels. *See* Plywood
Vermiculite, 43, 45, 46
Vibration, 106, 115, 145
Viscosity, 390
asphalt cement, 358, 362, 364, 370–72, 389–91

asphalts, 358, 362–64, 367–68, 388–99
plastics, 436
Viscosity, coefficient of, 363, 390
Void/cement ratio. *See* Void content
Void content, 44, 53–54, 68–69, 131
bulking, 51
gradation and, 56–57, 72–73
soils, 448, 450, 452, 453–56
test, 72–73
Volatilization, 365, 369

Waferboard. *See* Wood products
Water resistance. *See* Permeability
Water retentivity. *See* Absorption
Water-to-cement ratio (w/c)
admixtures, 176, 177–80
bleeding and, 106
rupture, 136–37
strength and, 128–31, 137, 200
Wear resistance. *See* Abrasion resistance; Durability
Weathering. *See* Durability; Freeze-thaw cycle
West Coast Lumber Inspection Bureau (WCLIB), 314
Western Wood Products Association (WWPA), 314, 321
Wet conditions, 50
Wood, 2–9, 12, 14, 292. *See also* Lumber; Wood products
absorption, 305
adsorption, 298
cell structure, 293–95
classifications, 295–97, 311–21
in construction, 112, 224, 306, 344–48
decay, 308–10
defects, 302–4, 306
density, 299–302
durability, 30, 302, 308–10
elasticity, 25, 297–98, 301, 306–7
examples, 302
grain, 294, 303–4
grain and strength, 319–23, 326, 327–28
growth, 292–93
moisture content, 298–301, 304–8, 318, 319, 349
permeability, 85
rupture, 321, 351–52
saturation point, fiber, 298, 301, 305
sawdust, 43, 46–47
seasoning, 307–8, 312–13, 318, 319
shrinkage, 298, 301, 305–7
specific gravity, 300–302, 349–50
standards, 349–51
strength, 30, 297–98, 301–2, 305, 319–23, 326–28, 350–51
tests, 349–52
treatment, 308–9
Wood products, 14, 207, 331–33, 339–44, 348, 352–53. *See also* Glulam; Plywood
Workability, 47, 67, 100–106, 217. *See also* Consistency
Wythe, 11, 225–26

Yield of concrete mix, 209–10
Yield point, 23, 24–25, 426
steel, 23–24, 403–4, 408–9, 418–19
Yield strength, 24, 188, 427
steel, 188, 404–6, 418–20, 422–23
Young's modulus. *See* Elasticity, modulus of
Young, Thomas, 25